Exploring Pedigree

Handicapping's Newest Frontier

by Mike Helm

CITY MINER BOOKS

Some of the material in this book has been adapted from articles appearing in the *Cramer-Olmsted Report* and *Horseplayer Magazine*. The author would also like to thank Golden Gate Fields and Sam Spear, the *Daily Racing Form* for permission to reproduce their past performances and results charts and last, but not least, Mark Cramer for his critical reading of the manuscript and invaluable advice.

Sire Rating System™ 1994 Michael Helm

Library of Congress Catalog Card Number: 94-79186

Publishing Services: Penn&Ink
Design & Layout: Dayna Goforth Schippmann
Cover Design: Don McCartney

Published by City Miner Books, 1995
P.O. Box 176
Berkeley, CA 94701

ISBN 0-933944-16-0

Printed in the United States of America

10 9 8 7 6 5 4 3 2 1

*"If one were to await the possession
of absolute truth, one must either be a fool or a mute."*

— Jose Clemente Orozco

Table of Contents

Foreword

by Mark Cramer

Mike Helm, author of *A Breed Apart* and *Bred to Run* has reached an inevitable point in his career as writer and horseplayer. For this his first hard-core handicapping book, he has chosen to venture into uncharted territories. With all the volumes out there on class, form cycle, speed and pace, it is remarkable that this is the first book to base its methodology on pedigree, considering that breeding is the oldest factor in handicapping tradition.

In Helm's new book, for the first time, handicapping based on breeding is objectified and to a large extent quantified. While Helm was writing this book, he asked me to test his methods at the eastern tracks as a sort of control to what he was doing out west. At the windows, I got paid for my job.

The book includes a vintage chapter on the history of the breed. For handicappers, this section introduces a major contemporary issue, the tug-of-war between breeding for speed and stamina, and how this has affected the carding of races, which in turn has an impact on breeding methods.

Helm also explains how he derived a series of sire ratings, covering: (1) class; (2) ability to produce first-time-starter winners; (3) turf; (4) wet and tiring tracks; (5) stamina index (ability to stretch out); (6) age bias (if sire is clearly better with 2-year-olds or 3- to 4-year-olds).

Heretofore, pedigree class ratings have been derived from average earnings. Helm is acutely aware of how numbers crunchers tend to be overly trusting of one-dimensional statistics. He explains that averages are often distorted by one big earner and instead uses median earnings along with other elegant factors to project the class of a sire's offspring. Similarly his stamina index is derived in a more subtle fashion than the misleading average-distance-of-winners stat, which is distorted by the fact that most racing venues don't provide a whole lot of opportunities for routers to run at their ideal distances.

8   In the realm of turf, Helm illustrates how knowledge of pedigree, sire and dam, enables the handicapper to make objective investments on horses that have never before raced on the grass, often horses that show terrible dirt form. The average mutuel of this type of bet is unusually high, and yet, the how-to mechanics of this investment are readily applicable.

Helm's stamina index has helped me in the weakest area of my play: the stretchout. I now give less credence to the running style of the sprint when evaluating a young horse's ability to stretch out for the first time. I find that Helm's stamina index is a far more objective measure.

Helm's class ratings provide a remarkably simple way to eliminate lightly-raced horses in maiden races; but in order to achieve simplicity without being one-dimensionally simplistic, someone has to struggle first with complexity, and Helm has done that for us.

The balance and core of Helm's book involves the handicapping of specific races and how the breeding factor is integrated into the process. Helm differentiates between situations where breeding may be an only factor and the majority of times when pedigree is integrated with other factors. In each case, he gives the reader a lucid walk-through of the process, illustrating how and where the pedigree factor comes into play.

With the prior absence of guides as to how pedigree knowledge is integrated objectively in the handicapping process and the paucity of three-dimensional data on the class and stamina capability of sires, Helm's new book figures to offer a quantam leap in handicapping consciousness for most players.

Let me make it clear that Helm's approach is not simplistic. For example, he explains why a first-time-starter whose sire has gotten a high percentage of first-time winners is not an automatic bet. Toteboard and trainer specialty are also incorporated, and if that's not enough, the author introduces the concept of sire cycles.

For example, just because Mr. Prospector has shown a high percentage of first-time winners doesn't mean his debut horses are a good bet today, for as a horse progresses in his stud career, he may be hooked up with fewer brilliant, precocious mares because he no longer has to prove anything in the short run. With sire cycles to consider, fixed-percentage stats can sure be misleading.

With a real edge tougher to find within other areas of handicapping, due to a glut of information, the primitive factor of pedigree has been refined into a revolutionary new frontier. Helm explores previously uncharted territory methodically and with a fine sense for precision and nuances.

Mike Helm's *Exploring Pedigree: Handicapping's Newest Frontier* is a fresh alternative for horseplayers who have been attempting to grind it out by using the same tired factors as everyone else in the crowd.

The best news is that this book is not going to hurt the average mutuel for pedigree bets because the culture engrained in handicappers is to look to running lines in the past performances, and they will find it difficult to bet on information that projects future occurrences without past races to go on. First-time starters, first time stretching out, first time on the grass and various other pedigree gems will continue to be best bets for those who are ready and willing to apply the methods in this book.

Preface

After the high rolling mutuels that skillful use of Beyer and other speed figures enjoyed in the 1980s, winning with that approach has become much more difficult in the 1990s. One reason for this development is that the track takeout has generally remained constant while the top figure horses have increasingly been identified in the *Daily Racing Form* (as well as through various subscription services) and thus lost much of their pari-mutuel value.

Another reason for declining mutuels is the fact that fewer horses are being bred. With the reduction in the annual crop from 50,000 to 35,000 foals, the average number of horses entered in most races has also declined. This in turn has resulted in the odds and potential payouts for each horse being correspondingly shaved across the board. These days it is not unusual to see five and six-horse fields, not only at the stakes and allowance level, but also in mid-range claiming races. The upshot is that top figure horses which used to pay a minimum of six and seven dollars now often pay less than four in the win pool and little more than ten in the exactas.

Given this harsh reality, numbers players have increasingly been reduced to a tedious grind-it-out approach that is not only, as Mark Cramer has shown, barely profitable on a dollar-per-hour basis, but also destroys much of the fun of handicapping.

Fortunately, what the pari-mutuel system takes away with one hand, it inexorably gives back with another and those handicappers flexible enough to adjust their thinking are the ones most likely to succeed in the meaner and leaner '90s.

Within this context pedigree analysis — the oldest of handicapping approaches — is ironically making a comeback, albeit with a strong assist from the information gathering capacity of the computer revolution. Whereas knowledge of pedigree used to be the exclusive province of breeders, today anyone willing to make a modest effort can apprise themselves of how the progeny of various sires are likely to run, particularly in Maiden races and NW1 turf events which often have the largest fields and highest payoffs. Interestingly, it is precisely in these kinds of races — where an appropriately bred horse is trying a sprint, route or turf for the first time — that the top figure horse is most vulnerable and worth taking a shot against.

12 This is not to say that pedigree analysis, by itself, is sufficient. Sometimes it is, but most of the time it isn't. You can beat a race with it, but the overall art of handicapping will never be reduced to one monotheistic factor. An understanding of pedigree, however, when integrated with other factors such as trainer and workout analysis, toteboard and paddock inspection, or a knowledge of winning par times, can make a significant difference in a handicapper's bottom line.

What follows here is the outgrowth of several years of study and work. The race analysis and stallion ratings included are offered, not as the final word, but as points of departure for your own creativity in trying to beat a difficult game. Best of luck!

Mike Helm
October 1994
Berkeley, California

1
Origins of the Ratings

Since much of the pedigree analysis in *Exploring Pedigree* depends upon the accuracy of the sire ratings in the back of the book, handicappers have every reason to ask how they were derived and why they should prove useful. First off, handicappers should know that each rating for the established stallions is based on interpreting statistical information available from such sources as the *Daily Racing Form*, *Thoroughbred Times*, *The Blood-Horse* and Bloodstock Research Information Services.

Interpreting the statistical data on each sire is crucial because breeding is a dynamic and not a static process. Most of the information made available to handicappers is based on lifetime averages, even though the performance of a stallion's get can vary considerably from one crop to the next depending on where he is in his stud cycle and what kind of mares are being bred to him.

Stud managers, for example, are under tremendous pressure to "prove" their young stallions as soon as possible. Consequently, they tend to initially "book" them to the most precocious mares available in the hopes of getting a large number of juvenile winners. Young stallions whose initial 2-year-old crops fail to achieve good results are typically dismissed by commercial breeders and have trouble commanding continued patronage.

Once a young stallion has proven to be precocious, however, a stud manager may change the way he breeds him in order to further increase his stud value. He may be bred to stouter mares in order to increase his chances of getting a Classic winner and his percentage of juvenile and first-time winners will correspondingly go down. Additionally, if a sire has a grass pedigree, his turf ratings will often improve dramatically after his fourth and fifth crops because the turf favors older runners. Once he's proven himself on the turf, a stallion may be bred even more that way.

On the other end of the stud cycle, as most stallions age, their productivity — compared to their lifetime averages — typically declines because the class of mare breeders are willing to send to them generally drops. One of the strengths of *Exploring Pedigree* is that variables generated by the stud cycle are taken into account in arriving at a stallion's precocity, turf and class ratings.

Another useful feature in the *Exploring Pedigree* ratings is that handicappers are apprised of not only each sire's sire, but also of the sire of his dam. This can be helpful with a young stallion because knowledge of his broodmare sire can shed significant light as to whether his get are likely to be precocious or run well on the grass. Handicappers who knew, for example, that Gone West is out of a Secretariat mare were not surprised to see his progeny often run well on the grass, despite his dirt-oriented Mr. Prospector breeding.

Since a sire contributes only half of the genetic makeup of each foal, it is obviously important to pay attention to each dam. In a book of this size, unfortunately, it is impossible to list and rate every mare. Furthermore, the relatively small sample of foals each dam produces makes any rating system suspect. Interested handicappers can, however, supplement *Exploring Pedigree* by purchasing the most recent edition of *Maiden Stats* which is annually published by Bloodstock Research Information Services. Besides giving its sire's lifetime averages, *Maiden Stats* annually lists for each 2-year-old foal the number of winners to starters its dam has produced. Everything else being equal, a handicapper who knows one dam has four winners from five starters should prefer wagering on her foal rather than on the foal of another dam who is zero for five.

Another important factor handicappers should take note of in assessing the potential of a dam is the quality of her sire. Toward this end *Exploring Pedigree* has put in bold-faced type the names of those stallions that are important broodmare sires and whose daughters are more likely than usual to produce a quality foal. In choosing between contenders in races, handicappers should give extra consideration to young horses whose dams are by one of these sires.

Finally, some sires seem to have a pronounced gender bias when it comes to the sex of their winners. Those sires that currently have a 2-1 or better ratio in favor of either colt or filly winners have been respectively identified by placing a c or an f after their names.

With this as background, let's take a closer look at the criteria used for each rating category, remembering all the time that they usually need to be integrated with other handicapping factors for maximum effectiveness:

FTS - First-time starters: This category deals with how well a sire's progeny do as first-time starters. Given the vagaries of the stud cycle and the need for high-quality handicapping information, I have been extremely selective before giving a sire an A or B rating. While most industry pedigree analysts consider 15 percent debut winners to be the minimum standard for a precocious sire, *Exploring Pedigree* generally requires 18-25 percent debut winners before a sire is assigned a B rating and 26 percent or above for an A rating.

If there is a minus sign after an A or B rating, such as Mr. Prospector has, that means a sire's recent crops have been less precocious than his lifetime average indicate and that his progeny are less likely to win first time out than their ratings would otherwise indicate. Nevertheless, I have continued to assign them high FTS ratings because their loss in precocity may be a function of their being bred to stouter mares and, in any event, they remain important as broodmare sires whose daughters are now passing on their speed and precocity.

I have also sometimes upgraded a stallion, such as Its Freezing — who nominally falls within the 10-17 percent C rating — to a B rating because he has proven his consistency over a much larger number of starters than many of his more lightly bred, but higher-rated competitors.

Employing the same logic, I have frequently lowered the ratings of a lightly-bred stallion, such as Aras An Uachtarain, from an A to a B because his limited number of starters, in my opinion, has given him a higher percent of success than he is likely to sustain with subsequent crops.

Similarly, some lightly-bred sires that marginally qualified as a B have been lowered to C+ to caution handicappers from prematurely jumping on them. In fact, handicappers are well advised, especially with young stallions, to limit their initial bets until they see how well their progeny perform at their tracks. A number of cheaper regional sires have inflated ratings based on local class advantages that won't be duplicated when their progeny are shipped elsewhere.

Handicappers will also note that a number of the stallions listed in the ratings section have an asterisk * after their name. These are generally stallions whose progeny have only started between thirty and fifty times. Stallions in this category have been provisionally assigned A-F ratings. While helpful in identifying a lightly-bred sire's proclivities, these ratings should be viewed with caution because they can change dramatically in the light of only a few additional starts.

Stallions with less than thirty starters are also identified with an asterisk. In most cases, unless they have shown phenomenal early percentages, they have been assigned one of three ratings: L for Likely; M for Maybe; or U for Unlikely. These ratings involved the most guesswork and were determined by analyzing the race record and pedigree of each stallion. They should be viewed with caution except when a sire receives an underlined L̲ rating. In this context, L's should initially be treated as the equivalent of a B- rating, then adjusted upward or downward depending upon actual race results.

Those stallions whose first crop is racing in 1994 or will race in 1995 have been identified with a delta ^ after their names. These sires have also been provisionally assigned L, M, or U ratings and should be played with extreme caution. Only those runners with an underlined L̲ rating should be seriously considered. Dayjur, for example, received an L under the FTS category because, given his extremely precocious pedigree, he is quite likely to get a high number of debut winners from his first crop. Ls, in general, are only worth considering when all the experienced starters are proven losers, and all the other first-time starters have D, F or U ratings.

Finally, handicappers should note that every sire has been assigned an age bias (AB). Sires that get 25 percent or better 2-year-old winners to starters have been placed in the 2-3 category. If the 2 is underlined, that means the sire is extremely precocious and gets better than 30 percent juvenile winners to starters. Alternatively, if a sire gets a 3-4 rating, that means his progeny are more likely to break their maidens at three. If the 3 is underlined that means that less than 13 percent of their runners break their maidens at two. An underlined 3-4, such as Cox's Ridge, Devil's Bag, Nijinsky II, Pleasant Colony and Private Account receive, means a sire's progeny are late to mature and tend not to break their maidens until the late spring and summer of their third year. Horses by these sires, such as Cardmania, Region, Twilight Agenda and Valley Crossing, often do their best running at the age of four and beyond.

CL - Class: While a sire's FTS rating is important, so too is a knowledge of the class level at which his progeny typically win. Consequently, each sire in *Exploring Pedigree* has been given a class rating on a scale of 1-10, with 1 being the highest and 10 the lowest class rating. Assigning class ratings allows handicappers to combine them with their FTS ratings and thus "predict", for example, that a Slewdledo, with an A8 combined rating, is unlikely to be competitive with a Forty Niner and his A1 combined rating.

Though racing is full of examples of horses that transcend and outrun their class levels, the progeny of sires with class ratings between 1-5 will generally be competitive at the Maiden Special Weight, Allowance and stakes level; those with 5-7 class ratings will generally run well in high maiden claiming, claiming and allowance events; and sires at the 8 -10 level are most likely to produce runners that win at the bottom maiden and claiming levels. **Once a young horse has a race record, however, it is important to emphasize that his actual performance on each surface supercedes whatever class and other ratings his sire may have initially given him.**

While the breeding industry tends to define each stallion's class and stud fee by looking at his Average Earnings Index (AEI), I have used each sire's adjusted lifetime *median* earnings as the major criteria for establishing his class. This is because from a handicapping point of view a sire's median earnings are a far better indication of how his progeny *typically* run, than his average earnings which can be skewed by a few large earners.

Stud fees can also be misleading guides. A sire with a high stud fee, such as Capote who is at the $40,000 level, can have median earnings no greater than those of a sire such as Copelan whose stud fee is only $10,000.

Nevertheless, I have included the estimated 1994 stud fees for more than two hundred of the top stallions at the back of the book because a knowledge of them can be useful when handicapping for breeder intent in Maiden Claiming races. Knowing that a breeder, besides all the other expenses, has $45,000 invested in a stud fee in a Storm Bird colt, for example, and is running him for a Maiden Claiming $32,000 tag his first start, should make one cautious about backing it. Since the breeder is, in effect, publicly stating that he is willing to lose money on the colt, why should handicappers be enthusiastic at the windows?

Because of the nature of the stud cycle, even using a sire's lifetime median earnings to determine his class rating has its complexities. Where a stallion has a high lifetime median earning, but a suspiciously low stud fee and low last crop median earnings I have, in most cases, adjusted his class downward to reflect the fact that he is now getting lower quality mares. With a number of proven older and deceased stallions whose last crop median earnings were low (such as Mr. Prospector and Nijinsky II) I have, however, used their lifetime median earnings to determine their class ratings because their influence will now be primarily as a broodmare sire and their better running daughters are the ones most likely to be bred. These stallions have been identified with an asterisk after their class rating.

Establishing class ratings for those sires whose first crop of 2-year-olds will be running in 1994 and 1995 was more problematic. Their class ratings should be viewed as provisional and were determined by considering their stud fee, race record and pedigree. Stallions that earned more than $2,000,000 on the track usually received a class rating of 4, those that earned more than a million got a 5, and those that earned between $500,00 and a million received a 6. Some young stallions with good racetrack earnings, however, were given lower class ratings because they come from female families with unproductive histories and breeders are unlikely to send their best mares to them.

TRF - Turf: This is a rating for how well a sire's progeny generally do on the turf: Sires whose progeny win 20 percent or better of their starts have been assigned an A rating; 15-19 percent a B rating; 10-14 percent a C rating; 5-9 percent a D rating; and 0-4 percent an F rating. If an A or B is underlined, that means a sire's runners have been particularly effective their first time on the turf. When a sire is given an A- or B- the minus sign means his progeny may need a race or two on the turf before being good bets. Since opportunities vary considerably around the country for first-timers on the turf, the best rule of thumb is to initially use the overall rating and invoke the minus caveat primarily when a horse is competing against horses that have already run well or won on the turf. Similarly, if a horse is stretching out for the first time, caution is in order, unless it figures to be the lone speed and control the pace or has a stamina rating that projects a successful stretchout. As was the case with maiden first-time starters the progeny of unproven sires with L, M, or U turf ratings should be viewed with caution, unless the L is underlined.

OT - Off-Track: Establishing useful sire ratings for off-track performance is extremely difficult. This is partly because the nature of an off-track can vary considerably — depending not only upon its composition and texture, but also on its pitch, the rate and amount of rainfall, the temperature, and amount of wind and sunshine present as it is drying out. Sloppy surfaces, for example, seem to favor speed sires (that aren't out of Buckpasser mares) from the Mr. Prospector and In Reality lines, while tiring surfaces tend to favor sires from the Damascus line whose progeny seem to have the fortitude and physical conformation to get up from off the pace. To reflect

20 these different styles, *Exploring Pedigree* has two types of off-track ratings: The first, and by far the largest category, is primarily for wet tracks listed as sloppy and muddy; The second is for slow and tiring tracks where running times are typically seven or more ticks below par and/or the track variant published in *Daily Racing Form* is above 22.

Sires whose progeny are running on wet tracks with fast or good times have been rated as follows: A+ = 22 percent and above winners; A = 20-21 percent; B = 19 percent; C = 13-18 percent; D = 12 percent; and F = 11 percent or below winners. The breakoff points for each rating have been adapted from Mark Cramer's analysis of the bell curve for off-track winners published in the *Cramer/Olmsted Report*.

In order to maintain the integrity and utility of the highest off-track ratings, I have generally downgraded those stallions with a limited number of starters, which otherwise qualified for an A or A+ off-track rating, to a B rating. On the lower end, I have upgraded those sires with a large number of starters which fell in the 17-18 percent range to a C+ rating. As with the other L, M and U ratings, only those stallions with an underlined L should be considered for off-tracks and then only when none of the experienced horses have run well in the slop and none of the other first-time mudders are bred for it.

The T rating has been assigned to those sires whose progeny seem to move up on tiring off-tracks, especially in northern and southern California. These ratings are more impressionistic, based primarily on the author's own experience, and handicappers in other parts of the country should view them with caution until they are locally corroborated.

SI - Stamina Index: This index indicates how far the progeny of a particular sire are likely to be able to run. The stamina ratings in *Exploring Pedigree* are unique in that they are based on a detailed look at how far a sire's progeny have actually run. This approach is far more useful for handicapping purposes than employing either average winning distances or Dosage numbers. Average winning distance can be deceptive because of the fact that sprints dominate American racing. This not only results in shorter average winning distances for each sire, but also blurs important stamina distinctions beyond a mile. Dosage is limited by its theoretical bent and the fact that the sons and grandsons of *chef de races* don't necessarily run to type.

The criteria used for establishing the stamina index ratings in *Exploring Pedigree* are as follows: An SI rating of 1 means that a sire's progeny have demonstrated the ability to run 1 1/4 mile or more; a rating of 2 indicates a distance capacity of up to 1 1/8 mile; a rating of 3 suggests a range of up to 1 1/16 mile; a stamina rating of 4 means a stallion gets runners able to both sprint and win up to a mile; and an SI rating of 5 indicates a distance capacity of 6-7 furlongs.

If a stamina number is underlined, that indicates a sire's progeny are particularly effective at that distance. If the 1 is underlined that means the a sire is an extreme source of stamina. If the 4 is underlined, that means that a mile is the most frequent winning distance of a stallion's progeny. An underlined <u>5</u> indicates a stallion's progeny are especially brilliant and less likely to win beyond 6 furlongs.

A stamina number of 3 or 4 followed by an asterisk means the stallion is a mixer who can contribute enough speed to his progeny to get a sprinter if the broodmare was one or, alternatively, enough stamina to get a stayer if she was one. In most cases, a sire with a 4* rating (such as Mr. Prospector) will be more precocious than stout. He will get lots of sprint and mile winners, but need a mare by a very stout sire to produce a runner that can go further than 1 1/8 mile. Alternatively, a sire with a stamina rating of 3* (such as Seattle Slew) will generally be less precocious, but more likely to get a runner able to stay a 1 1/4 mile. A stamina index of 1* (such as Halo, Lyphard, Majestic Light and Silver Hawk have been assigned) indicates a stallion who is primarily a source of stamina, but who can get a debut sprint winner if bred to a precocious mare. While stallions with SI's of 1* sometimes get 6 furlong winners, their progeny are generally more effective in 6 1/2 and 7 furlong sprints.

It is important to note that the likely distance capacity of a sire's progeny must be adjusted by the stamina index of the broodmare sire, with that index counting roughly half that of the sire's because it is one generation further back in the pedigree. Also a stamina index of 2 or 3 for both sire and broodmare sire increases the potential stamina of the foal so that it might run further than either of its parents.

In conclusion, I want to reiterate that these ratings are not set in stone. The goal was to create ratings that transcend single misleading statistics and that incorporate vital nuances previously ignored by most pedigree analysts. While I believe the ratings accurate and state-of-the-art, all the while remaining mercifully simple to use, it is important to keep in mind that there will always be an element of mystery when it comes to pedigree and that no ratings can be infallible. The study of pedigree is not now, nor will it ever be a "deterministic" science.

Furthermore, because of the dynamic nature of the stud cycle, some of the ratings have had to be adjusted during the writing of this book as important changes in the way some stallions are being bred became more apparent. Similarly, because of possible local track biases, handicappers should feel free to adjust the ratings based on their own actual experience. I would, of course, be pleased to receive any information or ideas that would make future printings or editions of *Exploring Pedigree* more accurate and useful. Please send any comments care of *City Miner Books* whose address is listed on the copyright page at the front of the book. The sire ratings may also be periodically updated. If you want to be on that mailing list, drop a note to the publisher.

2

The Dynamic Between Speed and Stamina in the Evolution of the Thoroughbred

Involved in the immediate drama of each race most handicappers give little thought to the fact they are participating in an ancient spectacle. For human beings have been breeding, racing, and no doubt, wagering on horses since shortly after they were first domesticated some six thousand years ago. In the long history of the sport, the development of the pari-mutuel system and the art of handicapping are but recent episodes.

Within the broad four-and-half million year sweep of equine history, the Thoroughbred is also a relatively modern creation. Its evolution as a distinct breed began little more than three centuries ago during the reign of Charles II when English nobility first fused the explosive Quarter Horse-like speed and agility of Hobby and Galloway mares with the courage and stamina of a handful of imported "Oriental" stallions, the majority of which had been specifically bred for long-distance nocturnal raids by desert sheiks.

Though Charles II wouldn't have called himself the sport's first racing secretary, he essentially became that when he proclaimed the initial King's Plate (or Cup) at Newmarket in 1665. Charles's proclamation was significant because prior to that time English aristocrats exclusively bred their "blooded" stock to compete over hill and dale in six to twelve mile "hunting" events while the more common folk concentrated on "flat" racing at sprint distances of up to six hundred yards.

A radical innovation, the King's Plate revealed Charles II's preference for a faster-paced action. Rather than competing in twelve mile hunting events, Charles II wrote a series of heats for fully mature 6-year-old horses that required them to carry their speed and twelve stone (168 pounds) over the shorter distance of only four miles.

Scandalous as the shorter distance of the King's Plate must have seemed to traditionalists, Charles II was, after all, the King and English nobles and breeders quickly took to the challenge of trying to inject more speed into their studs.

While early on it was accurate to say that the fillies couldn't beat the boys, this was a distance-biased observation that gradually became obsolete as more stamina was bred into the female descendants of the original Hobby mare population and more speed into the male descendants of the original Oriental stallions. Ironically, today the thoroughbred's gene pool has been so recycled and transformed that most breeders, especially in North America, now look for speed from their stallions and stamina from their mares. Yet, even so, it should come as no surprise when a modern filly such as a Safely Kept or Meafera harks back to her Hobby ancestors and outsprints the best of the boys.

By the middle of the eighteenth century, with the gradual decline of the aristocracy and the corresponding rise of a wealthy, mercantile class in England, the bias towards more speed, shorter distances and earlier racing of horses was accelerated. Less patient than the nobility, few self-respecting burghers, it seems, were prepared to buy high-priced horseflesh and then wait six years for a little action and a possible return on their investment.

Not long after the American Declaration of Independence, at the behest of commercial breeders, the first of the classic English 3-year-old races the St. Leger was instituted in 1776, followed by the Oaks in 1779 and the Derby in 1780. While Great Britain might well lose its colonies, English breeders damned well intended to protect their investment in bloodstock.

The English or Epsom Derby at a 1 1/2 mile soon became the definition of a Classic race and the distance at which breeders aspired to to win. With the creation of the English stud book in 1791 as its linchpin, English Thoroughbred bloodstock dominated the export market throughout the world during the entire nineteenth and well into the twentieth century.

Toward the middle of the twentieth century, however, the primacy of English bloodstock began to wane when American A.B. "Bull" Hancock and Canadian E. P. Taylor imported Nasrullah and Nearctic. These two stallions respectively founded the speed-oriented Bold Ruler and Northern Dancer lines whose best contemporary male descendants include such prepotent sires as Seattle Slew, Storm Cat and Danzig.

By the 1970s the 1 1/2 mile Belmont Stakes was increasingly viewed by most American commercial breeders as an antiquated marathon and winning the Kentucky Derby at 1 1/4 mile became the most coveted prize.

In conjunction with the rise to prominence at stud of Bold Ruler in the 1970s, Northern Dancer in the early 1980s and Mr. Prospector and his sons in the 1990s, the international center of bloodstock moved from Newmarket in England to the Keeneland auctions in Lexington, Kentucky.

With the establishment of the Breeders' Cup Classic in 1984, proponents of speed and precocity further institutionalized the 1 1/4 mile definition of a Classic distance, winning yet another battle against those who favor greater patience in racing and breeding for stamina and soundness.

Yet, in the pursuit of ever more speed and a quicker turnover of dollars, even the 1 1/4 mile Classic distance did not long remain sacrosanct. Chagrined by the fact that no Breeders' Cup Juvenile winner has gone on to become a Classic champion at three, the more aggressive commercial breeders began to publicly suggest that shortening the Classic distance to 1 1/8 mile might not be a bad idea. Along with such horsemen as ex-quarter horse trainer D. Wayne Lukas, this vocal minority is on record as favoring the reduction of the Kentucky Derby to 1 1/8 mile so that precociously bred 2-year-olds have a better chance to win the most important Classic at three.

The bias toward shorter distances and ever more speed, however, may have reached a genetic limit. After all, the Thoroughbred is not a Quarter Horse. While the average winning times of Thoroughbreds have continued to improve over the past thirty years, at the top of the game the best horses are not running faster than their predecessors. It may be more than coincidental that Dr. Fager's Arlington mile record (1:32 1/5) set in 1968 while carrying 134 pounds and Secretariat's 1973 Kentucky Derby (1:59 2/5) and Belmont (2:24) times still stand.

Despite the financial pressure to train and race Thoroughbreds early, it may well be that shortening Classic distances and emphasizing 2-year-old racing is bad not only for the horses but also ultimately for the economics of the game. In the short run commercial breeders and some trainers may make out, but what about all the owners who, staggered by their juvenile losses, subsequently leave the game and warn off their friends?

The great majority of 2-year-olds, critics point out, are like teenage athletes in that only a few of them are physically and mentally mature enough to compete professionally. Under the stress of early training and racing, a lot of young horses are physically and psychologically traumatized in a way that could be avoided if they were given more time to mature. Some of them never get to the races, while others, compromised by early injury, either break down or begin to lose their speed, desire and economic value as they descend the claiming ladder.

Furthermore, its an open question whether premature racing of juveniles, along with widespread use of anti-inflammatory drugs, tarnishes racing's image and lessens the possibility of recruiting new fans to the sport. If owners, breeders and trainers, in their own self-interest, don't soon show more self-restraint, then there's the prospect of animal rights activists mobilizing public opinion and imposing it from without.

Still, it's easy to hold breeders to a higher moral standard than the rest of the culture, especially when the cost would come out of their pockets. The hard truth is that breeding and racing Thoroughbreds is an excruciatingly expensive business and there are no simple answers. Whatever the future of 2-year-old racing, it is important to acknowledge (in line with the historical pattern) that early speed and precocity have a huge edge in the contemporary game. Horses that don't have them don't win many races. And handicappers that don't bet on them, don't cash many tickets. With that in mind, let's turn our attention to handicapping a variety of maiden races.

3

The Case for Playing Maiden Races

Probably nothing is more disparaged by conventional gambling wisdom than playing maiden races. Yet, from a handicapping point of view, betting first-time starters by precocious sires in maiden races can be among the best plays at the track, especially when they are entered in Maiden Special Weight events. While as we shall see, opportunities also exist for cashing in on the progeny of cheaper precocious sires at the Maiden Claiming level, the Maiden Special Weight category is generally more reliable because none of the horses entered there are for sale. They are all being protected from the claiming box by their connections who obviously believe they have some ability and are likely to run to their breeding.

Conversely, a young horse with an expensive pedigree that is first started in a Maiden Claiming event is suspect. The fact that it is for sale suggests its connections lack confidence in its ability. It almost always has some hole in it and is often overbet compared to the get of cheaper precocious sires that are running where they belong.

Another thing that makes Maiden Special Weight races attractive, from a handicapping perspective, is that horses with vastly different chances of winning first time out are regularly entered into these events. Some of the first-time starters are by highly-touted, precocious sires such as Danzig, Forty Niner, Seeking The Gold and Storm Cat, while others are by less publicized sires such as Copelan, Dixieland Band and Saratoga Six that can sneak off at odds considerably greater than their genetic chances of winning.

Simultaneously, Maiden Special Weight races frequently also attract three or more first-time starters by sires whose genetic record strongly suggest they *won't* win early. Typically these are young horses sired by the likes of His Majesty, Pleasant Colony, Private Account, Cox's Ridge, Nijinsky II, Halo and Devil's Bag and/or out of daughters by stamina broodmare sires such as Al Hattab, Buckpasser, Damascus, Graustark, Hatchet Man, Roberto, Secretariat, No Robbery and Le Fabuleux.

Even though horses by these sires are not likely to win at first asking, they are entered into Maiden Special Weight sprints because they need the seasoning as a base

26 from which to subsequently stretch out to their best winning distance on either dirt or turf. This seasoning factor gives pedigree handicappers a tremendous edge in Maiden Special Weight sprints because they can throw out the non-precociously bred horses while concentrating on the ones that are. And they can do so with the ironic knowledge that, precisely because the non-precociously bred horses will have not performed well at first, they will often go off as juicy overlays when stretched out later.

First-time-starters in Maiden Special Weight events are further attractive because the past performances and speed figures of the horses that have run already are notoriously unreliable. This is so because most lightly-raced young horses are still in the process of discovering their talent and deciding whether or not they even want to compete on the track. One young horse may run the race of its life first time out, decide it doesn't like the stress of racing, and run progressively worse from then on. Another may run well and then "bounce" next time out from the soreness generated by a competitive initial effort. A third, less intelligent horse may race greenly for several races before understanding what is being asked of it.

This uncertainty over whether a young horse will continue to develop or regress often means that the maiden with the best recent speed figure in a race will be overbet proportionate to its chances of winning. The public forgets that maiden races are, by definition, events where all of the horses that have run have yet to demonstrate they have the will to win. Mesmerized by the top speed figure the public generally prefers to bet on the best past losing performance rather than on a first-time starter by a precocious sire with a winning genetic record. This is so even when the past performances of the horses that have run are below par and the genetic record of a particular sire of a first-time starter is superior.

Pedigree handicappers realize, of course, that in most maiden races a number of factors must be weighed before coming to a value-oriented betting decision. Even precociously-bred young horses are unlikely to break their maidens first time out of the box when they are in the hands of incompetent trainers or ones who don't try to win early. Correctly estimating the relative importance of the pedigrees of all the first-time starters in a race versus the race record, workout patterns, toteboard action and connections of the rest of the field is an ongoing challenge. With this in mind let's take a close look at a variety of juvenile maiden races with one or more precociously bred first-time starters.

4

Maiden Special Weight Races For 2-Year-Olds

Each year, beginning in May and heating up in June and July, the racing season begins in earnest for the most precocious members of the current 2-year-old crop. Owners and trainers have already selected the most mature of the lot and been training them at the track since February and March. Typically, "baby" races are first written at 3-and 4-1/2 furlongs, and then gradually extended up to 1 1/16 mile by Breeders' Cup day.

The vast majority of juvenile races, however, are maiden sprints where quick gate works, early speed and inside posts (except for the rail) are generally advantageous. The younger and less experienced the horse and the shorter the distance, the more these biases come into play. This is because young horses who fall behind tend to flinch from the dirt being kicked in their face and those parked on the outside haven't yet learned how to negotiate a wide turn. Keeping these factors in mind let's turn to some examples where identifying a precocious pedigree is a crucial element in finding a live horse. Included in our analysis is information from two excellent resources, Bloodstock Research Information Systems' *Maiden Stats 1994* and Bill Olmsted's *Trainer Pattern Pocket Guide*.

The fifth race at Belmont on June 23rd is an example of a playable debut longshot. Going over the six-horse field, the first thing to notice is that there is no clear favorite in this 5 furlong sprint for very young 2-year-olds. Dances In Gold has the highest speed figure and is the top choice at 2-1, but has already lost twice. At the short price, it's worth seeing if anyone can beat her.

Wilson's Courage is by Known Fact, an A-rated sire, but she's already lost to the favorite and actual race experience supercedes pedigree. While she might improve with blinkers, she hasn't worked in two weeks and might also bounce for her cold trainer.

This leaves us with the four debut horses to consider. Of these, Joe's Workingirl and Pretty Discreet are both getting bet at less than 3-1 and must be considered despite their mediocre precocity ratings. While Joe's Workingirl has a bullet work, Michael Daggett is not noted as a debut trainer and Air Forbes Won horses, according to his age bias of

5 Furlongs. (:56¹) MAIDEN SPECIAL WEIGHT. Purse $25,000. Fillies, 2–year–olds. Weight, 115 lbs.

START ▼
5 FURLONGS
▲ FINISH

Joe's Workingirl

Own: Singer Ann G
MCCAULEY W H (55 5 11 7 .09)

B. f. 2 (Apr)
Sire: Air Forbes Won (Bold Forbes)
Dam: Joe's Lil Girl (Sunrise Flight)
Br: Joseph B. Singer (Ky)
Tr: Daggett Michael H (13 2 1 0 .15)

C5/4

115

	Lifetime Record :		0 M 0 0		$0	
1994	0 M 0 0		Turf	0 0 0 0	
1993	0 M 0 0		Wet	0 0 0 0	
Bel	0 0 0 0		Dist	0 0 0 0	

WORKOUTS: ● Jun 21 Bel tr.t 4f fst :48³ H 1/13 Jun 2 Bel 5f fst 1:03 Bg 15/19 May 27 Bel 4f gd :49 Bg 4/10 May 22 Bel tr.t 5f fst 1:02¹ B 13/24 May 11 Bel 5f fst 1:01³ Hg 10/33 May 4 Bel 4f fst :49³ Bg 33/56
Apr 29 Bel 4f fst :48⁴ B 22/49 Apr 22 Bel 3f fst :37³ B 11/17 Apr 12 Bel tr.t 3f fst :39 B 15/17

Wilson's Courage

Own: Paraneck Stable
DAVIS R G (174 17 18 33 .10)

Dk. b or br f. 2 (Mar)
Sire: Known Fact (In Reality)
Dam: Jane Arnold (Damascus)
Br: Dr & Mrs R Smiser West & Mr & Mrs M Miller (Ky)
Tr: Sweigert Laura (9 0 0 1 .00)

A5/4 *C2/2* *BL*

115

	Lifetime Record :		1 M 0 1		$3,000	
1994	1 M 0 1		Turf	0 0 0 0	
1993	0 M 0 0		Wet	0 0 0 0	
Bel	1 0 0 1	$3,000	Dist	1 0 0 1	$3,000

9Jun94-3Bel fst 5f :22² :45⁴ :58² ⓕMd Sp Wt 66 4 3 31 21 32½ 34½ Davis R G 115 17.10 84 – 14 Pent Up Kiss115² DancesInGold115²½ Wilson'sCourage115⁴ Lacked rally 6
WORKOUTS: ● Jun 8 Aqu 3f fst :36¹ H 1/6 Jun 5 Aqu 5f fst 1:01² Hg 5/6 May 30 Aqu 5f fst :49⁴ Hg 2/5 May 25 Aqu 3f fst :37³ B 4/6

Love Tunnel

Own: Kinsman Stable *OB*
CHAVEZ J F (207 31 35 21 .15)

Ro. f. 2 (Mar)
Sire: Mining (Mr. Prospector)
Dam: Leveraged Buyout (Spectacular Bid)
Br: Kinsman Stud Farm (Ky)
Tr: Domino Carl J (12 1 2 1 .08)

BT4/4 2-3 *P6/2* *DAN 2⁸ᵗ* *debut*

115

	Lifetime Record :		0 M 0 0		$0	
1994	0 M 0 0		Turf	0 0 0 0	
1993	0 M 0 0		Wet	0 0 0 0	
Bel	0 0 0 0		Dist	0 0 0 0	

WORKOUTS: Jun 14 Bel 5f fst 1:00 H 2/19 Jun 9 Bel 5f fst 1:01⁴ B 12/21 May 28 Bel 4f fst :51 B 76/93

Doppio Espresso

Own: R F T Roses Stable
SANTOS J A (166 33 24 27 .20)

Ch. f. 2 (Mar)
Sire: Java Gold (Key to the Mint)
Dam: Eastern Dawn (Damascus)
Br: Dr. & Mrs. R. Smiser West & Mr. & Mrs. M. Miller (Ky)
Tr: Rice Linda (10 1 4 2 .10)

115

	Lifetime Record :		1 M 1 0		$5,500	
1994	1 M 1 0		$5,500	Turf	0 0 0 0	
1993	0 M 0 0		Wet	0 0 0 0	
Bel	1 0 1 0	$5,500	Dist	1 0 1 0	$5,500

Entered 22Jun94- 8 BEL
9Jun94-5Bel fst 5f :22¹ :45⁴ :58 ⓕMd Sp Wt 75 3 2 1ʰᵈ 2ʰᵈ 21 24 Santos J A 115 *1.50 87 – 14 Whirl's Girl115⁴ Doppio Espresso115³½ Garden Secrets115⁶ Second best 5
WORKOUTS: Jun 16 Bel 4f fst :48¹ B 9/58 Jun 5 Bel 4f fst :47² Hg 3/60 May 22 Bel 4f fst :48² B 24/120 May 12 Bel 4f fst :48³ H 7/44 May 4 Bel tr.t 3f fst :36² B 4/22 Apr 28 Bel tr.t 3f fst :37⁴ B 4/11

Dances In Gold

Own: Eberhard Johanne D
DOUGLAS R R (2 0 1 0 .00)

B. f. 2 (Mar)
Sire: Pentelicus (Fappiano)
Dam: Gold Edition (Dance in Time)
Br: Nada Kingwell (Fla)
Tr: Hine Hubert (2 1 1 0 .50)

L7/5 2-3 *D8/4*

115

	Lifetime Record :		2 M 1 0		$6,200	
1994	2 M 1 0		$6,200	Turf	0 0 0 0	
1993	0 M 0 0		Wet	0 0 0 0	
Bel	1 0 1 0	$5,500	Dist	1 0 1 0	$5,500

9Jun94-3Bel fst 5f :22² :45⁴ :58² ⓕMd Sp Wt 74 6 2 1½ 11 11½ 12 Douglas R R 115 3.50 87 – 14 Pent Up Kiss115² Dances In Gold115²½ Wilson's Courage115⁴ Weakened 6
21Apr94-3Hia fst 3f :21³ :33² ⓕMd Sp Wt — 7 4 3ⁿᵏ 41 Douglas R R 116 3.90 94 – 12 Fanny's Image116ⁿᵒ Quick Brew116ʰᵈ Singular Pleasure116¾ Weakened 10
WORKOUTS: Jun 18 Mth 5f fst 1:03 B 12/31 Jun 2 Mth 4f fst :48² B 4/29 May 23 Mth 3f fst :35⁴ B 1/3 Apr 17 GP 3f fst :36² B 3/7 Apr 13 GP 3f fst :38 B 2/2 Apr 4 GP 3f fst :39³ B 6/7

Pretty Discreet

Own: Robsham Einar P
BAILEY J D (168 40 24 32 .24)

B. f. 2 (Apr)
Sire: Private Account (Damascus)
Dam: Pretty Persuasive (Believe It)
Br: E. Paul Robsham (Ky)
Tr: Terrill William V (37 6 3 3 .16)

C+2/2 3-4 *D6/3* *3/4* *debut*

115

	Lifetime Record :		0 M 0 0		$0	
1994	0 M 0 0		Turf	0 0 0 0	
1993	0 M 0 0		Wet	0 0 0 0	
Bel	0 0 0 0		Dist	0 0 0 0	

WORKOUTS: Jun 20 Bel 4f fst :51² B 32/37 Jun 14 Bel 5f fst 1:02¹ Bg 13/19 Jun 8 Bel 4f fst :47³ Hg 11/66 Jun 2 Bel 4f fst :47² H 4/41 May 29 Bel 3f fst :36 H 5/25 May 22 Bel 4f fst :52 B 119/120

Ross Creek

Own: Hawkins Gregory D
SMITH M E (196 42 27 27 .21)

Dk. b or br c. 2 (Apr)
Sire: Gulch (Mr. Prospector)
Dam: Promising Times (Olden Times)
Br: Dr. & Mrs. Kirk A. Shiner (Ky)
Tr: Freeman Willard C (9 0 1 3 .00)

C5/4*

115

	Lifetime Record :		0 M 0 0		$0	
1994	0 M 0 0		Turf	0 0 0 0	
1993	0 M 0 0		Wet	0 0 0 0	
Bel	0 0 0 0		Dist	0 0 0 0	

WORKOUTS: Jun 19 Bel 4f fst :53 B 45/46 Jun 14 Bel 4f fst :50² Bg 41/68 Jun 8 Bel 5f fst 1:00² H 7/22 Jun 3 Bel 5f fst 1:02⁴ B 13/15 May 28 Bel 4f fst :49¹ B 34/93 May 21 Bel 4f gd :50¹ B 22/26

Towering Princess

Own: Sigel Marshall E *OB*
BECKNER D V (128 10 9 14 .08)

B. f. 2 (Apr)
Sire: Irish Tower (Irish Castle)
Dam: Ward's Princess (Wardlaw)
Br: Marshall Sigel (Ky)
Tr: Picou James E (9 0 1 0 .00)

B5/4 *C6/4* *DAN 0/1 2yr*

108⁷

	Lifetime Record :		0 M 0 0		$0	
1994	0 M 0 0		Turf	0 0 0 0	
1993	0 M 0 0		Wet	0 0 0 0	
Bel	0 0 0 0		Dist	0 0 0 0	

WORKOUTS: Jun 20 Bel 3f fst :36⁴ B 8/15 Jun 13 Bel 4f fst :48 H 3/31 Jun 8 Bel 5f fst 1:00³ H 9/22 May 30 Bel 4f fst :51 Bg 39/47 May 26 Bel 4f gd :52⁴ B 15/17 May 12 Bel 4f fst :51 Bg 29/44
May 3 Bel 3f fst :37 Bg 9/31

3-4, don't generally break their maidens at two. Furthermore, he's on the rail and likely to get pinned inside Wilson's Courage and Dances In Gold's speed.

Pretty Discreet, one the other hand (according to *Olmsted's Trainer Guide*) does have an excellent debut trainer in William Terrill. But, both Private Account and Believe It, her sire and broodmare sire, do better with 3-year-olds and their progeny normally need at least 6 1/2 furlongs. Pretty Discreet might beat us, but since this is a 5 furlong race for baby 2-year-olds, we're going to toss her out on pedigree grounds alone.

Love Tunnel (B+4) and Towering Princess (B5), on the other hand, are both bred to win early. Of the two, Love Tunnel has the higher precocity and better class rating and is also being bet some at 6-1, compared to Towering Princess's 22-1. Furthermore, Mining's underlined 2-3 age bias indicates he gets lots of juvenile winners and Love Tunnel's dam (according to *Maiden Stats 1994*) won with her only 2-year-old starter.

So how to bet? Given her 6-1 odds, Love Tunnel is definitely worth a Win bet. Since almost anyone can finish second and this is a short field, two alternative backup bets also make sense: One can either bet Place or make a small backwheel with Love Tunnel under the rest of the field in the exacta.

FIFTH RACE 5 FURLONGS. (.56¹) MAIDEN SPECIAL WEIGHT. Purse $25,000. Fillies, 2-year-olds. Weight, 115 lbs.

Belmont
JUNE 23, 1994

Value of Race: $25,000 Winner $15,000; second $5,500; third $3,000; fourth $1,500. Mutuel Pool $157,411.00 Exacta Pool $326,251.00

Last Raced	Horse	M/Eqt. A.Wt	PP	St	1/16	3/8	Str	Fin	Jockey	Odds $1
	Love Tunnel	2 115	3	3	5³	5⁶	4ʰᵈ	1ⁿᵏ	Chavez J F	6.20
	Pretty Discreet	2 115	5	5	3²½	3²	1ʰᵈ	2²	Bailey J D	2.20
9Jun94 ³Bel²	Dances In Gold	2 115	4	2	1¹½	1ʰᵈ	2²½	3⁵	Douglas R R	2.00
	Joe's Workingirl	2 115	1	4	4³½	4³	5⁹	4³	McCauley W H	2.80
9Jun94 ³Bel³	Wilson's Courage	b 2 115	2	1	2½	2¹	3½	5⁵½	Davis R G	6.40
	Towering Princess	b 2 108	6	6	6	6	6	6	Beckner D V⁷	21.90

OFF AT 2:56 Start Good. Won driving. Time, :22¹, :46¹, :59 Track fast.

$2 Mutuel Prices:
3-(C)-LOVE TUNNEL	14.40	6.60	2.60	
5-(F)-PRETTY DISCREET		4.00	2.20	
4-(E)-DANCES IN GOLD			2.20	

$2 EXACTA 3-5 PAID $65.40

Ro. f, (Mar), by Mining–Leveraged Buyout, by Spectacular Bid. Trainer Domino Carl J. Bred by Kinsman Stud Farm (Ky).

Not all Maiden races, of course, payoff in double digit figures. Frequently, an obvious pedigree contender is a low-priced overlay that is nevertheless worth playing. The third race at Belmont on May 24th is a case in point.

Of the eight first-time starters in this 5 furlong sprint, Susan's Choice is truly a standout. Not only does her sire Storm Cat have an A1 rating with an underlined (better-than-thirty percent) 2-3 age bias, but no other horse in the field has even a B rating. Furthermore, her dam Elopement (according to *Maiden Stats 1994*) already has four winners out of six starters and her class rating of 1 towers above the field. Susan's Choice also has a top jockey up in Craig Perret and a :48 breezing bullet work. The only negative is that she has been shipped in from Pimlico by an obscure trainer.

5 Furlongs. (:56¹) **MAIDEN SPECIAL WEIGHT. Purse $25,000. Fillies, 2-year-olds. Weight: 115 lbs.**

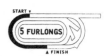

START ▼
5 FURLONGS
▲ FINISH

Finito   Ch. f. 2 (Mar)   *C7/1*

Own: Gracie Hill Stable

Sire: Ends Well (Lyphard)
Dam: Orteen (Believe the Queen)
Br: Edward L. Shapoff (NY)
Tr: Shapoff Stanley R (16 0 5 4 .00)

SANTAGATA N (62 5 14 6 .08) **115**

Lifetime Record :	1 M 0 1		$3,000
1994	1 M 0 1	$3,000	Turf 0 0 0 0
1993	0 M 0 0		Wet 0 0 0 0
Bel	1 0 0 1	$3,000	Dist 1 0 0 1 $3,000

13May94– 3Bel fst 5f :23² :47³ 1:00 ⒻMd Sp Wt 31 5 3 3½ 41¼ 3² 38¼ Santagata N 115 *1.00e 73–14 All Fast For You115⁶¼ Forever Proud115² Finito115¹½   No late bid 8

WORKOUTS: May 21 Bel tr.t 5f gd 1:02¹ B 3/10   May 9 Bel tr.t 4f gd :47⁴ H 2/9   May 4 Bel 4f fst :52 B 54/56   Apr 29 Bel 4f fst :50 B 33/49   Apr 24 Bel 4f fst :52 Bg 24/32   Apr 20 Bel 3f fst :38⁴ Hg 11/13

Lakespeed   B. f. 2 (Mar)   *D5/3*

Own: Bayne M

Sire: Slewpy (Seattle Slew)
Dam: Heart of America (Northern Jove)
Br: Arthur I. Appleton (Fla)
Tr: Contessa Gary C (15 2 3 3 .13)

LUTTRELL M G (69 5 7 8 .07) **110⁵**

Lifetime Record :	0 M 0 0		$0
1994	0 M 0 0		Turf 0 0 0 0
1993	0 M 0 0		Wet 0 0 0 0
Bel	0 0 0 0		Dist 0 0 0 0

WORKOUTS: May 15 Bel 4f fst :51 B 46/57   Apr 24 Hia 4f gd :51¹ B 14/16   Apr 16 Hia 3f fst :37 Bg 10/29   Apr 9 Hia 3f fst :37 Bg 10/22

Emperator Mistress   B. f. 2 (Mar)   *U8/1*

Own: Perez Robert

Sire: Hamza (Northern Dancer)
Dam: Heather V. (Empery)
Br: Robert Perez (NY)
Tr: Callejas Alfredo (6 1 1 0 .17)

CHAVEZ J F (82 11 15 9 .13) **115**

Lifetime Record :	0 M 0 0		$0
1994	0 M 0 0		Turf 0 0 0 0
1993	0 M 0 0		Wet 0 0 0 0
Bel	0 0 0 0		Dist 0 0 0 0

WORKOUTS: May 19 Bel 4f sly :51 B (d) 11/22   May 1 Bel 3f fst :37³ B 7/18   Apr 9 Bel tr.t 4f fst :50⁴ B 60/72   Apr 3 Bel tr.t 3f fst :39⁴ B 16/16

Heather Hawk   Dk. b or br f. 2 (Feb)   *C4/1*

Own: Live Oak Plantation

Sire: Silver Hawk (Roberto)
Dam: Ersatz (Clever Trick)
Br: Live Oak Stud (Fla)
Tr: Kelly Patrick J (12 1 0 3 .08)

MAPLE E (37 2 5 6 .05) **115**

Lifetime Record :	0 M 0 0		$0
1994	0 M 0 0		Turf 0 0 0 0
1993	0 M 0 0		Wet 0 0 0 0
Bel	0 0 0 0		Dist 0 0 0 0

WORKOUTS: May 20 Bel 4f sly :48⁴ H 4/13   May 15 Bel 3f fst :36⁴ Bg 7/19   May 10 Bel 3f fst :37 B 17/53   May 3 Bel 4f fst :50 Bg 32/47   Apr 29 Bel 3f fst :37³ Bg 9/23

Susan's Choice   *oK*   Dk. b or br f. 2 (Apr)   *A1/3   2–3*   *D4/6*

Own: Blusiewicz Leon J

Sire: Storm Cat (Storm Bird)
Dam: Elopement (Gallant Romeo)
Br: River Bend Farm & Mrs. W. L. Lyons Brown (Ky)
Tr: Blusiewicz Leon J (—)

PERRET C (8 0 1 2 .00) **115**

Lifetime Record :	0 M 0 0		$0
1994	0 M 0 0		Turf 0 0 0 0
1993	0 M 0 0		Wet 0 0 0 0
Bel	0 0 0 0		Dist 0 0 0 0

WORKOUTS: May 20 Pim 4f fst :49³ Bg 7/18   May 14 Pim 5f fst 1:01 H 6/14   May 7 Pim 5f fst 1:02² B 5/17   ●May 1 Pim 4f fst :48² B 1/5   Apr 25 Pim 4f fst :48 H 2/9

Fight Over Beauty   Gr. f. 2 (Apr)   *C7/4*

Own: Two Sisters Stable

Sire: Fight Over (Grey Dawn II)
Dam: Mighty Molina (Black Molo)
Br: Robert Price (Fla)
Tr: Pierce Joseph H Jr (5 1 0 0 .20)

DAVIS R G (68 8 4 18 .12) **115**

Lifetime Record :	0 M 0 0		$0
1994	0 M 0 0		Turf 0 0 0 0
1993	0 M 0 0		Wet 0 0 0 0
Bel	0 0 0 0		Dist 0 0 0 0

WORKOUTS: Apr 28 GS 4f fst :49⁴ B 6/11   Apr 12 PBD 4f gd :52 B 1/2   Apr 7 PBD 3f fst :38 B 1/2   Mar 2 Hia 3f fst :36¹ H 4/17   Feb 20 PBD 3f gd :38 B 4/8

Journey Proud   Ch. f. 2 (Apr)   *M7/3*

Own: Loblolly Stable

Sire: Idabel (Mr. Prospector)
Dam: Mrs. Beeton (Temperence Hill)
Br: Loblolly Stable (Ky)
Tr: Bohannan Thomas (2 0 0 1 .00)

VELAZQUEZ J R (73 9 12 12 .12) **115**

Lifetime Record :	0 M 0 0		$0
1994	0 M 0 0		Turf 0 0 0 0
1993	0 M 0 0		Wet 0 0 0 0
Bel	0 0 0 0		Dist 0 0 0 0

WORKOUTS: May 11 Bel 4f fst :49² B 21/49   May 5 Bel 4f fst :49⁴ B 16/22

Kathey's Attitude   B. f. 2 (Mar)   *M6/5*

Own: Weber Larry A

Sire: Glitterman (Dewan)
Dam: Allaise (Superbity)
Br: Lucas Farm, Inc. (Fla)
Tr: Rice Linda (2 0 1 0 .00)

SANTOS J A (76 13 16 12 .17) **115**

Lifetime Record :	1 M 1 0		$3,000
1994	1 M 1 0	$3,000	Turf 0 0 0 0
1993	0 M 0 0		Wet 1 0 1 0 $3,000
Bel	0 0 0 0		Dist 0 0 0 0

15Apr94– 4Kee sly 4½f :23³ :48² :55 ⒻMd Sp Wt 40 4 1 2hd 1hd 2¾ Sellers S J 117 5.10 83–25 KristyPig117¾ Kthy'sAtttud117² ProsprousUpstrt117²   Bid, led, 2nd best 8

WORKOUTS: May 12 Bel 4f fst :49 B 11/44   May 4 Bel 4f fst :48² B 10/56   Apr 28 Bel tr.t 4f fst :50 B 9/11   Apr 8 Kee 4f fst :50 Bg 3/28   Apr 6 GP 3f fst :38 B 5/9   Mar 29 GP 3f fst :38³ B 14/15

But, the trainer also owns the horse (usually a positive sign) and thinks enough of Susan's Choice's chances to ship her all the way to Belmont for the $25,000 purse. Last, but not least, she has been bet down to 3-2.

My own line makes Susan's Choice even-money which means that, if she ran against this weak field a hundred times, she should win at least half the time. At 3-2 her odds are low, but lest we get too greedy, it is important to emphasize that Susan's Choice does represent a 50 percent overlay on my line!

The only other horse that might contend is Kathey's Attitude who showed speed in her first race at Keeneland, but is now parked on the outside. She is by the cheap, but promising freshman sire Glitterman who should get a lot of sprint winners as indicated by his underlined stamina rating of 5. Still Storm Cat is a proven precocious sire with a huge class edge and that makes it hard to like Kathey's Attitude for anything but the Place position.

Results: Susan's Choice drew off to win by six with Kathy's Attitude securing the place by another seven lengths and the exacta paying $17.00. Interestingly, Journey Proud by Idabel (a cheap son of speed sire Mr. Prospector), closed for the show. If it had been a trifecta race, she would have been the logical third pedigree play.

THIRD RACE   5 FURLONGS. (.56¹) MAIDEN SPECIAL WEIGHT. Purse $25,000. Fillies, 2-year-olds. Weight: 115 lbs.

Belmont
MAY 24, 1994

Value of Race: $25,000 Winner $15,000; second $5,500; third $3,000; fourth $1,500. Mutuel Pool $141,425.00 Exacta Pool $334,635.00

Last Raced	Horse	M/Eqt. A.Wt	PP	St	3/16	3/8	Str	Fin	Jockey	Odds $1
	Susan's Choice	2 115	5	3	3½	3²½	1½	1⁶	Perret C	1.50
15Apr94 ⁴Kee²	Kathey's Attitude	b 2 115	8	5	2²½	2¹½	2²	2⁷	Santos J A	2.50
	Journey Proud	2 115	7	7	4²	4²½	4²½	3¹	Velazquez J R	16.90
	Lakespeed	2 110	2	2	1¹	1½	3³	4³	Luttrell M G⁵	14.00
	Heather Hawk	2 115	4	8	7⁹	6½	5¹	5³½	Maple E	6.30
	Fight Over Beauty	2 115	6	4	5½	5ʰᵈ	6²½	6⁴	Davis R G	19.40
13May94 ³Bel³	Finito	2 115	1	1	6³½	7⁸	7⁶	7⁶	Santagata N	6.00
	Emperator Mistress	2 115	3	6	8	8	8	8	Chavez J F	10.60

OFF AT 2:00 Start Good. Won driving. Time, :22¹, :46², :59 Track fast.

$2 Mutuel Prices:

5-(E)-SUSAN'S CHOICE	5.00	3.00	2.80
8-(H)-KATHEY'S ATTITUDE		3.00	2.60
7-(G)-JOURNEY PROUD			5.60

$2 EXACTA 5-8 PAID $17.00

Dk. b. or br. f, (Apr), by Storm Cat–Elopement, by Gallant Romeo. Trainer Blusiewicz Leon J. Bred by River Bend Farm & Mrs. W. L. Lyons Brown (Ky).

The third race at Hialeah on May 12th is another example with all the ingredients of a low-priced overlay. It is a restricted 3 furlong sprint for colts and fillies bred in Florida where speed and precocity are even more at a premium.

Looking at the past performances, Title Holder, the lone filly in the field, is a standout. Her sire Copelan's B5 rating with an underlined 2-3 age bias, not only makes

3

3 Furlongs. (:32²) MAIDEN SPECIAL WEIGHT. Purse $14,000 (plus $1,400 FOA). 2-year-olds. Weight, 116 lbs.

Christina Prince

L 7|4

Own: David W R

TORIBIO A R (134 17 19 25 .13)

116

WORKOUTS: May 7 Crc 3f fst :38³ Bg47/69

			Sire: Prince of Fame (Fappiano)
			Dam: Lady Christina (Welsh Saint)
			Br: William R. Davis (Fla)
			Tr: White William P (46 11 6 6 .24)

B. c. 2 (Apr)

Lifetime Record :	0 M 0 0		$0		
1994	0 M 0 0		Turf	0 0 0 0
1993	0 M 0 0		Wet	0 0 0 0
Hia	0 0 0 0		Dist	0 0 0 0

Graig Diamond Star

M 8|4

Own: Woodson Joanna

NO RIDER (—)

116

B. b or br c. 2 (Apr)
Sire: Racing Star (Baldski)
Dam: Fast Diamond (Diamond Prospect)
Br: Joanne Woodson & Irvin Woodson (Fla)
Tr: Woodson Kenneth J (—)

Lifetime Record :	0 M 0 0		$0		
1994	0 M 0 0		Turf	0 0 0 0
1993	0 M 0 0		Wet	0 0 0 0
Hia	0 0 0 0		Dist	0 0 0 0

WORKOUTS: May 8 Hia 3f fst :37 B 4/10 • May 3 Hia 3f fst :37¹ Bg6/14 • Apr 27 Hia 3f sly :52¹ B 3/3 • Apr 19 Hia 3f fst :37⁴ Bg7/12 • Apr 16 Hia 3f fst :36¹ H 3/29 • Apr 9 Hia 3f fst :37 B 10/22

Title Holder

ol B   *B 5|4*

Own: Hooper Fred W

RAMOS W S (222 37 28 37 .17)

113

B. f. 2 (Apr)
Sire: Copelan (Tri Jet)
Dam: Ski Title (Baldski)
Br: F. W. Hooper (Fla)
Tr: White William P (46 11 6 6 .24)

Lifetime Record :	0 M 0 0		$0		
1994	0 M 0 0		Turf	0 0 0 0
1993	0 M 0 0		Wet	0 0 0 0
Hia	0 0 0 0		Dist	0 0 0 0

WORKOUTS: May 7 Crc 3f fst :37 Bg9/69

Spotter Man

C 7|4

Own: Smith Robert G

FERRER J C (126 12 15 12 .10)

116

B. c. 2 (Apr)
Sire: Hooched (Danzig)
Dam: National Spot (National)
Br: Marilyn Lewis (Fla)
Tr: Smith Robert G (1 0 1 0 .00)

Lifetime Record :	0 M 0 0		$0		
1994	0 M 0 0		Turf	0 0 0 0
1993	0 M 0 0		Wet	0 0 0 0
Hia	0 0 0 0		Dist	0 0 0 0

Leap With Joy

C 6|4

Own: Lewis James Jr

RUSS M L (16 1 2 2 .06)

116

B. c. 2 (Mar)
Sire: Premiership (Exclusive Native)
Dam: Bouncing Joy (Giboulee)
Br: Ocala Stud Farm (Fla)
Tr: Tortora Emanuel (70 6 9 11 .09)

Lifetime Record :	0 M 0 0		$0		
1994	0 M 0 0		Turf	0 0 0 0
1993	0 M 0 0		Wet	0 0 0 0
Hia	0 0 0 0		Dist	0 0 0 0

WORKOUTS: May 8 Crc 4f fst :49² B 4/19 • May 3 Crc 3f fst :36³ H 4/19 • Apr 29 Crc 3f sly :37² B (d)5/18 • Apr 21 Crc 3f fst :36¹ H 2/24 • Apr 15 Crc 3f fst :37² B 6/13 • Apr 2 Crc 3f fst :36⁴ Hg20/61

Classic Quixote

U 9|4

Own: Foxcroft Farms

RIVERA J A II (135 15 18 26 .11)

116

B. c. 2 (Mar)
Sire: Dawn Quixote (Grey Dawn II)
Dam: Henrietta Hat (Tarboosh)
Br: F. Berens & S. Garazi (Fla)
Tr: Stanchfield Glenn M (11 0 4 3 .00)

Lifetime Record :	0 M 0 0		$0		
1994	0 M 0 0		Turf	0 0 0 0
1993	0 M 0 0		Wet	0 0 0 0
Hia	0 0 0 0		Dist	0 0 0 0

WORKOUTS: May 8 Crc 5f fst 1:04 B 10/19 • May 5 Crc 3f fst :36³ Bg2/24 • May 1 Crc 4f fst :49² B 6/28 • Apr 28 Crc 3f gd :38³ Bg5/11 • Apr 24 Crc 4f fst :50³ B 11/27 • ● Apr 17 Crc 3f fst :36 H 1/23
Apr 10 Crc 3f fst :36² H 3/25 • Apr 3 Crc 4f fst :48¹ H 7/52

Andrews Buddy

N|A

Own: X Bar Ranch

BAIN G W (51 0 5 2 .00)

116

B. c. 2 (Apr)
Sire: Buddy Breezin (J. O. Tobin)
Dam: Beth's Rushing (Whirling Saucer)
Br: Bruce Remsburg (Fla)
Tr: Nazareth John (12 1 3 0 .08)

Lifetime Record :	0 M 0 0		$0		
1994	0 M 0 0		Turf	0 0 0 0
1993	0 M 0 0		Wet	0 0 0 0
Hia	0 0 0 0		Dist	0 0 0 0

her the only precociously-bred horse but also the classiest one in the race. Furthermore, her dam Ski Title already has produced a $126,000 winner that broke her maiden at two. Title Holder's connections are also excellent. Trainer William White has a 24 percent win record and (according to Olmsted's *Trainer Guide*) excels with first-time starters. The icing on the cake is that she has been bet down to 6-5 from her morning line of 3-1.

Heavy toteboard action of this kind on debut runners is particularly revealing because it usually doesn't come from the public, but rather from people connected to the backstretch who have been able to assess the horse's talent during its morning works. In fact, early toteboard action that sees a first-time starter open considerably below its morning line is probably more significant in maiden races than any other kind of event.

Results: Given Title Holder's superior pedigree, connections and her toteboard action, she probably should have gone off at 4-5. Instead, she was another low-priced overlay at 6-5 and paid $4.40 to win. Even though horses like Title Holder are psychologically hard for value-oriented handicappers to play, a 50 percent overlay on your line is a 50 percent overlay, whether it means betting a horse that should be 4-5 at 6-5 or one that should be 4-1 at 6-1. The important thing to remember about short-priced overlays is that their probability of coming in is greater than with longer-priced overlays and they can psychologically bolster your confidence by minimizing your losing streaks or stopping them before they begin.

Considerably more complex to analyze is the seventh race at Hollywood Park on June 19th which was a 5 furlong Maiden Special Weight event consisting of a field of ten juvenile starters.

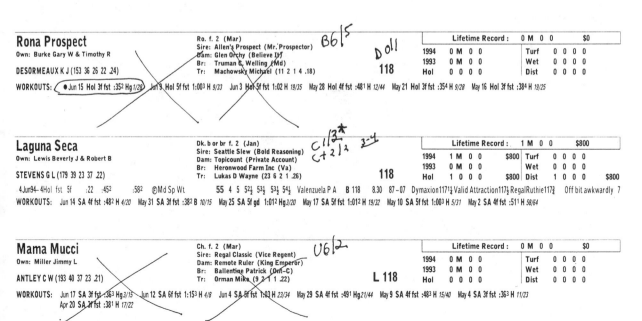

Hollywood Park

7

5 Furlongs. (:56²) MAIDEN SPECIAL WEIGHT. Purse $32,000. Fillies 2–year–olds. Weight, 118 lbs

continued

Chordette
Own: Moss Mr & Mrs J S

DELAHOUSSAYE E (89 16 18 18 .18)

Ch. f. 2 (Apr)
Sire: Dixieland Band (Northern Dancer)
Dam: Buda Lady (Crimson Satan)
Br: Russell S. Davis (Ky)
Tr: Mayberry Brian A (23 2 5 2 .09)

handwritten: B+3/3 B3L5 R3/7 M1/3 2y

118

	Lifetime Record :		0 M 0 0		$0	
1994	0 M 0 0			Turf	0 0 0 0	
1993	0 M 0 0			Wet	0 0 0 0	
Hol	0 0 0 0			Dist	0 0 0 0	

WORKOUTS: ●Jun 12 Hol 3f fst :35 H 1/26   Jun 5 Hol 4f fst :474 Hg12/61   May 27 Hol 4f fst :472 H 5/40   ●May 20 Hol 3f fst :344 H 1/19   May 13 Hol 3f fst :371 H 8/18   May 1 Hol 3f fst :353 H 7/26
Apr 2 Hol 3f fst :364 H 10/38

How So Oiseau
Own: Siegel Jan & Mace & Samantha

PEDROZA M A (177 16 33 28 .09)

B. f. 2 (Feb)
Sire: Saratoga Six (Alydar)
Dam: Deauville Dove (No Robbery)
Br: Bruce Kline (Ky)
Tr: Mayberry Brian A (23 2 5 2 .09)

handwritten: Bt 5/4   R 6L6 2/4

118

	Lifetime Record :		0 M 0 0		$0	
1994	0 M 0 0			Turf	0 0 0 0	
1993	0 M 0 0			Wet	0 0 0 0	
Hol	0 0 0 0			Dist	0 0 0 0	

WORKOUTS: Jun 10 Hol 5f fst :594 Hg3/45   Jun 1 Hol 4f fst :48 H 14/39   May 22 Hol 3f fst :351 H 5/28   ●May 15 Hol 3f fst :334 H 1/27   Apr 29 Hol 3f fst :354 H 2/17   Apr 17 Hol 3f fst :344 Hg2/28
Apr 10 Hol 3f fst :361 H 11/31   Apr 2 Hol 3f fst :384 H 35/38

It's Autumn
Own: Shapiro & Stratton & Vizvary

SORENSON D (54 3 5 7 .06)

Ch. f. 2 (Mar)
Sire: It's Freezing (T. V. Commercial)
Dam: Steppin Early (Belted Earl)
Br: Steve Shapiro & Susie Vizvary, et al. (Fla)
Tr: Jumps Kenneth J (3 0 0 2 .00)

handwritten: B4/4   D0/1

L 118

	Lifetime Record :		1 M 0 1		$4,800	
1994	1 M 0 1		$4,800	Turf	0 0 0 0	
1993	0 M 0 0			Wet	0 0 0 0	
Hol	1 0 0 1		$4,800	Dist	1 0 0 1	$4,800

22May94–3Hol fst 5f :22 :451 :574 ⒻMd Sp Wt   46 7 6 743 571 310 Sorenson D   B 117   10.30   85–08 DancinAtTheWire1175 Artica1175 It'sAutumn1171   Stumbled start, wide 7
WORKOUTS: Jun 15 Hol 4f fst :513 H 39/43   May 30 Hol 4f fst :462 H 2/63   May 14 Hol 4f fst :492 Hg24/48   May 7 Hol 5f fst :593 H 18/55   Apr 30 Hol 4f fst :49 H 18/41   Apr 23 Hol 4f fst :472 H 2/51

Texinadress
Own: Pegram Mike

NAKATANI C S (57 8 12 8 .14)

B. f. 2 (Mar)
Sire: Copelan (Tri Jet)
Dam: Something Unique (Secretariat)
Br: Janal Farms Inc (Fla)
Tr: Baffert Bob (24 4 5 2 .17)

handwritten: B5/4

118

	Lifetime Record :		0 M 0 0		$0	
1994	0 M 0 0			Turf	0 0 0 0	
1993	0 M 0 0			Wet	0 0 0 0	
Hol	0 0 0 0			Dist	0 0 0 0	

WORKOUTS: Jun 15 SA 4f fst :48 H 6/37   Jun 9 SA 5f fst 1:00 Hg4/26   Jun 2 SA 5f fst 1:003 Hg2/20   May 27 SA 5f fst 1:001 H 3/32   ●May 20 SA 4f fst :462 H 1/35   May 15 SA 4f fst :49 H 14/21
May 10 SA 3f fst :373 H 13/18   May 4 SA 3f fst :362 H 9/23   Apr 11 SA 3f fst :373 H 12/20   Apr 5 SA 3f fst :372 H 16/25

Wild Spice
Own: Rice & Seawind Stables

BAZE G (5 1 1 0 .20)

B. f. 2 (Feb)
Sire: Once Wild (Baldski)
Dam: Fame and Spice (Fame and Power)
Br: Cheryl A. Curtin (Fla)
Tr: Ward Wesley A (31 5 3 4 .16)

handwritten: M6/4   O0/2

118

	Lifetime Record :		1 M 0 0		$2,400	
1994	1 M 0 0		$2,400	Turf	0 0 0 0	
1993	0 M 0 0			Wet	0 0 0 0	
Hol	1 0 0 0		$2,400	Dist	1 0 0 0	$2,400

4Jun94–4Hol fst 5f :22 :452 :582 ⒻMd Sp Wt   61 1 6 41 411 311 421 Pedroza M A   B 117   3.40   89–07 Dymaxion1171¾ Valid Attraction1171½ Regal Ruthie1171¾   Inside trip 7
WORKOUTS: May 28 Hol 3f fst :36 Hg6/26   May 22 SA 5f fst 1:001 H 8/25   May 12 SA 4f fst :502 B 23/26   May 3 SA 3f fst :361 H 4/32

Rockaroller
Own: Takahashi Richard K

BLACK C A (190 19 23 26 .10)

Ch. f. 2 (Apr)
Sire: Dixieland Band (Northern Dancer)
Dam: So Tempted (Exclusive Native)
Br: Joan C. Johnson (Ky)
Tr: Garcia Eddie (8 0 0 1 .00)

handwritten: Bt3/3   Dam 1/3 Jr 7a

118

	Lifetime Record :		0 M 0 0		$0	
1994	0 M 0 0			Turf	0 0 0 0	
1993	0 M 0 0			Wet	0 0 0 0	
Hol	0 0 0 0			Dist	0 0 0 0	

WORKOUTS: Jun 14 Hol 3f fst :362 H 2/15   Jun 7 Hol 4f fst :474 H 8/22   Jun 1 Hol 5f fst 1:023 H 22/26   May 25 Hol 4f fst :49 H 18/51   May 18 Hol 4f fst :49 H 21/57   May 12 Hol 3f fst :353 Hg4/23
May 6 Hol 3f fst :353 H 2/16   May 1 Hol 3f fst :374 H 20/26

Sovereign Dawn
Own: Bonilla & Weissman

PINCAY L JR (157 20 18 20 .13)

B. f. 2 (Apr)
Sire: Sovereign Dancer (Northern Dancer)
Dam: Soft Dawn (Grey Dawn II)
Br: W. R. Hawn (Ky)
Tr: Conlon Melody (—)

handwritten: C4/2

118

	Lifetime Record :		0 M 0 0		$0	
1994	0 M 0 0			Turf	0 0 0 0	
1993	0 M 0 0			Wet	0 0 0 0	
Hol	0 0 0 0			Dist	0 0 0 0	

WORKOUTS: Jun 15 Hol 3f fst :362 Hg6/28   Jun 9 Hol 4f fst :491 Hg20/38   Jun 3 Hol 3f fst :483 H 10/29   May 29 Hol 5f fst 1:001 H 17/46   May 21 Hol 4f fst :493 H 36/42   May 16 Hol 4f fst :50 H 33/42
May 10 Hol 4f fst :48 H 6/34   May 4 Hol 4f fst :502 H 34/48   Apr 28 Hol 3f fst :361 H 11/30   Apr 23 Hol 3f fst :37 H 8/19

The first thing to note is that the experienced horses, Laguna Seca (55), It's Autumn (46) and Wild Spice (61) all have below-par Beyer numbers for the Maiden Special Weight level at Hollywood Park. Leguna Seca may improve but she is out of a Private Account mare and will probably, according to her stamina index (2) and age bias (3-4), need more time and distance before breaking her maiden. Still, she has been bet down to 9-2 from her debut odds of 8-1 and shouldn't be totally dismissed.

It's Autumn stumbled out of the gate her first start, a bad sign, has mediocre connections and is not being bet. Wild Spice has a good debut trainer and the highest Beyer, but she didn't win first time out when she was bet and is now going off at nearly 7-1. So the most crucial insight about this race is that it wouldn't take much ability for one or more of the first-timer starters to beat the experienced horses.

After scratching Rona Prospect and Mama Mucci and assigning precocity, class and stamina ratings to the rest of the field, there appear to be four legitimate pedigree contenders: Chordette, How So Oiseau, Texindress and Rockaroller all qualify with a B or better rating.

In terms of toteboard action, any first-time starter that opens at 6-1 or less should generally be viewed as a live horse. Since Rockaroller is dead on the board at 42-1 and has a cold trainer, she can be thrown out. The remaining three highly-rated debut runners, however, are all being bet.

Looking at their connections Chordette and How So Oiseau are an uncoupled entry trained by Brian Mayberry who excels with debut runners. While Chordette has a classier sire in Dixieland Band than How So Oiseau has in Saratoga Six, the latter sire, looking at his shorter stamina index, probably has injected more speed, especially at this distance. The workouts tend to confirm this. Both horses have bullet works, but How So Oiseau's May 15th gate work of :33.4 is exceptional! Even better, her dam (according to *Maiden Stats 1994*) has thrown six winners from as many starters, a rare success rate, and How So Oiseau is going off at nearly 9-2.

Texinadress, the other live first-time starter, is trained by Bob Baffert who rivals Mayberry as a debut trainer. Her sire Copelan is slightly less precocious than Saratoga Six, but has the same class rating and gets a lot of sprint winners. Interestingly, she is the actual betting favorite at 2-1.

So how to bet? My own line makes How So Oiseau 2-1, Chordette 5-2 and Texindress 7-2 as the prime contenders, with the rest of the field collectively grouped at 6-1. Since How So Oiseau is going off at better than 4-1, she is an overlay and the only horse worth betting in the Win pool. Players who prefer not to make personal odds lines would not be excluded from this investment opportunity. A handicapper giving each of the three primary contenders an equal chance would simply bet the horse with the highest odds. But, I also keyed How So Oiseau, top and bottom, with Chordette, Texindress and Laguna Seca (the bet down experienced horse) in the exacta.

Results: Unexpectedly, Chordette was the speed of the race and How So Oiseau had to make two moves from slightly off the pace to get up. She paid $10.80 to win and a healthy $41.60 for the exacta. Bolder exotic players will note that this was a trifecta race and that Chordette, the other pedigree play, would have completed it!

SEVENTH RACE   5 FURLONGS. (.56²) MAIDEN SPECIAL WEIGHT. Purse $32,000. Fillies 2-year-olds. Weight, 118 lbs.

Hollywood
JUNE 19, 1994

Value of Race: $32,000 Winner $17,600; second $6,400; third $4,800; fourth $2,400; fifth $800. Mutuel Pool $309,277.00 Exacta Pool $258,239.00 Trifecta Pool $190,340.00 Quinella Pool $32,102.00

Last Raced	Horse	M/Eqt. A.Wt	PP	St	¼	¾	Str	Fin	Jockey	Odds $1
	How So Oiseau	B 2 118	3	3	2¹	2hd	2½	1nk	Pedroza M A	4.40
	Texinadress	Bb 2 118	5	1	4³	4³½	3½	2nk	Nakatani C S	2.00
	Chordette	B 2 118	2	2	1hd	1hd	1hd	3²½	Delahoussaye E	2.80
4Jun94 4Hol⁵	Laguna Seca	B 2 118	1	4	3hd	3¹	4⁵	4⁴½	Stevens G L	4.60
22May94 3Hol³	It's Autumn	LB 2 118	4	7	7⁸	5½	5½	5¹½	Sorenson D	14.00
	Rockaroller	B 2 118	7	6	6¹	7⁶	6¹½	6¹½	Black C A	42.50
4Jun94 4Hol⁴	Wild Spice	B 2 118	6	5	5hd	6hd	7¹⁰	7¹⁴	Baze G	6.90
	Sovereign Dawn	B 2 118	8	8	8	8	8	8	Pincay L Jr	40.40

OFF AT 3:59 Start Good. Won driving. Time, :21⁴, :45, :57³ Track fast.

$2 Mutuel Prices:

5–HOW SO OISEAU	10.80	4.80	3.40
7–TEXINADRESS		3.80	2.60
4–CHORDETTE			2.80

$2 EXACTA 5–7 PAID $41.60 $2 TRIFECTA 5–7–4 PAID $94.00 $2 QUINELLA 5–7 PAID $22.80

B. f, (Feb), by Saratoga Six–Deauville Dove, by No Robbery. Trainer Mayberry Brian A. Bred by Bruce Kline (Ky).

It would be misleading, of course, to see pedigree as always determinant. The seventh race at Bay Meadows on December 18th represents another kind of playable Maiden Special Weight configuration, where there is a justifiable favorite with proven ability running against known competition that is suspect and several promising debut runners.

Looking first at the runners with racing experience, Reality's Conquest is a standout. While his sire Cutlass Reality has a modest C6 rating, he is coming off two very fast sprints with par figures good enough to contend at the allowance and 2-year-old stakes level in northern California. **It is crucial to emphasize here that once a horse has proven it can run, its experience supercedes any pedigree rating.** In this case, Reality's Conquest's impressive 1:09.3 running time and 80 Beyer rating make him an automatic "A". His only major drawback is that he has been the beaten favorite twice and may lack the will to win. Still, his known competition is weak:

Log Buster is by the important broodmare, but non-precocious, sire Hatchet Man and has already lost twice. Even though Log Buster improved in his last sprint, the pace was slow and his stamina index (1) indicates he is bred to route. A pass for now.

Dinadidit is a by Baron O'Dublin (C7) and appears to be a need-to-lead type that probably cannot go with Reality's Conquest. She also has a poor recent work suggesting she may bounce off her comeback race. Given her two close finishes, she is likely to be overbet. Her class 7 rating also indicates she probably needs a drop to the Maiden Claiming level.

6 Furlongs. (1:07¹) **MAIDEN SPECIAL WEIGHT. Purse $15,000. 2-year-olds. Weight, 118 lbs.**

START ▼
6 FURLONGS
▲ FINISH

Clever Remark
Own: Wolf Willow Farms

KAENEL J L (327 49 35 39 .15)

Dk. b or br c. 2 (May)
Sire: Regal Remark (Vice Regent)
Dam: Irish Comet (Irish Castle)
Br: Wolf Willow Farms (Alb-C)
Tr: Wig Janet C (6 1 1 0 .17)

B+7/4

118

	Lifetime Record :		0 M 0 0			$0		
1993	0 M 0 0		Turf	0 0 0 0
1992	0 M 0 0		Wet	0 0 0 0
BM	0 0 0 0		Dist	0 0 0 0

WORKOUTS: Dec 5 BM 6f fst 1:12³ Hg2/17 • Nov 28 BM 5f fst 1:05 H 40/40 Nov 21 BM 5f fst 1:02³ H 57/81 Nov 16 BM 4f fst :49⁴ H-18/35 Oct 31 StP 4f fst :50³ H 4/6 Oct 19 StP 3f fst :38¹ H 1/2
● Oct 3 StP 4f fst :48² Hg1/5 Sep 26 StP 3f fst :36³ Hg2/11

Reality's Conquest "A"
Own: Hubler Pete Or Vi

BAZE R A (497 129 89 67 .26)

C6f2

Ch. c. 2 (Apr)
Sire: Cutlass Reality (Cutlass)
Dam: L' Wonder (L'Natural)
Br: Pete Hubler & Vi Hubler (Cal)
Tr: Martin R L (47 8 6 7 .17)

L 118

	Lifetime Record :		2 M 2 0			$6,000		
1993	2 M 2 0	$6,000	Turf	0 0 0 0
1992	0 M 0 0		Wet	0 0 0 0
BM	2 0 2 0	$6,000	Dist	2 0 2 0	$6,000

4Dec93-4BM fst 6f :22¹ :44³ :56⁴ 1:09² Md Sp Wt 80 7 2 3⁴ 2⁴ 2¹¹½ 2¹ Baze R A LB 118 *1.10 88-16 Trekkor118¹ Reality's Conquest118⁶ Native Blast118⁵ Mild late bid 8
6Nov93-4BM fst 6f :22³ :45³ :58 1:10⁴ Md Sp Wt 74 3 3 1hd 1hd 2hd 2²¹½ Baze R A LB 118 *.70 79-21 SaratogaBndit118²¼ Relity'sConquest118⁵ WrningLbel118¹½ Second best 8
WORKOUTS: Nov 28 BM 5f fst :59² H 3/40 ● Nov 20 BM 4f fst :45² H 1/32 Oct 31 BM 4f fst :47² Hg3/60 ● Oct 25 BM 5f fst :58¹ H 1/81 Oct 18 BM 5f fst 1:01 H 16/87 Oct 9 BM 6f fst 1:15¹ H 11/18

Log Buster
Own: Franks John R

JUDICE J C (183 15 32 17 .08)

C3/1

Ro. c. 2 (Apr)
Sire: Hatchet Man (The Axe II)
Dam: Aunt Stel (Advocator)
Br: John R Franks (Ky)
Tr: Hollendorfer Jerry (247 46 44 38 .19)

L 118

	Lifetime Record :		2 M 0 0			$1,350		
1993	2 M 0 0	$1,350	Turf	0 0 0 0
1992	0 M 0 0		Wet	0 0 0 0
BM	2 0 0 0	$1,350	Dist	2 0 0 0	$1,350

20Nov93-1BM fst 6f :22³ :45³ :58 1:10⁴ Md Sp Wt 60 4 6 7⁵½ 6⁷ 5⁸½ 4³¼ Judice J C L 118 22.80 79-14 WriterToPass118² ParlayAegean118² GardenShdow118½ Very wide trip 7
4Nov93-4BM fst 6f :22³ :45³ :58 1:10³ Md 20000 28 8 8 8⁷½ 8¹⁰ 7¹⁴ 5¹⁷ Chapman T M L 118 16.30 66-16 LFbuluxFort118⁶ SuprfnWn118hd Notnthslftm113⁵ Broke out start, wide 8
WORKOUTS: Dec 13 BM 6f gd 1:17 H 14/18 ● Dec 7 BM 4f fst :48 H 1/9 Dec 2 BM 4f gd :49 H 12/41 Nov 26 BM 4f fst :47⁴ H 5/26 Nov 16 BM 5f fst 1:01² H 10/23 Nov 10 BM 5f fst 1:01 H 10/24

Dinadidit
Own: Mastrangelo William

GONZALEZ R M (344 32 33 43 .09)

C7/4

Dk. b or br g. 2 (May)
Sire: Baron O'Dublin (Irish Ruler)
Dam: Eaglesis (Terresto)
Br: Mastrangelo William (Cal)
Tr: Mastrangelo William (24 2 4 2 .08)

L 118

	Lifetime Record :		2 M 2 0			$6,600		
1993	2 M 2 0	$6,600	Turf	0 0 0 0
1992	0 M 0 0		Wet	0 0 0 0
BM	1 0 1 0	$3,000	Dist	1 0 1 0	$3,000

4Dec93-1BM fst 6f :22⁴ :45⁴ :58² 1:11² Md Sp Wt 73 7 3 1hd 1¹ 1¹½ 2¹ Gonzalez R M LB 118 b 3.10 78-16 Emperor Jared118¹ Dinadidit118² Warning Label118¹ Second best 8
12Jun93-3GG fst 5f :21² :45 :58 Md Sp Wt 58 4 2 1¹ 1hd 2hd 2¹ Patterson A B 118 b 11.80 91-10 Gold Bet118¹ Dinadidit118no Florida Wings118¹¼ Held gamely 12
WORKOUTS: Dec 13 BM 5f gd 1:03³ H 31/48 ● Nov 28 BM 6f fst 1:14 H 1/10 Nov 23 BM 6f fst 1:16¹ H 1/3 Nov 17 BM 5f fst 1:03 Hg12/28 Nov 11 BM 5f my 1:06⁴ H (d)3/6 Nov 6 BM 5f fst 1:03³ H 30/33

Luthier Fever
Own: Cuadra Tyt

BOULANGER G (493 99 74 71 .20)

B+

Ch. c. 2 (May)
Sire: Mt. Livermore (Blushing Groom)
Dam: Warfever (Luthier)
Br: Cuadra T Y T Inc. (Cal)
Tr: Silva Jose L (127 14 21 18 .11)

B+4/4
M2/2

L 118

	Lifetime Record :		0 M 0 0			$0		
1993	0 M 0 0		Turf	0 0 0 0
1992	0 M 0 0		Wet	0 0 0 0
BM	0 0 0 0		Dist	0 0 0 0

WORKOUTS: Dec 10 BM 5f my 1:03¹ Hg(d)8/40 Dec 3 BM 6f gd 1:15⁴ Hg5/15 Nov 27 BM 5f fst 1:01⁴ Hg25/53 Nov 20 BM 5f fst 1:05¹ H 44/45 Nov 7 BM 5f fst 1:03 H 49/61 Oct 31 BM 4f fst :48² H 14/60
Oct 25 BM 3f fst :36³ H 4/18 Jly 4 GG 4f fst :49⁴ H 28/56 Jun 20 GG 3f fst :38⁴ H 18/25

Lt. Donte
Own: Goodman Ed & Karolyn

NOGUEZ A M (35 4 5 2 .11)

D8/4

Ch. c. 2 (May)
Sire: Laomedonte (Raise a Native)
Dam: Lady Natural (Lord Avie)
Br: Karolyn Goodman (WA)
Tr: Goodman Karolyn (16 1 0 1 .06)

118

	Lifetime Record :		0 M 0 0			$0		
1993	0 M 0 0		Turf	0 0 0 0
1992	0 M 0 0		Wet	0 0 0 0
BM	0 0 0 0		Dist	0 0 0 0

WORKOUTS: Dec 9 BM 4f sly :51⁴ H (d)13/21 Dec 2 BM 4f gd :48³ Hg8/41 Nov 26 BM 7f fst 1:31 H 3/3 Nov 19 BM 5f fst 1:02² H 18/24 Oct 28 BM 4f fst :49⁴ Hg18/26 Oct 16 BM 4f fst :50 H 11/20
Oct 8 GG 4f fst :50¹ H 8/11 Sep 18 YM 3f fst :36² H 4/10

Ballazzi
Own: Walter Mr & Mrs Robert H

WARREN R J JR (366 29 53 64 .08)

F 8/4

B. g. 2 (Apr)
Sire: Ballydoyle (Northern Dancer)
Dam: Dazzling Daffodil (Secretariat)
Br: Mr. & Mrs. Robert H. Walter (Cal)
Tr: Hay Noble III (29 2 0 5 .07)

L 118

	Lifetime Record :		1 M 0 0			$0		
1993	1 M 0 0		Turf	0 0 0 0
1992	0 M 0 0		Wet	0 0 0 0
BM	1 0 0 0		Dist	1 0 0 0

4Dec93-4BM fst 6f :22¹ :44³ :56⁴ 1:09² Md Sp Wt 34 5 5 6⁷¼ 6¹⁰ 6¹⁴ 7¹⁹ Meza R Q B 118 17.60 70-16 Trekkor118¹ Reality's Conquest118⁶ Native Blast118⁵ Outrun 8
WORKOUTS: Dec 13 GG 4f gd :48² H 3/14 Nov 23 GG 6f fst 1:12¹ H 1/3 Nov 16 GG 6f fst 1:15 H 2/3 Nov 8 GG 6f fst 1:15 H 6/10 Nov 3 GG 5f fst 1:04 Hg7/8 Oct 25 GG 6f fst 1:16⁴ H 5/12

Thirtyfive Black B
Own: Chaiken & Chaiken & Heller & Heller

BELVOIR V T (468 83 85 61 .18)

B6/3

Dk. b or br c. 2 (Feb)
Sire: Desert Wine (Damascus)
Dam: Near Star (Star Envoy)
Br: Mr. & Mrs. Barton D. Heller (Cal)
Tr: Morey William J Jr (56 7 9 10 .13)

118

	Lifetime Record :		0 M 0 0			$0		
1993	0 M 0 0		Turf	0 0 0 0
1992	0 M 0 0		Wet	0 0 0 0
BM	0 0 0 0		Dist	0 0 0 0

WORKOUTS: Dec 13 BM 5f gd 1:01¹ H 6/48 Dec 4 BM 6f gd 1:13² H 2/21 Nov 27 BM 3f fst 1:00 H 4/53 Nov 20 BM 5f fst 1:01³ H 31/45 Nov 13 BM 4f fst :49 H 33/59 Nov 7 BM 4f fst :48³ H 18/72
Oct 28 BM 3f fst :35⁴ H 4/21 Oct 20 BM 3f fst :38 H 9/11 ● Oct 8 Pln 5f fst 1:02 H 1/4 Sep 28 Pln 4f fst :48¹ H 4/13 Sep 17 Pln 4f fst :50 H 6/9

continued

Lypheor Castle

Own: Takitani Aiko

JUDICE J C (183 15 32 17 .08)

Dk. b or br c. 2 (Apr)
Sire: Vernon Castle (Seattle Slew)
Dam: Maui Lyphear J.*Jp (Lypheor)
Br: Aiko Takitani (Cal)
Tr: Delima Clifford (39 4 4 3 .10)

Art's speed

L 118

	Lifetime Record :		1 M 0 0	$1,125					
1993	1 M 0 0	$1,125	Turf	0 0 0 0					
1992	0 M 0 0		Wet	0 0 0 0					
BM	1 0 0 0	$1,125	Dist	1 0 0 0	$1,125				

4Dec93- 1BM fst 6f :22⁴ :45⁴ :58² 1:11² Md Sp Wt   66 4 6 5³ 31½ 31½ 44 Judice J C   LB 118 11.00 75–16 Emperor Jared118¹ Dinadidit118² Warning Label118¹   Off slowly, wide 8

WORKOUTS: Nov 26 BM 5f fst :59³ H 2/17 Nov 17 BM 5f fst 1:02 Hg3/28 Nov 10 BM 5f fst 1:01² H 14/24 Nov 3 BM 4f fst :48⁴ Hg11/30 Oct 28 BM 4f fst :48³ H 6/26 Oct 22 BM 4f fst :50³ H 17/23

Just Trust Me

Own: Jollo Ralph

EIDE B M (214 24 22 25 .11)

Ch. c. 2 (Mar)
Sire: Tonzarun (Arts and Letters)
Dam: Brillant*Ar (Faridoon)
Br: Bluestem Farm Inc. (Cal)
Tr: Moger Ed Jr (63 10 7 9 .16)

c f/q

L 118

	Lifetime Record :		1 M 0 0	$375					
1993	1 M 0 0	$375	Turf	0 0 0 0					
1992	0 M 0 0		Wet	0 0 0 0					
BM	1 0 0 0	$375	Dist	1 0 0 0	$375				

4Dec93- 4BM fst 6f :22¹ :44³ :56⁴ 1:09² Md Sp Wt   43 1 8 7¹⁵ 7¹⁶ 7¹⁶ 5¹⁵½ Eide B M   118 47.30 73–16 Trekkor118¹ Reality's Conquest118⁶ Native Blast118⁵   No rally 8

WORKOUTS: Dec 15 BM 4f my :49 H 8/36 Dec 1 BM 3f my :40¹ H 13/14 Nov 21 BM 6f fst 1:15² H 8/12 Nov 14 BM 6f fst 1:15 H 5/17 Nov 7 BM 5f fst 1:00⁴ H 16/61 Oct 31 BM 5f fst 1:01³ H 29/64

Lypheor Castle is by the even cheaper sire Vernon Castle. Still, he was off slowly in his debut, showed some ability, and could improve. But his sire's stamina index of 2 indicates he is bred to route and unlikely to keep up with the pace Reality's Conquest and Dinadidit will likely set.

Ballazzi and Just Trust Me, the last two experienced runners, seem hopeless. They just lost to the favorite by over 14 lengths and their class rating of 8 indicates they need a serious drop.

If Reality's Conquest is going to be beat, it looks like a first-time starter will have to do it. Let's look at the possibilities:

Clever Remark by Regal Remark is precociously bred on top, but not on his Irish Castle bottom. Furthermore, this colt is stuck on the rail and his class rating of 7 indicates that he is more likely to break his maiden when dropped into the claiming level.

Luthier Fever is a very interesting possibility. As the only horse by a Kentucky sire in the race, Luthier Fever merits a closer look on that factor alone, because Kentucky sires generally get better quality mares. Furthermore, his sire Mt. Livermore is the sire of Housebuster and Eliza. His B+4 rating is also the precocious class of this field. Jose Silva's competence as a debut trainer adds to the interest. But, on the downside, the horse is not being heavily bet.

Two, perhaps related, reasons for the light action on Luthier Fever may be the slow workouts and the possibility that the stamina influence of his broodmare sire Luthier (2) may dull the Mt. Livermore speed influence first time out in a sprint. When it comes to assessing a broodmare sire's influence on a foal, a good rule of thumb is that he is estimated to have half the influence of the sire because he is one generation further removed in the pedigree.

Despite the relative lack of action, Luthier Fever has to be considered a contender. He might not win a 3 furlong or 5 furlong sprint, but this one is for 6 furlongs where a little extra stamina might be helpful. He's got an excellent sire, trainer Jose Silva wins races, and there is that "sneaky" :36.3 work on October 25.

Lt. Donte, the next first-time starter, is by a cheap, unprecocious sire from

Washington with a private stud fee. Except with Kentucky stallions, "private" usually means a stud manager is having trouble getting mares to his stallion and is willing to deal cheap or even foal share in lieu of a stud fee. Lt. Donte also has uninspiring connections and probably needs a severe drop in class and added real estate before being competitive.

Thirtyfive Black is another possible debut contender. He is by Desert Wine who has a precocious B6 rating. Trainer Bill Morey is decent with first-time starters and jockey Vann Belvoir being up is also a good sign. On the down side, Thirtyfive Black isn't getting bet and his dam (according to *Maiden Stats 1993*) has thus far primarily gotten route winners.

So how to play the race? Reality's Conquest looks tough to beat. He's coming up to his third race, often a horse's best, with an impressive 80 Beyer figure. Still, he has been beaten twice. I make him 6-5. The public jumps on him and he opens at 1-5, then goes off at 3-5. The colt may well win, but according to my line, he is definitely underlayed.

Looking at the challengers, Log Buster, Lt. Donte, Ballazi. Lypheor Castle and Just Trust Me are not by precocious sires and can be thrown out. Regal Remark may have some speed, but he's stuck on the rail in his first start, opens at 18-1, and is a throwout.

Dinadidit will challenge Reality's Conquest for the early lead. However, the gelding is coming off a tough race after a layoff and its December 13th five-furlong work of 1:03.3 is not inspiring. Also, its trainer doesn't win many races. Furthermore, at 7-1 it's way above its 2-1 morning line. Some risk, but I'm throwing him out.

Luthier Fever continues to loom as the most precocious debut play. He is not only likely to show speed, but his sire Mt. Livermore is the pedigree class of the field with his $15,000 stud fee which is twice that of Reality's Conquest. While stud fee isn't the primary basis for establishing the class ratings in *Exploring Pedigree*, it is a good barometer of potential class in a foal because the quality of the mare sent to a sire tends to be directly proportional to the size of his stud fee. Breeders don't normally invest an expensive stud fee in an inferior mare.

Because Mt. Livermores with their B+ rating are genetically 7-2 to win their debuts, I automatically make all Mt. Livermores 5-1 on my morning line when they are first entered in Maiden Special Weight sprints. From there I adjust my odds according to the strength of their connections and quality of the competition. In this field, except for the favorite, the competition looks weak and I make Luthier's Fever 3-1. He opens at 13-1, drops sharply to 8-1, then drifts back up to 12-1. Positive betting action according to Mark Cramer's toteboard research and an obvious value play!

Thirtyfive Black is the other first-time starter that must be respected. But Morey's horses usually get bet and Thirtyfive Black is dead on the board. He opens at 10-1, then drifts up to 18-1. Given the heavy action on Reality's Conquest and Mt. Livermore's better record as a precocious sire, I toss Thirtyfive Black out too.

40   So, based on a pedigree analysis of the race and Reality's Conquest 3-5 odds, the only value way to play this race is a Win bet on Luthier Fever at 13-1, with him behind Reality's Conquest in the exacta as a place bet.

Results: The actual race ran pretty much as plotted. Dinadidit took the early lead and then wilted when pressured by Reality's Conquest and Luthier Fever. Those two dueled into the stretch where Luthier Fever tired, but still managed to hold on for second. The $2 exacta paid $21.80 (compared to the Place price of $6.00) which was close to what Luthier Fever's Win price would have been at 12-1!

Another example of playing pedigree in a race with an experienced standout is the first race on May 4th at Churchill Downs. Perusing the past performances, Grand R. J. is clearly the best of the four horses that have already run and is being bet below his morning line of 3-1. Still, he is by a cheap sire and even though his 56 Beyer is not bad for May, he looks beatable.

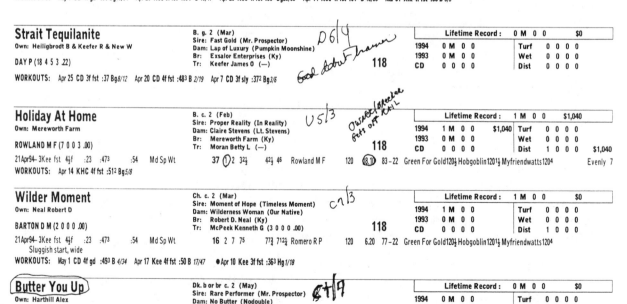

Churchill Downs

1

4½ Furlongs. (:51²) **MAIDEN SPECIAL WEIGHT.** Purse $18,000 (plus $5,800 KTDF supplement and $1,820 from the KDWOTB fund). 2-year-olds. Weight, 118 lbs.

Coupled – Wilder Moment and Village Baron

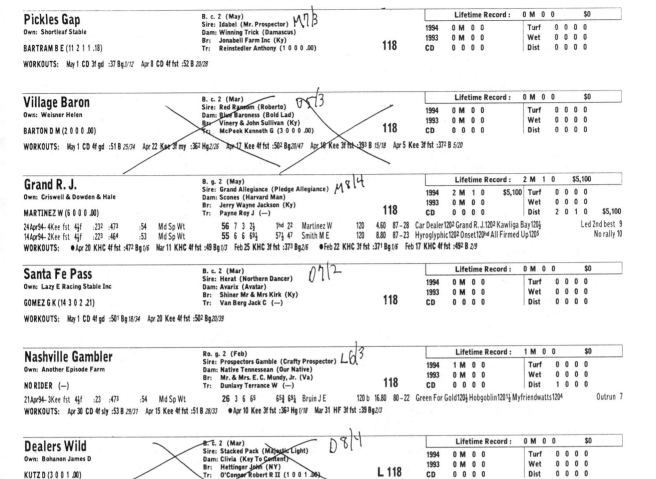

Pickles Gap — M7B

Own: Shortleaf Stable

B. c. 2 (May)
Sire: Idabel (Mr. Prospector)
Dam: Winning Trick (Damascus)
Br: Jonabell Farm Inc (Ky)
Tr: Reinstedler Anthony (1 0 0 0 .00)

BARTRAM B E (11 2 1 1 .18)

118

	Lifetime Record:	0 M 0 0	$0				
1994	0 M 0 0			Turf	0 0 0 0		
1993	0 M 0 0			Wet	0 0 0 0		
CD	0 0 0 0			Dist	0 0 0 0		

WORKOUTS: May 1 CD 3f gd :37 Bg3/12 • Apr 8 CD 4f fst :52 B 20/28

Village Baron — 07/3

Own: Weisner Helen

B. c. 2 (Mar)
Sire: Red Ransom (Roberto)
Dam: Blue Baroness (Bold Lad)
Br: Vinery & John Sullivan (Ky)
Tr: McPeek Kenneth G (3 0 0 0 .00)

BARTON D M (2 0 0 0 .00)

118

	Lifetime Record:	0 M 0 0	$0				
1994	0 M 0 0			Turf	0 0 0 0		
1993	0 M 0 0			Wet	0 0 0 0		
CD	0 0 0 0			Dist	0 0 0 0		

WORKOUTS: May 1 CD 4f gd :51 B 25/34 • Apr 22 Kee 3f my :36² Hg2/26 • Apr 17 Kee 4f fst :50² Bg20/47 • Apr 10 Kee 3f fst :39³ B 15/18 • Apr 5 Kee 3f fst :37² B 5/20

Grand R. J. — M8/4

Own: Criswell & Dowden & Hale

B. g. 2 (May)
Sire: Grand Allegiance (Pledge Allegiance)
Dam: Scones (Harvard Man)
Br: Jerry Wayne Jackson (Ky)
Tr: Payne Roy J (—)

MARTINEZ W (6 0 0 0 .00)

118

	Lifetime Record:	2 M 1 0	$5,100				
1994	2 M 1 0	$5,100		Turf	0 0 0 0		
1993	0 M 0 0			Wet	0 0 0 0		
CD	0 0 0 0			Dist	2 0 1 0	$5,100	

24Apr94– 4Kee fst 4½f :23² :47³ :54 Md Sp Wt 56 7 3 2½ 1hd 22 Martinez W 120 4.60 87–28 Car Dealer120² Grand R. J.120² Kawliga Bay120½ Led 2nd best 9
14Apr94– 2Kee fst 4½f :22³ :46⁴ :53 Md Sp Wt 55 6 6 6⁹½ 57½ 47 Smith M E 120 8.80 87–23 Hyroglyphic120² Onset120hd All Firmed Up120⁵ No rally 10

WORKOUTS: • Apr 20 KHC 4f fst :47² Bg1/6 • Mar 11 KHC 4f fst :49 Bg1/3 • Feb 25 KHC 3f fst :37³ Bg2/6 • Feb 22 KHC 3f fst :37¹ Bg1/6 • Feb 17 KHC 4f fst :49² B 2/9

Santa Fe Pass — 07/2

Own: Lazy E Racing Stable Inc

B. c. 2 (Mar)
Sire: Herat (Northern Dancer)
Dam: Avarix (Avatar)
Br: Shiner Mr & Mrs Kirk (Ky)
Tr: Van Berg Jack C (—)

GOMEZ G K (14 3 0 2 .21)

118

	Lifetime Record:	0 M 0 0	$0				
1994	0 M 0 0			Turf	0 0 0 0		
1993	0 M 0 0			Wet	0 0 0 0		
CD	0 0 0 0			Dist	0 0 0 0		

WORKOUTS: May 1 CD 4f gd :50¹ Bg 18/34 • Apr 20 Kee 4f fst :50² Bg20/39

Nashville Gambler — Ld3

Own: Another Episode Farm

Ro. g. 2 (Feb)
Sire: Prospectors Gamble (Crafty Prospector)
Dam: Native Tennessean (Our Native)
Br: Mr. & Mrs. E. C. Mundy, Jr. (Va)
Tr: Dunlavy Terrance W (—)

NO RIDER (—)

118

	Lifetime Record:	1 M 0 0	$0				
1994	1 M 0 0			Turf	0 0 0 0		
1993	0 M 0 0			Wet	0 0 0 0		
CD	0 0 0 0			Dist	1 0 0 0		

21Apr94– 3Kee fst 4½f :23 :47³ :54 Md Sp Wt 26 3 6 6⁵ 6⁵½ 6⁹½ Bruin J E 120 b 16.80 80–22 Green For Gold120½ Hobgoblin120½ Myfriendwatts120⁴ Outrun 7

WORKOUTS: Apr 30 CD 4f sly :53 B 29/31 • Apr 15 Kee 4f fst :51 B 28/33 • • Apr 10 Kee 3f fst :36³ Hg1/18 • Mar 31 HF 3f fst :39 Bg2/3

Dealers Wild — D8/4

Own: Bohanon James D

B. c. 2 (Mar)
Sire: Stacked Pack (Majestic Light)
Dam: Clivia (Key To Content)
Br: Hettinger John (NY)
Tr: O'Connor Robert R II (1 0 0 1 .00)

KUTZ D (3 0 0 1 .00)

L 118

	Lifetime Record:	0 M 0 0	$0				
1994	0 M 0 0			Turf	0 0 0 0		
1993	0 M 0 0			Wet	0 0 0 0		
CD	0 0 0 0			Dist	0 0 0 0		

Of Grand R.J.'s. experienced competition, Holiday At Home seems most likely to improve. He ran a decent first race, beating Nashville Gambler by 3 lengths, and is getting off the rail. On the downside, however, Proper Reality's progeny seem to need more than 4 1/2 furlongs to break their maidens. Of even more concern is that Holiday At Home is dead on the board at 19-1.

Among the first-time starters, Squadron Leader looks like a standout. He has the only B or better rating, the highest class in the race and is trained by Peter Vestal who is excellent with first-time starters. With Storm Bird as his sire and his strong connections Squadron Leader is a definite threat. On the downside, he's stuck on the rail, has very mediocre works and at 6-1 is not being bet as heavily as he should be.

Strait Tequilanite, despite Fast Gold's unprecocious D6 rating, has been made the second choice in the morning line at 4-1, an odds line with which the public basically agrees. But this horse is probably getting overbet because of the Pat Day-at-Churchill Downs factor as well as trainer James Keefer's decent record with first-time starters.

42   Still, based on his sire's poor debut record, Strait Tequilante looks like a horse worth taking a stand against.

Of the rest of the debut runners Butter You Up and Pickles Gap, because of his connections, are also worth considering. On the positive side for Butter You Up is his sire Rare Performer, a hard-knocking son of Mr. Prospector who gets his share of debut winners as his C+ rating testifies. He also has that nifty :36.3 bullet gate work and has been bet down a couple of ticks below his morning line of 8-1. On the downside Don Kemper isn't known as a debut trainer.

Pickles Gap is part of freshman sire Idabel's first crop. His "M" rating is based on the fact that he didn't race at two or win at three. Still, breeding is unpredictable and Idabel, another son of Mr. Prospector, might pass on some of his sire's precocious genes. Less speculatively, Pickles Gap's trainer Anthony Reinstedler is identified in *Olmsted's Trainer Guide* as excelling with debut runners.

So how to bet? Grand R.J. has already shown some ability and is being bet. But with that 56 Beyer he is no sure thing. Furthermore, at less than 2-1, he doesn't have any value in the Win pool. On the surface Squadron Leader is the obvious pedigree play. But, despite his breeding and connections, he is not being bet. He opens at 7-1 and stays above his morning line of 5-1. He can't be a prime bet.

Butter You Up, on the other hand, continues to be bet and his C+ rating is the second most precocious in the field. The best strategy here is to bet both Butter You Up and Squadron Leader to Win and put them both behind Grand R.J. in the exacta. While Strait Tequilanite and Pickles Gap might also win, this is the kind of race where you have to be careful not to overextend yourself and they are best thrown out because of the weaker arguments on their behalf.

Results: Butter You Up broke alertly from the gate and opened a 2 length lead before being caught by Grand R. J. in deep stretch. The $2 exacta returned a nice $44.20. Squadron Leader reflected his lack of action and never made a move.

The second race at Santa Anita on December 26th is another Maiden race that was worth playing because it included a vulnerable favorite. The first thing to note about this juvenile race is that the heavy favorite at 3-2 is Whatawoman. While she has the highest Beyers and might win, she has already lost four races, has a cold trainer and is losing Gary Stevens. I make her 2-1. Given her post-time odds, she has no value and seems worth taking a stand against.

But, who can beat her? Of the experienced competition Stop The Croquet and Gliding Lark have already lost to Whatawoman and are throwouts. Gondwanaland, Taxquenya and Contraindication are not being bet which leaves Meadow Moon at 5-2 as the only experienced runner the public thinks can improve enough to win in this spot.

Santa Anita Park

2

6 Furlongs. (1:071) **MAIDEN SPECIAL WEIGHT. Purse $32,000. Fillies, 2-year-olds. Weight, 117 lbs.**
(Non-starters for a claiming price of $32,000 or less in their last three starts preferred.)

START ▾ 6 FURLONGS ▴FINISH

Stop The Croquet

Own: Evans Edward P
FLORES D R (—)

Ch. f. 2 (May)
Sire: Stop the Music (Hail to Reason)
Dam: Merci Croquet (Jacinto)
Br: Anunzio Stanchieri (Ky)
Tr: Becerra Rafael (—)

D5|4 C

L 117

Lifetime Record :	2 M 0 0	$2,100			
1993	2 M 0 0	$2,100	Turf	0 0 0 0	
1992	0 M 0 0		Wet	0 0 0 0	
SA	1 0 0 0	$2,100	Dist	2 0 0 0	$2,100

31Oct93- 4SA fst 6f :213 :444 :572 1:10 ⒻMd Sp Wt 65 8 7 85¼ 73¾ 54 48½ Flores D R L 117 b 15.10 78-12 PrincessMitterand117hd Whatawomn1174 GlidingLrk1174½ Steadied early 9
14Aug93- 5Dmr fst 6f :214 :451 :572 1:101 ⒻMd Sp Wt 20 6 9 74¾ 89½ 816 823½ Pincay L Jr LB 117 b 12.50 64-13 Stellar Cat117nk Sophisticatedcielo1172½ Magical Avie1175 Broke slowly 10
WORKOUTS: Dec 23 SA 3f fst :362 H 8/16 Dec 13 SA 5f fst :594 H 6/36 Dec 7 SA 6f fst 1:151 H 10/18 Dec 1 SA 5f fst 1:014 H 28/50 Nov 24 Hol 4f fst :47 H 4/40 Nov 18 Hol 4f fst :484 H 12/28

Gliding Lark

Own: Team Valor & Roncari & Tsujimoto
STEVENS G L (—)

Ch. f. 2 (Mar)
Sire: Woodman (Mr. Prospector)
Dam: Eighty Lady (Flying Lark)
Br: Robert S West Jr & London Throughbrd Services Ltd. (Ky)
Tr: Hennig Mark (—)

Ct 5|

L 117

Lifetime Record :	3 M 0 1	$5,700			
1993	3 M 0 1	$5,700	Turf	0 0 0 0	
1992	0 M 0 0		Wet	0 0 0 0	
SA	1 0 0 1	$4,200	Dist	2 0 0 1	$5,700

26Nov93- 9Hol fst 6½f :213 :442 1:093 1:16 ⒻMd Sp Wt 70 2 9 79½ 67½ 64½ 66 Valenzuela F H B 118 5.60 83-09 Jacodra's Devil1132¾ Sliced Twice1181¼ DevotedDanzig118no Saved ground 10
31Oct93- 4SA fst 6f :213 :444 :572 1:10 ⒻMd Sp Wt 76 1 8 42¼ 41¾ 42 34 Valenzuela F H B 117 9.30 83-12 Princess Mitterand117hd Whatawoman1174 Gliding Lark1174½ No late bid 9
1Oct93- 5Bel fst 6f :223 :463 :591 1:121 ⒻMd Sp Wt 53 7 8 3½ 3nk 43 46 Davis R G 117 3.50 73-20 Aly's Conquest117nk Annie Bonnie1174¼ Vibelle1171½ Dueled, tired 10
WORKOUTS: Dec 17 Hol 5f fst 1:01 H 7/44 Dec 9 Hol 4f fst :472 H 6/21 ●Nov 22 Hol 4f fst :463 H 1/30 Nov 16 Hol 5f fst 1:021 H 11/19 Nov 10 Hol 4f fst :474 H 2/21 ●Oct 23 Hol 5f fst :583 H 1/16

Queen Gen

Own: Hersh Philip
GRYDER A T (—)

Ch. f. 2 (Jan)
Sire: General Assembly (Secretariat)
Dam: Queen Breeze (Breezing On)
Br: Pillar Stud, Inc. (Ky)
Tr: Bernstein David (—)

D6|3

117

Lifetime Record :	0 M 0 0	$0		
1993	0 M 0 0		Turf	0 0 0 0
1992	0 M 0 0		Wet	0 0 0 0
SA	0 0 0 0		Dist	0 0 0 0

WORKOUTS: Dec 21 SA 7f fst 1:272 H 8/12 Dec 17 SA 6f fst 1:133 H 22/45 Dec 11 SA 6f fst 1:134 H 17/27 ●Dec 6 SA 5f fst :594 H 1/29 Nov 30 SA 5f gd 1:013 H 6/12 Nov 24 SA 4f fst :48 H 14/40
Nov 18 SA 4f fst :48 H 7/27 Nov 12 SA 3f gd :382 H 17/24 Aug 29 Dmr 4f fst :482 H 21/46 Aug 5 Dmr 4f fst :50 H 39/63

Meadow Moon

Own: Headley Bruce
SOLIS A (—)

Dk. b or br f. 2 (Feb)
Sire: Meadowlake (Hold Your Peace)
Dam: Shamrock Reality (In Reality)
Br: Brylynn Farm (Ky)
Tr: Headley Bruce (—)

A+4|4

117

Lifetime Record :	1 M 0 0	$2,100			
1993	1 M 0 0	$2,100	Turf	0 0 0 0	
1992	0 M 0 0		Wet	0 0 0 0	
SA	1 0 0 0	$2,100	Dist	1 0 0 0	$2,100

14Nov93- 6SA fst 6f :214 :443 :563 1:084 ⒻMd Sp Wt 72 8 4 3nk 43½ 45 48½ Solis A B 117 *.80 84-06 C'monLetsDance1173½ CasualMeeting117½ Jacodr'sDevil1124½ Weakened 9
WORKOUTS: Dec 21 SA 5f fst :59 H 5/51 ●Dec 14 SA 5f fst :592 H 1/41 Dec 7 SA 5f fst 1:001 H 7/38 ●Dec 1 SA 5f fst :593 H 1/50 Nov 25 SA 5f fst 1:011 H 13/39 Nov 5 SA 6f fst 1:151 Hg 19/32

Gondwanaland

Own: Recachina Laura
BLACK C A (—)

B. f. 2 (Apr)
Sire: Chief's Crown (Danzig)
Dam: Sue Babe (Mr. Prospector)
Br: Regent Farms (Ky)
Tr: Lukas D Wayne (—)

C5|3 improve?

117

Lifetime Record :	1 M 0 0	$0		
1993	1 M 0 0		Turf	0 0 0 0
1992	0 M 0 0		Wet	0 0 0 0
SA	1 0 0 0		Dist	1 0 0 0

14Nov93- 6SA fst 6f :214 :443 :563 1:084 ⒻMd Sp Wt 57 1 7 77¼ 79½ 811 714 Stevens G L B 117 11.50 79-06 C'monLtsDnc1173½ CsulMting117½ Jcodr'sDvil1124½ Stumbled after start 9
WORKOUTS: Dec 16 SA 4f fst :474 H 6/46 Dec 9 SA 5f fst 1:001 H 2/20 Dec 2 SA 5f fst 1:023 H 28/33 Nov 24 SA 4f fst :482 H 16/40 Nov 8 SA 5f fst 1:021 H 32/42 Nov 2 SA 5f fst :594 H 9/44

Emerite

Own: Haras Santa Maria & Oppenheimer
PINCAY L JR (—)

B. f. 2 (Apr)
Sire: Mogambo (Mr. Prospector)
Dam: Tough Tender (Bold Bidder)
Br: Haras Santa Maria de Araras & Bridget Oppenheimer (Fla)
Tr: McAnally Ronald (—)

D7|3

117

Lifetime Record :	0 M 0 0	$0		
1993	0 M 0 0		Turf	0 0 0 0
1992	0 M 0 0		Wet	0 0 0 0
SA	0 0 0 0		Dist	0 0 0 0

WORKOUTS: Dec 20 Hol 6f fst 1:123 H 3/13 Dec 14 Hol 6f fst 1:142 H 5/17 ●Dec 8 Hol 7f fst 1:254 H 1/5 ●Dec 2 Hol 6f fst 1:121 H 1/29 Nov 26 Hol 6f fst 1:124 H 3/23 Nov 20 Hol 5f fst 1:014 Hg 40/63
Nov 8 Hol 5f fst 1:02 H 21/31 Nov 1 Hol 5f fst 1:011 H 14/30 Oct 26 Hol 5f fst 1:024 H 12/18

Taxquenya

Own: Butterfield Interest Inc
PEDROZA M A (—)

Dk. b or br f. 2 (Apr)
Sire: Kris S. (Roberto)
Dam: Crown Silver (Spectacular Bid)
Br: Keswick Stables (Va)
Tr: Threewitt Noble (—)

C 2|2

117

Lifetime Record :	1 M 0 0	$0		
1993	1 M 0 0		Turf	0 0 0 0
1992	0 M 0 0		Wet	0 0 0 0
SA	0 0 0 0		Dist	0 0 0 0

7Aug93- 6Dmr fst 5½f :213 :443 :561 1:03 ⒻMd Sp Wt 20 2 6 32 45½ 715 726¾ Pedroza M A B 117 6.50 69-09 Sardula11710 My Fling1171½ Princess Leia M D1176½ Stopped 8
WORKOUTS: ●Dec 22 SA 4f fst :463 H 1/49 Dec 16 SA 5f fst 1:001 H 6/53 Dec 9 SA 6f fst 1:123 H 2/23 Dec 2 SA 5f fst 1:001 H 11/33 Nov 27 SA 4f fst :511 H 36/37 Nov 19 SA 4f fst :50 B 29/35

Contraindication

Own: Milch & Silverman & Silverman
MCCARRON C J (—)

Ch. f. 2 (Feb)
Sire: Dixieland Band
Dam: Contrafaire (Sham)
Br: Greely John J III & Hill Herbert L (Ky)
Tr: Vienna Darrell (—)

B+3|3 D4|4 Bred to route

117

Lifetime Record :	1 M 0 0	$2,100		
1993	1 M 0 0	$2,100	Turf	0 0 0 0
1992	0 M 0 0		Wet	0 0 0 0
SA	0 0 0 0		Dist	0 0 0 0

26Nov93- 9Hol fst 6½f :213 :442 1:093 1:16 ⒻMd Sp Wt 74 9 7 910 89¾ 54¼ 44 McCarron C J B 118 10.20 85-09 Jacodra's Devil1132¾ Sliced Twice1181¼ Devoted Danzig118no Finished well 10
WORKOUTS: Dec 23 SA 4f fst :481 H 13/40 Dec 18 SA 5f fst 1:022 H 50/61 Dec 13 SA 4f fst :513 H 37/38 Nov 20 SA 6f fst 1:123 H 2/20 Nov 14 SA 5f fst 1:004 B 25/55 Nov 7 SA 5f fst 1:012 H 40/58

continued

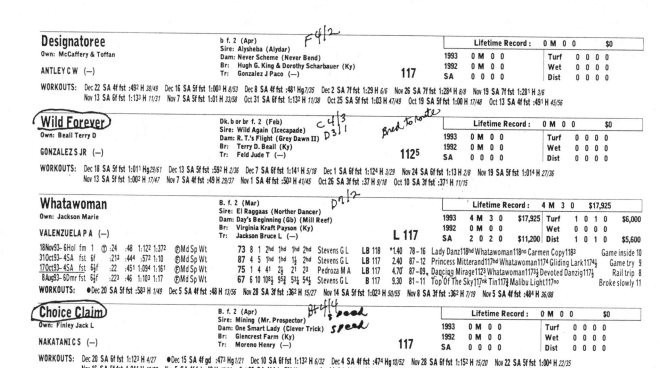

Designatoree F4/2
Own: McCaffery & Toffan
ANTLEY C W (—)

b f. 2 (Apr)
Sire: Alysheba (Alydar)
Dam: Never Scheme (Never Bend)
Br: Hugh G. King & Dorothy Scharbauer (Ky)
Tr: Gonzalez J Paco (—)

117

Lifetime Record:	0 M 0 0			$0
1993	0 M 0 0	Turf	0 0 0 0	
1992	0 M 0 0	Wet	0 0 0 0	
SA	0 0 0 0	Dist	0 0 0 0	

WORKOUTS: Dec 22 SA 4f fst :49^2 H *38/49* Dec 16 SA 5f fst 1:00^3 H *8/53* Dec 8 SA 4f fst :48^1 Hg *7/35* Dec 2 SA 7f fst 1:29 H *6/6* Nov 26 SA 7f fst 1:28^4 H *8/8* Nov 19 SA 7f fst 1:28^1 H *3/6*
Nov 13 SA 6f fst 1:13^3 H *11/31* Nov 7 SA 5f fst 1:01 H *33/58* Oct 31 SA 6f fst 1:13^2 H *11/38* Oct 25 SA 5f fst 1:03 H *47/49* Oct 19 SA 5f fst 1:00 H *17/48* Oct 13 SA 4f fst :49^1 H *45/56*

Wild Forever C4/3 D3/1 Bred to route
Own: Beall Terry D
GONZALEZ S JR (—)

Dk. b or br f. 2 (Feb)
Sire: Wild Again (Icecapade)
Dam: R. T.'s Flight (Grey Dawn II)
Br: Terry D. Beall (Ky)
Tr: Feld Jude T (—)

112^5

Lifetime Record:	0 M 0 0			$0
1993	0 M 0 0	Turf	0 0 0 0	
1992	0 M 0 0	Wet	0 0 0 0	
SA	0 0 0 0	Dist	0 0 0 0	

WORKOUTS: Dec 18 SA 5f fst 1:01^1 Hg *29/61* Dec 13 SA 5f fst :59^2 H *2/36* Dec 7 SA 6f fst 1:14^1 H *5/18* Dec 1 SA 6f fst 1:12^4 H *3/29* Nov 24 SA 6f fst 1:13 H *2/8* Nov 19 SA 5f fst 1:01^4 H *27/36*
Nov 13 SA 5f fst 1:00^2 H *17/47* Nov 7 SA 4f fst :49 H *29/37* Nov 1 SA 4f fst :50^3 H *41/45* Oct 26 SA 3f fst :37 H *9/18* Oct 10 SA 3f fst :37^1 H *11/15*

Whatawoman D7/2
Own: Jackson Marie
VALENZUELA P A (—)

B. f. 2 (Mar)
Sire: El Raggaas (Norther Dancer)
Dam: Day's Beginning (Gb) (Mill Reef)
Br: Virginia Kraft Payson (Ky)
Tr: Jackson Bruce L (—)

L 117

Lifetime Record:	4 M 3 0	$17,925			
1993	4 M 3 0	$17,925	Turf	1 0 1 0	$6,000
1992	0 M 0 0		Wet	0 0 0 0	
SA	2 0 2 0	$11,200	Dist	1 0 1 0	$5,600

18Nov93- 6Hol	fm	1 Ⓣ	:24	:48	1:12^2	1:37^2	ⒻMd Sp Wt	73	8	1	2^{hd}	1^{hd}	1^{hd}	2^{hd}	Stevens G L	LB 118	*1.40	78-16	Lady Danz118^{hd} Whatawoman118^{no} Carmen Copy118^3	Game inside 10
31Oct93- 4SA	fst	6f	:21^3	:44^4	:57^2	1:10	ⒻMd Sp Wt	87	4	5	1^{hd}	1^{hd}	1½	2^{hd}	Stevens G L	LB 117	2.40	87-12	Princess Mitterand117^{hd} Whatawoman117^4 Gliding Lark117^{4½}	Game try 9
17Oct93- 4SA	fst	6½f	:22	:45^1	1:09^4	1:16^1	ⒻMd Sp Wt	75	1	4	4^1	2½	2^1	2^3	Pedroza M A	LB 117	4.70	87-09	Dancing Mirage112^3 Whatawoman117^{3½} Devoted Danzig117½	Rail trip 8
8Aug93- 6Dmr	fst	6½f	:22^3	:46	1:10^3	1:17	ⒻMd Sp Wt	67	6	10	10^{8½}	9^{5¾}	5^{3¼}	5^{4½}	Stevens G L	B 117	9.30	81-11	Top Of The Sky117^{nk} Tin117¾ Malibu Light117^{no}	Broke slowly 11

WORKOUTS: ●Dec 20 SA 5f fst :58^3 H *1/49* Dec 5 SA 4f fst :48 H *13/56* Nov 28 SA 3f fst :36^2 H *15/27* Nov 14 SA 5f fst 1:02^3 H *50/55* Nov 8 SA 3f fst :36^2 H *7/19* Nov 5 SA 4f fst :48^4 H *36/88*

Choice Claim B+4/4 speed speed
Own: Finley Jack L
NAKATANI C S (—)

B. f. 2 (Apr)
Sire: Mining (Mr. Prospector)
Dam: One Smart Lady (Clever Trick)
Br: Glencrest Farm (Ky)
Tr: Moreno Henry (—)

117

Lifetime Record:	0 M 0 0			$0
1993	0 M 0 0	Turf	0 0 0 0	
1992	0 M 0 0	Wet	0 0 0 0	
SA	0 0 0 0	Dist	0 0 0 0	

WORKOUTS: Dec 20 SA 6f fst 1:12^3 H *4/27* ●Dec 15 SA 4f gd :47^3 Hg *1/21* Dec 10 SA 6f fst 1:13^2 H *6/32* Dec 4 SA 4f fst :47^4 Hg *10/52* Nov 28 SA 6f fst 1:15^2 H *15/20* Nov 22 SA 5f fst 1:00^4 H *22/35*
Nov 16 SA 5f fst 1:01^1 H *16/39* Nov 5 SA 4f fst :49 H *45/88* Oct 30 SA 4f fst :50^3 H *62/69* Oct 24 SA 3f fst :37 H *16/27* Oct 16 SA 5f fst 1:02 H *9/15* Oct 10 SA 5f fst 1:02^3 H *41/49*

Of the first-time starters Queen Gen (D6), Emerite (D7), Designatoree (F4) and Wild Forever (C4) are all by unprecocious sires and above their morning lines. All of them, except Designatoree, are probably being prepped for routes as can be seen by the stamina indexes of their broodmare sires. Designatoree may be out of a fast Never Bend mare, but Alysheba has such a poor debut rating that his progeny will probably be best stretching out at three.

This leaves Choice Claim (B+4) as the only precociously-bred first-time starter in the race! She has Mining/Mr. Prospector speed on top and Icecapade/Clever Trick speed on the bottom of her pedigree. Adding interest is Henry Moreno who is competent with debut runners. Also that :47.3 December 15th gate work on a good track is eye-catching. Best of all, Choice Claim has been bet down below her 12-1 morning line, has Corey Nakatani up, and has drawn the outside post where she can stay out of trouble, yet use her speed to get good tactical position before hitting the turn. While her outside post would be a disadvantage in a shorter sprint, here she has more time to get over. Furthermore, being the last to load in the gate, she is more likely to break alertly than if she was stuck on the inside and had to wait for all the other horses.

So, how to bet? Choice Claim's morning line is 12-1. My line 6-1. Post time odds 10-1. She has the breeding, connections and has gotten some action, so she is a definite Win bet. Of the experienced contenders, Meadow Moon is bred, top and bottom, to sprint. She has been rested off a tough first race loss where she faded as the 4-5 favorite and comes back with a bullet work. I make her 5-2 which are her post-time odds (no advantage in the Win hole). Whatawoman has run two good races in a row, but at 3-2 is an underlay and may throw in a clunker. Since I already have Choice Claim to Win, I decide to key her in the exacta, top and bottom, with Meadow Moon and Whatawoman.

Results: Choice Claim broke alertly from the outside and outran Meadow Moon in the stretch to pay $21.40 on the Win, $9.60 to Place and $4.40 for the Show. The $2 exacta returned $90.80 with Whatawoman finishing third.

The fourth race on December 19th at Hollywood Park is a classic illustration that the more precociously-bred horses there are in a race, the more difficult it becomes to handicap from a pedigree perspective. Looking at the eight horse field for this shorter 5 1/2 furlong sprint, My Isabella and Rustica Ridge are the only automatic pedigree throwouts. Their sires Ferdinand and Cox's Ridge just don't get 2-year-old debut winners in sprints. The sires of Lorez, Social Style and Espadrille, however, do.

continued

Wild Impulse — C4|3

Own: L M J Stables & Ross & Ross

BAILEY J D (1 0 0 0 .00)

B. f. 2 (Apr)
Sire: Wild Again (Icecapade)
Dam: Lyphia (Lyphard)
Br: Mr. & Mrs. Al Ross (Ky)
Tr: Van Berg Jack C (62 5 7 9 .08)

118

	Lifetime Record:	3 M 0 0	$180
1993	3 M 0 0	$180	Turf	0 0 0 0
1992	0 M 0 0		Wet	0 0 0 0
Hol	0 0 0 0		Dist	0 0 0 0

29Oct93–3Kee fst 6½f :22² :46 1:11² 1:17⁴ ⒻMd Sp Wt 20 3 2 2hd 21½ 8¹⁴ 9²¹½ Rowland M F L 120 45.60 66–11 Altar Guild120½ Rainbowsnlollipops120²½ LostLove120² Used inside early 10

10Aug93–4EIP fst 6f :22⁴ :47 1:00¹ 1:13⁴ ⒻMd Sp Wt 22 4 1 2½ 3² 45½ 7¹⁰½ Trotter M A L 118 10.60 66–20 Show Star118³ Apache Pines118½ Lifes A Party118½ 9
Check, in tight, 5/16's, weakened

18Jly93–1EIP fst 6f :22² :46¹ :59² 1:13¹ ⒻMd Sp Wt 39 3 3 98½ 45¼ 34½ 55½ Trotter M A L 118 18.70 74–13 Miss Woodchuck118¾ Redheaded Honey118⁴ Ancy Dancy118nk 9
Fractious paddock, flattened out

WORKOUTS: Dec 9 Hol 5f fst 1:00¹ H 9/38   Dec 1 Hol 6f fst 1:14² Hg 18/25   Nov 24 Hol 5f fst 1:00² H 10/42   Nov 18 Hol 5f fst 1:02 H 25/38   Nov 11 Hol 4f fst :47³ H 4/10   Oct 24 Kee 4f fst :51 B 29/32

⚡ **Espadrille** — A+4|4

Own: Haras Santa Maria De Araras

PINCAY L JR (114 15 20 13 .13)

B. f. 2 (Feb)
Sire: Meadowlake (Hold Your Peace)
Dam: La Ole Review (Reviewer)
Br: Haras Santa Maria de Araras (Fla)
Tr: McAnally Ronald (47 5 10 6 .11)

118

	Lifetime Record:	0 M 0 0	$0
1993	0 M 0 0		Turf	0 0 0 0
1992	0 M 0 0		Wet	0 0 0 0
Hol	0 0 0 0		Dist	0 0 0 0

WORKOUTS: Dec 14 Hol 5f fst 1:00⁴ H 2/53   Dec 8 Hol 4f fst :47³ H 4/39   Dec 2 Hol 5f fst 1:01 H 8/38   Nov 26 Hol 5f fst 1:00¹ H 6/42   Nov 20 Hol 4f fst :49³ H 20/26   Nov 14 Hol 3f fst :35¹ H 3/16
Nov 8 Hol 3f fst :35² H 3/12   ●Oct 14 Hol 6f fst 1:12 H 1/13   Oct 8 Hol 5f fst 1:00⁴ Hg 10/31   Oct 2 Hol 5f fst :59² H 3/40   Sep 26 Hol 4f fst :48 H 6/24   Sep 20 Hol 4f fst :51¹ H 21/28

It's Ben Freezing — Better Post — B4|4

Own: Triple Dot Dash Stable Trust

STEVENS G L (85 16 16 14 .19)

Ch. f. 2 (May)
Sire: It's Freezing (T. V. Commercial)
Dam: Benzina (Jaazeiro)
Br: Joyce Vandervoort (Ky)
Tr: Vienna Darrell (15 3 1 1 .20)

118

	Lifetime Record:	1 M 0 0	$0
1993	1 M 0 0		Turf	0 0 0 0
1992	0 M 0 0		Wet	0 0 0 0
Hol	0 0 0 0		Dist	0 0 0 0

24Oct93–4SA fst 6f :21⁴ :45 :57¹ 1:09¹ ⒻMd Sp Wt 48 1 6 5¹½ 75¾ 78½ 7¹⁷ McCarron C J B 117 35.70 74–12 Lakeway117nk Madder ThanMad117² DesertStormette1177½ Checked 5/16 7

WORKOUTS: Dec 14 SA 6f fst 1:14² H 19/26   Dec 9 SA 5f fst 1:01¹ H 10/20   Dec 4 SA 6f fst 1:16² H 18/20   Nov 29 SA 4f fst :49⁴ B 35/42   Nov 13 SA 6f fst 1:15 H 28/31   Nov 9 SA 5f fst :50² H 39/50

On a pedigree basis, Espadrille by Meadowlake (A+4) is a debut standout. But, at 4-5 she is hard to bet. Lorez (B5) and Social Style (B6) are also precociously bred and at the shorter 5 1/2 furlong distance are advantageously drawn inside Espadrille, but she probably outclasses them.

Of the experienced runners Fresh Berries and It's Ben Freezing could improve. Fresh Berries, typical of the Capotes, showed lots of speed in her first start, setting the pace with splits of :21 and change, and Clifford Sise is putting the blinkers on for her second try. It's Ben Freezing was checked in her first start and is getting off the rail and could also move up. Even Wild Impulse, who is back after a seven-week layoff, can't be thrown out. She showed more speed in her last race in Kentucky and has been working fairly regularly at Hollywood Park (with one quick move of :47.3) for the always dangerous Jack Van Berg.

So how to bet this sprint which is loaded with speed? Even though Espadrille has the best breeding, good debut connections, and is being heavily backed, at 4-5 she is no bargain in the Win pool. Rather than pass the race, I part wheel Espadrille on top in the exacta with everyone, except My Isabella and Rustica Ridge, and hope for a big price.

FOURTH RACE   5½ FURLONGS. (1.021) MAIDEN SPECIAL WEIGHT. Purse $27,00. Fillies, 2–year–olds. Weight, 118 lbs.

Hollywood
DECEMBER 19, 1993

Value of Race: $27,000 Winner $14,850; second $5,400; third $4,050; fourth $2,025; fifth $675. Mutuel Pool $290,392.00 Exacta Pool $270,119.00 Trifecta Pool $178,791.00 Quinella Pool $35,460.00

Last Raced	Horse	M/Eqt.	A.Wt	PP	St	¼	⅜	Str	Fin	Jockey	Odds $1
29Oct93 3Kee9	Wild Impulse	LB	2 118	6	3	3^3	2^1	1^{hd}	$1^{\frac{1}{2}}$	Bailey J D	15.80
	Espadrille		2 118	7	6	6^4	$6^{3\frac{1}{2}}$	5^1	$2^{2\frac{1}{2}}$	Pincay L Jr	0.80
5Dec93 4Hol7	Fresh Berries	LBb	2 118	1	2	1^1	$1^{1\frac{1}{2}}$	2^4	$3^{\frac{3}{4}}$	Flores D R	7.40
24Oct93 4SA7	It's Ben Freezing	B	2 118	8	4	4^6	4^{hd}	$4^{4\frac{1}{2}}$	$4^{5\frac{1}{2}}$	Stevens G L	11.30
	Social Style	B	2 118	5	7	5^1	$5^{1\frac{1}{2}}$	6^3	5^1	Gryder A T	13.40
	Rustica Ridge	B	2 118	4	5	8	8	8	6no	Delahoussaye E	13.70
	Lorez	B	2 118	2	1	2^1	3^3	3^{hd}	$7^{1\frac{1}{2}}$	McCarron C J	3.90
	My Isabella	B	2 118	3	8	7^1	$7^{1\frac{1}{2}}$	7^{hd}	8	Antley C W	22.90

OFF AT 2:07 Start Good. Won driving. Time, :22¹, :45², :57⁴, 1:04 Track fast.

$2 Mutuel Prices:

6–WILD IMPULSE	33.60	8.80	4.80
7–ESPADRILLE		3.20	2.60
1–FRESH BERRIES			4.00

$2 EXACTA 6–7 PAID $110.00 $2 TRIFECTA 6–7–1 PAID $460.80 $2 QUINELLA
6–7 PAID $33.00

B. f, (Apr), by Wild Again–Lyphia, by Lyphard. Trainer Van Berg Jack C. Bred by Mr. & Mrs. Al Ross (Ky).

One reason smart handicappers demand an overlay on a horse is the ever present possibility of bad racing luck. The best horse doesn't always win. Espadrille broke slowly on the outside and fell back 12 lengths behind Fresh Berries who had the early lead over Wild Impulse and Lorez. At the top of the stretch Wild Impulse grabbed the lead and just held off Espadrille's bold and prolonged closing move. If this had been a 6 furlong sprint, Espadrille's slow start and outside post wouldn't have hurt her and she would have won easily. As it was, Wild Impulse returned $33.30 to Win and $110 for the $2 exacta. Interestingly, Espadrille paid $3.20 for Place, which considering that odds-on favorites often run second, would have been another way of playing this race.

5

Maiden Special Weight Races for 3-year-olds

While the progeny of most precocious sires are expected to make their debut at two, not all of them do. Some first-time starters, especially those out of stouter mares, simply don't mature quickly enough to race before they are three. These later bloomers, however, can be excellent bets, particularly during the first two months of the new year when the racing industry annually shifts its focus from looking for promising 2-year-olds in maiden races to evaluating the proven 3-year-old performers who are now competing in the stakes races that lead up the Kentucky Derby.

Late maturing, but talented 3-year-old maidens that debut in January and February tend to get lost in the shuffle and can go off at surprisingly attractive odds. The second race on January 16th at Santa Anita is one good example.

Meadow Moon with her 81 Beyer rating ran an improved race second time out and looks like a legitimate favorite at 2-1. Her experienced competition — Gondwanaland, Eau De Vivre and Masteful Dawn — look outgunned. Still, she has already lost two races while being heavily bet and is facing some potentially talented first-time starters.

Of these Paradisa (A+3) is a possible upsetter that fits the pattern of the promising 3-year-old debut runner. She is by the extremely precocious sire Seeking the Gold (A+3), but out of a stouter Lyphard mare on the bottom half of her pedigree. The Lyphard influence is particularly interesting in that while he is a good debut sire (B2), he is also an important (bold-faced) broodmare sire whose low stamina index suggests his daughters will throw later-maturing foals. Given her breeding, it makes sense that Paradisa has not made her debut until now.

Paradisa's connections are also encouraging. Fabio Nor, according to Olmsted's *Trainer Guide*, is good with debut runners. Furthermore, Nor and his wife own the filly (often a sign of good intent) and Paradisa has that bullet gate work of :46.4 on January 8th.

On the downside, Paradisa is stuck in the two hole inside Meadow Moon's speed and is not being heavily bet. Still, Nor has a small stable, the kind that doesn't usually draw automatic action, and Paradisa with Eddie D up does open at 9-1, two points below her morning line.

Santa Anita Park

2

6½ Furlongs. (1:14) MAIDEN SPECIAL WEIGHT. Purse $32,000. Fillies, 3-year-olds. Weight, 117 lbs.
(Non-starters for a claiming price of $32,000 or less in their last three starts preferred.)

START ▼ 6½ FURLONGS ▲ FINISH

Tee Pee Colony
Own: Hibbert R E
STEVENS G L (85 8 16 13 .09)

B. f. 3 (Jan)
Sire: Cherokee Colony (Pleasant Colony) D6/2 Router
Dam: June Showers (Topsider)
Br: Robert Hibbert (Ky)
Tr: Rash Rodney (22 4 3 3 .18)

117

Lifetime Record :	0 M 0 0	$0		
1994	0 M 0 0		Turf	0 0 0 0
1993	0 M 0 0		Wet	0 0 0 0
SA	0 0 0 0		Dist	0 0 0 0

WORKOUTS: Jan 12 SA 4f fst :47¹ Hg 9/59 Jan 7 SA 7f fst 1:26⁴ H 7/13 ●Dec 28 SA 7f fst 1:25 H 1/14 Dec 22 SA 6f fst 1:11⁴ H 3/45 Dec 17 SA 6f fst 1:12³ H 11/45 Dec 11 SA 5f fst :59 H 3/51
Nov 22 SLR tr.t 4f fst :49 H 1/2 Aug 18 Dmr 5f fst 1:01 H 19/53 Aug 13 Dmr 4f fst :48 H 14/57 Aug 6 SLR tr.t 4f fst :49 B 7/8 Jly 25 SLR tr.t 3f fst :37¹ H 1/2

Paradisa
Own: Nor Joanne H
DELAHOUSSAYE E (60 7 8 9 .12) Bullet

B. f. 3 (Apr)
Sire: Seeking the Gold (Mr. Prospector) A+3/4
Dam: Purusha (Lyphard) B2/2*
Br: Joanne Nor (Ky)
Tr: Nor Fabio (2 0 1 1 .00)

117

Lifetime Record :	0 M 0 0	$0		
1994	0 M 0 0		Turf	0 0 0 0
1993	0 M 0 0		Wet	0 0 0 0
SA	0 0 0 0		Dist	0 0 0 0

WORKOUTS: ●Jan 8 Hol 4f fst :46⁴ Hg 1/20 Dec 31 Hol 5f fst 1:00³ H 13/56 Dec 24 Hol 4f fst :48² H 10/37 Dec 17 Hol 4f fst :48 H 5/51 Dec 4 Hol 3f fst :37⁴ H 15/20 Nov 27 Hol 5f fst 1:02² H 37/46
Nov 24 Hol 3f fst :37⁴ H 28/34 Nov 15 Hol 5f fst 1:00¹ H 3/29 Nov 11 Hol 3f fst :36³ B 3/3 Nov 4 Hol 5f fst 1:00⁴ H 6/26 Oct 27 Hol 5f fst 1:02⁴ H 16/25 Oct 20 Hol 4f fst :47¹ Hg 3/25

Meadow Moon
Own: Headley Bruce
SOLIS A (84 14 9 17 .17)

Dk. b or br f. 3 (Feb)
Sire: Meadowlake (Hold Your Peace) A+4/4
Dam: Shamrock Reality (In Reality)
Br: Brylynn Farm (Ky)
Tr: Headley Bruce (7 1 2 1 .14)

117

Lifetime Record :	2 M 1 0	$8,500		
1993	2 M 1 0	$8,500	Turf	0 0 0 0
1992	0 M 0 0		Wet	0 0 0 0
SA	2 0 1 0	$8,500	Dist	0 0 0 0

26Dec93- 2SA fst 6f :21⁴ :44⁴ :57 1:09³ ⒻMd Sp Wt 81 4 3 3¹ 4²¾ 4²½ 2² Solis A B 117 2.90 87-09 Choice Claim117² Meadow Moon117½ Whatawoman117³ Finished well 12
14Nov93- 6SA fst 6f :21⁴ :44³ :56³ 1:08⁴ ⒻMd Sp Wt 72 8 4 3nk 4³½ 4⁵ 4⁸½ Solis A B 117 *.80 84-06 C'monLetsDance117³½ CasualMeeting117½ Jacodr'sDevil112⁴½ Weakened 9
WORKOUTS: Jan 13 SA 3f fst :35² H 4/18 Jan 7 SA 4f fst :50 H 35/43 Dec 21 SA 5f fst :59 H 5/51 ●Dec 14 SA 5f fst :59² H 1/41 Dec 7 SA 5f fst 1:00¹ H 7/38 ●Dec 1 SA 5f fst :59³ H 1/50

Gondwanaland
Own: Recachina Laura
BLACK C A (66 4 3 4 .06)

B. f. 3 (Apr)
Sire: Chief's Crown (Danzig) C5/3
Dam: Sue Babe (Mr. Prospector)
Br: Regent Farms (Ky)
Tr: Lukas D Wayne (24 1 2 4 .04)

117

Lifetime Record :	2 M 0 0	$0		
1993	2 M 0 0		Turf	0 0 0 0
1992	0 M 0 0		Wet	0 0 0 0
SA	2 0 0 0		Dist	0 0 0 0

26Dec93- 2SA fst 6f :21⁴ :44⁴ :57 1:09³ ⒻMd Sp Wt 44 5 4 5²¾ 10⁶¾ 12¹⁴ 12¹⁶¼ Black C A B 117 33.40 73-09 Choice Claim117² Meadow Moon117½ Whatawoman117³ Brief speed 12
14Nov93- 6SA fst 6f :21⁴ :44³ :56³ 1:08⁴ ⒻMd Sp Wt 57 1 7 7⁷¼ 7⁹½ 8¹¹ 7¹⁴ Stevens G L B 117 11.50 79-06 C'monLtsDnc117³½ CsulMting117½ Jcodr'sDvil112⁴½ Stumbled after start 9
WORKOUTS: Jan 9 SA 6f fst 1:14¹ H 22/37 Jan 2 SA 5f fst 1:03⁴ H 43/47 Dec 16 SA 4f fst :47⁴ H 6/46 Dec 9 SA 5f fst 1:00¹ H 2/20 Dec 2 SA 5f fst 1:02³ H 28/33 Nov 24 SA 4f fst :48² H 16/40

Emerald Express
Own: Keith Diane & Harold
DESORMEAUX K J (54 14 6 4 .26) 3 Bullets

B. f. 3 (Apr) D7/5
Sire: Falstaff (Lyphard)
Dam: Emerald Green (Sir Ivor)
Br: Eclipse Investments (Cal)
Tr: Dupuis Jean-Pierre (1 0 0 0 .00)

L 117

Lifetime Record :	0 M 0 0	$0		
1994	0 M 0 0		Turf	0 0 0 0
1993	0 M 0 0		Wet	0 0 0 0
SA	0 0 0 0		Dist	0 0 0 0

WORKOUTS: ●Jan 8 Hol 5f fst :58² Hg 1/23 Dec 31 Hol 5f fst 1:01³ H 29/56 Dec 10 Hol 6f fst 1:14³ H 2/12 Dec 4 Hol 4f fst :48¹ Hg 10/30 Nov 21 Hol 6f fst 1:12² H 3/8 Nov 13 Hol 5f fst 1:00³ H 4/26
●Nov 5 Hol 5f fst :58⁴ H 1/30 Oct 31 Hol 4f fst :48³ H 14/30 Oct 25 Hol 4f fst :47⁴ H 2/15 ●Oct 19 Hol 3f fst :34² H 1/8 Oct 12 Hol 3f fst :36¹ H 6/13 Oct 5 Hol 3f fst :39⁴ H 4/5

Eau De Vivre
Own: Haras Santa Maria De Araras
PINCAY L JR (84 12 12 6 .14)

Dk. b or br f. 3 (Feb)
Sire: Gulch (Mr. Prospector) C5/4
Dam: Harvard's Bay (Halpern Bay)
Br: Haras Santa Maria De Araras (Fla)
Tr: McAnally Ronald (30 3 3 6 .10)

L 117

Lifetime Record :	1 M 0 0	$675		
1993	1 M 0 0	$675	Turf	0 0 0 0
1992	0 M 0 0		Wet	0 0 0 0
SA	0 0 0 0		Dist	0 0 0 0

5Dec93- 4Hol fst 6f :21⁴ :44⁴ :57¹ 1:09² ⒻMd Sp Wt 56 1 7 8⁸¾ 6¹¹ 6¹⁰ 5¹²¾ Pincay L Jr B 118 49.60 80-10 Madder Than Mad118¹¼ My Fling118⁸ Tulgey Wood118no Off bit slow 10
WORKOUTS: Jan 13 Hol 5f fst 1:01² H 15/31 Jan 8 Hol 6f fst 1:14 H 8/22 Jan 2 Hol 5f fst 1:00² H 7/21 Dec 27 Hol 5f fst 1:01¹ H 7/25 Dec 20 Hol 5f fst 1:00² H 3/20 Dec 14 Hol 4f fst :49⁴ H 11/31

Henie
Own: Paulson Allen E
GRYDER A T (43 1 3 2 .02)

B. f. 3 (Jan)
Sire: Theatrical (Nureyev) B4/1
Dam: Different Worlds (Cornish Prince)
Br: Paulson Allen E (Ky)
Tr: Hassinger Alex L Jr (3 1 2 0 .33)

L 117

Lifetime Record :	0 M 0 0	$0		
1994	0 M 0 0		Turf	0 0 0 0
1993	0 M 0 0		Wet	0 0 0 0
SA	0 0 0 0		Dist	0 0 0 0

WORKOUTS: Jan 10 SA 5f fst 1:01² Hg 16/41 Jan 5 SA 3f fst :35³ Hg 5/39 Dec 29 SA 6f fst 1:12² H 3/25 Dec 23 SA 6f fst 1:12³ H 14/39 Dec 16 Hol 5f fst 1:03¹ H 19/34 Dec 6 Hol 4f fst :49³ H 22/35
Aug 13 SLR tr.t 3f fst :36³ H 2/7 ●Aug 8 SLR tr.t 3f fst :37¹ H 1/4 Aug 3 SLR tr.t 3f fst :35² H 3/6 ●Jly 22 SLR tr.t 3f fst :35⁴ H 1/8

Foggy Rose
Own: N & M Boyce Racing Stable Inc
VALENZUELA F H (84 10 5 9 .12)

Ro. f. 3 (Mar)
Sire: Relaunch (In Reality) B2/4*
Dam: Josette (Top Command)
Br: J. Allison & N & M Boyce Racing Stable, Inc (Cal)
Tr: Boyce Neil B (1 0 0 0 .00)

117

Lifetime Record :	0 M 0 0	$0		
1994	0 M 0 0		Turf	0 0 0 0
1993	0 M 0 0		Wet	0 0 0 0
SA	0 0 0 0		Dist	0 0 0 0

WORKOUTS: Jan 9 SA 6f fst 1:16² Hg 34/37 Jan 3 SA 5f fst 1:00⁴ Hg 34/60 Dec 26 Haw 5f fst 1:02 B 1/2 ●Dec 20 Haw 5f gd 1:00 H 1/12 Dec 14 Spt 4f fst :51³ B 2/3 Dec 9 Spt 5f fst 1:02¹ B 1/3
●Dec 4 Haw 4f sl :48 H 1/13 Nov 29 Haw 4f fst :50³ B 19/41 Nov 24 Haw 3f fst :36⁴ B 5/26 Nov 16 Haw 3f gd :38 B 8/12

continued

Of the other first-time starters Emerald Express (D7), despite her poor debut rating, opens at 7-2 and must therefore be considered. She has three bullet works and leading rider Kent Desormeux in the irons. Henie (B4), Foggy Rose (B2) and Made of Jade (B3)

Made Of Jade

Own: McNall Racing Ltd

NAKATANI C S (82 4 9 16 .05)

B. f. 3 (Mar)
Sire: Jade Hunter (Mr. Prospector)
Dam: Rogatien (Storm Bird)
Br: Summa Stable (Ky)
Tr: Becerra Rafael (17 4 1 4 .24)

B3/4

117

		Lifetime Record:	0 M 0 0			$0
1994	0 M 0 0		Turf	0	0 0 0
1993	0 M 0 0		Wet	0	0 0 0
SA	0 0 0 0		Dist	0	0 0 0

WORKOUTS: Jan 10 SA 6f fst 1:14⁴ H 12/22 Jan 4 SA 6f fst 1:14² H 11/21 Dec 29 SA 6f fst 1:14 Hg 11/25 Dec 24 SA 5f fst 1:00³ H 43/114 Dec 13 Hol 5f gd 1:02⁴ Hg 9/16 Dec 6 Hol 5f fst 1:00³ H 5/18
●Nov 26 Hol 3f fst :34³ H 1/21 Nov 21 Hol 4f fst :51⁴ Hg 37/38 Nov 16 Hol 4f fst :50⁴ H 19/23

Masterful Dawn

Own: Ortega Charles & Jim & Joseph S

PEDROZA M A (53 6 7 2 .11)

Dk. b or br f. 3 (Mar)
Sire: Masterful Advocate (Torsion)
Dam: After the Dawn (Africanus)
Br: Ortega James & Joseph (Cal)
Tr: Bernstein David (10 1 2 1 .10)

U8/4

L 117

		Lifetime Record:	3 M 1 1			$12,500
1994	1 M 1 0		$7,000	Turf	0	0 0 0
1993	2 M 0 1		$5,500	Wet	0	0 0 0
SA	1 0 1 0		$7,000	Dist	0	0 0 0

1.Jan94–6SA	fst 6f	:21² :44 :55³ 1:08	Ⓢ Md Sp Wt	72	7 2 2½ 2½ 2⁴ 2¹⁰	Pedroza M A	LB 117	6.00	87–03	Flying In The Lane117¹⁰ Masterful Dawn117²¼ Taj Aire117³	No match 10
24Nov93–6Hol	fst 6f	:22 :45 :57³ 1:10²	Ⓢ Md Sp Wt	60	9 2 4½½ 5³ 52½ 5⁴	Desormeaux K J	LB 118	*2.00	84–09	Lovely Music118½ Linda Lou B.118½ Fresh Parsley118ⁿᵏ	Wide trip 10
11Jly93–6Hol	fst 5½f	:22 :45¹ :57³ 1:04¹	Ⓢ Md Sp Wt	60	10 2 1¹ 1¹ 22½ 38½	Flores D R	LB 117	9.30	82–12	Choobloo117⁸¼ Top Of The Sky117ⁿᵒ Masterful Dawn117½½	12
	Bumped start, greenly lane										

WORKOUTS: Jan 10 SA 4f fst :47² H 6/53 Dec 27 SA 4f fst :47³ H 12/43 Dec 21 SA 6f fst 1:11⁴ H 3/18 Dec 10 SA 5f fst :59¹ H 4/66 ●Dec 3 SA 4f fst :45² H 1/31 Nov 18 Hol 5f fst 1:00 H 3/38

also have precocious sires but their connections do not excel with debut runners and all of them are dead on the board. While Tee Pee Colony has shown some speed in her morning works and is getting bet, she is bred to route and a tossout on pedigree grounds.

So how to bet? Paradisa, despite the *Daily Racing Form* telling the public that Seeking the Gold gets 43 percent first-time winners, is going off at an extremely generous 11-1. Normally this isn't a good sign. But Nor is a low-profile trainer and Paradisa has everything else going for her — great breeding, good connections and a bullet work. The race is also 6 1/2 furlongs which tactically gives the ever patient Delahousaye a little more time to get Paradisa up if Meadow Moon breaks on top. I'm tempted to make Paradisa a prime Win bet, but am still intimidated by her relative lack of action. I decide to bet her moderately to Win and box her in the exacta with Meadow Moon and Emerald Express, the two action horses.

Results: Paradisa dueled for the early lead with Tee Pee Colony, fought off Masterful Dawn in the stretch, and prevailed by 2 lengths. She paid $24.80 to Win, $10 to Place and $7 to Show. Emerald Express (the big action debut horse) hopped at the start and steadied at the quarter pole, but still got up for second, with the $2 exacta paying $99.00. Masterful Dawn held on for third while Meadow Moon finished a distant fourth.

The fifth race at Santa Anita on January 2nd is another example of a 3-year-old debut runner by a precocious sire that has been bred to a stouter mare with an interesting additional twist. This is a Maiden Special Weight sprint restricted to Cal-breds.

The first thing to note about this race is that one of the entrants Phone Prince is by Phone Trick who is not only a highly precocious sire, but also the only Kentucky stallion in the race. Owner/Breeder James Biller shrewdly sent his stout Kris S. mare all the way to Kentucky for her "date" with Phone Trick and then shipped her back so she could foal in California and thus technically qualify as a Cal-bred and be eligible to run for some of the attractive state-bred purses. While the definition of what constitutes a

5 6 Furlongs. (1:07¹) MAIDEN SPECIAL WEIGHT. Purse $35,000 (includes $3,000 from California–bred race fund if race fills Cal–bred otherwise purse $32,000.) 3–year–olds (Foals of 1991) bred in California. Weight, 118 lbs. (Non–starters for a claiming price of $32,000 or less in their last three starts preferred.)

START ▼
6 FURLONGS
▲ FINISH

Phone Prince
(circled) BMA 1'05 Fire (foot sire) B4/3 Ky Stallion
Own: Biller James M
Dk. b or br c. 3 (Feb)
Sire: Phone Trick (Clever Trick) $5,000
Dam: Kris's Nickle (Kris S.) 3/2
Br: James Biller (Cal)
Tr: Harte Michael G (3 1 0 0 .33) speed
VALENZUELA F H (20 4 1 2 .20) 118

	Lifetime Record :	0 M 0 0	$0
1994	0 M 0 0	Turf	0 0 0 0
1993	0 M 0 0	Wet	0 0 0 0
SA	0 0 0 0	Dist	0 0 0 0

WORKOUTS: Dec 24 SA 6f fst 1:14³ H 47/68 Dec 16 SA 3f fst :36² Hg9/39 Dec 6 SA 5f fst 1:00¹ B 3/29 Nov 24 SA 5f fst :59³ H 4/50 Nov 14 SA 4f fst :47² H 7/39 Nov 5 SA 3f fst :36⁴ H 17/41

Muirfield Village
B4/2 $5,600 half broke while speed clle 2lep
Own: Oak Cliff Stable
Dk. b or br c. 3 (Feb)
Sire: Skywalker (Relaunch)
Dam: Kankam (Minera)
Br: Thomas P. Tatham (Cal)
Tr: Speckert Christopher (—)
DELAHOUSSAYE E (19 3 2 2 .16) 118

	Lifetime Record :	1 M 1 0	$4,900
1993	1 M 1 0	$4,900	Turf	0 0 0 0
1992	0 M 0 0		Wet	0 0 0 0
SA	1 0 1 0	$4,900	Dist	0 0 0 0

10Nov93–4SA fst 6½f :22 :45 1:09³ 1:15⁴ ⑤Md Sp Wt 83 4 8 89½ 89¼ 54½ 21½ Delahoussaye E B 118 14.30 91–12 SomeDrmr118½ DH WindwoodLd118 DH MuirfldVillg118¹ (Off slow, wide) 8
WORKOUTS: Dec 31 SA 3f fst :37¹ B 16/29 Dec 27 SA 5f fst 1:00 H 4/41 Dec 21 SA 6f fst 1:12² H 7/18 Dec 14 SA 5f fst 1:02¹ H 30/41 Dec 8 SA 5f fst 1:02³ B 41/48 Nov 24 SA 5f fst 1:01² H 35/50

Heezfor Gramps
(circled) B4'6/4 $5,000
Own: D & V Enterprises Inc
Ch. g. 3 (Apr)
Sire: Al Mamoon (Believe It)
Dam: Grandpa's Kelly (Ruffinal)
Br: D & V Enterprises (Cal)
Tr: Gregson Edwin (3 0 0 0 .00)
ANTLEY C W (17 2 3 1 .12) 118

	Lifetime Record :	0 M 0 0	$0
1994	0 M 0 0	Turf	0 0 0 0
1993	0 M 0 0	Wet	0 0 0 0
SA	0 0 0 0	Dist	0 0 0 0

WORKOUTS: Dec 29 SA 4f fst :47³ Hg 11/5 Dec 23 SA 6f fst 1:11² H 5/39 Dec 17 SA 6f fst 1:12⁴ H 7/45 Dec 10 SA 3f fst :35² Hg4/37 Dec 5 SA 6f fst 1:12³ H 4/27 Nov 30 SA 5f gd 1:02 H 9/12 Nov 24 SA 5f fst :59⁴ H 74/50 Nov 19 SA 4f fst :48⁴ H 14/35 Nov 14 SA 5f fst :48¹ H 15/39 Nov 9 SA 3f fst :35⁴ H 3/13 Nov 4 SA 3f fst :36⁴ H 6/16

Kleven's Best
B4'9/4 fate
Own: Craft Ben
Ch. g. 3 (Jan)
Sire: Kleven (Alydar)
Dam: Ima Early Blossom (Imacornishprince)
Br: Silverio Martinez (Cal)
Tr: Moreno Henry (1 1 0 0 1.00)
GONZALEZ S JR (18 3 2 2 .17) 113⁵

	Lifetime Record :	0 M 0 0	$0
1994	0 M 0 0	Turf	0 0 0 0
1993	0 M 0 0	Wet	0 0 0 0
SA	0 0 0 0	Dist	0 0 0 0

WORKOUTS: Dec 28 SA 7f fst 1:26 H 4/14 Dec 22 SA 4f fst :46³ Hg 1/49 Dec 16 SA 6f fst 1:15¹ H 22/37 Dec 10 SA 6f fst 1:13³ H 8/32 Dec 4 SA 5f fst 1:01⁴ H 29/57 Nov 29 SA 4f fst :48² H 10/42

Superfluously
D8/4
Own: Risdon Arthur G & Larry G
B. g. 3 (Apr)
Sire: Dance in Time (Northern Dancer)
Dam: Another Cute One (Briar Bend)
Br: A. G. Risdon & L. G. Risdon (Cal)
Tr: Risdon Larry G (—)
VALENZUELA P A (21 0 3 4 .00) 118

	Lifetime Record :	2 M 1 1	$9,650
1993	2 M 1 1	$9,650	Turf	0 0 0 0
1992	0 M 0 0		Wet	0 0 0 0
SA	0 0 0 0		Dist	2 0 1 1	$9,650

16Dec93–2Hol 6f :22 :45³ :57⁴ 1:10¹ ⑤Md Sp Wt 83 7 1 4 1½ 3½ 41 3⁴ Lopez A D B 118 4.60 85–15 Halloween Treat118⁴ Camera Ready118hd Superfluously118³ (4 wide trip) 8
21Nov93–4Hol 6f :22¹ :44⁴ :56³ 1:09 ⑤Md Sp Wt 77 6 5 4 1½ 3 1½ 2⁴ Lopez A D B 118 13.90 91–07 Argold118⁴ Superfluously118²½ Dry Tortuga118no Second best 7
WORKOUTS: Dec 24 SA 5f fst 1:00¹ H 31/114 Dec 14 Hol 3f fst :35 H 2/22 Dec 8 Hol 5f fst 1:00³ H 6/40 Dec 1 Hol 4f fst :49⁴ H 27/38 Nov 19 Hol 3f fst :36³ H 2/15 Oct 31 Hol 5f fst 1:01² H 7/15

Myceenote
D7/2 Barton Favourite
Own: Straub–Rubens Cecilia P
Ro. g. 3 (Feb)
Sire: Herat (Northern Dancer)
Dam: Furious Cee (Caro)
Br: Cecilia Straub–Rubens (Cal)
Tr: Robbins Jay M (3 0 0 0 .00)
PINCAY L JR (21 3 4 3 .14) 118

	Lifetime Record :	1 M 0 0	$0
1993	1 M 0 0	Turf	0 0 0 0
1992	0 M 0 0	Wet	0 0 0 0
SA	0 0 0 0	Dist	1 0 0 0

16Dec93–2Hol 6f :22 :45³ :57⁴ 1:10¹ ⑤Md Sp Wt 70 6 3 52½ 5 4 64½ Delahoussaye E B 118 *2.40 80–15 HalloweenTreat118⁴ CameraRedy118hd Superfluously118³ Bobbled start 8
WORKOUTS: Dec 26 SA 5f fst 1:00 H 10/52 Dec 11 SA 4f fst :59 H 3/51 Dec 5 SA 5f fst :58⁴ H 2/35 Nov 29 SA 5f fst :59² H 1/31 Nov 23 SA 5f fst 1:00² H 13/25 Nov 17 SA 4f fst :47³ H 9/27

Paster's Caper
C+3/3
Own: Auerbach Ernest
Dk. b or br c. 3 (May)
Sire: Flying Paster (Gummo)
Dam: Showy Cape (Snow Sporting)
Br: Ernest Auerbach (Cal)
Tr: Robbins Jay M (3 0 0 0 .00)
NAKATANI C S (27 4 0 6 .15) 118

	Lifetime Record :	0 M 0 0	$0
1994	0 M 0 0	Turf	0 0 0 0
1993	0 M 0 0	Wet	0 0 0 0
SA	0 0 0 0	Dist	0 0 0 0

WORKOUTS: Dec 30 SA 4f fst :48² Hg26/46 Dec 25 SA 7f fst 1:27 H 1/3 Dec 18 SA 6f fst 1:13 H 6/32 Dec 11 SA 6f fst 1:12⁴ H 9/27 Dec 4 SA 5f fst :58³ H 1/57 Nov 28 SA 5f fst :59⁴ H 5/44 Nov 22 SA 4f fst :47² H 9/49 Nov 16 SA 4f fst :48¹ H 7/42 Oct 3 SA 5f fst 1:00 H 3/4 Sep 28 SA 5f fst 1:00³ H 12/59 Sep 22 SA 4f fst :49 H 25/47 Sep 11 SLR tr.t 5f fst 1:01 Hg1/12

Fly With Ty
C 6/3
Own: Hi Card Ranch & Silver Horse Ranch
Ro. g. 3 (Apr)
Sire: Swing Till Dawn (Grey Dawn II)
Dam: La Cresta (Tell)
Br: Hi Card Ranch (Cal)
Tr: Becerra Rafael (3 1 0 1 .33)
DESORMEAUX K J (4 1 1 0 .25) 118

	Lifetime Record :	0 M 0 0	$0
1994	0 M 0 0	Turf	0 0 0 0
1993	0 M 0 0	Wet	0 0 0 0
SA	0 0 0 0	Dist	0 0 0 0

WORKOUTS: Dec 28 SA 5f fst :59³ H 6/55 Dec 23 SA 6f fst 1:12⁴ H 16/39 Dec 18 SA 6f fst 1:12⁴ H 2/32 Dec 13 SA 6f fst 1:15² H 13/19 Dec 8 SA 5f fst 1:02 H 36/48 Dec 3 SA 5f fst :59³ H 5/31 Nov 28 SA 5f fst 1:04² Hg43/44 Nov 23 SA 4f fst :47² H 8/38 Nov 18 SA 4f fst :49² H 15/27 Sep 11 SLR tr.t 3f fst :37² H 9/15

Cold Cool N'bold
C 6/5
Own: Atwell & Broberg & Jimac Stable
Dk. b or br c. 3 (May)
Sire: Bolger (Damascus)
Dam: Quick n' Cool (Ice Age)
Br: Running Luck, Inc. (Cal)
Tr: Dutton Jerry (3 1 0 0 .33)
STEVENS G L (24 3 7 0 .13) 118

	Lifetime Record :	0 M 0 0	$0
1994	0 M 0 0	Turf	0 0 0 0
1993	0 M 0 0	Wet	0 0 0 0
SA	0 0 0 0	Dist	0 0 0 0

WORKOUTS: Dec 24 SA 4f fst :46 H 1/85 Dec 17 SA 6f fst 1:14 Hg27/45 Dec 10 SA 6f fst 1:15 H 22/32 Dec 4 SA 4f fst :46¹ H 1/52 Nov 29 SA 5f fst 1:00² H 8/31 Nov 22 SA 4f fst :46² H 3/49 Nov 16 SA 4f fst :46⁴ H 1/42 Nov 10 SA 4f fst :48³ H 16/45 Nov 4 SA 3f fst :37 H 7/16

---

continued

College Town
Own: Brown David N
SOLIS A (26 5 2 3 .19)

Dk. b or br c. 3 (Mar)
Sire: Snow Chief (Reflected Glory)
Dam: Magnificent Crown (The Irish Lord)
Br: Dr. David N. Brown (Cal)
Tr: Stute Melvin F (8 1 0 2 .13)

118

Lifetime Record :	0 M 0 0	$0		
1994	0 M 0 0		Turf	0 0 0 0
1993	0 M 0 0		Wet	0 0 0 0
SA	0 0 0 0		Dist	0 0 0 0

WORKOUTS: Dec 30 SA 6f fst 1:12^1 H *3/34* Dec 23 SA 6f fst 1:13 Hg *19/39* Dec 18 Hol 5f fst 1:01^2 Hg *8/17* Dec 13 Hol 5f gd 1:02^1 H *4/16* Dec 8 Hol 4f fst :48^3 H *19/40* Nov 30 SLR tr.t 5f fst 1:00 H *1/2*
●Nov 22 SLR tr.t 5f fst 1:00 H *1/5* ●Nov 15 SLR tr.t 4f fst :47^4 H *1/4* Nov 8 SLR tr.t 4f fst :47^4 H *1/3* Nov 2 SLR tr.t 3f fst :36^1 H *1/3* Oct 26 SLR tr.t 3f fst :36^4 H *3/7* Jly 31 SLR tr.t 3f fst :38^3 H *15/20*

Camera Ready
Own: Golden Eagle Farm
BLACK C A (15 0 0 2 .00)

Ch. c. 3 (Apr)
Sire: Half a Year (Riverman)
Dam: Cute Show (Secretariat)
Br: Mabee Mr & Mrs John C (Cal)
Tr: Hofmans David (2 0 1 0 .00)

118

Lifetime Record :	1 M 1 0	$5,400.			
1993	1 M 1 0	$5,400	Turf	0 0 0 0	
1992	0 M 0 0		Wet	0 0 0 0	
SA	0 0 0 0		Dist	1 0 1 0	$5,400

16Dec93-2Hol fst 6f :22 :45^3 :57^4 1:10^1 [S]Md Sp Wt 84 1 4 1$\frac{1}{2}$ 1$\frac{1}{2}$ 1$\frac{1}{2}$ 2^4 Gryder A T B 118 36.80 85-15 Halloween Treat118^4 Camera Ready118^{hd} Superfluously118^3 Inside duel 8
WORKOUTS: Dec 27 SA 5f fst 1:00^1 H *7/41* Dec 13 Hol 4f gd :48^1 Hg *2/14* Dec 8 Hol 6f fst 1:15^3 H *22/22* Dec 1 Hol 6f fst 1:13^1 H *8/25* Nov 24 Hol 5f fst 1:00^2 H *10/42* Nov 16 SLR tr.t 5f fst 1:01^1 H *2/6*

Prince Quick
Own: Bettwy & Myers & Petruccione
ATHERTON J E (4 0 0 2 .00)

Ch. c. 3 (Mar)
Sire: Fast Forward (Pleasant Colony)
Dam: Silver Princess (Prize Silver)
Br: Ronald Petruccione & Joyce Petruccione (Cal)
Tr: Bacorn Herbert L (3 0 0 0 .00)

113^5

Lifetime Record :	1 M 0 0	$0		
1993	1 M 0 0		Turf	0 0 0 0
1992	0 M 0 0		Wet	0 0 0 0
SA	0 0 0 0		Dist	1 0 0 0

16Dec93-2Hol fst 6f :22 :45^3 :57^4 1:10^1 [S]Md Sp Wt 62 4 7 7^6 8$^{7\frac{3}{4}}$ 7^8 8$^{12\frac{1}{2}}$ Black C A 118 58.40 77-15 Halloween Treat118^4 Camera Ready118^{hd} Superfluously118^3 No rally 8
WORKOUTS: Dec 26 SA 4f fst :48^3 H *22/46* Dec 9 Hol 6f fst 1:14^3 Hg *9/13* Dec 1 Hol 6f fst 1:14^4 H *21/25* Nov 24 Hol 4f fst :48 H *15/40* Nov 17 Hol 5f fst 1:03^4 Hg *22/23* Nov 9 SA 5f fst 1:02^2 H *34/42*

state-bred varies from state to state, handicappers should always keep a sharp eye out for the progeny of Kentucky sires when they appear in races restricted to state-breds because Kentucky (and sometimes Florida) sires are almost always superior to the local stallions that typically service the state-bred mare population.

Looking at the experienced horses in the race, Muirfield Village, Superflously and Camera Ready all ran well in their debut and look to contend again. Muirfield Village broke poorly in his first start and is the favorite at 5-2 followed by Camera Ready and Superfluously who are the second and third lukewarm choices at nearly 6-1. The latter two come out of the same race and are respectively switching to better, but cold jockeys in Pat Valenzuela and Corey Black. Superflously went 4-wide last time and is drawn inside Camera Action for this sprint, with good gate jockey Pat V figuring to gun her from the start.

Of the first-time starters Phone Prince (B4), Heezfor Gramps (B+6) and Kleven's Best (B+7) are all precociously-bred. But the latter two are probably outclassed here and need either a weaker Maiden Special Weight race or to be dropped into the Maiden Claiming ranks. Cold Cool N'Bold (C6) is not precociously-bred or classy but, with several bullets and Gary Stevens in the irons, he is being bet, as is Fly With Ty (C6) with Kent Desormeux up.

Looking closer at Phone Prince, he has a series of decent works and is trained by Michael Harte who is competent with debut runners. Even though Phone Prince is stuck on the rail, he figures to have more speed than Muirfield Village has thus far shown and has just as much class. Furthermore, Phone Prince fits the profitable pattern of the late-maturing, 3-year-old out of a stoutly-bred mare (Kris S.) that makes its debut early in the year. On the downside, despite Phone Prince's considerable allure, it is worriesome that he is going off at 30-1.

So how to bet? While Phone Prince is an attractive play, there are three experienced horses with Beyers over 80 and several other precociously-bred, first-time starters that may outrun their breeding. Given all the possibilities, the best way, before the fact, is to eschew the exotics and play Phone Prince across the board. After the fact, a bolder wheel and backwheel, Phone Prince with the aforementioned contenders, looks like a better bet.

Results: Superfluously broke alertly and put away Kleven's Best and Myceenote on the backstretch. Phone Prince outbroke Muirfield Village and tracked Superflously to the wire, just holding off Muirfield Village for the Place. Superflously paid $13.80 to Win, while Phone Prince returned $25.60 to Place and $10 to Show. The $2 exacta paid (ouch) $461.00!

The fourth race at Bay Meadows on January 2nd is another type of potentially playable Maiden Special Weight event. This is a race where the horses that have run all have below par Beyers and the heavy favorite is a debut runner that can be dismissed on pedigree grounds alone.

The first thing to observe about this race is that all of the experienced runners come out of the same weak sprint. Among Lovely Krissy, Katelyn Alexis, Light The Gold, and Sonata Sky, only Katelyn Alexis showed speed her first-time out. She is getting bet at 4-1 and is likely to improve. As the lesser of evils she might win, but her initial 37 Beyer is not very impressive.

The heavy false favorite here is Aly's Dearest, a first-time starter by Alysheba that is trained by Neil Drysdale and ridden by leading rider Russell Baze. Besides having no pari-mutuel value, Aly's Dearest can be confidently bet against here because of Alysheba's F4 rating which indicates that less than 5 percent of his progeny win their debuts. In general descendants of Alydar have not been precocious sires. Aly's Dearest is also out of a stamina-oriented Speak John mare and is thus most likely bred to route.

Why then has the public made Aly's Dearest the 3-2 favorite when genetically she should be at least 20-1? The explanation, of course, lies in her connections. Neil Drysdale is good with debut runners and the northern California public obviously believes that he hasn't shipped this filly up for the exercise. The fact that he has hired Russell Baze to ride her is further indication of positive trainer intent. The public's logic seems to be further reinforced by the weakness of the field. There are no first-time starters by A or B sires and the horses that have already run have very low Beyers. If Aly's Dearest is going to win her debut, this field probably offers her the best shot.

Still, the fact remains that Alysheba is a terrible debut sire and the chances are excellent somebody is going to beat Aly's Dearest. But who? Of the other first-time starters Key To V.A. is one possibility. At least she has a C precocity rating. But with Key to the Mint on top and Prince John on the bottom she too looks like she is bred to route. Still, Jose Silva is a good debut trainer and she has been bet down to 4-1 from her morning line of 10-1.

Bay Meadows

4

6 Furlongs. (1:07¹) MAIDEN SPECIAL WEIGHT. Purse $16,500. Fillies 3-year-olds. Weight, 117 lbs

START ▼
6 FURLONGS
▲ FINISH

Spread speed to stretch

Rapa Lady
D5/4
Own: Fleur De Lis Stable
B. f. 3 (Apr)
Sire: Capote (Seattle Slew)
Dam: Just Say Whoa (Secretariat)
Br: L R French Jr., Barry Beal, et al. (Ky)
Tr: Klokstad Bud (87 23 17 12 .26)

WARREN R J JR (410 37 59 68 .09)   L 117

	Lifetime Record :	0 M 0 0		$0	
1994	0 M 0 0		Turf	0 0 0 0	
1993	0 M 0 0		Wet	0 0 0 0	
BM	0 0 0 0		Dist	0 0 0 0	

WORKOUTS: Dec 27 BM 5f fst 1:04 H 47/52   Dec 18 BM 6f gd 1:17¹ H 8/9   Dec 10 BM 4f my :51¹ H (d)3/26   Dec 3 BM 5f gd 1:02² H 11/45   Nov 26 BM 5f fst 1:02 H 14/17   Nov 18 BM 5f fst 1:02¹ Hg 12/23
Nov 12 BM 5f gd 1:02 H 14/33   Nov 6 BM 4f fst :48³ Hg 12/33   Oct 30 BM 4f fst :35³ H 2/20   • Oct 23 BM 3f fst :35³ H 1/10   Oct 16 BM 3f fst :39 H 3/3

Lovely Krissy
C 2/2   LATER
Own: Bonelli E & Jean
Dk. b or br f. 3 (Feb)
Sire: Kris S. (Roberto)
Dam: Daisy's Fling (Quiet Fling)
Br: Mr. & Mrs. Edward A. Bonelli (Cal)
Tr: Utley Doug (143 13 12 14 .09)

SCHVANEVELDT C P (83 8 9 6 .10)   L 117

	Lifetime Record :	4 M 1 0		$5,315	
1993	4 M 1 0	$5,315	Turf	0 0 0 0	
1992	0 M 0 0		Wet	0 0 0 0	
BM	3 0 1 0	$4,125	Dist	1 0 1 0	$3,000

5Dec93–6BM fst 6f   :22³ :46¹ :58² 1:11² (F)Md Sp Wt   43 6 7 8¹¹ 88¾ 49 26   Schvaneveldt C P LB 117 b 31.00 73–18 Woven Gold117⁶ Lovely Krissy117hd Sonata Sky117³½   Rallied wide 9
17Oct93–1BM gd 1   :22⁴ :47⁴ 1:13³ 1:40¹ (F)Md Sp Wt   40 3 4 5⁵ 46¼ 48 41²¼ Kaenel J L LB 118 b 9.70 56–28 Jilly's Halo117⁶ Vanilla Wafer117¹¼ Day Rate117⁵   Even try 5
6Sep93–6BM fst 5½f   :22 :45⁴ :58² 1:04⁴ (F)Md Sp Wt   37 1 7 99¾ 87¼ 79½ Gonzalez R M LB 117 b 9.00 78–14 Vail Link117½ Eishin Ohio117⁴ Bel Native117³½   No threat 9
29Jun93–9Pln fst 5½f   :21³ :46 :57³ 1:04² (F)Md Sp Wt   51 5 9 9¹² 7¹⁰ 6¹¹ 47   Gonzalez R M B 117 b 11.30 84–09 Saucy Trick117³ Gentle Rainfall117²¼ Jilly's Halo117½   No rally 10

WORKOUTS: Dec 29 BM 5f fst 1:01⁴ H 26/43   Dec 22 BM 5f fst 1:02 H 18/39   Dec 15 BM 5f my 1:03 H 13/17   Dec 2 BM 5f gd 1:01⁴ H 6/21   Nov 25 BM 5f fst 1:02 H 13/19   Nov 5 BM 5f fst 1:02⁴ H 19/31

Don't Dally Dear
D8/3   LATER
Own: Kjell Qvale H
B. f. 3 (Feb)
Sire: Variety Road (Kennedy Road)
Dam: Telling Blow (Silveyville)
Br: Kjell H. Qvale (Cal)
Tr: Cornell Linda (18 2 3 5 .11)

KAENEL J L (349 52 41 40 .15)   117

	Lifetime Record :	0 M 0 0		$0	
1994	0 M 0 0		Turf	0 0 0 0	
1993	0 M 0 0		Wet	0 0 0 0	
BM	0 0 0 0		Dist	0 0 0 0	

WORKOUTS: Dec 29 BM 5f fst 1:01 Hg 13/43   Dec 24 BM 6f fst 1:15² H 12/30   Dec 16 BM 6f gd 1:18³ H 7/7   Dec 7 BM 5f fst 1:05¹ H 10/10   Dec 2 BM 4f gd :49² H 15/41   Nov 24 BM 4f fst :48 H 4/36
Nov 19 BM 3f fst :39² H 13/14   Nov 12 BM 4f gd :52³ H 21/22   Nov 5 BM 3f fst :39¹ H 15/23

Refined Reality
U5/3   FULL TO MR. N.
Own: Wolf Willow Farms
Dk. b or br f. 3 (May)
Sire: Proper Reality (In Reality)
Dam: Fight for Gold (Search for Gold)
Br: Albert H. Cohen, Randy L. Cohen, et al. (Md)
Tr: Wig Janet C (7 1 1 0 .14)

BELVOIR V T (515 98 92 69 .19)   117

	Lifetime Record :	0 M 0 0		$0	
1994	0 M 0 0		Turf	0 0 0 0	
1993	0 M 0 0		Wet	0 0 0 0	
BM	0 0 0 0		Dist	0 0 0 0	

WORKOUTS: Dec 30 BM 4f fst :48⁴ H 18/43   Dec 18 BM 5f gd 1:04¹ H 41/47   Dec 5 BM 5f fst 1:02² H 39/52   Nov 21 BM 3f fst :36⁴ H 14/40   Oct 26 StP 3f fst :38² H 3/4   Oct 19 StP 3f fst :38¹ H 1/2

Aly's Dearest
F4/2   LATER
Own: Appleton Arthur I
B. f. 3 (Feb)
Sire: Alysheba (Alydar)
Dam: Middle Cornish (Speak John)
Br: W. S. Kilroy (Ky)
Tr: Drysdale Neil (10 2 1 0 .20)

BAZE R A (544 139 92 70 .26)   117

	Lifetime Record :	0 M 0 0		$0	
1994	0 M 0 0		Turf	0 0 0 0	
1993	0 M 0 0		Wet	0 0 0 0	
BM	0 0 0 0		Dist	0 0 0 0	

WORKOUTS: Dec 28 SA 5f fst 1:01 H 32/55   Dec 22 SA 6f fst 1:15² H 33/45   Dec 6 Hol 6f fst 1:14² H 5/12   Dec 1 Hol 5f fst 1:03² H 41/41   Nov 25 Hol 5f fst 1:03¹ H 14/18   Nov 20 Hol 3f fst :36⁴ H 5/22
Oct 29 SA 5f fst 1:04² H 44/47   Oct 24 SA 6f fst 1:16² H 24/26   Oct 18 SA 4f fst :48³ H 15/48   Oct 13 SA 5f fst 1:01¹ H 20/40   Oct 8 SA 6f fst 1:17¹ H 21/23   Oct 3 SA 6f fst 1:13⁴ H 14/28

Katelyn Alexis
[6f]   C4/3 3rd   BL speed
Own: Bonde Jeff & Cox Carl & Wallace Gle
Dk. b or br f. 3 (Apr)
Sire: Conquistador Cielo (Mr. Prospector)
Dam: Northern Dynasty (Northern Jove)
Br: Richard S. Kaster & Nancy R. Kaster (Ky)
Tr: Bonde Jeff (101 15 7 14 .15)

BOULANGER G (545 105 83 81 .19)   L 117

	Lifetime Record :	1 M 0 0		$1,125	
1993	1 M 0 0	$1,125	Turf	0 0 0 0	
1992	0 M 0 0		Wet	0 0 0 0	
BM	1 0 0 0	$1,125	Dist	1 0 0 0	$1,125

5Dec93–6BM fst 6f   :22³ :46¹ :58² 1:11² (F)Md Sp Wt   34 8 2 1hd 2hd 35 49¼ Baze R A LB 117 3.90 69–18 Woven Gold117⁶ Lovely Krissy117hd Sonata Sky117³½   Weakened 9

WORKOUTS: Dec 29 BM 3f fst :35³ H 2/17   Dec 22 BM 6f fst 1:14² H 2/6   Dec 16 BM 3f gd :35 Hg 2/31   Dec 1 BM 3f my :37 Hg 3/14   Nov 24 BM 6f fst 1:15³ Hg 8/11   Nov 17 BM 5f fst 1:02¹ H 4/28

Light The Gold
C5/2   LATER
Own: Golden Eagle Farm
B. f. 3 (May)
Sire: Slew o' Gold (Seattle Slew)
Dam: Candlelight Service (Blushing Groom)
Br: Mr. & Mrs. John C. Mabee (Ky)
Tr: Severinsen Allen (110 15 12 15 .14)

CHAPMAN T M (396 40 37 60 .10)   117

	Lifetime Record :	1 M 0 0		$375	
1993	1 M 0 0	$375	Turf	0 0 0 0	
1992	0 M 0 0		Wet	0 0 0 0	
BM	1 0 0 0	$375	Dist	1 0 0 0	$375

5Dec93–6BM fst 6f   :22³ :46¹ :58² 1:11² (F)Md Sp Wt   30 2 5 65¾ 56 6¹² 5¹¹ Chapman T M B 117 b 23.20 68–18 Woven Gold117⁶ Lovely Krissy117hd Sonata Sky117³½   No rally 9

WORKOUTS: Dec 30 BM 3f fst :37² H 7/14   Dec 23 BM 5f fst 1:02⁴ H 32/42   Dec 17 BM 4f gd :52¹ H 32/42   Nov 24 BM 5f fst 1:01⁴ H 23/46   Nov 13 BM 4f fst :49 Hg 33/59   Nov 5 BM 4f fst :50³ H 30/33

A Trifle To Spare
3 fast works   C6/3 speed
Own: Robocker Irene E
B. f. 3 (May)
Sire: Dixieland Brass (Dixieland Band)
Dam: Udder Delight (Raja Baba)
Br: Farish W S Jr (Ky)
Tr: Comiskey C A (42 0 4 4 .00)

MARTINEZ J C (70 3 3 4 .04)   117

	Lifetime Record :	0 M 0 0		$0	
1994	0 M 0 0		Turf	0 0 0 0	
1993	0 M 0 0		Wet	0 0 0 0	
BM	0 0 0 0		Dist	0 0 0 0	

WORKOUTS: Dec 29 BM 4f fst :47¹ Hg 2/44   Dec 23 BM 5f fst 1:01 H 14/42   Dec 16 BM 5f gd 1:02³ Hg 16/55   Dec 10 BM 4f my :54⁴ H (d)23/26   Dec 4 BM 4f gd :49⁴ H 23/40   Nov 28 BM 6f fst 1:18⁴ H 10/10
Nov 21 BM 5f fst 1:02³ Hg 57/81   Nov 14 BM 4f fst :51 H 51/59

Key To V. A.
Own: Cuadra Tyt

bcr   *C 3/3 LrTeR*

B. f. 3 (Mar)
Sire: Key to the Mint (Graustark)
Dam: Lady Fame (Prince John)
Br: Cuadra TYT Inc. (Cal)
Tr: Silva Jose L (141 16 24 18 .11)

MEZA R Q (299 36 39 42 .12)

L 117

Lifetime Record :		0 M 0 0		$0		
1994	0 M 0 0		Turf	0 0 0 0		
1993	0 M 0 0		Wet	0 0 0 0		
BM	0 0 0 0		Dist	0 0 0 0		

WORKOUTS: Dec 19 BM 5f fst 1:01² H 40/106   Dec 10 BM 5f my 1:03¹ Hg(d)8/40   Dec 3 BM 6f gd 1:15⁴ Hg5/15   Nov 26 BM 5f fst 1:00⁴ Hg8/17   Nov 18 BM 5f fst 1:03² H 18/23   Nov 7 BM 4f fst :48⁴ H 24/72
● Oct 31 BM 3f fst :35¹ H 1/29   Jly 11 GG 4f fst :48⁴ H 18/61   Jly 4 GG 4f fst :49³ H 26/56

Sonata Sky
Own: Tatham Larraine

*B 4/3 ***

Dk. b or br f. 3 (Apr)
Sire: Skywalker (Relaunch)
Dam: Banjo Melody (Apalachee)
Br: Lorraine Tatham (Cal)
Tr: Benedict Jim (86 17 14 11 .20)

TOHILL K S (291 22 27 36 .08)

L 117

Lifetime Record :		5 M 1 2		$9,675		
1993	5 M 1 2		$9,675	Turf	0 0 0 0	
1992	0 M 0 0			Wet	0 0 0 0	
BM	5 0 1 2		$9,675	Dist	2 0 1 1	$5,250

5Dec93–6BM fst 6f	:22³ :46¹ :58² 1:11²	ⓕMd Sp Wt	43 1 4 4²¾ 3² 2⁵ 3⁶	Tohill K S	LB 117	5.80	73–18	Woven Gold117⁶ Lovely Krissy117ʰᵈ Sonata Sky117³¾	Even late 9
14Nov93–3BM fst 1	:23⁴ :48² 1:13³ 1:39²	ⓕMd Sp Wt	46 4 3 3¹½ 3²½ 3⁴½ 3⁵½	Tohill K S	LB 117	1.60	68–15	Lucky Jo B.117ⁿᵏ Vanilla Wafer117⁵ Sonata Sky117¹⁰	5
	Broke in tangle, steadied start								
28Oct93–6BM fst 6f	:22¹ :46 :58² 1:11	ⓕMd Sp Wt	63 2 7 7⁶ 4⁵½ 3¹½ 2²	Tohill K S	LB 117	11.00	79–17	Catnip117² Sonata Sky117² Valnesian118³¾	Broke in tangle 7
6Sep93–6BM fst 5½f	:22 :45⁴ :58² 1:04⁴	ⓕMd Sp Wt	48 8 1 2¹½ 6³½ 4⁴ 4⁵½	Tohill K S	LB 117	10.60	82–14	Vail Link117½ Eishin Ohio117⁴ Bel Native117¾	Steadied sharply 1/4 9
18Aug93–9BM fst 5½f	:22 :45² :58¹ 1:05	ⓕMd Sp Wt	48 7 7 6³½ 4⁴½ 3⁵½ 4⁶¾	Tohill K S	LB 117	15.80	83–11	Delightful Genie117⁴½ Valnesian117ⁿᵏ My Blue Genes117²	Rail trip 7

WORKOUTS: Dec 27 BM 4f fst :50² H 46/58   Dec 1 BM 3f my :37⁴ H 6/14   Oct 23 BM 4f fst :48 H 8/43 '   Oct 10 BM 5f fst 1:00⁴ H 7/47'

A Trifle to Spare is another debut possibility. Her sire Dixieland Brass is from the Northern Dancer-line which generally imparts speed as does her broodmare sire Raja Baba. This filly's :47.1 gate work on December 29th is eye-catching and suggests she might pop at first asking. On the downside, her connections are ice cold and she is only getting lukewarm toteboard action.

Equally problematic is Rapa Lady. She is by Capote and out of a dam by premier broodmare sire Secretariat. On the basis of stud fees, she is the class of the race. On the downside, even though Capote is a precocious 2-year-old sire, his progeny tend to need several races before breaking their maidens, as his D5 rating testifies. Trainer Bud Klokstad, however, is excellent with first-time starters and there is that bullet :35.3 October work, suggesting Rapa Lady has some speed and might get loose on the lead. Since Bay Meadows generally plays to front-running speed, and there doesn't appear to be much of it in this field, Rapa Lady has a good chance of outrunning her 16-1 odds.

Don't Dally Dear is by a cheap California sire and is probably bred to route. At 28-1 she is dead on the board and a throwout which leaves Refined Reality as the last debut runner to consider.

Refined Reality is part of Proper Reality's first 2-year-old crop and his U5 rating is based on a very limited number of starters. So, it may be too early to give up on Proper Reality as a debut sire. As a racehorse he won his only race at two, but did his best running at three and four when he won the Metropolitan Mile. Proper Reality's first crop may similarly improve as they get older, with his 3-year-old debut runners being more precocious than his 2-year-olds. Also worth noting is Refined Reality's broodmare sire Search for Gold who is a full brother to Mr. Prospector and adds speed to the bottom half of her pedigree.

So how do you bet a race with a heavy false favorite but no clear pedigree standout? In Dick Mitchell felicitous phrase, this is a "chaos" race where anything can happen. The most logical thing to do is pass. But anytime you can throw out a horse like Aly's Dearest and 40 percent of the Win pool it's hard not to take a shot.

56   Since we can't bet every horse in the race, some eliminations are in order. Lovely Krissy, Don't Dally Dear and Light The Gold are all bred to route and easy to throw out. So too is Sonata Sky. Even though she has the highest Beyer, she's already had five tries and is not being bet. At nearly 7-1 she is nearly double her 7-2 morning line. Katelyn Alexis is a more difficult elimination. Even though she showed speed in her debut (and is now being bet), the pace of her first race was slow and she still couldn't break 1:13 for the six furlongs.

    Rapa Lady is more intriguing. Her :35.3 work looks to be faster than anything in the race. In fact, Rapa Lady should have enough speed to outrun the horses just outside her, who are bred to route, and avoid the pressure of being pinned to the rail. She is a wire-to-wire threat, if she doesn't quit like many young Capote's are prone to do. But with Secretariat as her broodmare sire, she may well have the extra stamina she needs. I make Rapa Lady 6-1, while the public has her at 16-1.

    Refined Reality looks great in the post parade, but is still dead on the board. Nevertheless, I am reluctant to throw her out because over the years I've found that body language with debut runners is especially important. The ones that look calm and confident in the post-parade often outrun their odds. Though I don't know Refined Reality's trainer Janet Wig, she's done well with a limited number of starters. I make Refined Reality 10-1 compared to the public's 37-1.

    A Trifle To Spare continues to be another possibility. She has shown some speed in her morning works and is being bet down a tick below her 12-1 morning line odds, despite her ice-cold trainer. The mystery horse here is Key To V.A. who is bet early and at 4-1 is way below her morning line of 10-1. What to believe? Her toteboard action and Jose Silva's good stats as a debut trainer? Or what looks like her route breeding? Despite the action and her connections, I decide to toss Key To V. A. out on pedigree and value grounds and swing for the fences.

    Since there are a lot of unknowns in this race, I decide to limit my bets and key on Rapa Lady's speed. I play her to Win and Place and box her in the exacta with A Trifle To Spare (the other good work horse) and Refined Reality (the post parade play). But what if A Trifle To Spare or Refined Reality win? I cover myself with a Win bet on the former and, because of the absence of action on Refined Reality, only a Place bet on her. My Place bets makes sense here here because, if the heavy favorite finishes third or worse, both Place bets will pay handsomely.

Results: The way I played this race is a classic example of good handicapping being compromised by bad money management. Rapa Lady, as I expected, rushed from up along the rail and took a two length lead down the backstretch. But, just when it looked like Rapa Lady would wire the field, she began to quit (Capote's speed-crazy influence with this filly was apparently stronger than Secretariat's calming, stamina one). Next Katelyn Alexis made a move and took a brief lead before she was passed by the action horse Key To V. A. who looked like a winner until Refined Reality caught her in deep stretch. Aly's Dearest made a late move and got up for third. Refined Reality paid

$76.40 to Win and $16.80 to Place. The $2 exacta paid $468.00. Even though I collected on my Place bet, I was dismayed because I hadn't put any Win money on Refined Reality.

This race was a painful reminder that the first rule in handicapping a false favorite race is to bet all your overlayed contenders to Win. There is nothing wrong with dutching three horses when you can throw out a 3-2 favorite like Aly's Dearest, strictly on pedigree grounds. While Refined Reality paid $76 to win, had Rapa Lady won she would have paid $34 and A Trifle To Spare would have paid $24. A $10 win bet on all three would have cost $30 and returned a minimum $120.00. As it was, my Place bet on Refined Reality barely covered my action.

Still, this was a precarious race to play because there wasn't an A or B debut contender, nor any other obvious handicapping angle to separate the horses. While in the short run playing against the false favorite in this kind of race can burn money, in the long run it's a positive expectation to bet this type of race, even randomly. Those players who can handle losing streaks without caving in should bet when the 3-2 favorite can be eliminated.

Ironically, Refined Reality would have been upgraded in my play had I been in possession of one more piece of information. Her dam, I belatedly discovered in *Maiden Stats 1993*, had already thrown eight winners from ten starters, including a couple of 2-year-old winners. Furthermore, she had a speed-leaning dosage of nearly 6 which made it likely that Refined Reality would also be precocious. That information, plus Refined Reality's positive body language in the post-parade, would have given me the confidence to more aggressively bet her.

Just as there are attractive short-priced overlays in 2-year-old races, the same holds true at three. The sixth race at Gulfstream on January 27th is a good example.

The first thing to notice about this Maiden Special Weight sprint is that it is for 7 furlongs — a distance racing secretaries write specifically for 3-year-old debut runners out of stouter mares who need more than 6 furlongs to break their maidens.

The second thing to notice is that none of the experienced runners look like world beaters. Retrospection has the highest Beyer and is the likely favorite, but he has already lost four times and his soundness is suspect because of his two layoffs. Carpet broke poorly in his debut and could improve in his second start. He is by the extremely precocious Danzig (A+1) and comes into the race, after a six-month layoff, with a bullet drill at Hialeah. Horses that have been lightly-raced at two and then rested for several months come back physically more mature and often show a dramatic improvement over their 2-year-old form. Carpet fits this pattern and his Claiborne/Mott connections don't hurt either. The rest of the experienced runners are proven losers and probably need class drops.

Gulfstream Park

6

7 Furlongs. (1:203) **MAIDEN SPECIAL WEIGHT. Purse $22,000 (plus $2,000 FOA). 3-year-olds. Weight, 120 lbs.**

START ▾ 7 FURLONGS ▴ FINISH

Satellite Photo

Own: Evans Edward P

BRAVO J (88 10 12 10 .11)

Dk. b or br c. 3 (May)
Sire: Crafty Prospector (Mr. Prospector)
Dam: Key Flight (Bates Motel)
Br: Edward P. Evans (Va)
Tr: Attfield Roger L (8 0 1 0 .00)

120

Lifetime Record: 2 M 0 1 $1,890

1994 2 M 0 1 $1,890	Turf 0 0 0 0		
1993 0 M 0 0	Wet 0 0 0 0		
GP 1 0 0 0 $210	Dist 1 0 0 0 $210		

15Jan94-6GP fst 7f :224 :46 1:103 1:241 Md Sp Wt 57 3 5 54 6¾ 6½ 512¼ Santos J A 120 4.80 70-19 Cool Bandit120¾ Youthful Legs1204¼ Corajudo120¼ Failed to menace 9
2Jan94-6Crc fst 6f :221 :454 :583 1:122 Md Sp Wt 58 8 3 86¼ 610 59 37 Ramos W S 120 2.20 78-20 Edward Thomas1203½ Turn West1203½ Satellite Photo120½ 9
Bumped start, passed tired rivals

WORKOUTS: ●Jan 10 Pay 5f sly 1:014 B 1/4 ●Dec 24 Pay 4f fst :50 Bg1/16 ●Dec 20 Pay 6f fst 1:181 B 1/5 Dec 13 Pay 5f fst 1:05 B 3/6 Nov 29 Pay 5f fst 1:042 B 3/4 Nov 22 Pay 4f fst :512 B 3/11

Retrospection

Own: Brophy B Giles

MAPLE E (34 6 6 7 .18)

Gr. c. 3 (Feb)
Sire: Relaunch (In Reality)
Dam: Nice Tradition (Search Tradition)
Br: Ronald K. Kirk (Ky)
Tr: Johnson Philip G (7 1 0 0 .14)

120

Lifetime Record: 4 M 2 1 $15,760

1993 4 M 2 1 $15,760	Turf 0 0 0 0
1992 0 M 0 0	Wet 1 0 1 0 $5,940
GP 0 0 0 0	Dist 0 0 0 0

2Dec93-5Aqu fst 6f ⊡ :222 :462 :584 1:114 Md Sp Wt 78 1 9 44½ 1hd 3nk 21½ Santos J A 118 4.00 82-17 Mr. Shawklit1181½ Retrospection1181½ Prank Call1181¼ Good effort 12
21Oct93-5Aqu sly 1 :22 :453 1:101 1:352 Md Sp Wt 70 2 3 31½ 41 23 210½ Samyn J L 118 4.80 76-22 Go For Gin11810 Retrospection1181¾ A Track Attack1183 Held place 7
30Sep93-5Bel fst 6f :222 :46 :582 1:112 Md Sp Wt 67 6 7 88½ 75 45½ 48 Samyn J L 118 3.50 75-22 Bermuda Cedar1186 Prank Call118nk Plutonius1181¾ Squeezed break 9
31Jly93-3Sar fst 6f :222 :46 :583 1:111 Md Sp Wt 64 6 7 52¾ 51½ 51¾ 33¼ Bailey J D 118 7.70 83-08 Gold Tower1181¼ Colonel Slade1182 Retrospection118nk Late gain 8

WORKOUTS: Jan 22 Crc 6f fst 1:154 B 4/7 Jan 16 Crc 6f fst 1:152 B 3/3 Jan 6 Crc 5f fst 1:012 H 2/21 Dec 29 Crc 4f fst :482 B 4/35 Dec 23 Crc 4f fst :50 B 4/25 Dec 10 Bel tr.t 4f fst :483 B 6/54

Torero

Own: Stelling C E

DOUGLAS R R (110 14 15 14 .13)

Dk. b or br c. 3 (Apr)
Sire: Capote (Seattle Slew)
Dam: Only a Rumour (Ela-Mana-Mou)
Br: Stelcar Stables, Inc. (Ky)
Tr: Azpurua Eduardo Jr (4 0 0 1 .00)

120

Lifetime Record: 0 M 0 0 $0

1994 0 M 0 0	Turf 0 0 0 0
1993 0 M 0 0	Wet 0 0 0 0
GP 0 0 0 0	Dist 0 0 0 0

WORKOUTS: Jan 22 Crc 5f fst 1:02 Bg9/42 Jan 19 Crc 3f sly :372 B (d)5/17 Jan 7 Crc 4f fst :484 H 7/32 Dec 31 Crc 3f fst :382 B 14/24 Dec 17 Crc 3f fst :39 B 16/18 Dec 10 Crc 3f fst :372 B 4/14
Oct 6 Crc 3f gd :392 Bg31/32 Oct 2 Crc 5f fst 1:03 B 15/44 Sep 25 Crc 5f sly 1:03 B (d) 10/32 Sep 19 Crc 4f fst :49 B 9/61 Sep 11 Crc 4f fst :511 B 43/64 Aug 28 Crc 3f fst :37 Bg14/54

Twining

Own: Darley Stud Management Inc

SANTOS J A (67 9 11 9 .13)

Ch. c. 3 (May)
Sire: Forty Niner (Mr. Prospector)
Dam: Courtly Dee (Never Bend)
Br: Alexander/Groves (Ky)
Tr: Schulhofer Flint S (17 3 2 3 .18)

120

Lifetime Record: 0 M 0 0 $0

1994 0 M 0 0	Turf 0 0 0 0
1993 0 M 0 0	Wet 0 0 0 0
GP 0 0 0 0	Dist 0 0 0 0

WORKOUTS: Jan 21 GP 4f fst :493 B 40/52 Jan 16 GP 5f fst 1:024 B 52/72 Jan 11 GP 3f gd :362 B 6/26 Jan 5 GP 5f fst 1:022 B 25/47 Dec 28 GP 5f fst 1:013 Bg11/46 Dec 21 GP 4f fst :482 H 2/34
Dec 16 GP 3f fst :362 Bg6/26 Dec 10 GP 3f fst :37 B 9/18 Dec 5 GP 4f fst :514 B 40/42 Sep 30 Sar tr.t 3f fst :402 B 6/7 Sep 25 Sar 3f fst :393 B 3/6

C. D. Prospect

Own: Condry L & Karpf L

SELLERS S J (57 9 8 8 .16)

Dk. b or br c. 3 (Feb)
Sire: Regal Search (Mr. Prospector)
Dam: Star Wish (Cutlass)
Br: Rothenberger Joseph (Fla)
Tr: Sanders Gregory E (11 0 2 3 .00)

120

Lifetime Record: 2 M 0 1 $2,390

1994 1 M 0 1 $2,280	Turf 0 0 0 0
1993 1 M 0 0 $110	Wet 0 0 0 0
GP 1 0 0 1 $2,280	Dist 1 0 0 1 $2,280

7Jan94-6GP fst 7f :22 :442 1:092 1:22 Md Sp Wt 64 2 8 65½ 78¼ 47 311¼ Castaneda M 120 12.70 82-12 Canaveral1208¾ Gallant Warfare1202½ C. D. Prospect1204½ Ducked in start 9
30Dec93-6Crc fst 6f :221 :46 :583 1:114 Md 50000 42 8 10 76½ 59½ 513 515¼ Castaneda M 120 4.10 73-15 Turf Star12011 Bomb Free1181 Nardo1181½ Broke in air 10

WORKOUTS: Jan 18 Crc 3f sly :382 Bg8/10 Jan 15 Crc 4f fst :491 B 14/47 Dec 26 Crc 4f sly :492 B (d)2/21 Dec 21 Crc 5f fst 1:001 Hg2/17 Dec 15 Crc 4f fst :502 Bg20/31 Nov 25 Crc 3f sly :373 B (d)2/10

Bonus Money (GB)

Own: Iron Country Farm Inc

RAMOS W S (100 9 13 14 .09)

B. c. 3 (Mar)
Sire: Chief's Crown (Danzig)
Dam: Christmas Bonus (Key to the Mint)
Br: G M Breeding Farms Inc (GB)
Tr: Vanier Harvey L (3 0 0 0 .00)

L 120

Lifetime Record: 7 M 0 2 $6,252

1993 7 M 0 2 $6,252	Turf 0 0 0 0
1992 0 M 0 0	Wet 1 0 0 1 $1,600
GP 0 0 0 0	Dist 1 0 0 0 $480

16Nov93-6CD fst 1 1/16 :234 :474 1:14 1:471 Md Sp Wt 72 3 5 511 47 43½ 36½ Sellers S J L 119 fb 8.20 71-25 Chief Deputy1195 Blazing Basque1191½ Bonus Money1192¼ No late threat 11
20Oct93-2Kee my 6½f :222 :454 1:104 1:171 Md Sp Wt 63 9 8 913 89½ 46½ 36½ Sellers S J L 120 fb *3.10 84-17 Ballybunion Bear1202½ Cowboy's Kid1204 Bonus Money120nk Mild gain 12
14Oct93-4Kee fst *7f :224 :453 1:11 1:263 Md Sp Wt 57 4 8 55½ 54 55 512 Arguello F A Jr L 118 f 13.80 84-08 College Station1185½ Exclusive Casino118hd Lake Ali1185½ 11
Check, in tight, 3/16's
29Sep93-5AP fst 6f :221 :461 :59 1:124 Md Sp Wt 62 3 5 68 56½ 57 43 Ramos W S L 120 fb 4.40 77-23 Spot Tv1202½ DH Sharp Drums120 DH Travel Claim120½ Mild rally 7
6Sep93-2AP fst 6f :224 :463 :591 1:122 Md Sp Wt 41 2 10 77 88½ 611 69½ Fires E 120 fb 9.70 73-19 Golden Gear1203 Spot Tv1201½ Sir Cognac120no No rally 11
25Aug93-7AP fst 6f :222 :462 :59 1:114 Md Sp Wt 55 6 7 89 85 55½ 43½ Torres F C 119 fb 18.70 81-13 Clev Er Irish1191½ My Quincy119hd Andover Again1192 Broke inside 9
12Aug93-7AP fst 5½f :223 :463 :59 1:052 Md Sp Wt 24 8 9 813 76½ 811 815 Fires E 119 fb 2.00e 71-15 Smart Enough119½ Spot Tv1195 Golden Gear119¾ Showed little 9

WORKOUTS: Jan 24 GP 5f fst 1:01 B 3/11 Jan 19 GP 3f gd :38 B 4/12 Jan 14 GP 5f sly 1:023 H (d) 1/1 Jan 10 GP 5f sly 1:033 B (d) 11/12 Nov 24 Kee 4f fst :512 B 4/5 Oct 27 Kee 4f fst :502 B 10/19

Chester's Gold

Own: Bowles C F & Scott J D

SIMPSON G B (1 0 0 0 .00)

Gr. g. 3 (Mar)
Sire: Interdicto (Grey Dawn II)
Dam: Lindsay's Gold (Strike Gold)
Br: C. Bowles & J. D. Scott (Fla)
Tr: Bowles Chester F (4 1 1 0 .25)

120

Lifetime Record: 3 M 0 0 $680

1994 1 M 0 0 $140	Turf 0 0 0 0
1993 2 M 0 0 $540	Wet 0 0 0 0
GP 0 0 0 0	Dist 1 0 0 0 $140

1Jan94-6Crc fst 7f :223 :46 1:12 1:253 Md Sp Wt 24 9 2 21 88 1020 1124½ Smith G S5 115 b 59.70 60-15 Critical Mass1201½ Fleet Eagle1202½ Plutonius1201¼ Brief speed 12
18Dec93-6Crc fst 6f :22 :454 :584 1:122 Md Sp Wt 47 3 10 97 79 511 512¼ Smith G S5 115 b 63.10 74-12 Copenquack1134¼ Cool Bandit120no My Magic Touch1204 In close start 12
16Nov93-6Crc fst 5f :223 :463 :593 Md 25000 27 2 4 97¾ 813 615 413½ Smith G S7 112 50.00 85-12 LightingForce1195 TrulyThBst1196 WisJudgmnt1192¼ Passed tired ones 9

WORKOUTS: Jan 23 Crc 4f fst :481 H 3/50 Dec 16 Crc 4f fst :51 B 19/21 Dec 11 Crc 3f fst :37 B 9/23 Nov 24 Crc 4f fst :493 Bg8/36 Nov 14 Crc 4f fst :49 B 8/41 Nov 9 Crc 4f sly :511 B (d) 14/22

Carpet
Own: Claiborne Farm & The Gamely Corp
SMITH M E (97 15 16 14 .15)

B. c. 3 (Feb)
Sire: Danzig (Northern Dancer)
Dam: Kashan (Damascus)
Br: Claiborne Farm & The Gamely Corp. (Ky)
Tr: Mott William I (27 7 4 4 .26)
120

Lifetime Record :	1 M 0 0		$0
1993	1 M 0 0	Turf	0 0 0 0
1992	0 M 0 0	Wet	0 0 0 0
GP	0 0 0 0	Dist	0 0 0 0

8Jly93- 3Bel fst 5f :221 :452 :581 Md Sp Wt 26 3 8 67½ 66½ 79 715½ Smith M E 118 4.10 74-09 QuickToFinsh118no ByouBrtholomw1181¼ Prsntly1182½ Broke slowly, wide 8

WORKOUTS: ●Jan 21 Hia 6f fst 1:143 Hg 1/12 Jan 15 Hia 5f fst 1:03 B 6/16 Jan 8 Hia 4f fst :50 B 15/22 Dec 30 Hia 4f fst :514 B 16/20 Dec 24 Hia 4f fst :524 B 38/38 Dec 16 Pay 4f fst :52 B 2/10

Plutonius
Own: Shields Joseph V Jr
PERRET C (26 4 4 5 .15)

Dk. b or br c. 3 (Apr)
Sire: Silver Buck (Buckpasser)
Dam: Chattooga (Ack Ack)
Br: J. V. Shields Jr. (Fla)
Tr: Alexander Frank A (3 1 0 0 .33)
120

Lifetime Record :	5 M 2 2	$15,400			
1994	1 M 0 1	$1,400	Turf	0 0 0 0	
1993	4 M 2 1	$14,000	Wet	0 0 0 0	
GP	0 0 0 0		Dist	3 0 1 1	$6,900

1Jan94- 6Crc fst 7f :223 :48 1:12 1:253 Md Sp Wt 67 4 11 75½ 54 44½ 34 Smith M E 120 fb 2.30 80-15 Critical Mass1201½ Fleet Eagle1202½ Plutonius1201¼ 12
Knocked back start, lacked rally
22Nov93- 5Aqu fst 7f :23 :463 1:123 1:261 Md Sp Wt 30 2 6 42½ 73¼ 86 820½ McCauley W H 118 b *2.40 53-22 King Of Kolchis1182 Alydawn118hd Poets Pistol1132½ 8
Checked turn, took up stretch
4Nov93- 5Aqu fst 7f :23 :461 1:113 1:244 Md Sp Wt 78 4 3 2½ 11 2hd 21¼ McCauley W H 118 b 2.20 79-17 I'm Very Irish1181¼ Plutonius1188 Signal Tap118½ Gamely 8
17Oct93- 5Bel fst 6f :222 :454 :582 1:114 Md Sp Wt 60 5 5 42½ 52 43½ 26 McCauley W H 118 11.90 75-14 Brush Full1186 Plutonius118½ Beasleyathisbest1182 Up for place 10
30Sep93- 5Bel fst 6f :222 :46 :582 1:112 Md Sp Wt 71 1 6 42½ 32 34 36¼ Antley C W 118 54.70 77-22 Bermuda Cedar1186 Prank Call118nk Plutonius1181½ Saved ground 9

WORKOUTS: Jan 21 Crc 5f fst 1:01 H 2/16 ●Jan 15 Crc 5f fst 1:012 H 1/31 Dec 26 Crc 5f sly 1:02 B (d) 2/16 Dec 19 Crc 5f fst 1:033 B 11/31 Dec 13 Crc 4f fst :503 B 5/13 Nov 15 Bel 5f fst 1:022 B 13/18

Alyjul
Own: Condren W J & Cornacchia J
BARTON D M (25 1 4 3 .04)

B. c. 3 (May)
Sire: Alydar (Raise a Native)
Dam: My Juliet (Gallant Romeo)
Br: Calumet-Gussin No. 1 (Ky)
Tr: Zito Nicholas P (18 1 2 2 .06)
120

Lifetime Record :	5 M 0 0	$200		
1994	1 M 0 0	$200	Turf	0 0 0 0
1993	4 M 0 0		Wet	1 0 0 0
GP	1 0 0 0	$200	Dist	2 0 0 0

4Jan94- 2GP gd 1 1/16 :233 :48 1:131 1:464 Md Sp Wt 48 9 10 96¾ 98¼ 910 914¼ Ferrer J C 120 b 47.30 62-21 [D]StonrCrk1202¾ [D]MyMgicTouch120¾ SilvrProfl120¾ Bumped hard start 11
27Nov93- 2Aqu fst 1 :233 :472 1:13 1:392 Md Sp Wt 53 5 5 53 64½ 77¾ 711½ Velazquez J R 118 16.40 55-31 Pleasant Dancer118nk Chockie Mountain1181¾ Signal Tap1186½ No threat 10
10Nov93- 6Aqu fst 1 :233 :474 1:133 1:401 Md Sp Wt 58 6 6 51¾ 42½ 45 59 Perret C 118 b 11.90 54-38 ATrackAttack1184 FinalClearance1184¼ MoscowMgic118hd Flattened out 8
8Oct93- 3Bel fst 7f :231 :47 1:12 1:243 Md Sp Wt 66 4 6 72½ 83¾ 76½ 611¼ Chavez J F 118 b 36.10 68-23 Our Emblem1181½ Final Clearance1183 Frisco Gold1182 No threat 8
19Sep93- 5Bel my 7f :23 :462 1:112 1:243 Md Sp Wt 30 2 8 813 812 810 821 Smith M E 118 11.40 59-20 Personal Escort1182½ Turnbull Creek1182 Red Mcfly1189½ Broke slowly 8

WORKOUTS: Jan 22 GP 5f fst 1:02 B 30/53 Dec 26 GP 5f fst 1:031 B 19/28 Dec 18 GP 5f fst 1:023 B 27/40 Dec 8 GP 5f fst 1:032 B 8/9 Nov 21 Bel 5f fst 1:001 H 2/13 Nov 5 Bel 5f gd 1:022 H 8/13

Dancer's Hideaway
Own: Opstein Kenneth J
TURNER T G (20 1 1 3 .05)

Gr. c. 3 (Mar)
Sire: Gate Dancer (Sovereign Dancer)
Dam: Giovanelli (Monteverdi-Ire)
Br: Sims Sonny & L W & Wallace J (Fla)
Tr: Sims Monti N (2 0 0 1 .00)
120

Lifetime Record :	1 M 0 0	$190		
1994	1 M 0 0	$190	Turf	0 0 0 0
1993	0 M 0 0		Wet	0 0 0 0
GP	1 0 0 0	$190	Dist	0 0 0 0

8Jan94- 6GP fst 6f :22 :45 :571 1:094 Md Sp Wt 65 4 5 65½ 79 78¾ 68½ Turner T G 120 79.90 83-09 Kyle's Code1202½ Meadow Monster1203½ Magic Caver120¾ No threat 11

WORKOUTS: Dec 28 Crc 6f fst 1:152 Hg 2/2 Dec 18 Crc 5f fst 1:042 B 24/29 Dec 15 Crc 4f fst :533 B 30/31 Dec 9 Crc 6f fst 1:162 B 2/2 Dec 2 Crc 4f fst :502 Bg 4/26 Nov 27 Crc 4f fst :523 B 49/52

Kris' Rainbow
Own: Evans R S
BAILEY J D (77 18 9 4 .23)

Dk. b or br c. 3 (May)
Sire: Kris S. (Roberto)
Dam: Ashley's Rainbow (Key to the Mint)
Br: Evans Robert S (Fla)
Tr: Schulhofer Flint S (17 3 2 3 .18)
120

Lifetime Record :	0 M 0 0		$0
1994	0 M 0 0	Turf	0 0 0 0
1993	0 M 0 0	Wet	0 0 0 0
GP	0 0 0 0	Dist	0 0 0 0

WORKOUTS: Jan 24 Hia 4f fst :512 B 3/4 Jan 19 Hia 3f fst :371 B 2/5 Jan 14 Hia 4f sly :491 Bg 4/10 Jan 8 Hia 4f fst :482 H 4/22 Jan 3 Hia 4f fst :491 B 3/9 Dec 29 Hia 4f fst :49 B 9/16
Dec 24 Hia 6f fst 1:152 B 1/2 Dec 19 Hia 4f fst :49 B 5/10 Dec 8 Bel tr.t 5f fst 1:033 B 25/36 Dec 2 Bel tr.t 4f fst :52 B 40/41 Nov 27 Bel tr.t 4f fst :504 B 57/82 Nov 22 Bel tr.t 4f fst :512 B 40/43

Also Eligible (Not in Post Position Order):

Gallant Warfare
Own: Roron Stable
CASTILLO H JR (42 3 2 5 .07)

B. g. 3 (Jan)
Sire: War (Majestic Light)
Dam: Gallant Libby (Gallant Knave)
Br: Appleton Arthur I (Fla)
Tr: Mazza John F (12 1 1 1 .08)
L 120

Lifetime Record :	8 M 1 1	$8,165			
1994	2 M 1 0	$3,630	Turf	2 0 0 1	$3,385
1993	6 M 0 1	$4,535	Wet	0 0 0 0	
GP	2 0 1 0	$3,630	Dist	2 0 1 0	$3,630

15Jan94- 6GP fst 7f :224 :46 1:103 1:241 Md Sp Wt 49 9 1 64½ 73¾ 810 816 Castillo H Jr L 120 4.20 66-19 Cool Bandit1202½ Youthful Legs1204¼ Corajudo1202¼ Faltered 9
7Jan94- 6GP fst 7f :22 :442 1:092 1:22 Md Sp Wt 70 4 6 32 32 24 28¾ Castillo H Jr L 120 2.90 84-12 Canaveral1208¾ Gallant Warfare1202¼ C. D. Prospect1204½ No match 9
7Dec93- 6Crc fm *1 1/16 Ⓣ 1:472 Md Sp Wt 59 8 2 21 21½ 43 54½ Ramos W S 119 *2.00 62-35 [D]Dance In The Ring119¾ Ensign MickeyB119no [DH]GottaScore119 Tired 8
3Nov93- 1Med fst 1 :233 :471 1:121 1:384 Md Sp Wt 57 4 1 21½ 25 25 58¼ Thomas D B 118 4.90 74-17 Jericho Blaise1183¾ Barge In1182 Pleasant Dancer1182 In tight early 9
9Sep93- 5Bel gd 1 Ⓣ :224 :463 1:12 1:371 Md Sp Wt 61 2 2 2½ 21 1hd 33 Castillo H Jr 118 17.30 73-22 Warm Wayne1183 Ocean Code118no Gallant Warfare1181¼ Willingly 12
15Aug93- 5Mth fst 1 :23 :464 1:13 1:423 Md Sp Wt 46 10 6 712 410 39 47¼ Bracho J A 118 76.60 52-25 Grey Chandon118nk Orestes1186½ National Kid118½ Flattened out 10
30Jun93- 5Mth fst 5f :222 :463 :59 Md Sp Wt 5 12 7 52¼ 54¾ 913 1023¼ Castillo H Jr 118 b 50.40 62-19 Dehere1184 Justinthefastlane1184¾ Luckie Peri1184 Gave way 12
18Jun93- 5Mth fst 5f :221 :461 :584 Md Sp Wt 26 9 2 32 42½ 56½ 517¼ Bravo J 118 15.80 70-20 Code Home1187¾ Mr Vincent1133¾ Ridan's Whirl1183 Lugged in stretch 9

WORKOUTS: ●Jan 6 GP 3f fst :35 H 1/28 Dec 31 GP 5f fst 1:01 B 5/51 Dec 24 GP 5f fst 1:013 B 17/63 ●Dec 18 GP 4f fst :48 B 1/42 ●Dec 5 GP 4f fst :47 H 1/42 ●Nov 30 GP 6f fst 1:132 H 1/4

Of the debut runners, Torero and Dancer's Hideaway have poor precocity ratings and are throwouts. Kris's Rainbow, the other half of Schulhofer's uncoupled entry, has some class and might be competitive at this longer debut distance. He also has Jerry Bailey up.

60   But Twining is the pedigree standout. His sire Forty Niner (A+1) gets better than 40 percent debut winners and Scotty Schulhofer is competent with first-time starters. Furthermore, his dam Courtly Dee has already thrown fourteen winners out of fifteen starters and is by the important broodmare sire Never Bend. While Never Bend can impart speed, his most successful sons Mill Reef and Riverman are sources of stamina. With Forty Niner speed on top and Never Bend stamina on the bottom, Twining is ideally bred to win this 7 furlong race and then stretch further out, maybe even on the grass.

Given the weak field, Twining's breeding and connections, and the fact that Carpet is no lock to improve, this son of Forty Niner should be 3-2 in this race. Instead, perhaps because Jose Santos is on him, Twining goes off at nearly 5-2. At that price he is a prime low-priced overlay Win bet. Since the exacta is difficult to figure, the best backup is probably a Place bet, although a case could also be made for a box with Carpet.

Results: Twining broke slowly, patiently made up ground, and then flew six-wide down the stretch to catch Kris's Rainbow (the other Schulhofer horse) at the wire. He paid $6.80 to Win and $5.40 to Place, while Kris's Rainbow returned $16 for the Place and $10 for the Show. The $2 exacta, for hidden-entry players, paid $144. Carpet ran out of the money.

Most races, of course, do not have single-debut contenders. The ninth race at Santa Anita on February 12th is a more complex 6 furlong sprint with an experienced favorite and two first-time starters with precocious pedigrees.

Wolf Bait is the heavy favorite here at 3-2 and on the surface he looks hard to go against. He has very respectable Beyers and is trained by Bob Baffert. On the downside, Wolf Bait did give up a two-length lead in his debut and was beaten at low odds last time out. He might be a money-burning horse who likes to finish second. Furthermore, Wolf Bait's 37.1 work on February 10th is suspiciously slow for a Baffert horse and comes in the wake of a declining Beyer.

Nevertheless, none of the other experienced horses look competitive, with the possible exception of Native Blast who helped set a fast pace last time out when he was defeated by the promising Strodes Creek. Still, Native Blast will have to contend with Blushing Victor early and is likely to fade once again.

The mystery horse here is Gold Miner's Slew, who showed speed and was bet in his second start. He's been rested for six months by John Sadler (who is excellent with layoff horses) and might dramatically improve over his 2-year-old form. Still at 38-1, he seems a reach.

Looking at the first-time starters, Slewpy Dewpy Doo and Clever Ricky have poor ratings and are throwouts. But Meadow Mischief (A+4) and Bedminster (A1+) are standouts. Their respective sires Meadowlake and Mr. Prospector are two of the best debut sires around, and that makes these two colts automatic contenders.

Santa Anita Park

9

6 Furlongs. (1:071) MAIDEN SPECIAL WEIGHT. Purse $32,000. 3-year-olds. Weight, 118 lbs. (Non-starters for a claiming price of $32,000 or less in their last three starts preferred).

START ▾ 6 FURLONGS ▴ FINISH

Meadow Mischief

Own: Emerald Meadows Ranch
DESORMEAUX K J (122 27 20 17 .22)

B. c. 3 (Feb)
Sire: Meadowlake (Hold Your Peace)
Dam: Bold 'n Determined (Bold and Brave)
Br: Robertson Corbin J (Ky)
Tr: Lewis Craig A (34 4 4 1 .12)

118

Lifetime Record :	0 M 0 0	$0		
1994	0 M 0 0		Turf	0 0 0 0
1993	0 M 0 0		Wet	0 0 0 0
SA	0 0 0 0		Dist	0 0 0 0

WORKOUTS: Feb 9 SA 4f gd :50 H *14/24* Feb 3 SA 7f fst 1:33 H *8/8* Jan 18 SA 5f fst 1:004 H *20/44* Jan 12 SA 5f fst 1:001 H *8/46* Jan 4 SA 5f fst 1:004 H *23/45* Dec 28 SA 4f fst :474 H *16/50* Dec 22 SA 4f fst :493 H *40/49* Dec 15 SA 3f gd :374 H *9/17* Dec 8 SA 3f fst :363 H *10/21* Nov 14 Hol 3f fst :354 H *4/16* Nov 2 Hol 3f fst :372 H *4/6*

Wolf Bait

Own: Earnhardt Hal
NAKATANI C S (188 22 22 30 .12)

Ro. g. 3 (Apr)
Sire: Wolf Power (Flirting Around)
Dam: Georgia More (George Lewis)
Br: Star Breeders (Ky)
Tr: Baffert Bob (34 2 11 4 .06)

L 118

Lifetime Record :	2 M 2 0	$12,800			
1994	2 M 2 0	$12,800	Turf	0 0 0 0	
1993	0 M 0 0		Wet	0 0 0 0	
SA	2 0 2 0	$12,800	Dist	2 0 2 0	$12,800

30Jan94- 4SA	fst	6f	:213	:443	:564	1:094	Md Sp Wt	76	1	3	42¾	41¾	2hd	2nk	Nakatani C S	LB 118	*1.60	88-07	Jabbawat118nk Wolf Bait1181¾ Superpak1181½	4 wide into lane	7
8Jan94- 4SA	fst	6f	:213	:442	:571	1:092	Md Sp Wt	85	1	5	1½	12	12	21¾	Nakatani C S	B 118	5.10	88-07	Honest Happiness1181¾ Wolf Bait1182¾ Stop The Fight1181½		10

Broke stride briefly near wire

WORKOUTS: Feb 10 SA 3f fst :371 H *7/19* Jan 25 SA 5f sl 1:032 H *5/7* Jan 19 SA 4f fst :492 H *46/60* Jan 4 SA 4f fst :474 H *13/49* Dec 30 SA 6f fst 1:14 H *19/34* Dec 16 SA 6f fst 1:134 Hg *7/37*

Gold Miner's Slew

Own: Fast Lane Farms & D'Silva & Watchorn
SOLIS A (170 27 27 28 .16)

Dk. b or br c. 3 (Apr)
Sire: Slew City Slew (Seattle Slew)
Dam: Pan D' Ore (Mr. Prospector)
Br: Polk Dr. Hiram & Richardson Dr. David (Ky)
Tr: Sadler John W (34 7 4 5 .21)

L 118

Lifetime Record :	2 M 0 0	$0		
1993	2 M 0 0		Turf	0 0 0 0
1992	0 M 0 0		Wet	0 0 0 0
SA	0 0 0 0		Dist	1 0 0 0

22Aug93- 3Dmr	fst	1	:22	:46	1:113	1:38	Md Sp Wt	24	3	1	12½	46¼	513	625½	Desormeaux K J	LB 117	3.90	51-16	Ocean Crest1173 Hot Number1173½ Mint Green1178	Hustled early	7
24Jly93- 4Hol	fst	6f	:214	:442	:563	1:093	Md Sp Wt	57	5	9	911	914	914	811	Desormeaux K J	LB 117	8.60	81-08	Feliz Hora117¾ Devil's Mirage1172 Subtle Trouble1173½	Off slowly, wide	9

WORKOUTS: Feb 6 SA 6f fst 1:134 H *9/49* Jan 31 SA 5f fst 1:011 H *20/34* Jan 16 SA 6f fst 1:163 H *26/29* Jan 10 SA 6f fst 1:141 H *8/22* Jan 4 SA 5f fst 1:011 H *30/45* Dec 29 SA 5f fst 1:013 H *35/52*

Slewpy Dewpy Doo

Own: Friendly Ed & Natalie
LOPEZ A D (47 3 1 3 .06)

B. g. 3 (Apr)
Sire: Slewpy (Seattle Slew)
Dam: Give Blood (What a Pleasure)
Br: Gerald Robins (Ky)
Tr: Hendricks Dan L (25 3 1 3 .12)

118

Lifetime Record :	0 M 0 0	$0		
1994	0 M 0 0		Turf	0 0 0 0
1993	0 M 0 0		Wet	0 0 0 0
SA	0 0 0 0		Dist	0 0 0 0

WORKOUTS: Feb 10 SA 3f fst :351 Hg *2/19* Feb 5 SA 4f gd :494 Hg *13/40* Jan 28 SA 5f fst 1:011 Hg *48/80* Jan 22 SA 6f fst 1:154 H *35/39* Jan 16 SA 6f fst 1:133 H *12/29* Jan 10 SA 5f fst 1:013 H *21/41* Jan 5 SA 4f fst :49 H *29/44* Dec 30 SA 3f fst :362 H *16/31* Dec 24 SA 4f fst :493 H *67/85* Dec 18 Hol 4f fst :484 H *15/29*

Blushing Victor

Own: Malmuth Mr & Mrs Marvin
VALENZUELA P A (112 13 13 18 .12)

Ch. c. 3 (Feb)
Sire: Blushing John (Blushing Groom)
Dam: Elegant Victress (Sir Ivor)
Br: Mr. & Mrs. Marvin Malmuth (Ky)
Tr: Jory Ian P D (28 7 1 0 .25)

L 118

Lifetime Record :	5 M 1 0	$9,500		
1994	1 M 0 0	$2,400	Turf	0 0 0 0
1993	4 M 1 0	$7,100	Wet	0 0 0 0
SA	3 0 0 0	$3,100	Dist	1 0 0 0

15Jan94- 2SA	fst	6½f	:213	:441	1:092	1:154	Md Sp Wt	53	6	1	3nk	32	43½	411½	Valenzuela P A	LB 118 b	4.60	81-08	Strodes Creek1185 Native Blast1181 Devon Dancer1185½	Weakened	7
26Dec93- 9SA	fst	1	:221	:45	1:101	1:363	Md Sp Wt	65	4	1	1½	3nk	43½	611	Delahoussaye E	LB 117 b	12.50	76-10	Almaraz1171¾ Dramatic Gold1171¾ Advantage Miles1176	Weakened	10
28Nov93- 4Hol	fst	6f	:214	:444	:57	1:091	Md Sp Wt	61	1	2	21	33	55¼	613	Delahoussaye E	LB 118	3.50	81-09	Al's River Cat1187½ Defrocker118hd Fraley1181¾	Rail trip	9
13Nov93- 6SA	fst	6½f	:222	:452	1:093	1:154	Md Sp Wt	70	4	1	2hd	2hd	21	55¼	Solis A	LB 118	3.50	87-11	Numerous118no Amarisingstar1182¼ Crowning Decision1182¾	Inside trip	7
12May93- 3Hol	fst	4½f	:22	:451		:511	Md Sp Wt	63	5	2	32½		31½	26	Pincay L Jr	B 117	2.20	94-08	Ramblin Guy1176 Blushing Victor1174½ Mi Profe1175	Second best	6

WORKOUTS: Feb 2 SA 6f fst 1:121 H *2/20* Jan 27 SA 4f fst :48 H *8/28* Jan 10 SA 5f fst 1:014 H *27/41* Jan 5 Hol 4f fst :48 H *3/17* Dec 17 Hol 6f fst 1:133 H *3/18* Dec 10 Hol 7f fst 1:304 H *7/9*

Clever Ricky

Own: Lambert Marjorie
PEDROZA M A (117 8 17 10 .07)

B. c. 3 (Apr)
Sire: Tank's Prospect (Mr. Prospector)
Dam: Ante (Buckfinder)
Br: Lambert Marjorie (Cal)
Tr: Lukas D Wayne (44 2 5 8 .05)

118

Lifetime Record :	0 M 0 0	$0		
1994	0 M 0 0		Turf	0 0 0 0
1993	0 M 0 0		Wet	0 0 0 0
SA	0 0 0 0		Dist	0 0 0 0

WORKOUTS: Feb 10 SA 4f fst :493 Hg *33/67* Feb 3 SA 5f fst 1:011 H *25/62* Jan 28 SA 5f fst :591 H *5/80* Jan 22 SA 5f fst 1:021 H *53/62* Jan 14 SA 3f fst :362 H *5/28* Jan 8 SA 3f fst :351 H *5/20* Jan 2 SA 3f fst :361 H *4/11* Aug 14 Dmr 4f fst :493 H *45/58*

continued

Native Blast B2/4*
Own: Nor Joanne H
ANTLEY C W (156 22 15 8 .14)

Ch. g. 3 (Apr)
Sire: Relaunch (In Reality)
Dam: Native Whisper (Exclusive Native)
Br: Nor Joanne H (Ky)
Tr: Nor Fabio (6 1 2 1 .17)
118

Lifetime Record :	4 M 1 2	$12,700		
1994	1 M 1 0	$6,400	Turf	0 0 0 0
1993	3 M 0 2	$6,300	Wet	0 0 0 0
SA	2 0 1 0	$6,400	Dist	3 0 0 2 $6,300

15Jan94-2SA fst 6½f :213 :441 1:092 1:154 Md Sp Wt 68 7 2 2hd 2hd 2hd 25 Flores D R B 118 3.00 88-08 Strodes Creek1185 Native Blast1181 Devon Dancer1185½ No match 7
20Dec93-1Hol fst 6f :221 :444 :564 1:09 Md Sp Wt 75 8 6 62¾ 52½ 43½ 35¾ Flores D R B 118 12.10 89-09 Pollock's Luck1181¾ Superpak1184 Native Blast118½ 4 wide turn 8
4Dec93-4BM fst 6f :221 :443 :564 1:092 Md Sp Wt 65 4 6 46 49 48½ 37 Chapman T M B 118 6.50 82-16 Trekkor1181 Reality's Conquest1186 Native Blast1185 Even late 8
10Oct93-9SA fst 6f :214 :443 :57 1:094 Md Sp Wt 50 5 5 2hd 2½ 53 813¼ McCarron C J 117 26.10 75-16 Sky Kid117hd Gracious Ghost1175 Al's River Cat1171¼ Dueled, tired 11

WORKOUTS: Feb 7 Hol 4f sly :484 H 1/2 Jan 27 Hol 3f fst :364 H 5/7 Jan 7 Hol 5f fst 1:013 H 13/27 Dec 29 Hol 3f fst :354 H 6/16 Dec 13 Hol 4f gd :491 H 4/14 Nov 27 Hol 5f fst 1:01 H 18/46

De Vito B3/3
Own: West Gary L & Mary E
PINCAY L JR (175 26 24 22 .15)

B. c. 3 (Apr)
Sire: Topsider (Northern Dancer)
Dam: Ripley (Believe It)
Br: Claiborne Farm & The Gamely Corp. (Ky)
Tr: Bradshaw Randy (15 2 0 2 .13)
118

Lifetime Record :	1 M 0 0	$0		
1993	1 M 0 0		Turf	0 0 0 0
1992	0 M 0 0		Wet	0 0 0 0
SA	0 0 0 0		Dist	0 0 0 0

18Jly93-2AP gd 5f :223 :454 :582 Md Sp Wt 31 1 7 56 55½ 610 612½ Silva C H 119 b 6.00 84-03 Brass Jacks1192½ Parting Sea1191½ Long Suit1191¼ Broke slowly 7

WORKOUTS: ●Feb 9 Hol 3f fst :363 Hg 1/4 Feb 3 Hol 5f fst 1:011 H 8/35 ●Jan 27 Hol 5f fst :584 H 1/24 Jan 22 Hol 5f fst 1:004 H 8/30 Jan 16 Hol 4f fst :483 H 4/11 Dec 9 Hol 5f fst 1:00 Hg 7/38

Me Tonto B9/4
Own: Larson Mr & Mrs Melvin C
LINARES M G (11 0 3 1 .00)

Dk. b or br g. 3 (Jun)
Sire: Inherent Star (Pia Star)
Dam: Dress Me Up's Girl (Aczay)
Br: Larson Mr & Mrs M C (Cal)
Tr: Cosme Ruben G (7 0 0 0 .00)
118

Lifetime Record :	1 M 0 0	$0		
1994	1 M 0 0		Turf	0 0 0 0
1993	0 M 0 0		Wet	0 0 0 0
SA	1 0 0 0		Dist	0 0 0 0

20Jan94-2SA fst 1 :231 :472 1:121 1:374 Ⓢ Md 40000 29 8 8 815 812 815 821½ Castanon A L B 117 b 56.10 58-20 Two Minute Drill1151½ Dash Of Vanilla1153 Heavenly Star117hd Never close 8

WORKOUTS: Feb 9 Fpx 3f gd :36 Hg 1/2 Jan 29 Fpx 4f fst :484 Hg 3/12 Jan 17 Fpx 3f fst :361 H 2/7 Jan 9 Fpx 6f fst 1:144 H 1/2 Jan 2 Fpx 5f fst 1:014 Hg 2/2 Dec 27 Fpx 6f fst 1:16 H 1/3

Bedminster A+/4*
Own: Wygod Mr & Mrs Martin J
DELAHOUSSAYE E (152 21 23 19 .14)

Dk. b or br c. 3 (Mar)
Sire: Mr. Prospector (Raise a Native)
Dam: One of a Klein (Danzig)
Br: Wygod Mr & Mrs Martin J (Ky)
Tr: Hendricks Dan L (25 3 1 3 .12)
118

Lifetime Record :	0 M 0 0	$0		
1994	0 M 0 0		Turf	0 0 0 0
1993	0 M 0 0		Wet	0 0 0 0
SA	0 0 0 0		Dist	0 0 0 0

WORKOUTS: Feb 10 SA 3f fst :351 Hg 2/19 Jan 28 SA 6f fst 1:142 H 28/48 Jan 22 SA 6f fst 1:152 H 33/39 Jan 15 SA 5f fst 1:003 H 16/49 Jan 9 SA 5f fst 1:01 H 24/49 Jan 3 SA 4f fst :48 H 15/38
Dec 29 SA 4f fst :50 H 39/52 Dec 24 SA 3f fst :374 H 27/39 Dec 17 Hol 3f fst :371 H 14/28 Dec 9 Hol 3f fst :37 H 11/21

Jade Master B3/4
Own: Lee China
ATKINSON P (64 6 4 5 .09)

Ch. c. 3 (May)
Sire: Jade Hunter (Mr. Prospector)
Dam: Mama Coca (Secretariat)
Br: Belford David (Ky)
Tr: Van Berg Jack C (91 15 21 8 .16)
118

Lifetime Record :	1 M 0 0	$0		
1994	1 M 0 0		Turf	0 0 0 0
1993	0 M 0 0		Wet	0 0 0 0
SA	1 0 0 0		Dist	0 0 0 0

28Jan94-3SA fst 6½f :22 :45 1:093 1:16 Md c-50000 45 3 4 2hd 2½ 67¼ 617 Gryder A T B 118 b 7.00 75-12 Sacrifice1183¼ Island Sport1182½ Capo1181¾ Dueled, tired 6
Claimed from Suffolk Racing Partnership Ltd, Gregson Edwin Trainer

WORKOUTS: Feb 6 Hol 5f fst 1:014 H 17/38 Jan 22 SA 4f fst :491 H 40/50 Jan 16 SA 6f fst 1:132 H 9/29 Jan 10 SA 4f fst :48 Hg 18/53 Jan 4 SA 5f fst :594 H 10/45 Dec 23 SA 6f fst 1:131 H 21/39

Sixth Of March D3/2
Own: Garcia & Scanell
IAMMARINO M P (27 0 1 0 .00)

B. c. 3 (Apr)
Sire: Cryptoclearance (Fappiano)
Dam: Rolling Nest (Rollicking)
Br: Sondra Bender & Howard M. Bender (Md)
Tr: Garcia Victor L (2 2 0 0 1.00)
118

Lifetime Record :	0 M 0 0	$0		
1994	0 M 0 0		Turf	0 0 0 0
1993	0 M 0 0		Wet	0 0 0 0
SA	0 0 0 0		Dist	0 0 0 0

WORKOUTS: Feb 9 Hol 4f fst :48 Hg 2/8 Feb 2 Hol 7f fst 1:282 H 2/3 Jan 27 Hol 6f fst 1:152 H 10/13 Jan 21 Hol 6f fst 1:154 Hg 5/7 Jan 15 Hol 5f fst 1:004 Hg 7/28 Jan 9 Hol 5f fst 1:004 H 14/23
Jan 3 Hol 4f fst :473 H 2/19 Dec 28 Hol 3f fst :362 H 3/14 Oct 6 Hol 4f fst :49 H 7/17 Sep 27 Fpx 6f fst 1:151 H 1/1 Sep 21 Fpx 4f fst :511 H 8/9 Sep 15 Fpx 4f fst :492 H 5/16

Also Eligible (Not in Post Position Order):

Sooper Night B6/5
Own: Petitt Karlene & Richard
PINCAY L JR (175 26 24 22 .15)

Ch. c. 3 (Mar)
Sire: Knights Choice (Drum Fire)
Dam: Filly Will Fly (Flying Lark)
Br: Dr. & Mrs. Alex Ryncarz (WA)
Tr: Penney Jim (18 2 0 0 .11)
118

Lifetime Record :	4 M 1 0	$8,835		
1994	1 M 0 0		Turf	0 0 0 0
1993	3 M 1 0	$8,835	Wet	0 0 0 0
SA	1 0 0 0		Dist	3 0 1 0 $8,835

22Jan94-4SA fst 6f :212 :441 :56 1:082 Md Sp Wt 37 2 1 22 43 711 723½ Antley C W B 118 b 17.70 71-09 Fly'n J. Bryan1184½ Laabity1183½ Paster's Caper1182 Gave way 7
20Sep93-10YM fst 1$\frac{1}{16}$:231 :462 1:112 1:433 JGottstn Fut100k — 5 3 3nk 3½ 1017 — Cooper B 120 b 27.20 — 06 GLJnor1203 TooGoodForWords1201½ Rghtpgrmpsly1202½ Stopped, eased 10
11Sep93-10YM fst 6f :223 :451 :58 1:112 L Knowles SC35k 63 6 1 33 32 1hd 2hd Baze G 120 b 8.30 86-10 Mr.EsyMoney120hd SooprNight1201½ TooGoodForWords1153 Just missed 7
27Aug93-10YM fst 6f :232 :471 1:00 1:122 Washtn Brds15k 28 1 7 1½ 1hd 2hd 47½ Cooper B 120 b 3.20 73-12 ⒹFiery Skies1171 Bold Crew1163½ Fortunes Rib1183 Drifted in 1/8 7
Disqualified and placed 5th

WORKOUTS: Feb 6 SA 5f fst 1:014 H 36/108 Feb 2 SA 4f fst :493 H 34/43 Jan 19 SA 6f fst 1:131 H 7/25 Jan 13 SA 6f fst 1:142 H 5/17 Jan 7 SA 6f fst 1:14 Hg 15/30 Jan 2 SA 5f fst :593 H 3/47

Meadow Mischief, however, is going off at 13-1 despite his breeding and the fact that leading rider Kent Desormeux is in the irons. Maybe that's because, while Craig Lewis is competent with first-time starters, he's only won four races at the Santa Anita meet. Another possible reason for the absence of action is that Meadow Mischief is stuck on the rail and Desormeux will have to use his speed early to avoid being trapped. Given the other speed in the race, Meadow Mischief might not have anything left with which to finish the race. Still, Meadowlake gets better than 34 percent debut winners which genetically makes his progeny 2-1. Even if we make him 6-1 in this race, he is definitely worth a play.

Bedminster, meanwhile, is getting some action at 7-2. But that isn't as heavy as you would expect for a horse with Mr. Prospector speed on top and Danzig speed on the bottom. After all, those two are the most expensive stallions in the world. Add that :35.1 gate work and the fact he has a good debut trainer in Dan Hendricks and you'd think Bedminster would be no worse than 8-5.

So how to play the race? By my line both Meadow Mischief and Bedminster are overlays and worth betting to Win. Since Wolf Bait certainly has the ability to win, a backup exacta with Meadow Mischief and Bedminster in the second hole also seems in order.

Results: Meadow Mischief broke alertly, took the lead away from Blushing Victor after a quarter mile and pulled away from the field. He paid $29.60 to Win. Gold Miner's Slew, the Sadler layoff horse, came flying in the stretch to return $25.60 for the Place, while Wolf Bait got up for the Show. Bedminster, meanwhile, was off slowly and never put in a run. The $2 exacta, which I didn't have, returned $561.

The second race at Santa Anita on March 6th is a contentious affair with seven first-time starters running against what looks like a vulnerable favorite in De Vito. While he made a quantum leap in his second race, De Vito's only had one work in twenty-four days and may regress. Still, the public has bet him down to 5-2 from his morning line of 6-1 and he can't be dismissed.

Of the debut runners, only Hello Chicago, Minero and Corridors of Power have precocious sires. Corridors of Power can be thrown out because he is out of a Le Fabuleux mare and is definitely bred to route. Minero is the obvious pedigree play and has a good debut trainer. While he is getting some action at 4-1, he doesn't have any exceptional works. Since he's out of a stout Gallant Man mare, he might be better suited to 7 furlongs or even a mile.

Hello Chicago gets early action at 5-1, then gradually drifts above his 20-1 morning line. Broad Brush is a precocious sire, but this colt is also out of a stamina-oriented (Speak John) mare. On the other hand, Hess is a decent debut trainer. So there are mixed signals here. Mint Chocolate is getting some action too, but Key to the Mint is a considerably less precocious sire than either Forty Niner or Broad Brush.

Of the experienced runners, Swank is by Topsider the same precocious sire as De Vito. Though Swank lost his first start, it might be thrown out because of his poor break. Some horses that have trouble in their first start improve in their second race, while for others a bad break is the sign of a bad actor. At better than 5-1 Swank is rightfully nearly double his morning line. Even though Kent Desormeux continues to take the call, Swank hasn't raced in nearly a year, didn't show any speed his first time out, and has a non-precocious broodmare sire.

MARCH 6th (handwritten)

2

6 Furlongs. (1:07¹) **MAIDEN SPECIAL WEIGHT. Purse $32,000. 3–year–olds. Weight, 118 lbs.** (Non–starters for a claiming price of $32,000 or less in their last three starts preferred).

Loveyawhenyawin
Own: Barbara & Lapera & McKean

PINCAY L JR (233 35 32 28 .15)

Gr. c. 3 (Mar)
Sire: Kris S. (Roberto)
Dam: Bishop's Delight (Sawbones)
Br: Hobeau Farm (Fla)
Tr: Spawr Bill (84 15 12 18 .18)

D2|1 3 (handwritten)

118

Lifetime Record :	0 M 0 0		$0				
1994	0 M 0 0			Turf	0 0 0 0		
1993	0 M 0 0			Wet	0 0 0 0		
SA	0 0 0 0			Dist	0 0 0 0		

WORKOUTS: Mar 4 SA 3f fst :36³ Bg6/23 • Feb 27 SA 6f fst 1:13² Hg21/46 • Feb 23 SA 6f fst 1:15¹ H 17/26 • Feb 16 SA 6f fst 1:12¹ H 5/57 • Feb 11 SA 5f fst 1:00³ H 21/62 • Jan 31 SA 5f fst 1:02¹ H 27/34
Jan 15 Hol 5f fst 1:02² H 19/28 • Jan 8 Hol 5f fst 1:04¹ H 21/23 • Dec 31 Hol 4f fst :50² H 21/31 • Dec 24 Hol 4f fst :49³ H 28/37 • Dec 17 Hol 3f fst :38¹ H 23/28

Pocomo
Own: Hibbert R E

ANTLEY C W (230 29 25 14 .13)

B. c. 3 (Apr)
Sire: Cherokee Colony (Pleasant Colony)
Dam: Old Fashioned Lady (Olden Times)
Br: R. E. Hibbert (Ky)
Tr: Rash Rodney (57 6 7 10 .11)

D6|2 2 (handwritten)

118

Lifetime Record :	0 M 0 0		$0				
1994	0 M 0 0			Turf	0 0 0 0		
1993	0 M 0 0			Wet	0 0 0 0		
SA	0 0 0 0			Dist	0 0 0 0		

WORKOUTS: Mar 3 SA 4f fst :48² H 10/40 • Feb 24 SA 5f fst :59⁴ Hg7/31 • Feb 16 SA 7f fst 1:27⁴ H 18/19 • Feb 10 SA 7f fst 1:30³ H 11/14 • Feb 3 SA 6f fst 1:15⁴ H 21/35 • Jan 29 SA 6f fst 1:12¹ H 4/31
Jan 23 SA 6f fst 1:15⁴ H 15/25 • Dec 27 SLR tr.t 4f fst :48⁴ H 4/4 • Dec 20 SLR tr.t 4f fst :47⁴ H 1/1 • Dec 13 SLR tr.t 4f fst :49 H 2/3 • Dec 6 SLR tr.t 3f fst :37 H 2/4

Flying Marfa
Own: Gleis Josephine T

STEVENS G L (269 41 56 47 .15)

Gr. g. 3 (Mar)
Sire: Marfa (Foolish Pleasure)
Dam: Flying Fortress (Carwhite)
Br: Gleis Mrs Josephine T (Ky)
Tr: Vienna Darrell (56 7 11 5 .13)

D6|3 (handwritten)

118

Lifetime Record :	0 M 0 0		$0				
1994	0 M 0 0			Turf	0 0 0 0		
1993	0 M 0 0			Wet	0 0 0 0		
SA	0 0 0 0			Dist	0 0 0 0		

WORKOUTS: Feb 28 SA 6f fst 1:14¹ H 15/25 • Feb 16 SA 6f fst 1:13² H 17/57 • Feb 10 SA 5f fst 1:01³ H 21/70 • Feb 2 SA 5f fst 1:03⁴ B 41/42 • Jan 28 SA 5f fst 1:02³ H 72/80 • Jan 23 SA 4f fst :48² H 12/37
Jan 18 SA 4f fst :51 H 45/52 • Jan 12 SA 3f fst :36³ H 12/29 • Jan 7 SA 3f fst :38 H 12/17 • Nov 10 SA 5f fst 1:02⁴ H 44/45 • Nov 5 SA 4f fst :48⁴ H 36/88 • Oct 31 SA 4f fst :51 H 54/59

Hello Chicago *EARLY ACTION* (handwritten)
Own: Sloan Mike H

FLORES D R (75 4 12 14 .05)

B. c. 3 (May)
Sire: Broad Brush (Ack Ack)
Dam: Party Worker (Speak John)
Br: Manning Family Trust & Honeagle Farm (Ky)
Tr: Hess R B Jr (75 8 8 7 .11)

B+3| 2 2 (handwritten)

118

Lifetime Record :	0 M 0 0		$0				
1994	0 M 0 0			Turf	0 0 0 0		
1993	0 M 0 0			Wet	0 0 0 0		
SA	0 0 0 0			Dist	0 0 0 0		

WORKOUTS: Mar 3 Hol 4f fst :47⁴ Hg2/10 • Feb 25 Hol 5f fst 1:00² H 4/28 • Feb 18 Hol 5f fst 1:02¹ H 10/20 • Feb 11 Hol 5f fst 1:02¹ Hg27/41 • Feb 5 Hol 5f gd 1:01¹ Hg3/12 • Jan 30 Hol 6f fst 1:13² H 3/9
Jan 24 Hol 6f fst 1:14 H 6/8 • Jan 18 Hol 6f fst 1:15³ H 6/10 • Jan 12 Hol 5f fst 1:01 H 15/28 • Jan 5 Hol 5f fst 1:00⁴ H 4/24 • Dec 31 Hol 4f fst :50² H 21/31 • Dec 16 SA 4f fst :48² H 14/46

Fit And Soxy
Own: Kirby & Schow

BLACK C A (228 20 21 25 .09)

Ch. g. 3 (Apr)
Sire: Secreto (Northern Dancer)
Dam: Fit and Fancy (Vaguely Noble)
Br: Marriott Woodrow D (Ky)
Tr: Jackson Declan A (2 0 0 0 .00)

C6|3 (handwritten)

118

Lifetime Record :	0 M 0 0		$0				
1994	0 M 0 0			Turf	0 0 0 0		
1993	0 M 0 0			Wet	0 0 0 0		
SA	0 0 0 0			Dist	0 0 0 0		

WORKOUTS: Feb 25 SA 3f fst :38¹ Hg14/19 • Feb 10 SA 4f fst :47¹ Hg4/67 • Jan 31 SA 4f fst :49 H 21/34 • Jan 30 SA 3f fst :35¹ Hg2/40 • Jan 24 SA 6f fst 1:12³ H 6/23 • Jan 19 SA 5f fst :59³ H 7/46
Jan 13 SA 5f fst 1:02 H 33/46 • Jan 2 SA 5f fst 1:02¹ H 32/47 • ●Dec 27 SA 3f fst :34³ H 1/25 • Dec 21 SA 4f fst :47⁴ H 9/34 • Dec 16 SA 4f fst :47⁴ H 6/46 • Dec 10 SA 4f fst :53⁴ H 42/43

Falcon Bid
Own: Philip & Sophie Hersh Trust

PEDROZA M A (166 13 21 16 .08)

Gr. c. 3 (May)
Sire: Imperial Falcon (Northern Dancer)
Dam: Qui Bid (Spectacular Bid)
Br: Hedgestone Management (Ont–C)
Tr: Bernstein David (30 5 6 4 .17)

D6|3 (handwritten)

L 118

Lifetime Record :	2 M 0 0		$800				
1994	2 M 0 0		$800	Turf	0 0 0 0		
1993	0 M 0 0			Wet	0 0 0 0		
SA	2 0 0 0		$800	Dist	1 0 0 0		$800

5Feb94–4SA fst 6½f :22¹ :45 1:09² 1:15⁴ Md Sp Wt — 6 2 3½ — — — McCarron C J LB 118 4.80 — 11 Irgun118⁴ Scenic Route118nk Devon Dancer118¹³ 6
Lost his action, fell 3/8
22Jan94–4SA fst 6f :21² :44¹ :56 1:08² Md Sp Wt 71 7 4 66½ 65½ 57½ 511¾ Pedroza M A LB 118 14.40 83–09 Fly'n J. Bryan118⁴½ Laabity118³½ Paster's Caper118² Wide into lane 7
WORKOUTS: Feb 28 SA 5f fst 1:01² B 47/63 • Feb 22 SA 4f fst :49³ H 62/88 • ●Feb 1 SA 4f fst :46² H 1/27 • Jan 14 SA 5f fst 1:00⁴ Hg17/38 • ●Jan 8 SA 7f fst 1:25³ H 1/7 • Jan 2 SA 6f fst 1:13¹ H 9/24

Mint Chocolate *EARLY ACTION* (handwritten)
Own: Stenger Mr & Mrs Richard

VALENZUELA F H (262 28 16 24 .11)

Dk. b or br c. 3 (Jan)
Sire: Key to the Mint (Graustark)
Dam: So Tenderlee (Forceten)
Br: Stenger Cally & Richard (Ky)
Tr: Fanning Jerry (50 7 4 9 .14)

C5|4 (handwritten)

118

Lifetime Record :	0 M 0 0		$0				
1994	0 M 0 0			Turf	0 0 0 0		
1993	0 M 0 0			Wet	0 0 0 0		
SA	0 0 0 0			Dist	0 0 0 0		

WORKOUTS: Mar 1 SA 5f fst 1:00¹ H 10/48 • Feb 23 SA 6f fst 1:12⁴ H 4/26 • Feb 13 SA 5f fst :59³ H 3/60 • Feb 3 SA 5f fst 1:00² H 7/62 • Jan 28 SA 5f fst :59 H 3/80 • Jan 21 SA 4f fst :48¹ H 12/40
Jan 15 SA 4f fst :49 H 27/46 • Jan 7 SA 3f fst :35¹ H 3/17

Corridors Of Power
Own: Gainesway Farm

DELAHOUSSAYE E (242 41 35 32 .17)

Ch. c. 3 (Mar)
Sire: Lyphard (Northern Dancer)
Dam: Fabuleux Jane (Le Fabuleux)
Br: Gainesway Thoroughbreds Ltd. (Ky)
Tr: Drysdale Neil (22 6 3 5 .27)

B 3|1 3 (handwritten)
U2|1 (handwritten)

118

Lifetime Record :	0 M 0 0		$0				
1994	0 M 0 0			Turf	0 0 0 0		
1993	0 M 0 0			Wet	0 0 0 0		
SA	0 0 0 0			Dist	0 0 0 0		

WORKOUTS: Feb 27 SA 7f fst 1:27³ H 4/5 • Feb 22 SA 5f fst 1:01 H 37/77 • Feb 11 SA 5f fst 1:02 H 47/62 • Jan 5 SA 3f fst :35⁴ B 10/39 • Dec 31 SA 6f fst 1:16⁴ H 35/37 • Dec 26 SA 6f fst 1:15⁴ H 20/24
Dec 20 SA 6f fst 1:16³ H 24/27 • Dec 14 SA 5f fst 1:02⁴ H 36/41 • Dec 4 Hol 4f fst :51 H 28/30 • Nov 24 Hol 3f fst :37² B 23/34

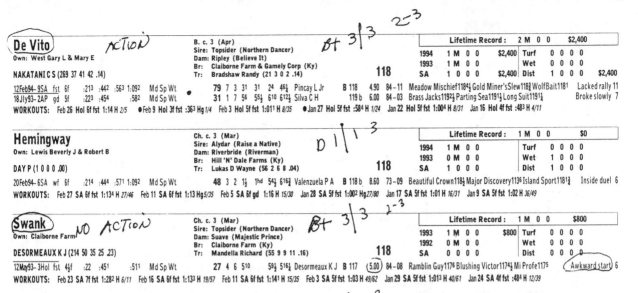

Of the rest of the field, Hemingway showed speed in his debut, but with Alydar on top and Riverman on the bottom he is bred to route. Falcon Bid and Sixth of March have already been beaten and are dead on the board.

So how to play the race? Bet Minero and Hello Chicago to Win and put them both behind De Vito in the exacta as a Place bet. While Mint Chocolate might beat us, he loses McCarron and is suspect on pedigree grounds.

Results: De Vito, the heaviest action horse, wired the field and paid $7.60 to Win. Minero finished second six lengths in front of Hello Chicago and returned $4.60 to Place. The $2 De Vito/Minero exacta paid $34.40, a 16-1 return. Interestingly, if this had been a trifecta race, Hello Chicago would have completed it at nearly 25-1. Combining the action experienced horse with the two highest-rated debut runners in maiden races can be a very lucrative trifecta play.

The second race at Santa Anita on April 9th is another example of a combination action and pedigree play. It's a race that includes four experienced runners and seven first-time starters. The first thing to note once again is that all the horses that have run

Santa Anita Park

2

6 Furlongs. (1:07^{1}) **MAIDEN SPECIAL WEIGHT. Purse $32,000. Fillies, 3-year-olds. Weight, 117 lbs. (Non-starters for a claiming price of $32,000 or less in their last three starts preferred.)**

START ▼ 6 FURLONGS ▲ FINISH

Cathy's Dynasty
Own: Young Cathy
MCCARRON C J (219 38 38 33 .17)

Br. f. 3 (Feb)
Sire: Gate Dancer (Sovereign Dancer)
Dam: Fashion Dynasty (Well Decorated)
Br: Opstein Kenneth (Fla)
Tr: Van Berg Jack C (210 28 42 24 .13)
117

Lifetime Record :	0 M 0 0	$0		
1994	0 M 0 0		Turf	0 0 0 0
1993	0 M 0 0		Wet	0 0 0 0
SA	0 0 0 0		Dist	0 0 0 0

WORKOUTS: ●Apr 6 SA 4f fst :45^{3} Hg *1/31* Mar 29 SA 5f fst 1:00 H *13/52* Mar 23 SA 6f fst 1:16^{4} Hg *58/62* Mar 15 SA 5f fst :59^{1} H *2/59* Mar 9 SA 5f fst 1:00^{2} H *9/59* Mar 2 SA 5f fst 1:00^{4} H *9/38* Feb 23 SA 4f fst :48^{1} H *18/49* Feb 12 SA 4f fst :50^{3} H *40/49*

Empress Sissi
Own: Haras Santa Maria De Araras
FLORES D R (141 8 19 20 .06)

B. f. 3 (Feb)
Sire: Ogygian (Damascus)
Dam: Quip Mask (Waldmeister)
Br: Haras Santa Maris De Araras (Fla)
Tr: McAnally Ronald (126 19 16 17 .15)
117

Lifetime Record :	1 M 0 0	$0		
1993	1 M 0 0		Turf	0 0 0 0
1992	0 M 0 0		Wet	0 0 0 0
SA	1 0 0 0		Dist	1 0 0 0

31Oct93- 4SA fst 6f :21^{3} :44^{4} :57^{2} 1:10 ⒻMd Sp Wt 54 5 2 3nk 6$3\frac{1}{4}$ 8$7\frac{1}{2}$ 8$12\frac{3}{4}$ Pincay L Jr 117 31.30 74-12 PrincessMitterand117hd Whatawoman117^{4} GlidingLark117$4\frac{1}{2}$ Steadied 1/4 9

WORKOUTS: Apr 2 Hol 6f fst 1:14^{1} H *7/13* Mar 28 Hol 5f fst 1:01^{1} H *13/33* Mar 22 Hol 6f fst 1:16^{1} H *18/21* Mar 15 Hol 4f fst :51^{4} H *18/18* Mar 9 Hol 5f fst 1:01 H *12/28* Mar 2 Hol 4f fst :47^{3} H *2/18*

Border Tramp
Own: Beal B & French L R Jr
DAY P (4 1 0 0 .25)

Ch. f. 3 (Jan)
Sire: Tejano (Caro)
Dam: Bo K. (Raise a Native)
Br: Beal Barry & French L R (Ky)
Tr: Lukas D Wayne (78 3 11 13 .04)
117

Lifetime Record :	2 M 0 0	$1,600			
1994	2 M 0 0	$1,600	Turf	0 0 0 0	
1993	0 M 0 0		Wet	0 0 0 0	
SA	2 0 0 0	$1,600	Dist	2 0 0 0	$1,600

26Mar94- 3SA fst 6f :21^{4} :44^{4} :57 1:09^{2} ⒻMd Sp Wt 65 4 2 2$\frac{1}{2}$ 3$1\frac{1}{2}$ 4$4\frac{1}{2}$ 5$11\frac{1}{2}$ Valenzuela F H B 117 15.40 79-08 Magical Avie117nk Masterful Dawn117^{9} Prematurely Gray117hd Gave way 6
5Mar94- 6SA fst 6f :22 :44^{3} :57 1:09^{4} ⒻMd Sp Wt 60 10 3 6$2\frac{1}{2}$ 3^{3} 4$4\frac{1}{2}$ 5$8\frac{1}{2}$ McCarron C J B 117 9.60 79-08 Palestrina117^{1} Magical Avie117$3\frac{1}{2}$ Jackie Ramos117$\frac{1}{2}$ 5 wide to turn 12

WORKOUTS: Apr 3 SA 4f fst :49 H *16/39* Mar 22 SA 4f fst :48^{2} H *8/62* Mar 14 SA 5f fst :59^{3} H *3/54* Feb 27 SA 6f fst 1:12^{1} Hg *7/46* Feb 19 SA 5f fst 1:01^{3} Hg *51/75* Feb 12 SA 5f fst 1:00^{1} H *16/86*

Locate
Own: Claiborne Farm
PINCAY L JR (349 48 50 40 .14)

B. f. 3 (May)
Sire: Cox's Ridge (Best Turn)
Dam: Find (Mr. Prospector)
Br: Claiborne Farm (Ky)
Tr: Mandella Richard (82 14 13 14 .17)
117

Lifetime Record :	1 M 0 0	$2,400			
1994	1 M 0 0	$2,400	Turf	0 0 0 0	
1993	0 M 0 0		Wet	0 0 0 0	
SA	1 0 0 0	$2,400	Dist	1 0 0 0	$2,400

5Mar94- 6SA fst 6f :22 :44^{3} :57 1:09^{4} ⒻMd Sp Wt 69 1 8 9^{4} 7$5\frac{1}{4}$ 5^{5} 4^{5} Pincay L Jr 117 2.40 83-08 Palestrina117^{1} Magical Avie117$3\frac{1}{2}$ Jackie Ramos117$\frac{1}{2}$ Steadied 5/8 12

WORKOUTS: Apr 7 SA 3f fst :37^{2} H *18/24* Mar 31 SA 6f fst 1:13^{2} H *3/15* Mar 22 SA 5f fst 1:02^{4} H *77/93* Mar 16 SA 3f fst :37^{3} H *18/22* Mar 4 SA 3f fst :37 H *8/23* Feb 27 SA 5f fst 1:02^{4} H *41/44*

Just Tops
Own: Biszantz & Greenman & Vandeweghe
DESORMEAUX K J (329 79 57 42 .24)

Dk. b or br f. 3 (May)
Sire: Topsider (Northern Dancer)
Dam: Justicara (Rusticaro)
Br: Echo Valley Horse Farm Inc (Ky)
Tr: Greenman Walter (38 9 5 3 .24)
117

Lifetime Record :	0 M 0 0	$0		
1994	0 M 0 0		Turf	0 0 0 0
1993	0 M 0 0		Wet	0 0 0 0
SA	0 0 0 0		Dist	0 0 0 0

WORKOUTS: Apr 4 SA 5f fst 1:02 H *44/53* ●Mar 28 SA 6f fst 1:11^{2} Hg *1/28* Mar 22 SA 6f fst 1:12^{3} H *4/36* Mar 16 SA 5f fst 1:01^{3} H *23/44* Mar 9 SA 6f fst 1:13 Hg *3/42* Mar 3 SA 4f fst :48 H *8/40* Feb 22 SA 6f fst 1:14 H *23/44* Feb 15 SA 5f fst 1:03^{2} H *32/33* Feb 10 SA 4f fst :49 H *24/67* Feb 2 SA 4f fst :47^{1} H *6/43* Jan 28 SA 4f fst :49^{4} H *42/55* Jan 22 SA 3f fst :36^{4} H *10/27*

Crystal Hailey
Own: Seabaugh & Thomason & Whittingham C
ATKINSON P (152 12 8 16 .08)

Ch. f. 3 (May)
Sire: Greinton (Green Dancer)
Dam: Little Hailey (Blushing Groom)
Br: William Seabaugh & Charlie Whittingham (Ky)
Tr: Whittingham Charles (97 9 17 11 .09)
117

Lifetime Record :	0 M 0 0	$0		
1994	0 M 0 0		Turf	0 0 0 0
1993	0 M 0 0		Wet	0 0 0 0
SA	0 0 0 0		Dist	0 0 0 0

WORKOUTS: ●Apr 6 SA 3f fst :34^{1} H *1/18* Apr 1 SA 6f fst 1:14^{4} H *10/17* Mar 27 SA 5f fst 1:02^{1} H *45/68* Mar 22 SA 6f fst 1:12^{4} H *7/36* Mar 16 SA 4f fst :47^{2} Hg *4/56* Mar 10 SA 6f fst 1:13 H *2/30* Feb 27 SA 6f fst 1:14^{3} H *34/46* Feb 22 SA 5f fst 1:02 H *56/77* Feb 15 SA 5f fst 1:03 H *30/33* Feb 10 SA 4f fst :49^{3} H *33/67* Feb 5 SA 4f gd :48^{1} H *2/40* Jan 28 SA 3f fst :38^{3} H *25/28*

Masterful Dawn
Own: Ortega Charles & Jim & Joseph S
STEVENS G L (386 60 72 66 .16)

Dk. b or br f. 3 (Mar)
Sire: Masterful Advocate (Torsion)
Dam: After the Dawn (Africanus)
Br: Ortega James & Joseph (Cal)
Tr: Bernstein David (43 5 7 6 .12)
L 117

Lifetime Record :	7 M 3 2	$33,325			
1994	5 M 3 1	$27,825	Turf	0 0 0 0	
1993	2 M 0 1	$5,500	Wet	0 0 0 0	
SA	5 0 3 1	$27,825	Dist	4 0 3 0	$21,100

26Mar94- 3SA fst 6f :21^{4} :44^{4} :57 1:09^{2} ⒻMd Sp Wt 93 3 1 1$\frac{1}{2}$ 1$1\frac{1}{2}$ 1hd 2nk Antley C W LB 117 4.10 90-08 MagicalAvie117nk MasterfulDawn117^{9} PrematurelyGry117hd Game inside 6
26Feb94- 6SA fst 6f :21^{3} :44^{4} :57 1:09^{4} ⒻⓈMd Sp Wt 81 5 1 2hd 2$\frac{1}{2}$ 2$\frac{1}{2}$ 2^{2} Stevens G L LB 117 2.80 86-11 Serena's World118^{2} Masterful Dawn117$3\frac{1}{2}$ La Frontera117no Outfinished 9
30Jan94- 2SA fst 6$\frac{1}{2}$f :22^{1} :45 1:08^{4} 1:15^{2} ⒻⓈMd Sp Wt 79 7 2 1$1\frac{1}{2}$ 1$\frac{1}{2}$ 2^{1} 4$4\frac{1}{2}$ Pedroza M A LB 117 3.50 90-07 Ballerina Gal117$1\frac{1}{2}$ Pirate's Revenge117$2\frac{3}{4}$ Krispy117nk Weakened 7
16Jan94- 2SA fst 6$\frac{1}{2}$f :21^{2} :44^{3} 1:09^{3} 1:16^{1} ⒻMd Sp Wt 78 10 1 3^{2} 3$\frac{1}{2}$ 2hd 3$3\frac{1}{4}$ Pedroza M A LB 117 6.10 88-09 Paradisa117^{2} Emerald Express117$1\frac{1}{4}$ Masterful Dawn117$5\frac{1}{2}$ Best of rest 10
1Jan94- 6SA fst 6f :21^{2} :44 :55^{3} 1:08 ⒻⓈMd Sp Wt 75 7 2 2$1\frac{1}{2}$ 2$1\frac{1}{2}$ 2^{4} 2^{10} Pedroza M A LB 117 6.00 87-03 Flying In The Lane117^{10} Masterful Dawn117$2\frac{1}{2}$ Taj Aire117^{3} No match 10
24Nov93- 6Hol fst 6f :22 :45 :57^{3} 1:10^{2} ⒻⓈMd Sp Wt 60 9 2 4$1\frac{1}{2}$ 5^{3} 5$2\frac{1}{2}$ 5^{4} Desormeaux K J LB 118 *2.00 84-09 Lovely Music118$\frac{1}{2}$ Linda Lou B.118$\frac{1}{2}$ Fresh Parsley118nk Wide trip 10
11Jly93- 6Hol fst 5$\frac{1}{2}$f :22 :45^{1} :57^{3} 1:04^{1} ⒻⓈMd Sp Wt 60 10 2 1^{1} 1^{1} 2$2\frac{1}{2}$ 3$8\frac{1}{2}$ Flores D R LB 117 9.30 82-12 Choobloo117$8\frac{1}{2}$ Top Of The Sky117no Masterful Dawn117$1\frac{3}{4}$ 12
Bumped start, greenly lane

WORKOUTS: ●Mar 16 SA 6f fst 1:12^{2} H *1/23* Mar 8 SA 4f fst :46^{3} H *2/42* Feb 16 SA 5f fst 1:00^{3} B *23/76* Feb 11 SA 4f fst :46^{3} H *4/57* Jan 10 SA 4f fst :47^{2} H *6/53*

Turkomine
Own: McCaffery & Toffan
GONZALEZ S JR (328 38 30 40 .12)

Ch. f. 3 (May)
Sire: Turkoman (Alydar)
Dam: Let's Get Pinned (Dewan)
Br: Brereton C. Jones (Ky)
Tr: Gonzalez J Paco (54 11 10 3 .20)
112^{5}

Lifetime Record :	1 M 0 0	$2,400			
1994	1 M 0 0	$2,400	Turf	0 0 0 0	
1993	0 M 0 0		Wet	1 0 0 0	$2,400
SA	1 0 0 0	$2,400	Dist	0 0 0 0	

19Mar94- 4SA sly 6$\frac{1}{2}$f :22^{1} :45^{4} 1:12^{4} 1:20^{1} ⒻMd Sp Wt 43 5 6 5$3\frac{1}{2}$ 3$4\frac{1}{2}$ 3^{6} 4^{9} Gonzalez S^{5} B 112 b *1.40 62-21 Capote Peak117^{6} Gondwanaland117$\frac{1}{2}$ She's A D. A.117$2\frac{1}{2}$ Broke in air 6

WORKOUTS: Apr 4 SA 5f fst :59^{2} H *3/53* Mar 29 SA 5f fst 1:00^{3} H *20/52* Mar 15 SA 4f fst :47^{4} H *12/41* ●Mar 9 SA 6f fst 1:12^{3} H *1/42* Mar 2 SA 5f fst 1:00^{3} B *6/38* Feb 24 SA 6f fst 1:15 Hg *19/22*

Mischievous Ways ~~C4/4*~~

Own: Golden Eagle Farm

Ch. f. 3 (Feb)
Sire: Blushing Groom (Red God)
Dam: Waterlot (Buckpasser)
Br: Mabee Mr & Mrs John C (Ky)
Tr: Mandella Richard (82 14 13 14 .17)

117

ANTLEY C W (347 57 33 22 .16)

		Lifetime Record :	0 M 0 0		$0	
	1994	0 M 0 0		Turf	0 0 0 0	
	1993	0 M 0 0		Wet	0 0 0 0	
	SA	0 0 0 0		Dist	0 0 0 0	

WORKOUTS: Mar 27 SA 6f fst 1:13⁴ Hg 11/29   Mar 18 Hol 6f fst 1:15³ H 10/10   Mar 10 Hol 6f fst 1:13² H 2/8   ●Mar 3 Hol 5f fst :59² H 1/16   Feb 25 Hol 3f fst :36² H 4/8   Feb 19 Hol 4f fst :48 H 8/24
Jan 29 SLR tr.t 4f fst :48³ H 10/19   Jan 18 SLR tr.t 3f fst :35³ H 4/6   Nov 16 Hol 4f fst :48⁴ H 5/23   Nov 8 Hol 4f fst :46⁴ H 2/28   Nov 1 Hol 3f fst :36⁴ H 4/6   Oct 25 Hol 3f fst :36² H 7/12

Fast A Foot ~~D B/4~~

Own: Kirby & Schow

Ch. f. 3 (Apr)
Sire: Afleet (Mr. Prospector)
Dam: Reassert Yourself (Caucasus)
Br: Bud Boschert's Stables Inc (Ky)
Tr: Jackson Declan A (4 0 0 0 .00)

117

BLACK C A (328 26 37 37 .08)

		Lifetime Record :	0 M 0 0		$0	
	1994	0 M 0 0		Turf	0 0 0 0	
	1993	0 M 0 0		Wet	0 0 0 0	
	SA	0 0 0 0		Dist	0 0 0 0	

WORKOUTS: Apr 2 SA 5f fst 1:01 Hg 18/43   Mar 28 SA 5f fst 1:01 H 20/41   Mar 13 SA 4f fst :49² Hg 40/56   Mar 4 SA 5f fst 1:02³ H 38/49   Feb 27 SA 4f fst :47⁴ H 20/51   Feb 22 SA 5f fst 1:02² H 60/77
Feb 11 SA 4f fst :50¹ H 49/57   Feb 5 SA 3f gd :39² H 17/21   Jan 28 SA 4f fst :51² H 51/55   Jan 22 SA 3f fst :39⁴ H 27/27   Jan 18 SA 3f fst :37¹ H 12/19   Jan 12 SA 3f fst :39³ H 29/29

Deanna's Sunshine ~~M 4/2~~

Own: Jiles Deanna & E W

B. f. 3 (Apr)
Sire: Sunshine Forever (Roberto)
Dam: Deanna's Special (Storm Bird)
Br: Jiles E W (Ky)
Tr: Moreno Henry (16 2 0 0 .13)

117

VALENZUELA P A (290 33 35 48 .11)

		Lifetime Record :	2 M 0 0		$800	
	1994	2 M 0 0	$800	Turf	0 0 0 0	
	1993	0 M 0 0		Wet	1 0 0 0	
	SA	2 0 0 0	$800	Dist	1 0 0 0	$800

19Mar94-4SA sly 6½f   :22¹ :45⁴ 1:12⁴ 1:20¹   ⑤Md Sp Wt   31 4 2 4½ 4⁵ 5⁸ 6¹⁴½ Nakatani C S   B 117b 2.20 57-21 Capote Peak117⁶ Gondwanaland117½ She's A D.A.117²½   Gave way 6
23Jan94-4SA fst 6f   :21³ :44² :57 1:10¹   ⑤Md Sp Wt   71 7 4 5½ 5³¾ 5⁴½ 5⁴ Nakatani C S   B 117b 3.80 82-10 Taj Aire117ʰᵈ Neeran117ⁿᵏ Prematurely Gray117½   4 wide into lane 7

WORKOUTS: ●Apr 7 SA 3f fst :34³ H 1/24   Apr 1 SA 5f fst 1:01 H 11/32   Mar 26 SA 4f fst :48³ H 14/27   Mar 16 SA 4f fst :47 H 2/56   Mar 11 SA 6f fst 1:14 H 6/13   Mar 5 SA 5f fst 1:01¹ H 19/57

Accountable Lady ~~B 5/3~~

Own: Gainesway Farm

B. f. 3 (Mar)
Sire: The Minstrel (Northern Dancer)
Dam: Issues n' Answers (Jacinto)
Br: Heronwood Farm Inc. (Va)
Tr: Drysdale Neil (34 6 3 9 .18)

L 117

DELAHOUSSAYE E (344 56 54 55 .16)

		Lifetime Record :	0 M 0 0		$0	
	1994	0 M 0 0		Turf	0 0 0 0	
	1993	0 M 0 0		Wet	0 0 0 0	
	SA	0 0 0 0		Dist	0 0 0 0	

WORKOUTS: Apr 4 Hol 5f fst 1:00 H 5/33   Mar 30 Hol 6f fst 1:15⁴ H 4/5   Mar 21 Hol 6f fst 1:12³ H 3/18   Mar 15 SA 6f fst 1:13³ H 9/19   Mar 10 SA 6f fst 1:15² H 20/30   Mar 4 SA 6f fst 1:14¹ B 14/36
Feb 27 SA 6f fst 1:13² H 21/46   Feb 3 SA 3f fst :37² Hg 16/28   Jan 29 SA 5f fst :59¹ H 3/61   Jan 24 SA 4f fst :51¹ H 38/39   Jan 18 SA 7f fst 1:28³ B 4/6   Jan 11 SA 7f fst 1:27² H 7/11

Also Eligible (Not in Post Position Order):

Top Rung OWNER/breeder ~~C1/1 C6/2~~

Own: Glen Hill Farm

B. f. 3 (Mar)
Sire: Seattle Slew (Bold Reasoning)
Dam: Feature Price (Quack)
Br: Glen Hill Farm (Fla)
Tr: Proctor Willard L (20 2 3 2 .10)

117

STEVENS G L (386 60 72 66 .16)

		Lifetime Record :	0 M 0 0		$0	
	1994	0 M 0 0		Turf	0 0 0 0	
	1993	0 M 0 0		Wet	0 0 0 0	
	SA	0 0 0 0		Dist	0 0 0 0	

WORKOUTS: Apr 3 SA 5f fst 1:02 H 38/56   Mar 27 SA 5f fst 1:00² Hg 13/68   Mar 18 SA 5f fst 1:00¹ Hg 8/36   Mar 11 SA 6f fst 1:13 H 2/13   Mar 4 SA 6f fst 1:14¹ H 14/36   Feb 26 SA 3f fst :36 H 7/27
Feb 19 SA 5f fst 1:00³ H 25/75   Feb 12 SA 5f fst 1:00³ H 28/86   Feb 5 SA 5f gd 1:03² H 37/59   Jan 30 SA 4f fst :47² H 13/44   Jan 23 SA 4f fst :50² H 36/37   Jan 17 SA 3f fst :37⁴ H 18/22

© 1994 DRF, Inc.

— Border Tramp, Locate, Turkomine and Deanna's Sunshine — have below par Beyers and are by route sires. Even though Locate is being bet at around 7-2, he is not the favorite. Clearly this is a race where a precociously-bred debut runner has a big shot.

The two highest debut qualifiers are Just Tops and Accountable Lady, both of whom have B ratings. Between the two, Just Tops has the class edge and an excellent debut trainer in Walter Greenman who is also part-owner of the horse. Throw in Kent Desormeux and that 1:11.2 bullet gate work at Santa Anita on March 28th and Just Tops becomes even more attractive. On the downside, this is a sprint and Rusticaro is a route sire that might dull Just Tops' speed. Also, given all her positives, it is of some concern that she is only the third choice at 4-1.

Accountable Lady, on the other hand, despite her precocious sire and highly competent connections, is dead on the board at 19-1. Either she's not well meant here or she's being ignored because Drysdale's had her working over at Hollywood Park. The out of sight, out of mind factor may be skewing her odds. Still, even factoring that angle in, it seems that she shouldn't be any worse than 6-1.

68   Top Rung is the real surprise debut horse here. He's the heavy favorite at 3-2, even though Seattle Slew is not normally a precocious sire and having Quack as his broodmare sire is likely to dull Top Rung's speed. Nevertheless, commercial breeder Glen Hill Farm has thought enough of Top Rung to keep him and Willard Proctor can certainly get them ready. And that action is hard to ignore.

So how to bet? Top Rung may be for real, but at 3-2 has no value in the Win pool. Best way to play him is on top of Just Tops and Accountable Lady in the exacta. Then bet those two for the Win.

Results: Top Rung, the action horse, was well meant. He broke alertly and led to midstretch before being caught by Accountable Lady. Locate finished third, while Just Tops never made a move and finished eight lengths back. Accountable Lady paid $40.20 for the Win, $12.20 to Place and $6.80 for the Show. The $2 exacta returned $134.60.

While the preceding races have mostly been examples of winning pedigree plays, it would be a great disservice to leave the impression that the ratings in *Exploring Pedigree* always work. They often don't. Even under optimum conditions — where a debut runner is highly-rated, has the right connections and is being bet below its morning line — he will lose more than half the time. The only way to combat this is to demand higher odds the more the contention and pass those races where there isn't a clear play.

Furthermore, it is important for pedigree handicappers to remember that racing is a seasonally-attuned sport and that precociously-bred horses are supposed to make their debuts anywhere from the middle of their juvenile year on up through March of their third year. Precociously bred horses that haven't gotten to the races by then often have something wrong with them and this can seriously affect the reliability of their sire's ratings. My own handicapping experience seems to confirm this. Once the racing calendar moved into April this past year, betting 3-year-old debut runners by precocious sires in Maiden Special Weight events at Golden Gate Fields, Santa Anita and Hollywood Park became a flat bet loss for me. With their 4-year-old debut runners, I suspect, the record would be even worse.

But having said this, the slow-to-mature progeny of a few select sires — particularly Cox's Ridge, Devil's Bag, Nijinsky II, Pleasant Colony and Private Account — can be good plays when they debut later in the year. Typically these are young horses that are out of stouter mares or they were late (May and June) foals that have chronologically just turned three. Debut runners that fit this profile are worth a serious look, especially in 7 furlong events like the third race at Belmont on June 5th and the sixth race at Churchill Downs on June 22nd.

The first thing to notice about the June 5th race at Belmont is the distance and conditions of the race. Typically, 7 furlong sprints are written by Racing Secretaries for classy, but late-maturing, 3-year-old horses whose breeding suggests they will need more than 6 furlongs to break their maiden.

Belmont Park

3

7 Furlongs. (1:202) MAIDEN SPECIAL WEIGHT. Purse $25,000. 3-year-olds and upward. Weights: 3-year-olds, 114 lbs. Older, 122 lbs.

START — 7 FURLONGS — FINISH

Coupled - Chrys and Send 'm Home

Convince
Own: Stone Farm
KRONE J A (26 3 3 7 .12)

Dk. b or br c. 3 (May)
Sire: Halo (Hail to Reason)
Dam: Persuadable (What a Pleasure)
Br: Hancock Arthur (Ky)
Tr: Schulhofer Flint S (46 11 8 7 .24)

114

Lifetime Record :	0 M 0 0	$0		
1994	0 M 0 0		Turf	0 0 0 0
1993	0 M 0 0		Wet	0 0 0 0
Bel	0 0 0 0		Dist	0 0 0 0

WORKOUTS: Jun 3 Bel 4f fst :471 H 3/39 May 28 Sar 4f fst :501 H 1/3 May 23 Sar 5f fst 1:032 H 1/1 May 18 Sar tr.t 5f gd 1:042 B 1/1 May 12 Sar tr.t 4f fst :51 B 1/1 Feb 9 GP 4f fst :48 Hg 4/35 Feb 4 GP 4f gd :523 B (d) 15/16 Jan 26 GP 4f gd :483 H (d) 4/48 Jan 21 GP 4f fst :49 B 29/52 Jan 16 GP 3f fst :37 B 12/46 Jan 7 GP 3f fst :383 B 14/24

Vigan
Own: Paulson Allen E
LUZZI M J (130 21 17 15 .16)

B. c. 3 (Apr)
Sire: Blushing John (Blushing Groom)
Dam: Northern Aspen (Northern Dancer)
Br: Paulson Allen E (Ky)
Tr: Zito Nicholas P (25 5 3 0 .20)

114

Lifetime Record :	0 M 0 0	$0		
1994	0 M 0 0		Turf	0 0 0 0
1993	0 M 0 0		Wet	0 0 0 0
Bel	0 0 0 0		Dist	0 0 0 0

WORKOUTS: May 29 Bel 5f fst 1:03 B 20/32 May 21 Bel 5f gd 1:003 H 13/25 May 14 Bel 5f fst 1:02 B 10/26 Apr 30 Bel 5f fst 1:05 B 31/31 ●Apr 18 Bel tr.t 5f fst 1:001 H 1/23 Apr 10 Bel tr.t 5f fst 1:02 B 9/19 Apr 2 Bel tr.t 5f fst 1:011 H 13/37 Mar 18 GP 5f fst 1:023 Bg 10/15 Mar 5 GP 3f fst :393 Bg 27/27 Feb 21 GP 5f fst 1:033 B 50/58 Feb 12 GP 4f fst :503 B 33/44 Feb 6 GP 5f fst 1:033 B 46/55

Jayayen
Own: Nester John
DAVIS R G (119 12 13 24 .10)

B. c. 3 (Apr)
Sire: Cryptoclearance (Fappiano)
Dam: Rare Sparkling (Cannonade)
Br: Boyd Mike (Fla)
Tr: Monaci David (6 0 1 0 .00)

114

Lifetime Record :	6 M 3 0	$13,190			
1994	5 M 3 0	$13,190	Turf	0 0 0 0	
1993	1 M 0 0		Wet	0 0 0 0	
Bel	2 0 1 0	$5,500	Dist	2 0 1 0	$4,910

22May94- 5Bel fst 6f :223 :453 :574 1:102 Md Sp Wt 74 4 1 11 1½ 21½ 26 Davis R G 122 f 7.20 83-19 Saratoga Shark1226 Jayayen1221 Count On Broadway1221¼ Held place 6
27Mar94- 3Hia fst 1⅛ :471 1:12 1:383 1:53 Md Sp Wt 61 6 1 15 16 23½ 212¼ Ramos W S 122 f 5.60 67-13 Alybro12212 Jayayen1222½ Brother Jerry12213 Second best 9
13Mar94- 4GP fst 6f :213 :444 :574 1:112 Md Sp Wt 49 7 5 75 75½ 64¼ 66¾ Ramos W S 122 f 6.40 77-14 Heroic Pursuit122no Spanos1221¾ Dixie Storm1223½ Failed to menace 10
6Mar94- 6GP fst 7f :222 :45 1:10 1:232 Md Sp Wt 74 5 4 2hd 1hd 2hd 22¾ Ramos W S 122 f 46.30 83-11 Youthful Legs1222¾ Jayayen1221 Chasin Gold122nk Gamely 11
5Feb94- 6GP fst 7f :22 :45 1:103 1:242 Md Sp Wt 56 1 3 1hd 2½ 49 715 Bailey J D 120 f 5.70 66-16 Theater Of War1205½ Colonel Slade1201 Expansionist120¾ 9
Set early pace, three wide, tired
13Sep93- 5Bel fst 6f :221 :454 :582 1:11 Md Sp Wt 19 2 5 3nk 74¼ 1114 1129½ Davis R G 118 22.30 57-23 You And I1181½ Palance1181¼ Hussonet1182¾ Used early 11

WORKOUTS: Apr 23 Hia 4f sly :483 H 3/13 Apr 15 Hia 3f fst :38 B 7/12 Mar 23 Hia 4f fst :492 B 5/10

Classic Arbitrage
Own: Klaravich Stables
CHAVEZ J F (133 18 20 14 .14)

B. c. 3 (Feb)
Sire: Pleasant Colony (His Majesty)
Dam: Royal Partner (Northern Dancer)
Br: Evans T M (Ky)
Tr: Sciacca Gary (52 6 7 8 .12)

114

Lifetime Record :	1 M 0 0	$1,440		
1994	1 M 0 0	$1,440	Turf	0 0 0 0
1993	0 M 0 0		Wet	0 0 0 0
Bel	0 0 0 0		Dist	0 0 0 0

30Apr94- 9Aqu fst 6f :214 :452 :581 1:104 3↑ Md Sp Wt 60 8 4 45½ 32 42½ 49 Bravo J 115 *2.30 79-15 Bold Spector1153½ King Protea1155 Dixie Reef117½ Lugged in stretch 10

WORKOUTS: May 30 Bel 4f fst :492 H 24/47 May 24 Bel 4f fst :483 H 18/51 Apr 28 Bel 3f fst :354 Hg 3/10 Apr 23 Bel 5f fst 1:012 H 12/41 Apr 6 Bel tr.t 6f fst 1:14 H 1/3 Apr 1 Bel tr.t 4f fst :47 Hg 2/41

Holy Mountain
Own: Lazer Two Stable
MIGLIORE R (86 13 17 11 .15)

Dk. b or br c. 3 (May)
Sire: Devil's Bag (Halo)
Dam: Regal Gal (Viceregal)
Br: Way-Oakcliff (Ky)
Tr: Klesaris Robert P (7 1 1 1 .14)

114

Lifetime Record :	3 M 1 1	$8,280		
1994	1 M 1 0	$5,280	Turf	0 0 0 0
1993	2 M 0 1	$3,000	Wet	0 0 0 0
Bel	1 0 0 0		Dist	1 0 0 0

1May94- 5Aqu fst 6f :22 :452 :582 1:12 Md Sp Wt 68 2 2 1½ 2hd 1hd 2no Migliore R 122 *1.50 82-19 Timeless Endeavor122no Holy Mountain1221½ Plutonius1223 Gamely 8
30Oct93- 5Aqu gd 6f :224 :471 :593 1:122 Md Sp Wt 66 8 1 2½ 2hd 31½ 37½ Migliore R 118 *1.20 72-24 MnOfThMnd1185½ CptnMnlght1182 HlMntn1183 Forced pace, weakened 12
8Oct93- 3Bel fst 7f :231 :47 1:12 1:243 Md Sp Wt 66 2 8 51½ 51¾ 65 711¾ Migliore R 118 3.20 68-23 OurEmblem1181½ FinalClearance1183 FriscoGold1182 Off slow, in traffic 8

WORKOUTS: May 31 Bel 5f fst 1:013 B 8/17 May 11 Bel 4f fst :483 B 13/49 Apr 25 Bel tr.t 5f fst 1:012 H 3/7 ●Apr 18 Bel tr.t 6f fst 1:121 H 1/4 Apr 9 Bel tr.t 5f fst 1:00 H 3/39 Apr 1 Bel tr.t 5f fst 1:022 B 23/41

Party Manners
Own: Phipps Ogden Mills
SMITH M E (109 20 15 12 .18)

B. c. 3 (May)
Sire: Private Account (Damascus)
Dam: Duty Dance (Nijinsky II)
Br: Ogden Mills Phipps (Ky)
Tr: McGaughey Claude III (17 4 2 5 .24)

114

Lifetime Record :	0 M 0 0	$0		
1994	0 M 0 0		Turf	0 0 0 0
1993	0 M 0 0		Wet	0 0 0 0
Bel	0 0 0 0		Dist	0 0 0 0

WORKOUTS: Jun 2 Bel 4f fst :484 Hg 14/40 May 25 Bel 5f fst 1:004 Hg 6/18 May 19 Bel 4f sly :501 B (d) 7/22 May 13 Bel 4f fst :491 Bg 16/25 May 6 Bel 4f fst :474 H 4/21 May 1 Bel 4f fst :50 B 19/28 Apr 27 Bel 4f fst :492 B 15/31 Apr 19 Bel tr.t 4f fst :501 B 18/32 Apr 5 Bel tr.t 4f fst :482 H 10/26 Mar 24 Bel tr.t 3f fst :394 B 12/12 Mar 19 Bel tr.t 3f fst :39 B 10/14 Mar 13 Bel tr.t 3f fst :361 H 7/19

Royce Joseph
Own: Double Oaks Stable
CRUGUET J (19 0 1 0 .00)

B. c. 3 (Feb)
Sire: Turkoman (Alydar)
Dam: Solicitous (Cutlass)
Br: Lion Crest Stable (Ky)
Tr: Brida Dennis J (26 2 3 4 .08)

114

Lifetime Record :	5 M 0 2	$7,140			
1994	3 M 0 1	$4,320	Turf	0 0 0 0	
1993	2 M 0 1	$2,820	Wet	1 0 0 0	$1,440
Bel	2 0 0 0		Dist	2 0 0 1	$2,880

5May94- 3Bel fst 1 1/16 :232 :463 1:112 1:421 3↑ Md Sp Wt 48 1 2 41¼ 76 615 625½ Davis R G 115 12.70 65-13 Unaccounted For1156 Telly's Kris115½ Sophie's Friend11510 Gave way 7
21Apr94- 3Aqu fst 7f :224 :462 1:122 1:253 Md Sp Wt 52 2 5 57½ 62¼ 42 34½ Leon F 122 13.10 71-19 All Clear122½ Bet On Rain1224 Royce Joseph1224 Boxed in 7
7Apr94- 3Aqu my 6f :222 :463 :583 1:112 3↑ Md Sp Wt 46 3 4 47½ 46 49½ 413¾ Leon F 115 b 8.10 71-23 Palance1157½ Timeless Endeavor115¾ Star Thief1195½ No threat 5
4Sep93- 5Bel fst 7f :231 :462 1:103 1:231 Md Sp Wt 47 3 6 42¼ 42½ 48 515½ Antley C W 118 fb 5.60 71-14 Changing Breeze118¾ Colonel Slade11813 Alydawn1181 Gave way 6
20Aug93- 4Sar gd 6f :221 :46 :583 1:112 Md Sp Wt 58 3 7 818 815 44 38½ McCauley W H 118 b 16.70 76-15 Royal Minister1186 Speedy Harry1182½ Royce Joseph1183 Belated rally 8

WORKOUTS: Jun 2 Bel 4f fst :482 H 9/40 May 29 Bel 4f fst :512 B 51/63 May 24 Bel 6f fst 1:16 H 3/7 Apr 18 Bel tr.t 5f fst 1:032 B 23/23 Apr 1 Bel tr.t 5f fst 1:04 B 35/41 Mar 24 Bel tr.t 4f fst :50 Hg 30/42

Send 'm Home
Own: Dark Hollow Farm
BAILEY J D (101 27 13 20 .27)

Dk. b or br c. 3 (Mar)
Sire: Private Account (Damascus)
Dam: Safely Home (Winning Hit)
Br: Hayden Mr & Mrs David (Md)
Tr: Arnold George R II (20 2 1 2 .10)

114

Lifetime Record :	1 M 0 0	$1,500		
1994	1 M 0 0	$1,500	Turf	0 0 0 0
1993	0 M 0 0		Wet	0 0 0 0
Bel	1 0 0 0	$1,500	Dist	0 0 0 0

22May94- 5Bel fst 6f :223 :453 :574 1:102 Md Sp Wt 68 1 6 610 67½ 56 48¼ Smith M E 122 1.90 81-19 Saratoga Shark1226 Jayayen1221 Count On Broadway1221¼ No factor 6

WORKOUTS: May 31 Bel 5f fst :59 H 2/17 ●May 20 Bel 3f sly :364 B 1/5 May 15 Bel 4f fst :474 Hg 2/58 May 10 Bel 4f fst :49 Bg 28/59 May 2 Bel 5f fst :594 H 2/12 Apr 26 Bel 5f fst 1:01 H 4/17

continued

Plutonius	Dk. b or br c. 3 (Apr)	Lifetime Record : 9 M 2 4 $22,750
Own: Shields Joseph V Jr	Sire: Silver Buck (Buckpasser)	1994 5 M 0 3 $8,750 Turf 0 0 0 0
	Dam: Chattooga (Ack Ack)	1993 4 M 2 1 $14,000 Wet 1 0 0 0 $220
	Br: J. V. Shields Jr. (Fla)	
SANTOS J A (118 21 20 20 .18)	Tr: Alexander Frank A (17 3 5 4 .18) **114**	Bel 3 0 1 2 $11,500 Dist 4 0 1 1 $7,120

13May94- 1Bel	fst	6f	:231 :461 :581 1:101 3↑	Md Sp Wt	**82**	6	6	31	3½	2½	3¾	Santos J A	115 b	2.90e	89-14	Kerfoot Corner115¾ Count On Broadway115no Plutonius1151½	Willingly	8
1May94- 5Aqu	fst	6f	:22 :452 :582 1:12	Md Sp Wt	**64**	3	6	32½	43	44½	31½	Bailey J D	122	7.00	80-19	Timeless Endeavor122no Holy Mountain1221½ Plutonius1223	Willingly	8
21Feb94- 6GP	fst	1$\frac{1}{16}$	:232 :47 1:113 1:434	Md Sp Wt	**63**	8	7	711	79¾	510	410¾	Perret C	120 f	10.90	80-09	Bonus Money1204½ Majestico1202 Iron Mountain1204¼	Belated bid	10
27Jan94- 6GP	sly	7f	:221 :451 1:10 1:223	Md Sp Wt	**64**	8	4	72½	54¼	66	612¼	Perret C	120 fb	2.60	78-15	Twining120nk Kris' Rainbow1203 Satellite Photo1203¾	Failed to menace	12
1Jan94- 6Crc	fst	7f	:223 :46 1:12 1:253	Md Sp Wt	**67**	4	11	75½	54	44½	34	Smith M E	120 fb	2.30	80-15	Critical Mass1201½ Fleet Eagle1202½ Plutonius1201¼		12
Knocked back start, lacked rally																		
22Nov93- 5Aqu	fst	7f	:23 :463 1:123 1:261	Md Sp Wt	**30**	2	6	42½	73¼	86	820½	McCauley W H	118 b	*2.40	53-22	King Of Kolchis1182 Alydawn118hd Poets Pistol1132½		8
Checked turn, took up stretch																		
4Nov93- 5Aqu	fst	7f	:23 :461 1:113 1:244	Md Sp Wt	**78**	4	3	2½	11	2hd	21¼	McCauley W H	118 b	2.20	79-17	I'm Very Irish1181¼ Plutonius1188 Signal Tap118½	Gamely	8
17Oct93- 5Bel	fst	6f	:222 :454 :582 1:114	Md Sp Wt	**60**	5	5	42½	52	43½	26	McCauley W H	118	11.90	75-14	Brush Full1186 Plutonius118½ Beasleyathisbest1182	Up for place	10
30Sep93- 5Bel	fst	6f	:222 :46 :582 1:112	Md Sp Wt	**71**	1	6	42½	32	34	36¼	Antley C W	118	54.70	77-22	Bermuda Cedar1186 Prank Call118nk Plutonius1181¾	Saved ground	9

WORKOUTS: May 28 Bel 4f fst :491 H *34/93* Apr 23 Bel 5f fst :594 H *5/41* Apr 16 Bel tr.t 5f my 1:02 H *10/30* Apr 9 Bel tr.t 5f fst 1:004 H *12/39* Apr 2 Bel tr.t 4f fst :482 H *11/63*

Crafty Mist	Dk. b or br c. 3 (Mar)	Lifetime Record : 0 M 0 0 $0
Own: Kaufman Robert	Sire: Crafty Prospector (Mr. Prospector)	1994 0 M 0 0 Turf 0 0 0 0
	Dam: Blanche Du Bois (Green Dancer)	1993 0 M 0 0 Wet 0 0 0 0
	Br: Kinderhill Select Bloodstock Inc. (NY)	
PERRET C (24 5 4 5 .21)	Tr: Serpe Philip M (25 4 1 9 .16) **114**	Bel 0 0 0 0 Dist 0 0 0 0

WORKOUTS: Jun 2 Bel 4f fst :493 B *24/40* May 28 Bel 6f fst 1:152 B *5/9* May 22 Bel 5f fst 1:023 B *51/70* May 10 Bel 6f fst 1:16 Bg *12/16* May 3 Bel 6f fst 1:164 B *5/8* Apr 29 Bel tr.t 4f fst :50 B *12/20* Apr 21 GP 5f fst 1:021 B *3/7* Apr 13 GP 4f fst :501 B *5/8* Apr 8 GP 3f fst :371 B *5/10* Mar 7 GP 1 fst 1:46 B *1/1* Mar 1 GP 5f fst 1:022 B *14/21* Feb 21 GP 7f fst 1:31 B *1/1*

The second thing to note is that the actual betting favorite Crafty Mist is a first-time starter, which suggests all the experienced horses are vulnerable. Of these, Plutonius comes out of a race with Beyers good enough to win at this level, but he is not being bet and along with Jayayen and Royce Joseph is a proven loser. The more lightly-raced Classic Arbitrage, Holy Mountain and Send'm Home are respectively by Pleasant Colony, Devil's Bag and Private Account (the right sires for the conditions of this race) but they also have already lost at low odds with undistinguished Beyers. And, of these horses, only Send'm Home at 3-1 is getting any action.

While Crafty Mist, with his B+ rating and excellent debut connections, has been hammered down below 2-1, he may be a vulnerable favorite for several reasons. First off, the progeny of Crafty Prospector, according to his 2-3 age bias are supposed to win at two and early in their third year. Why then is Crafty Mist making his debut this late in the year? Secondly, Crafty Mist has an unimpressive string of slow works. Even though he is out of a Green Dancer mare, Crafty Mist should have shown more of Crafty Prospector's speed than he has. Given these two negatives, it's worth seeing if any of the other debut runners might beat Crafty Mist.

Of the other debut horses, Vigan with his sire's C debut rating and 53-1 odds is an automatic throwout. While Convince also only has a C rating, he can't be dismissed because of his Krone/Schulhofer connections, that :47.1 work, and the fact he is getting bet at 6-1. His major negative is that (according to *Maiden Stats*) his dam is 0-3.

Party Manners, with his Phipps/McGaughey connections and Mike Smith in the irons, is a more attractive possibility. He was a late May foal and has just turned three. The fact that he is debuting relatively early for a Private Account and that McGaughey has asked him for speed in his works suggest he might be a precocious runner. Furthermore, apart from Crafty Mist, Party Manners' C+ debut ratings, on the top and bottom of his pedigree, are the highest in the race. And his dam Duty Dance has already thrown two 3-year-old winners at the same seven-furlong distance Party Manners is being asked to run today. On the downside, Party Manners is going off at 12-1.

So how to bet? While Send'm Home has fast works, he showed no speed in his debut, has mediocre Beyers, and (probably because the Form touted him) has been bet down to an unattractive 3-1. On value grounds, he's a toss out. This leaves us with three first-time starters in Convince, Party Manners and Crafty Mist. Of these colts, Party Manners is the only horse without a pedigree negative and worth betting to Win and Place at 12-1. Alternatively, he could be bet to Win and boxed in the exacta with Convince and Crafty Mist, although my vision was not at the same level before the race as it is now.

THIRD RACE
Belmont
JUNE 5, 1994

7 FURLONGS. (1.20²) MAIDEN SPECIAL WEIGHT. Purse $25,000. 3-year-olds and upward. Weights: 3-year-olds, 114 lbs. Older, 122 lbs.

Value of Race: $25,000 Winner $15,000; second $5,500; third $3,000; fourth $1,500. Mutuel Pool $285,621.00 Exacta Pool $533,201.00

Last Raced	Horse	M/Eqt. A.Wt	PP	St	¼	½	Str	Fin	Jockey	Odds $1
	Party Manners	3 114	6	7	6¹½	5¹	2²	1⁵	Smith M E	12.40
	Convince	3 114	1	9	1½	1½	1½	2²	Krone J A	6.50
5May94 ³Bel⁶	Royce Joseph	3 114	7	4	8ʰᵈ	8½	5¹	3ʰᵈ	Cruguet J	26.50
1May94 ⁵Aqu²	Holy Mountain	f 3 114	5	3	4½	3½	4½	4¾	Migliore R	7.20
22May94 ⁵Bel²	Jayayen	f 3 114	3	1	2²	2¹	3²	5¹¹½	Davis R G	12.00
	Crafty Mist	3 114	10	5	7¹	6½	6½	6²½	Perret C	1.90
13May94 ¹Bel³	Plutonius	b 3 114	9	6	5½	4ʰᵈ	7¹½	7²½	Santos J A	7.60
22May94 ⁵Bel⁴	Send 'm Home	b 3 114	8	8	9⁵	9³½	8¹	8³½	Bailey J D	2.90
30Apr94 ⁹Aqu⁴	Classic Arbitrage	f 3 114	4	2	3¹½	7½	9²½	9¹¾	Chavez J F	22.50
	Vigan	3 114	2	10	10	10	10	10	Luzzi M J	53.40

OFF AT 2:05 Start Good. Won ridden out. Time, :22², :45³, 1:10¹, 1:22² Track fast.

$2 Mutuel Prices:
8-(H)-PARTY MANNERS	26.80	11.00	6.60
2-(A)-CONVINCE		8.20	8.00
9-(I)-ROYCE JOSEPH			5.20

$2 EXACTA 8-2 PAID $185.20

B. c, (May), by Private Account-Duty Dance, by Nijinsky II. Trainer McGaughey Claude III. Bred by Ogden Mills Phipps (Ky).

© 1994 DRF, Inc.

The sixth race at Churchill Downs on June 22nd is another example of a late-maturing horse winning its debut. Going over the field, the first thing to notice is that the two favorites, Save A Slice at 9-5 and Smart Shopper at 7-2, have both already lost with less than overwhelming Beyers. Furthermore, both of them figure to be pressed by Kelly's Katie and may tire once again, especially Smart Shopper who hasn't worked in nearly a month.

Delightful Linda is getting some action at 9-2, but she is coming off a nine-month layoff, has slow works and her stamina indexes both suggest route breeding. The rest of the experienced runners are proven losers with low Beyers and can also be thrown out.

So, this leaves us with Time To Dance and Jesi's Promise as the two first-time starters to consider. Jesi's Promise is 126-1, by an obscure sire, and can be eliminated. Time to Dance, however, is very interesting. She is by Nijinsky II, one of the classy sires whose progeny mature more slowly, and out of a precociously-bred (B5) Timeless Moment mare. Furthermore, her dam Moment To Buy's first two foals are winners and James E. Baker, according to Olmsted's *Trainer Guide* is competent as a debut trainer. With all these positives, it's incredible that Time To Dance is going off at almost 10-1!

Churchill Downs

6

7 Furlongs. (1:21 1) MAIDEN SPECIAL WEIGHT. Purse $19,000 (plus $5,800 KTDF supplement and $2,100 from the KDWOTB fund). Fillies and mares, 3-year-olds and upward. Weights: 3-year-olds, 112 lbs. Older, 121 lbs.

START / 7 FURLONGS / FINISH

Lone Star Gale

Own: Team Valor Stables
GOMEZ G K (206 25 29 28 .12)

B. f. 3 (Mar)
Sire: Marfa (Foolish Pleasure)
Dam: Swept Off Her Feet (Forceten)
Br: Springtide Inc (NY)
Tr: Hennig Mark (14 2 3 1 .14)

L 112

Lifetime Record:	3 M 0 0	$260		
1994	3 M 0 0	$260	Turf	0 0 0 0
1993	0 M 0 0		Wet	0 0 0 0
CD	1 0 0 0		Dist	0 0 0 0

19May94-5CD fst 6f :22 :47 :592 1:132 3+ ⒻMd Sp Wt 53 2 6 21½ 3nk 31½ 54¼ Woods C R Jr L 110 10.90 75-15 Ferd's Nymph110nk Scrape1101 Taylor's Fireworks1102 Bid, weakened 12
13Apr94-6Kee fst 6½f :23 :473 1:141 1:21 ⒻMd Sp Wt 42 1 2 1hd 42½ 58½ 611¼ Woods C R Jr L 120 20.60 59-23 Episode1201½ PreaknessLdy120½ TokenOfEsteem1204 Done early inside 10
14Mar94-6GP fst 6f :223 :461 :582 1:11 ⒻMd Sp Wt 17 5 6 3½ 56½ 1018 1025¼ McCauley W H 121 11.50 61-13 Saxuality12110 Mistletoe And Ivy121nk Our Miz Waki1211½ 11
Drifted out top str, faltered

WORKOUTS: Jun 18 CD 4f fst :49 B 8/34 Jun 12 CD 4f fst :51 B 30/41 Jun 6 CD 4f fst :481 B 3/38 May 31 CD 4f fst :491 B 14/39 May 2 CD 4f fst :484 B 3/52 ●Apr 26 CD 4f fst :48 H 1/23

Save A Slice

Own: Ebert Barry & Hartlage Gary G
JOHNSON J M (60 9 7 8 .15)

Ch. f. 3 (Feb)
Sire: Savings (Buckfinder)
Dam: Slicing (Cutlass)
Br: Ebert Barry F (Ky)
Tr: Hartlage Gary G (47 8 6 8 .17)

112

Lifetime Record:	4 M 2 1	$11,302			
1994	4 M 2 1	$11,302	Turf	0 0 0 0	
1993	0 M 0 0		Wet	1 0 0 0	$1,260
CD	2 0 1 1	$7,742	Dist	1 0 1 0	$5,180

29May94-5CD fst 7f :224 :453 1:11 1:243 3+ ⒻMd Sp Wt 77 7 2 1hd 3nk 11 2¾ Johnson J M 111 *1.60 86-11 O.K.Mom111¾ SveASlice111½ Blondinthshowr1111½ Dueled, led, 2nd best 11
10May94-7CD fst 6½f :224 :463 1:122 1:192 3+ ⒻMd Sp Wt 63 3 4 12 1½ 11 32½ Johnson J M 111 *1.70 80-16 Mylittlegeneral1211½ Balanda107¾ Save A Slice1112 Pace, weakened 12
18Mar94-3OP fst 6f :22 :463 :591 1:112 ⒻMd 50000 74 4 3 1hd 11½ 11 22½ Johnson J M 120 *1.30 81-19 Scrutiny1162½ Save A Slice1207 Suspicious Nature1116 No match 6
11Feb94-2OP sly 6f :214 :454 :583 1:114 3+ ⒻMd Sp Wt 45 2 7 52½ 41½ 46½ 411½ Johnson J M 114 *1.50 70-18 Star Of My Eye1202½ Real Royalty1141 Caro's Beauty1208 Broke slow 7

WORKOUTS: Jun 17 CD 5f fst 1:012 B 3/34 Jun 10 CD 5f fst 1:013 B 10/34 May 25 CD 4f fst :493 B 15/52 May 4 CD 6f fst 1:142 H 1/3 Apr 29 CD 5f my 1:022 B 5/29 Apr 23 OP 5f fst 1:004 B 3/31

Kelly's Katie

Own: Finnway Farm
MCKNIGHT R E (15 1 0 0 .07)

B. f. 3 (Mar)
Sire: Trapp Mountain (Cox's Ridge)
Dam: Forest Dawn (Green Forest)
Br: Finney Ben (Ky)
Tr: Parks Robin (4 1 0 0 .25)

L 112

Lifetime Record:	5 M 0 1	$3,375			
1994	2 M 0 0	$1,295	Turf	0 0 0 0	
1993	3 M 0 1	$2,080	Wet	0 0 0 0	
CD	4 0 0 0	$1,295	Dist	2 0 0 0	$1,295

29May94-5CD fst 7f :224 :453 1:11 1:243 3+ ⒻMd Sp Wt 73 10 1 2hd 1hd 21 42¾ McKnight R E L 114 f 18.50 84-11 O.K.Mom111¾ SveASlice111½ Blondeintheshower1111½ Dueled, weakened 11
19May94-3CD fst 6f :214 :462 :584 1:123 3+ ⒻMd Sp Wt 58 5 2 41½ 53 77 56¼ McKnight R E L 114 f 16.10 77-15 Bunch1151½ Preakness Lady112nk Kahala Aloha1204½ No late threat 10
27Nov93-1CD gd 7f :23 :464 1:13 1:263 ⒻMd Sp Wt 19 9 2 31 75¾ 1223 1227 Miller D A Jr L 121 35.80 50-19 Halo's Gleam1211½ Handle With Care1211½ BlushingMaggie1217 Gave way 12
9Nov93-1CD fst 6f :221 :47 :594 1:131 ⒻMd Sp Wt 14 1 8 1hd 3½ 1114 1118¾ Miller D A Jr L 121 4.60 61-17 Lost Love121¾ Ajax Mountain121hd Wild Stark1214 Duel, tired 12
23Oct93-3Kee fst 6f :214 :453 :574 1:111 ⒻMd Sp Wt 57 9 1 1hd 1½ 23 34 Miller D A Jr L 121 14.50 78-14 Wildboutwinnie121½ DoubleArrow1213½ Kelly'sKtie1212 Duel, weakened 11

WORKOUTS: Apr 30 KHC 4f fst :52 B 8/8

Time To Dance

G/B

Own: Greentree Stable
ROWLAND M F (125 8 6 9 .06)

B. f. 3 (Mar)
Sire: Nijinsky II (Northern Dancer)
Dam: Moment to Buy (Timeless Moment)
Br: Mrs. John Hay Whitney (Ky)
Tr: Baker James E (14 2 1 1 .14)

Ct 1 1 BS 1 3 3-4 2-3 Dam 2/2

112

Lifetime Record:	0 M 0 0	$0		
1994	0 M 0 0		Turf	0 0 0 0
1993	0 M 0 0		Wet	0 0 0 0
CD	0 0 0 0		Dist	0 0 0 0

WORKOUTS: Jun 12 CD 5f fst 1:023 B 8/27 Jun 6 CD 4f fst :484 Bg 8/38 May 18 CD Ⓣ 6f fm 1:19 B (d) 3/4 May 11 CD 4f fst :492 Bg 14/45 May 6 CD 5f fst 1:023 B 21/50 Apr 27 CD 4f sly :51 B 7/20 Apr 22 CD 4f fst :50 B 8/31 Apr 18 CD 4f fst :492 B 5/23 Apr 9 CD 3f fst :38 B 7/13

Lawsey Ms. Mawsey

Own: Manor House Farm
BARTON D M (83 6 4 13 .07)

B. f. 3 (Feb)
Sire: Risen Star (Secretariat)
Dam: Northern Joke (Northern Dancer)
Br: Singer Craig B (Ky)
Tr: McPeek Kenneth G (37 2 2 4 .05)

L 112

Lifetime Record:	3 M 0 0	$0		
1994	2 M 0 0		Turf	1 0 0 0
1993	1 M 0 0		Wet	1 0 0 0
CD	1 0 0 0		Dist	1 0 0 0

25May94-7CD fm 1 1/16 Ⓣ :232 :464 1:12 1:44 3+ ⒻMd Sp Wt 36 6 5 64¾ 88½ 813 816 Barton D M L 111 54.10 70-14 Gotta Wear Shades1111½ Duda112½ First Alliance1122½ Swerved out start 10
29Apr94-2Kee sly 7f :223 :462 1:134 1:273 ⒻMd Sp Wt 21 5 8 87¾ 89½ 913 915 Barton D M 119 31.20 48-27 Carefree Flyer1192½ Rajana's Honor119nk Cahaba Dancer1195 Outrun 10
5Jun93-3CD fst 5½f :23 :471 :531 1:061 ⒻMd Sp Wt 4 7 11 115½ 109½ 1121 1121½ Martinez W 120 26.10 70-13 Miss Ra He Ra1202 Redheaded Honey1201½ Scrape1202 Broke in air 11

WORKOUTS: ●Jun 12 CD 3f fst :35 H 1/18 Jun 5 CD 6f fst 1:152 B 4/6 May 21 Kee 4f fst :512 B 13/24 Apr 22 Kee 5f my 1:05 B 16/17 Apr 17 Kee 4f fst :503 B 26/47 Apr 1 Kee 5f gd 1:041 B 3/3

Will Comply

Own: Phillips Mr & Mrs James W
BARTRAM B E (162 13 33 20 .08)

B. f. 4
Sire: Sovereign Dancer (Northern Dancer)
Dam: Laurie's Angel (Graustark)
Br: Galbreath Daniel M (Ky)
Tr: Morgan James E (13 0 2 2 .00)

C 4 1/2

121

Lifetime Record:	2 M 0 1	$2,010			
1994	2 M 0 1	$2,010	Turf	0 0 0 0	
1993	0 M 0 0		Wet	0 0 0 0	
CD	2 0 0 1	$2,010	Dist	1 0 0 1	$2,010

29May94-3CD fst 7f :23 :463 1:12 1:252 3+ ⒻMd Sp Wt 61 5 10 21 2½ 33½ 34 Bartram B E 121 29.00 79-11 Taylor'sFireworks1111½ CahbDncer1112½ WillComply121hd Weakened late 11
10May94-7CD fst 6½f :224 :463 1:122 1:192 3+ ⒻMd Sp Wt 42 5 10 108¾ 1010 912 811¼ Rowland M F 121 27.40 71-16 Mylittlegeneral1211½ Balanda107¾ Save A Slice1112 12
In tight start, no factor

WORKOUTS: Jun 11 CD 4f fst :503 B 39/54 May 21 CD 4f fst :492 B 12/43 Apr 22 CD 5f fst 1:022 H 3/21 ●Mar 30 Tam 4f fst :484 H 1/8 Mar 27 Tam 5f fst 1:031 Bg 10/14

Dancing Viking

Own: Sam-Son Farms
BARTON D M (83 6 4 13 .07)

Dk. b or br f. 3 (Apr)
Sire: Nureyev (Northern Dancer)
Dam: From Sea to Sea (Gregorian)
Br: Sam-Son Farm (Ont-C)
Tr: Day James E (18 4 0 3 .22)

C 3 1/3*

112

Lifetime Record:	5 M 2 2	$11,404			
1994	2 M 1 1	$5,750	Turf	4 0 2 2	$11,404
1993	3 M 1 1	$5,654	Wet	0 0 0 0	
CD	0 0 0 0		Dist	0 0 0 0	

23Mar94-6FG fm *1 Ⓣ :234 :481 1:132 1:384 ⒻMd Sp Wt 62 10 9 106 84¾ 73¾ 2½ Villeneuve F A 119 5.20 90-07 Promised Legacy119½ Dancing Viking119hd Pennywise1193½ Too late wide 11
15Feb94-4FG fm *7½f Ⓣ :253 :491 1:143 1:334 ⒻMd Sp Wt 58 7 7 2½ 2hd 21½ 35½ King R Jr 119 b 3.60 79-10 Oh So Sharp1194 Alytude1191½ Dancing Viking1191 Prominent, tired 10
31Oct93-4Grd fst 6½f :23 :471 1:133 1:202 ⒻMd Sp Wt 32 2 3 44 66¾ 611 818 Clark D 114 b 4.45 61-21 FleetWhine1157 ToMeBeTrue1142 ChrlottAccord1091½ Well placed early 10
6Oct93-3WO fm 1 Ⓣ :231 :474 1:132 1:392 ⒻMd Sp Wt 49 7 2 2hd 11 22 35¾ Clark D 114 b *1.15e 74-20 WickedMnners1145 SeymourMry115¾ DncingViking1142 Outside speed 10
25Sep93-3WO fm *7f Ⓣ :241 :474 1:144 1:273 ⒻMd Sp Wt 49 8 5 33 1hd 1hd 22½ Clark D 114 b 1.60 78-11 Fiesta Singer1192½ Dancing Viking114hd Malkia119¾ 9
Dueled three wide turn

WORKOUTS: Jun 19 CD 3f fst :361 B 2/20 Jun 11 CD 5f fst 1:012 H 2/18 Jun 2 CD 5f fst 1:023 B 9/28 May 21 CD 4f fst :492 B 12/43 May 13 CD 5f fst 1:022 B 12/25 Apr 24 Kee 5f fst 1:024 B 6/13

Smart Shopper
Own: Dee Pee Stable
SELLERS S J (221 42 33 32 .19)

Dk. b or br f. 3 (May)
Sire: Dynaformer (Roberto)
Dam: Sovereign Mistress (Sovereign Dancer)
Br: Wafare Farm (Ky)
Tr: Padgett Virginia (1 0 0 1 .00)

L 112

Lifetime Record:	1 M 0 1	$2,082		
1994	1 M 0 1	$2,082	Turf	0 0 0 0
1993	0 M 0 0		Wet	0 0 0 0
CD	1 0 0 1	$2,082	Dist	0 0 0 0

26May94- 5CD fst 1 :22⁴ :46² 1:12³ 1:39¹ 3↑ ⒻMd Sp Wt 71 6 7 31½ 1hd 1½ 31¼ Cooksey P J L 111 f 8.90 80–18 BluegrssLssie112nk LdyVencor108¹ SmrtShoppr111no Led, rail, weakened 12
WORKOUTS: May 18 Kee 6f fst 1:16⁴ Hg 3/7 Apr 24 Kee 4f fst :48³ Hg 3/25 Apr 19 Kee 5f fst 1:03² B 7/17

Jesi's Promise
Own: Galloway Joe
BARTRAM B E (162 13 33 20 .08)

B. f. 4
Sire: Leonardo Da Vinci (Brigadier Gerard)
Dam: Ruthie Root (Ruthie's Native)
Br: Hutchens Edward (Ky)
Tr: Galloway Harold (—)

121

Lifetime Record:	0 M 0 0	$0		
1994	0 M 0 0		Turf	0 0 0 0
1993	0 M 0 0		Wet	0 0 0 0
CD	0 0 0 0		Dist	0 0 0 0

WORKOUTS: Jun 18 KHC 5f fst 1:04¹ Bg 3/7 Jun 7 KHC 4f fst :51 Bg 2/7 May 23 KHC 3f fst :38⁴ B 2/3

Fall Foliage
Own: Mjaka IV Stable
BARTRAM B E (162 13 33 20 .08)

Ch. f. 3 (Feb)
Sire: It's Freezing (T. V. Commercial)
Dam: Asheville Blues (To the Quick)
Br: Lavin Allan G & Wedgewood Farm (Ky)
Tr: Reinstedler Anthony (29 3 3 6 .10)

112

Lifetime Record:	5 M 2 1	$11,561		
1994	5 M 2 1	$11,561	Turf	0 0 0 0
1993	0 M 0 0		Wet	0 0 0 0
CD	2 0 0 0	$1,281	Dist	0 0 0 0

4Jun94- 1CD fst 1 1/16 :24 :48 1:13² 1:46⁴ 3↑ ⒻMd Sp Wt 59 3 5 45½ 45½ 6⁸ 6⁸ Bartram B E 115 8.10 72–17 Rosy's Lorna111nk Get The Glory112² Scrape111⁵ 9
Came out, bumped start, tired
10May94- 7CD fst 6½f :22⁴ :46³ 1:12² 1:19² 3↑ ⒻMd Sp Wt 58 6 8 64¼ 74¾ 6⁶ 44¼ Martinez J R Jr 112 5.70 78–16 Mylittlegeneral1211½ Balanda107¾ Save A Slice111² No late response 12
2Apr94- 3TP fst 6f :22 :45¹ :58 1:11³ ⒻMd Sp Wt 64 5 8 6⁸ 79½ 5¹⁰ 3⁶ Martinez J R Jr 122 *1.60 79–16 Carats Please122⁴ Love Awaits122² Fall Foliage122¹ Improved position 12
2Mar94- 3OP gd 1 :23 :46⁴ 1:12³ 1:38⁴ ⒻMd Sp Wt 70 5 3 3⁴ 3³ 2⁵ 2¹⁰ Martinez J R Jr 118 *1.30 73–27 Fittin Purrfect118¹⁰ Fall Foliage118¹ Tressa Ann118⁷ Up for place 7
18Feb94- 1OP fst 6f :21³ :45 :58 1:11² ⒻMd Sp Wt 68 1 6 5⁷ 56½ 35½ 2⁴ Martinez J R Jr 120 13.30 80–13 Broad Avenue120⁴ Fall Foliage120³ Nevis120¹ Rallied 9
WORKOUTS: May 29 CD 4f fst :51 B 19/23 Apr 20 CD 4f fst :50³ B 11/19

Delightful Linda
Own: Hawn William R
DAY P (222 64 38 24 .29)

Ch. f. 3 (May)
Sire: Slew o' Gold (Seattle Slew)
Dam: Lovlier Linda (Vigors)
Br: W. R. Hawn (Ky)
Tr: Mott William I (31 10 3 2 .32)

L 112

Lifetime Record:	1 M 0 0	$0		
1993	1 M 0 0		Turf	0 0 0 0
1992	0 M 0 0		Wet	0 0 0 0
CD	0 0 0 0		Dist	1 0 0 0

5Aug93- 6Sar fst 7f :22³ :45² 1:11³ 1:25² ⒻMd Sp Wt 38 3 12 105½ 9⁸ 107½ 9¹⁰ Krone J A 117 3.10 68–15 British Bauble117hd Bunting117¾ Smart Connection117¾ Broke slowly 12
WORKOUTS: Jun 14 CD 6f fst 1:17 B 1/1 Jun 8 CD 5f fst 1:02⁴ Bg 7/17 May 31 CD 5f fst 1:06² B 14/14 May 24 CD 5f fst 1:03³ B 16/24 May 18 CD 4f fst :52³ B 32/36 Mar 24 KHC 5f fst 1:04³ B 2/7

Hurricane Judy
Own: Conrad Marion
ARGUELLO F A JR (173 15 20 22 .09)

Ch. f. 4
Sire: Storm Cat (Storm Bird)
Dam: Dream Harder (Hard Work)
Br: Conrad Marian (Ky)
Tr: Culp Harvey (7 3 1 0 .43)

L 121

Lifetime Record:	3 M 1 0	$5,850			
1994	3 M 1 0	$5,850	Turf	0 0 0 0	
1993	0 M 0 0		Wet	0 0 0 0	
CD	0 0 0 0		Dist	1 0 0 0	$140

14Apr94- 3Kee fst 6½f :23 :47³ 1:14² 1:20⁴ ⒻMd Sp Wt 31 3 2 2¹ 5⁶ 6⁸ 5¹⁴ Martinez W L 122 3.40 58–23 Twenty Eight Carat122⁶ Redial122² Queen Of The Spa122⁵ Tired 7
1Apr94-10Hia fst 7f :23³ :47³ 1:13¹ 1:26³ ⒻMd Sp Wt 48 6 2 2½ 2¹ 4² 57¼ Ferrer J C L 122 *1.70 69–17 Golden Autumn122¹ Signa122½ Genutech1151½ 9
Six wide top str, faded
11Mar94- 1GP fst 6f :23¹ :47² 1:00 1:13¹ ⒻMd Sp Wt 50 4 2 32½ 3² 2² 21¼ Ferrer J C L 121 2.30 74–17 Imlittlesowhat1211¼ Hurricane Judy1211½ Golden Autumn1212¼ Rallied 6
WORKOUTS: Jun 19 CD 3f fst :36² B 3/20 Jun 13 CD 4f fst :48² Hg 3/37 Jun 8 CD 5f fst 1:01⁴ B 2/17 May 29 CD 5f fst 1:04⁴ B 25/31 May 17 CD 4f fst :52² B 35/41 ●Apr 10 Kee 3f fst :36³ B 1/18

© 1994 DRF, Inc.

So how to bet? With both hands, Win and Place! The public has once again overbet the highest losing speed figures and ignored the possibility that an appropriately-bred, first-time starter might run faster. Since all the favorites are suspect and any horse could finish second, I would avoid the exacta. But, I know exotics players who would argue, for the same reasons, that a prime Win bet and a few selected exacta back and forth bets (keying Time To Dance) makes better sense.

SIXTH RACE
Churchill
JUNE 22, 1994

7 FURLONGS. (1.21¹) MAIDEN SPECIAL WEIGHT. Purse $19,000 (plus $5,800 KTDF supplement and $2,100 from the KDWOTB fund). Fillies and mares, 3–year–olds and upward. Weights: 3–year–olds, 112 lbs. Older, 121 lbs.

Value of Race: $25,160 Winner $17,485; second $4,220; third $2,110; fourth $1,345. Mutuel Pool $115,981.00 Exacta Pool $74,494.00 Trifecta Pool $72,216.00

Last Raced	Horse	M/Eqt.	A.	Wt	PP	St	¼	½	Str	Fin	Jockey	Odds $1
	Time To Dance		3	113	4	4	31½	3³	1hd	1²	Rowland M F	9.70
19May94 5CD5	Lone Star Gale	L	3	112	1	7	6¹	7²	3½	2³	Woods C R Jr	36.40
23Mar94 6FG2	Dancing Viking		3	112	7	8	9²	82½	6²	3¾	Barton D M	15.20
26May94 5CD3	Smart Shopper	Lf	3	113	8	3	7½	6hd	5²	41½	Sellers S J	3.30
29May94 5CD2	Save A Slice		3	112	2	9	1hd	1hd	21½	5no	Johnson J M	1.80
4Jun94 1CD6	Fall Foliage		3	115	10	1	10hd	9½	7³	6¹	Bartram B E	17.40
29May94 5CD4	Kelly's Katie	Lf	3	114	3	5	21½	2⁴	4²	7²	McKnight R E	8.80
5Aug93 6Sar9	Delightful Linda	L	3	112	11	10	11⁴	11²	9⁴	81½	Day P	4.60
14Apr94 3Kee5	Hurricane Judy	L	4	121	12	2	4hd	5hd	81½	9⁶	Arguello F A Jr	23.60
25May94 7CD8	Lawsey Ms. Mawsey	Lb	3	112	5	12	8hd	10¹	111½	10hd	Thompson T J	66.90
29May94 3CD3	Will Comply		4	121	6	6	51½	4hd	101½	111½	Bruin J E	32.00
	Jesi's Promise	bf	4	121	9	11	12	12	12	12	Thorwarth J O	126.70

OFF AT 5:30 Start Good. Won driving. Time, :22⁴, :46, 1:12, 1:25¹ Track fast.

$2 Mutuel Prices:				
	4–TIME TO DANCE	21.40	12.80	9.60
	1–LONE STAR GALE		39.00	20.40
	7–DANCING VIKING			7.60

EXACTA 4–1 PAID $659.40 TRIFECTA 4–1–7 PAID $8,999.20

B. f, (Mar), by Nijinsky II–Moment to Buy, by Timeless Moment. Trainer Baker James E. Bred by Mrs. John Hay Whitney (Ky).

© 1994 DRF, Inc.

6
Handicapping Maiden Claiming Races

Maiden Claiming events can be among the most difficult races to handicap on the card because of the tremendous mix of horses that are regularly entered in them and the fact that all of the contestants are still in the process of establishing their true racing class. Competing in these races for the first time are well-bred horses dropping in class, debut runners that have specifically been bred to win early at various levels of the Maiden Claiming ladder, and the usual quotient of proven losers and obscurely-bred pretenders.

While many Maiden Claiming races cannot be handicapped from a breeding perspective, there are a number of instances where a knowledge of a young horse's pedigree can be a powerful tool for identifying a potential winner. But, even here, pedigree and sire ratings, by themselves, should not be seen as determinant. They must be integrated with trainer and toteboard analysis, a knowledge of winning par times and stud fees, and a deeper understanding of the economics of the breeding industry. Many a Maiden Claiming bet has been lost by a handicapper who failed to appreciate that the entire Maiden Claiming structure is not only a place to win a purse, but also the court of last resort for a commercially-oriented breeder to sell his horse.

It is important when handicapping Maiden Claiming races to look at the connections of each of the horses and determine whether they are recreational owners, who have bought at auction and are in racing primarily for the excitement, or whether they are professional owner/breeders who are hoping to sell their horses in order to turn dollars and stay in the business. Because commercially-oriented, owner/breeders are in the business of selling horses and not just racing them, they are generally much more realistic about where they spot their horses than recreational owners who are inclined to get sentimentally attached to their horses. The last thing a commercial owner/breeder wants to do is to run a horse where it can't win while incurring the expensive day rates that will quickly eat up his profits. After all, a cheap maiden claimer that is in over its head eats as much as a John Henry or a Spectacular Bid while not earning any money. If they can "lose" their horse for a fair price, most owner/breeders are more than happy to do so because they will then have the capital to reinvest in another, potentially more lucrative breeding.

When professional breeders run a horse for a tag, their immediate goal is to recoup two-and-half times the stud fee they've invested plus the $5,000-$10,000 cost of raising and training it. Handicappers can often gauge how well meant a horse is in a particular spot by adding up an owner/ breeder's costs and seeing if, after expenses, he is at least breaking even. If, on the other hand, an owner/breeder initially runs a horse with a $30,000 stud fee in a Maiden $12,500 race, he is obviously taking a financial bath and handicappers should be extremely wary.

Recreational owners, for their part, are much more reluctant to lose their horses to the claiming box for several reasons: With their more expensive, but less able horses, they are loathed to admit they made a mistake at auction and tend, at least initially, to run them over their heads; with their cheaper horses, especially those that have shown some ability in their mornings works, they fear losing a potentially good one and shy away from running them for a realistic tag.

While some of the horses that recreational owners protect from the claiming box do win at the Maiden Special Weight level, the vast majority need to be dropped before they can break their maidens. Astute handicappers will let the ambitiously-placed exceptions win and wait for the majority to be dropped to where their breeding and class ratings suggest they have a more realistic shot of winning. In general, horses with class ratings of 1-5 will be competitive at the Maiden Special Weight level, those with 5-7 ratings at the high Maiden Claiming level, and those with 8-10 ratings in lower Maiden Claiming events.

Unorthodox as it might seem, it is extremely useful from a handicapping perspective to think of every maiden, each time it is dropped in class, as making its "debut" there: not only because it has never run at that level before, but more importantly because it has never lost there. Maidens that have shown speed against better and are now "debuting" at a lower Maiden Claiming level are among the best bets at the track. With these comments as a general framework, let's analyze a variety of Maiden Claiming races to identify those which are best worth playing.

Horses Dropping From the Maiden Special Weight Level

The second race at Santa Anita on February 16th is a good example of a well-spotted horse dropping in class to break its maiden for its owner/breeder connections. Looking at the conditions of the race, we notice that it is restricted to a bunch of "ancient" 4-year-old fillies that have yet to break their maidens. While obviously not a race with a stellar group of horses, it is imminently playable.

The first thing to do is throw out all the proven losers who have shown no ability at this level. This leaves us with Faceta, the heavy 4-5 favorite who has finished second three times in a row, My Sensation — who is dropping from the Maiden Special Weight level — and the first-time starters Mom's Alleged and Ruffmeup. Annaluce, who broke poorly in her debut, might also improve, but she is bred to route and isn't being bet.

Santa Anita Park

2

6½ Furlongs. (1:14) MAIDEN CLAIMING. Purse $20,000. Fillies and mares, 4–year–olds and upward. Weights: 4–year–olds, 119 lbs. Older, 120 lbs. Claiming price $40,000; if for $35,000, allowed 2 lbs.

START ▾ 6 1/2 FURLONGS ▴ FINISH

Pantalona

Own: Adams & Maestro Jr

IAMMARINO M P (29 0 1 0 .00) $40,000

Dk. b or br m. 5
Sire: Private Account (Damascus)
Dam: Corselette (Hoist the Flag)
Br: Keck Howard B (Ky)
Tr: Adams Craig (4 0 0 0 .00)

L 120

Lifetime Record :	12 M 0 2	$18,640			
1994	1 M 0 0		Turf	2 0 0 0	$2,250
1993	6 M 0 0	$8,015	Wet	0 0 0 0	
SA	3 0 0 0	$2,550	Dist	1 0 0 0	$1,650

28Jan94-9SA	fst	1	:223 :462 1:112 1:364	ⒻMd 40000	58	2	3	42	63¾	65½	68½	Iammarino M P	LB 120	13.90	76–15	Uncanny Ann1181¼ Dances With Wolves113hd [DH]Coral Isle118	Gave way	9
12Nov93-4SA	fst	6f	:214 :451 :574 1:11 3↑	ⒻMd 32000	60	4	8	89½	77¼	44½	43¾	Solis A	LB 120	3.50	78–16	Lady OfTheOpera1183 NeverQuicker118½ MissFortWorth120nk	Wide trip	8
27Oct93-4SA	fst	6f	:214 :452 :574 1:103 3↑	ⒻMd 32000	59	7	9	107½	107	67¾	47½	Solis A	B 120	9.10	76–14	NaskrasRapture117½ NativeRidge1174½ NeverQuicker1172½	Very wide trip	12
19Sep93-9Fpx	fst	1 1/16	:223 :462 1:113 1:442 3↑	ⒻMd Sp Wt	56	1	6	44½	37	49	413½	Pedroza M A	B 118	5.50	72–10	Future Guest11510 Noddy's Halo1181½ Lady Pancho1182	Off step slow	9
29Aug93-2Dmr	fm	1 Ⓣ	:224 :463 1:113 1:372 3↑	ⒻMd Sp Wt	67	4	4	43½	42	52½	63	Solis A	B 122	6.70	81–09	Gold Conde122nk Noddy's Halo122hd Sunday's Sis117½	4 Wide stretch	8
8Aug93-2Dmr	fst	6f	:221 :454 :581 1:104 3↑	ⒻMd 45000	61	2	8	89½	66½	65	45½	Fuentes J A5	B 114	26.90	79–11	Flying Safeera1151¼ Amity Ar1212 Future Guest1152¼		8

Shuffled back early, wide

23Jly93-2Hol	fst	6½f	:22 :443 1:103 1:172 3↑	ⒻMd 50000	34	6	4	55½	57½	58½	416	Fuentes J A5	B 116 f	6.30	66–13	TimToBLucky1112 FuturGust1167½ RoylEnggmnt1146½	Wide backstretch	6
16Dec92-2Hol	fm	1 1/16 Ⓣ	:24 :482 1:133 1:452 3↑	ⒻMd Sp Wt	67	5	8	86½	83¾	52½	41	Solis A	B 118	4.10	67–25	Air Sortie118½ La Tuna118hd Gold Conde118nk	Rallied	9
25Nov92-2Hol	fst	6f	:221 :454 :581 1:103 3↑	ⒻMd 40000	62	7	9	97½	77½	45	35¾	Solis A	119	4.40	81–13	Ravager1193 Test The Time1192¾ Pantalona119	Broke slowly	12
21Aug92-6Dmr	fst	1 1/16	:222 :46 1:111 1:442 3↑	ⒻMd Sp Wt	67	1	8	87	31½	41½	54½	Solis A	B 118	3.90	75–15	Chipeta Springs1181¾ Air Sortie118½ Letakia118	Weakened	9

WORKOUTS: Feb 12 SA 3f fst :364 Hg 15/26 Feb 6 SA 4f fst :491 H 37/83 Jan 22 SA 6f fst 1:133 H 15/39 Jan 8 SA 5f fst 1:004 H 19/43 Dec 31 SA 5f fst 1:014 H 45/66 Dec 26 SA 3f fst :37 H 10/13

Her Melody

Own: Allen Byron

TORRES R V (15 0 0 0 .00) $40,000

Ch. f. 4
Sire: Dance in Time (Northern Dancer)
Dam: Her Soir (Hekaton)
Br: Sokol Ernest I (Ky)
Tr: Allen Byron (8 0 0 0 .00)

L 119

Lifetime Record :	6 M 0 1	$2,800			
1994	2 M 0 0		Turf	0 0 0 0	
1993	4 M 0 1	$2,800	Wet	1 0 0 0	
SA	3 0 0 0		Dist	1 0 0 0	

17Jan94-4SA	fst	1 1/16	:234 :481 1:131 1:441	ⒻMd 35000	15	2	1	1hd	3½	612	633¾	Torres R V	LB 117	45.60	40–18	Smart Patch1191½ Raise The Runway1174 Perilous Flyer1173	Inside duel	7
5Jan94-2SA	fst	6f	:214 :442 :562 1:084	ⒻMd 35000	35	8	9	87	912	1019	1020½	Torres R V	LB 117	56.60e	72–09	Kalavrita11910 Smart Patch1192¼ Time For A Storm117½	No factor	12
11Dec93-6Hol	sly	6½f	:22 :453 1:131 1:202 3↑	ⒻMd 28000	33	2	6	3nk	45	78½	814¼	Torres R V	LB 117	23.60	53–19	NativeRidge119½ BbyTkeTheGold119¾ TimeForAStorm119½	Brief speed	9
1Dec93-2Hol	fst	6f	:214 :444 :572 1:104 3↑	ⒻMd 28000	56	8	1	23	24½	24	36	Torres R V	LB 117	185.80	80–13	Enchanted Beauty119½ Faceta1195½ Her Melody117½	Wide early	8
17Nov93-5Hol	fst	6f	:22 :451 :572 1:103 3↑	ⒻMd 28000	43	4	7	64¼	79	410	513¼	Navarro V G	L 117	113.20	74–14	Wild N Super1192 EnchantedBeauty1195½ NativeRidge1194½	By tired ones	10
27Oct93-4SA	fst	6f	:214 :452 :574 1:103 3↑	ⒻMd 32000	37	1	6	42½	64½	88¾	1016	Navarro V G	LB 117	200.50	68–14	Naskras Rapture117½ Native Ridge1174½ Never Quicker1172½	Steadied 3/8	12

WORKOUTS: Feb 12 Fpx 4f fst :48 H 4/9 Feb 6 Fpx 6f gd 1:171 H 2/6 ●Jan 29 Fpx 6f fst 1:14 H 1/12 ●Jan 23 Fpx 6f fst 1:142 H 1/6 ●Jan 12 Fpx 5f fst 1:01 H 1/7 Dec 30 Fpx 6f fst 1:144 H 1/3

Faceta

c5/4

Own: Mandella & Quinn Stables

OLDHAM D W (10 0 3 1 .00) $35,000

Dk. b or br f. 4
Sire: Groovy (Norcliffe)
Dam: Recital P. R. (High Tribute)
Br: Cosme Armando (Fla)
Tr: Mandella Richard (43 6 8 8 .14)

L 117

Lifetime Record :	4 M 3 0	$10,000			
1994	1 M 1 0	$4,000	Turf	0 0 0 0	
1993	3 M 2 0	$6,000	Wet	0 0 0 0	
SA	2 0 1 0	$4,000	Dist	1 0 1 0	$4,000

17Jan94-9SA	fst	6½f	:22 :451 1:101 1:163	ⒻMd 35000	75	1	3	2hd	1½	2hd	2no	Oldham D W	LB 117	3.70	89–09	BabyTakeTheGold119no Faceta1173 AlasknReveltion1182¾	Awkward start	12
20Dec93-4Hol	fst	6f	:221 :451 :574 1:101 3↑	ⒻMd 25000	72	12	3	31½	21½	21½	2½	Oldham D W	LB 119	*.80	88–09	Snow Petals119½ Faceta11911 Albert Queen1193	Floated out 1/16	12
1Dec93-2Hol	fst	6f	:214 :444 :572 1:104 3↑	ⒻMd 32000	70	5	3	53¾	37	34	2½	Oldham D W	LB 119	5.50	85–13	Enchanted Beauty119½ Faceta1195½ Her Melody117½	Rallied	8
27Oct93-4SA	fst	6f	:214 :452 :574 1:103 3↑	ⒻMd 32000	49	5	4	53	31½	46½	711¼	Oldham D W	LB 117	31.40	73–14	NaskrasRapture117½ NativeRidge1174½ NeverQuicker1172½	Steadied start	12

WORKOUTS: Feb 12 Hol 3f fst :362 H 9/19 Feb 6 Hol 4f fst :502 H 20/34 Jan 31 Hol 3f fst :382 H 6/7 Jan 10 Hol 5f fst 1:002 H 9/32 Dec 17 Hol 3f fst :391 H 25/28 ●Nov 24 Hol 6f fst 1:114 H 1/18

Mom's Alleged

c4/1

Own: Har Pen Jac Stables & Izumikawa

GONZALEZ S JR (180 23 15 22 .13) $40,000

Dk. b or br f. 4
Sire: Alleged (Hoist the Flag)
Dam: My Gallant Duchess (My Gallant)
Br: Arthur I. Appelton (Fla)
Tr: Hofmans David (21 5 4 4 .24)

114⁵

Lifetime Record :	0 M 0 0	$0		
1994	0 M 0 0		Turf	0 0 0 0
1993	0 M 0 0		Wet	0 0 0 0
SA	0 0 0 0		Dist	0 0 0 0

WORKOUTS: Feb 10 SA 6f fst 1:161 H 29/42 Feb 5 SA 6f gd 1:183 H 29/30 Jan 29 SA 5f fst 1:012 H 35/61 Jan 22 Hol 6f fst 1:18 H 9/10 Jan 16 Hol 5f fst 1:012 H 13/22 Jan 10 Hol 5f fst 1:024 H 28/32 Jan 5 Hol 5f fst 1:041 H 23/24 Dec 30 Hol 5f fst 1:042 H 9/9 Dec 24 Hol 4f fst :503 H 34/37 Dec 18 Hol 3f fst :354 H 7/28

Annaluce

c4/1 back to route

Own: Teakell Elizabeth

GULAS L L (14 0 0 0 .00) $35,000

B. m. 5
Sire: Majestic Light (Majestic Prince)
Dam: Tamil (Believe It)
Br: Howard B. Keck (Ky)
Tr: Bell Thomas R II (9 2 0 0 .22)

118

Lifetime Record :	1 M 0 0	$0		
1994	1 M 0 0		Turf	0 0 0 0
1993	0 M 0 0		Wet	0 0 0 0
SA	1 0 0 0		Dist	1 0 0 0

17Jan94-9SA	fst	6½f	:22 :451 1:101 1:163	ⒻMd 35000	19	11	12	1214	1218	1117	1024	Gulas L L	118	68.80	65–09	BabyTakeTheGold119no Faceta1173 AlaskanReveltion1182¾	Off very slow	12

WORKOUTS: Feb 12 SA 4f fst :483 H 23/49 Feb 6 SA 5f fst 1:021 H 54/108 Jan 30 SA 5f fst 1:012 H 46/58 Jan 14 SA 3f fst :38 Hg 20/28 Dec 29 SA 6f fst 1:152 H 18/25 Dec 22 SA 6f fst 1:164 H 44/45

Ode To Code

Own: Ciotti Tony J

CEDENO E A (4 0 1 0 .00) $35,000

Ch. f. 4
Sire: Lost Code (Codex)
Dam: Set for One (Banquet Table)
Br: Thomas James K (Ky)
Tr: Conner Kay (3 0 0 0 .00)

117

Lifetime Record :	3 M 0 0	$1,275			
1994	2 M 0 0		Turf	0 0 0 0	
1993	1 M 0 0	$1,275	Wet	0 0 0 0	
SA	2 0 0 0		Dist	0 0 0 0	

2Feb94-2SA fst 6f :21³ :44³ :57 1:10 ⒻMd 35000 34 10 6 10⁶¾ 8⁹½ 8¹⁵ 8¹⁵ Gulas L L B 117 f 14.60 72-13 Irish Maccool117no Intimacy1197 Princess Coup114 1¾ Raced wide 12
5Jan94-2SA fst 6f :21⁴ :44² :56² 1:08⁴ ⒻMd 35000 38 7 3 3²½ 3⁵½ 3⁸ 8¹⁹½ Gulas L L B 117 f 10.50 74-09 Kalavrita11910 Smart Patch1192¼ Time For A Storm117½ Gave way 12
20May93-6Hol fst 6f :21⁴ :45³ :58 1:10⁴ 3+ ⒻMd 32000 73 6 3 1¹½ 1hd 1hd 2¹¼ Garcia J A B 115 62.80 85-11 Datsdawayitis1221¼ DH OdeToCode1152 EscortHome120 5½ Impeded foe 3/8 12
Disqualified and placed 4th

WORKOUTS: Jan 28 SA 4f fst :47 H 8/55 Jan 21 SA 5f fst 1:01¹ H 14/48 Jan 14 SA 4f fst :46² H 3/31 Dec 29 SA 6f fst 1:13² H 7/25 Dec 22 SA 5f fst 1:00³ H 22/64 Dec 16 SA 4f fst :48⁴ H 26/46

B+4/4

My Sensation

10K STUD FEE

Own: Johnston & Stonebraker & Three M St

SOLIS A (193 30 31 31 .16) $35,000

B. f. 4
Sire: Mining (Mr. Prospector)
Dam: Some Sensation (Somethingfabulous)
Br: Old English Rancho & Stonebraker (Cal)
Tr: Warren Donald (17 3 2 1 .18)

L 117

Lifetime Record :	1 M 0 0	$0			
1994	1 M 0 0		Turf	0 0 0 0	
1993	0 M 0 0		Wet	0 0 0 0	
SA	1 0 0 0		Dist	1 0 0 0	

20Jan94-3SA fst 6½f :21⁴ :44³ 1:09¹ 1:15⁴ ⒻMd Sp Wt 39 4 5 3¹ 6⁹¼ 6¹³ 6¹⁹ Solis A LB 119 7.40 74-08 Dolce Amore1192 Really Tops114 1¾ Madrona1192 Done early 6

WORKOUTS: Feb 11 SA 5f fst 1:01 H 28/62 Feb 5 SA 5f gd 1:01⁴ H 13/59 Jan 30 SA 4f fst :48¹ H 23/44 Jan 13 SA 4f fst :46⁴ Hg 4/61 Jan 6 SA 6f fst 1:15⁴ Hg 20/25 Dec 30 SA 5f fst 1:00⁴ H 30/76

Lovely Explosion

Own: Castlebury Thrghbrds & Derby Co Ltd

PEDROZA M A (123 8 18 10 .07) $35,000

Ch. f. 4
Sire: Time to Explode (Explodent)
Dam: Feels Like Love (Northern Jove)
Br: Drakos Chris (Ky)
Tr: May Alan (5 0 0 0 .00)

L 117

Lifetime Record :	12 M 2 3	$15,005			
1994	2 M 0 0	$525	Turf	0 0 0 0	
1993	10 M 2 3	$14,480	Wet	0 0 0 0	
SA	5 0 0 2	$6,225	Dist	7 0 1 2	$8,980

28Jan94-9SA fst 1 :22³ :46² 1:11² 1:36⁴ ⒻMd 35000 63 5 7 7³¾ 4¹¼ 3³ 5⁶ Valenzuela F H L 117 b 12.00 79-15 UncnnyAnn1181½ DncesWithWolves113hd DH CorlIsle118 4 wide 2nd turn 9
5Jan94-2SA fst 6f :21⁴ :44² :56² 1:08⁴ ⒻMd 35000 54 9 8 9⁸½ 8¹¹ 7¹² 6¹³½ Pedroza M A LB 117 b 15.70 80-09 Kalavrita11910 Smart Patch1192¼ Time For A Storm117½ No rally 12
9Dec93-9Hol fst 6½f :21⁴ :45 1:11 1:17² 3+ ⒻMd 25000 57 6 9 8¹⁰ 6⁸ 4¹½ 3²¼ Valenzuela F H LB 118 b *2.70 80-10 Ashley Place118nk Madam Michelle1182 Lovely Explosion1188 11
Off bit slow, 4 wide turn
25Nov93-6Hol fst 7f :22¹ :46 1:11² 1:24 3+ ⒻMd 25000 55 11 2 9⁶ 4¾ 2½ 2³ Valenzuela F H LB 119 b *2.40 81-09 Lotta Reality1173 Lovely Explosion1193½ Ashley Place1192 4 Wide turn 11
3Nov93-4SA fst 1 :22⁴ :46⁴ 1:12¹ 1:38¹ 3+ ⒻMd 32000 49 5 8 9⁹ 7⁶½ 7¹⁰ 6¹⁴¾ Valenzuela F H LB 117 b 19.10 64-20 Charlierusse1173½ LarkInTheMeadow1202½ RaiseTheRunway1173½ No rally 10
15Aug93-3EP fst 6½f :22³ :45⁴ 1:11³ 1:18² ⒻMd Sp Wt 65 3 5 5⁴½ 3⁴½ 3³½ 2¹½ Loseth C L 120 b 6.75 86-13 AreYouTsingM1201½ LovlyExplosion1202½ VrySplshyLdy1152½ Closed well 5
4Aug93-6EP fst 6½f :21⁴ :45 1:11² 1:18¹ ⒻMd Sp Wt 51 3 5 6¹⁰ 5¹⁰ 5⁴½ 4⁶¾ Velasquez D W5 L 115 b 9.20 82-09 Briar N Brandy1151½ Very SplashyLady1152½ Carmanah1203 Showed little 7
18Jly93-6EP fst 6½f :22³ :46³ 1:13 1:19² ⒻMd Sp Wt 45 2 6 6⁵ 4³½ 4⁴ 4⁶¾ Velasquez D W5 L 115 b 7.15 76-20 WildPassions1202 BriarNBrandy1154 VerySplashyLdy115¾ Steadied 3/16 9
1Jly93-3EP fst 6½f :22³ :47 1:12⁴ 1:19³ ⒻMd Sp Wt 38 1 7 5²½ 6³ 7⁶½ 8⁹½ Loseth C L 120 b 8.75 73-19 Burning Lace120no Carmanah1203 Honor Garthorn1201 Steadied 1/2-7/16 8
15May93-4EP fst 6½f :22² :46 1:12² 1:19¹ ⒻMd Sp Wt 48 1 6 8⁶½ 7⁸½ 7¹⁰ 7¹¹½ Loseth C L 120 b *1.55 72-20 Donut Queen1203½ Very Splashy Lady1152½ Burning Lace1201½ Outrun 9

WORKOUTS: Feb 11 SA 4f fst :48² H 25/57 Dec 24 SA 5f fst 1:02 H 82/114 Nov 17 SA 4f fst :49³ H 23/27

Oreanna

Own: Elm Tree Farm & Jones

GRYDER A T (123 5 9 17 .04) $35,000

Dk. b or br f. 4
Sire: Kennedy Road (Victoria Park)
Dam: Devi (Twice Worthy)
Br: Cardiff Stud Farm (Cal)
Tr: Metz Jeff (1 0 0 0 .00)

L 117

Lifetime Record :	2 M 0 0	$0			
1994	1 M 0 0		Turf	0 0 0 0	
1992	1 M 0 0		Wet	0 0 0 0	
SA	1 0 0 0		Dist	2 0 0 0	

26Jan94-2SA fst 6½f :22¹ :45¹ 1:10² 1:17¹ ⒻⓈMd 35000 36 8 5 6⁵½ 7⁸ 8⁹¾ 8¹⁵ Atkinson P LB 117 fb 41.60 71-14 Kristys Lucky Lady1172¾ Under Full Sail117½ Missy'sHabit1193½ No threat 10
3Dec92-9Hol fst 6½f :21⁴ :44³ 1:10² 1:17 ⒻMd 28000 18 5 5 11¹⁷ 11¹⁷ 11¹⁸ 10¹⁸ Torres H B 117 b 104.90 69-08 Eurythmic1192½ Petite Wink119¾ Audra's Prospect1193 Wide, green 11

WORKOUTS: ●Feb 11 Fpx 5f fst 1:00⁴ H 1/9 ●Feb 6 Fpx 3f gd :36 H 1/11 ●Jan 19 Fpx 4f fst :48 H 1/5 Jan 12 SA 5f fst 1:00⁴ H 16/46 Jan 5 Fpx 6f fst 1:14⁴ H 1/1 Dec 29 SA 5f fst 1:03² H 46/52

Ruffmeup

D7/4

Own: Drakos Christopher

STEVENS G L (191 23 42 30 .12) $40,000

Dk. b or br f. 4
Sire: Native Royalty (Raise a Native)
Dam: Ruff Con (Ruffinal)
Br: Chris Drakos (Ky)
Tr: Smith Michael R (13 0 0 0 .00)

119

Lifetime Record :	0 M 0 0	$0			
1993	0 M 0 0		Turf	0 0 0 0	
1992	0 M 0 0		Wet	0 0 0 0	
SA	0 0 0 0		Dist	0 0 0 0	

WORKOUTS: Feb 11 SA 5f fst :59² H 6/62 Feb 2 SA 5f fst 1:00⁴ Hg 17/42 Jan 26 SA 5f gd 1:01 H 4/18 Jan 21 SA 5f fst 1:01² H 18/48 Jan 15 SA 4f fst :47³ H 4/46 Jan 9 SA 4f fst :48² H 24/55 Dec 27 SA 3f fst :37² H 18/25

Dancing Quoit

Own: Braeburn Farm

DELAHOUSSAYE E (176 28 26 24 .16) $35,000

B. f. 4
Sire: Waquoit (Relaunch)
Dam: Danzitup (Danzig)
Br: Franks John (Fla)
Tr: Speckert Christopher (15 1 2 2 .07)

117

Lifetime Record :	5 M 0 1	$5,025			
1994	2 M 0 0	$525	Turf	2 0 0 1	$4,500
1993	3 M 0 1	$4,500	Wet	0 0 0 0	
SA	2 0 0 0	$525	Dist	0 0 0 0	

17Jan94-4SA fst 1 1/16 :23⁴ :48¹ 1:13¹ 1:44¹ ⒻMd 40000 46 6 5 3¹ 5²½ 5⁸ 5¹⁵¾ Delahoussaye E B 119 b 5.20 58-18 SmartPatch1191¼ RaiseTheRunway1174 PerilousFlyer1173 Bid, gave way 7
5Jan94-4BM fm 1 Ⓣ :23 :47¹ 1:13³ 1:40⁴ Md Sp Wt 54 2 7 7¹³ 7⁵¼ 6³½ 6⁴ Boulanger G 114 *1.30 65-24 DeepInTime119½ GenuineCharity1191¼ HardBargin119½ Far wide stretch 8
25Nov93-4Hol fm 1 1/16 Ⓣ :23³ :47³ 1:12² 1:44¹ 3+ ⒻMd Sp Wt 59 8 8 7⁵½ 3¹½ 3³ 3⁴½ Solis A LB 118 10.20 70-25 Yerna1182 Erin Sweeney1182½ Dancing Quoit1185 No late bid 8
13Oct93-6SA fst 6f :21 :44¹ :56³ 1:09 3+ ⒻMd Sp Wt 55 7 12 12¹⁶ 12¹² 12¹⁰ 8¹⁴½ Lopez A D 117 237.40 77-11 Desert Stormer1171 Bawl For Beulah1172½ Hidden Dark1175 Outrun 12
30Aug93-6Dmr fst 6f :21¹ :44¹ :57² 1:10¹ 3+ ⒻMd Sp Wt 51 6 11 12¹⁷ 12¹⁸ 11¹¹ 9¹⁴ Delahoussaye E B 117 81.50 74-14 Cosmic Mirage1173½ Devil's Beware1171½ Slew All117nk Off slowly, wide 12

WORKOUTS: Feb 12 SA 5f fst 1:00⁴ H 36/86 Feb 5 SA 4f gd :50² H 21/40 Jan 14 SA 3f fst :38⁴ H 23/28 Jan 2 SA 5f fst 1:02² H 35/47 Dec 27 SA 5f fst 1:02¹ H 31/41 Dec 21 SA 4f fst :48 H 12/34

Faceta is owned and trained by the always dangerous Richard Mandella and must be respected. But at 4-5, she has no value. Furthermore, Faceta has already lost at this level and, given all those seconds, may lack a winning spirit.

But who can beat her? Of the first-time starters Mom's Alleged has a higher precocity and class rating than Faceta, but she also has a stamina index of 1 on the top and bottom of her pedigree. Competent as trainer David Hofmans is with debut runners, Mom's Alleged's 27-1 odds further confirm that she is probably being prepped for a route. Ruffmeup has Gary Stevens up and is being bet at 7-2, but her sire has a poor D7 rating and the filly is conditioned by a cold trainer.

My Sensation, with her superior B+4 rating and competent Donald Warren connections, is a much more intriguing possibility. She showed speed in her only start at the Maiden Special Weight level and is now making her Maiden Claiming "debut" for her Stonebraker owner/breeder connections. With her :44 and change fractions against better and Alex Solis back in the irons, My Sensation looks like a definite threat to control the pace at this Maiden Claiming level. In her only race, My Sensation's :22 first quarter mile equaled the more experienced Faceta's early fractions and maidens dropping in class have a great probability of improving their early fractions and final time. Furthermore, second-time starters, as a group, are more likely to improve than maidens like Faceta that have already run four times.

The owner/breeder economics also seem right. If My Sensation wins the race and gets claimed, she will earn 55 percent of the twenty grand purse plus the $40,000 claiming price for a total of some $52,000. On the expense side, Mining stood for $10,000 when she was bred. At two-and-half times his stud fee plus additional expenses of fifteen to twenty grand for four years, the Stonebraker connections have between $40-45,000 invested in the filly. This means that should they lose My Sensation to a claim, they will still make a modest profit if she wins, and thus, she is definitely well meant.

So how to bet? Given My Sensation's positive attributes, I make her 2-1 in this spot which means that a horse with her profile should win at least once out of every three times. At 9-2 odds she is a huge bargain and a definite Win bet. Rather than only putting her second to Faceta in the exacta, in this case, a box seems in order since it looks like strictly a two horse race and I like My Sensation better.

Results: The actual race unfolded with Faceta and Ode To Code contesting for the early lead and My Sensation being reserved in third on the outside. Faceta put away Ode To Code at the top of the stretch, but once again lacked the will to win. She was caught in deep stretch by My Sensation who paid $4.40 to Win with the exacta returning $29.40. Interestingly, Ruffmeup, the other action horse, got up for third which would have completed the trifecta had there been one.

Santa Anita Park

2

1 1/16 MILES. (1:39) **MAIDEN CLAIMING. Purse $21,000. Fillies and mares, 4-year-olds and upward. Weights: 4-year-olds, 119 lbs. Older, 120 lbs. Claiming price $40,000; if for $35,000, 2 lbs.**

1 1/16 MILES
START ▲ ▲ FINISH

Bel's Sue

Own: Aguilar & Meza
SILVA J G (5 0 1 0 .00) $35,000

B. m. 5
Sire: Bel Bolide (Bold Bidder)
Dam: Saratoga Sue (Minnesota Mac)
Br: Goppert & Kruljac Partnership (Ariz)
Tr: Schiewe Paul (1 0 0 0 .00)

C7/4

118

Lifetime Record :	0 M 0 0		$0
1993	0 M 0 0	Turf	0 0 0 0
1992	0 M 0 0	Wet	0 0 0 0
SA	0 0 0 0	Dist	0 0 0 0

WORKOUTS: Jan 21 SLR tr.t 6f fst 1:152 H 2/4 Jan 13 SLR tr.t 5f fst 1:011 H 3/3 Dec 17 SLR tr.t 5f fst 1:014 H 6/10 Dec 10 SLR tr.t 4f fst :501 H 6/8 Dec 2 SLR tr.t 4f fst :502 Hg5/7

Anelda

Own: Hibbert R E
ANTLEY C W (151 22 14 8 .15) $40,000

B. f. 4
Sire: Pancho Villa (Secretariat)
Dam: Roving Girl (Olden Times)
Br: Robert E. Hibbert (Ky)
Tr: Rash Rodney (39 4 5 6 .10)

BC/4

119

Lifetime Record :	4 M 1 2	$16,250			
1993	4 M 1 2	$16,250	Turf	0 0 0 0	
1992	0 M 0 0		Wet	0 0 0 0	
SA	0 0 0 0		Dist	2 0 1 1	$11,450

22Aug93-9Dmr fst 1 1/16 :223 :462 1:111 1:431 3+ ⒻMd Sp Wt 71 7 4 41½ 22 25 38¼ Stevens G L B 117 6.90 77-16 Sensational Eyes1174 Cindazanno1174½ Anelda1176 Weakened 9
5Jly93-4Hol fst 1 1/16 :231 :464 1:113 1:44 3+ ⒻMd Sp Wt 71 2 1 12½ 11½ 2½ 25 Stevens G L B 116 3.10e 76-18 Chickapee1175 Anelda1161½ Portugese Starlet116hd Bobbled start 10
15May93-6Hol fst 6½f :213 :441 1:094 1:164 3+ ⒻMd Sp Wt 39 4 6 42½ 35 58½ 716½ Delahoussaye E B 116 *1.60 69-09 Lydara1151½ Time For A Storm115½ Matter Of Fact1177 Gave way 10
25Apr93-6Hol fst 6½f :221 :452 1:101 1:163 3+ ⒻMd Sp Wt 69 8 5 62¾ 3nk 21½ 34¾ Valenzuela P A B 117 *1.30 81-11 HurryHomeHelen1153½ BbyTkeTheGold1161½ Anld1171¾ Wide backstretch 10

WORKOUTS: Feb 5 SA 5f gd 1:022 H 18/59 ● Jan 29 SA 3f fst :341 H 1/25 Jan 24 SA 1 fst 1:394 H 2/4 Jan 19 SA 5f fst :594 B 12/46 Jan 10 SA 4f fst :48 H 18/53 Jan 4 SA 5f fst :591 H 6/45

Fairly Flashy (GB)

Own: Mamakos James L
ATKINSON P (61 6 3 5 .10) $35,000

Ch. f. 4
Sire: Flash of Steel*GB (Kris-GB)
Dam: Fair Siobahn (Petingo)
Br: Mamakos Jason (GB)
Tr: Mamakos Jason (2 0 0 0 .00)

L 117

Lifetime Record :	2 M 0 1	$5,925			
1994	1 M 0 0	$1,575	Turf	0 0 0 0	
1993	1 M 0 1	$4,350	Wet	0 0 0 0	
SA	1 0 0 0	$1,575	Dist	2 0 0 1	$5,925

17Jan94-4SA fst 1 1/16 :234 :481 1:131 1:441 ⒻMd 35000 59 1 6 67 64 45½ 48¼ Gonzalez S5 LB 112 4.30 66-18 SmrtPtch1191½ RiseTheRunwy1174 PerilousFlyer1173 Improved position 7
17Dec93-2Hol fst 1 1/16 ⊗ :233 :464 1:11 1:441 3+ ⒻMd Sp Wt 63 3 6 613 613 39½ 310½ Atkinson P B 118 27.10 69-18 Wende1183 Perilous Flyer1187½ Fairly Flashy1188 Rail trip 6

WORKOUTS: Feb 6 Hol 4f fst :50 H 13/34 Jan 31 Hol 4f fst :502 H 17/20 Jan 15 Hol 3f fst :353 H 2/14 Nov 19 Hol 5f fst 1:014 H 18/33 Nov 12 Hol 3f fst :363 H 9/17

Dances With Wolves (Ire)

Own: McNall Racing Ltd
GONZALEZ S JR (156 21 14 19 .13) $35,000

B. m. 5
Sire: Woodman (Mr. Prospector)
Dam: Shahrood (Prince Tenderfoot)
Br: Ryan J (Ire)
Tr: Shoemaker Bill (27 5 2 3 .19)

Ct5/3*

L 113⁵

Lifetime Record :	10 M 2 1	$16,791			
1994	2 M 1 0	$4,200	Turf	3 0 1 0	$6,591
1993	5 M 0 1	$6,000	Wet	0 0 0 0	
SA	3 0 1 1	$6,750	Dist	1 0 0 0	$725

28Jan94-9SA fst 1 :223 :462 1:112 1:364 ⒻMd 35000 72 6 8 86½ 73¾ 53½ 21¼ Gonzalez S5 LB 113 4.70 84-15 UncnnyAnn1181¼ DncesWithWolvs113hd DH CorlIsl118 Far wide into lane 9
17Jan94-9SA fst 6½f :22 :451 1:101 1:163 ⒻMd 35000 56 3 11 119 109½ 88¾ 68 Gonzalez S5 LB 113 9.40 81-09 Baby Take The Gold119no Faceta1173 Alaskan Revelation1182¾ Off slow 12
17Dec93-2Hol fst 1 1/16 ⊗ :233 :464 1:11 1:441 3+ ⒻMd Sp Wt 44 5 3 42½ 47 511 521¼ Gonzalez S5 LB 116 3.60 58-18 Wende1183 Perilous Flyer1187½ Fairly Flashy1188 No rally 6
1Dec93-2Hol fst 6f :214 :444 :572 1:104 3+ ⒻMd 32000 54 4 8 810 715 710 46¼ Delahoussaye E B 121 *1.70 79-13 Enchanted Beauty119½ Faceta1195½ Her Melody117½ Wide into lane 8
10Nov93-2SA fst 6½f :22 :45 1:104 1:174 3+ ⒻMd 32000 52 8 12 98 88¾ 64½ 33¼ Delahoussaye E B 120 3.30 79-12 RightSmartGirl1182½ BestBerry111¾ DncesWithWolves120hd In tight 1/16 12
16Aug93-6Dmr fst 6½f :22 :45 1:101 1:163 3+ ⒻMd Sp Wt 59 10 3 74 57 67½ 510¼ Black C A B 122 9.80 78-11 Catch1171¾ Devil's Beware1173 Tuscan Tune1172½ Wide trip 11
4Jly93-10Hol fst 6½f :22 :444 1:10 1:162 3+ ⒻMd Sp Wt 60 11 5 41½ 21 55½ 59 Black C A B 121 14.90 78-11 Dreams At Night116no Secret Sight1163 Wild Express1172 Wide trip 11
24Sep92♦ Evry(Fr) yl *6f ⓉLH 1:111 Alw 27800 810¼ Black C A 124 15.00 Maritheus128nk Dauberval1283 Des Etoiles128½ 14
Tr: Nicolas Clement Prix de Milly *Chased in 5th, weakened 2f out*
17Aug91♦ Deauville(Fr) yl *7f ⓉStr 1:264 Stk 608000 14 Mosse G 122 16.00 Mariage Secret122¾ Cardoun1262 Lady Thynn122hd 16
Challenge d'Or Piaget (Listed) *Soon behind, never a factor*
12Jly91♦ M-Laffitte(Fr) gd *7f ⓉLH 1:284 ⒻMdn (FT)27600 21 Peslier O5 118 8.00 Verveine1231 Dances With Wolves118nk Local Heroine123½ 12
Prix Peniche *Tracked in 4th, rallied over 1f out, up for 2nd*

WORKOUTS: Jan 7 SA 5f fst :593 H 2/32 Dec 31 SA 5f fst 1:011 H 28/66 Dec 24 SA 3f fst :353 H 4/39 Dec 14 Hol 4f fst :491 H 8/31 Nov 25 Hol 4f fst :492 H 3/17 Nov 19 Hol 3f fst :364 H 3/15

Fager's Prospect

Own: Iron Country Farm Inc
DESORMEAUX K J (116 25 19 16 .22) $35,000

Ro. f. 4
Sire: Chief's Crown (Danzig)
Dam: Fager's Glory (Mr. Prospector)
Br: G M Breeding Farms Inc (Ky)
Tr: Lewis Craig A (32 4 4 1 .13)

C5/3

L 117

Lifetime Record :	10 M 0 1	$10,450			
1993	7 M 0 1	$10,450	Turf	1 0 0 0	
1992	3 M 0 0		Wet	0 0 0 0	
SA	4 0 0 0	$2,625	Dist	3 0 0 1	$7,825

31Dec93-9SA fst 1 :223 :462 1:112 1:37 3+ ⒻMd 35000 63 6 3 42 62¾ 62¾ 55 Desormeaux K J LB 116 b 4.80 80-07 Noddy's Halo121½ Raise The Runway1162 Uncanny Ann119nk Inside bid 10
4Sep93-2Dmr fm 1 1/16 Ⓣ :234 :49 1:132 1:442 3+ ⒻMd Sp Wt 54 4 4 32 42 67 69½ Solis A LB 118 b 9.80 74-12 Portugese Starlet118½ Chasse1184 Erin Sweeney118½ Gave way 8
22Aug93-9Dmr fst 1 1/16 :223 :462 1:111 1:431 3+ ⒻMd Sp Wt 61 8 6 63 63¾ 49½ 414¼ Solis A LB 117 b 7.30 71-16 Sensational Eyes1174 Cindazanno1174½ Anelda1176 Wide trip 9
5Aug93-6Dmr fst 1 1/16 :224 :462 1:113 1:443 3+ ⒻMd Sp Wt 71 12 11 1210 96¾ 74½ 31½ Solis A LB 115 b 5.30e 78-21 Indio Rose1161½ Sunday's Sis115no Fager's Prospect115hd Wide trip 12
5Jly93-4Hol fst 1 1/16 :231 :464 1:113 1:44 3+ ⒻMd Sp Wt 68 4 4 76½ 74¾ 55 57 Solis A LB 116 8.40 74-18 Chickapee1175 Anelda1161½ Portugese Starlet116hd No mishap 10
14Feb93-4SA fst 6½f :22 :45 1:101 1:164 ⒻMd Sp Wt 57 4 2 74½ 77¼ 68 610¼ Torres H LB 117 b 25.60 77-12 Kylie'sPride1171½ SweetSavanna1174 NughtyNineties1171½ Bumped start 7
31Jan93-4SA fst 5½f :214 :451 :581 1:044 ⒻMd Sp Wt 52 6 3 44 44½ 56 47¼ Torres H LB 117 b 56.60 79-12 Freezelin1172 TimeForAStorm117nk LongingToDnce1175 4 Wide stretch 10
26Dec92-4SA fst 6f :212 :443 :57 1:093 ⒻMd Sp Wt 45 1 5 64 96¼ 98½ 915 Torres H LB 117 b 91.10 73-09 Likeable Style1171 Aleyna's Love117no Malojen1176 Brushed start 12
13Jun92-4Hol fst 5f :214 :45 :573 Md Sp Wt 50 2 7 99 99¾ 89¼ 78½ Lopez A D LB 114 36.50 85-05 Super Season117no River Special1172½ Kaliman117 Outrun 10
7May92-4Hol fst 4½f :214 :451 :511 ⒻMd Sp Wt 33 5 7 108½ 99½ 914¼ Flores D R B 117 29.90e 88-09 Devil's Nell117½ Blue Moonlight1173½ Seattle Drama117 Outrun 10

WORKOUTS: Feb 6 SA 5f fst 1:034 H 98/108 Feb 1 SA 3f fst :381 H 13/13 Jan 8 SA 4f fst :494 H 30/39 Dec 24 SA 6f fst 1:163 H 64/68 Dec 17 SA 6f fst 1:131 H 19/45 Dec 10 SA 6f fst 1:154 H 25/32

continued

Riviere Miss

Own: Lurenz Steven

Ro. f. 4
Sire: Prince Bobby B (King Pellinore)
Dam: Vigors Miss (Vigors)
Br: Beckett James L (Cal)
Tr: Perez Mag (1 0 0 0 .00)

VALENZUELA F H (180 23 8 18 .13) $35,000   L 117

handwritten: D 7|4

Lifetime Record:	17 M 3 3	$26,325			
1994	2 M 0 0	$3,000	Turf	0 0 0 0	
1993	15 M 3 3	$23,325	Wet	1 0 0 1	$2,850
SA	9 0 1 2	$14,100	Dist	2 0 0 0	$1,600

26Jan94–2SA fst 6½f	:221 :451 1:102 1:171	⑤Md 35000	55 6 8 99 89½ 66¾ 46¾	Valenzuela F H	LB 117	7.80	79–14	Kristys Lucky Lady117¾ Under Full Sail117½ Missy's Habit119½	Mild bid 10
12Jan94–2SA fst 6f	:212 :441 :563 1:101	⑤Md 35000	61 4 7 78½ 511 58 45	Pedroza M A	LB 117	8.60	81–13	Date With Royalty117¾ Eyes Of Ashley117² Tameo117¾	Mild late bid 9
30Dec93–4SA fst 6½f	:213 :444 1:111 1:174 3+	⑤Md 35000	46 1 7 611 59½ 32 32½	Pedroza M A	LB 115	5.50	80–15	Old Teddy Bear117² Tameo115½ Riviere Miss115½	Outfinished 8
5Nov93–4SA fst 6f	:213 :452 :581 1:112 3+	⑤Md 32000	59 10 9 99 73½ 64½ 53½	Pedroza M A	LB 118	21.30	77–17	Kats Aly118² Under Full Sail118no Walking Liberty116½	Bumped start 12
22Oct93–4SA fst 6f	:214 :45 1:111 1:18 3+	⑤Md 32000	54 5 9 78½ 67 65½ 63½	Black C A	LB 117	*1.70	78–10	Shana Two112½ Point Position117hd Walking Liberty117no	Mild late bid 10
10Oct93–4Fpx fst 1¹⁄₁₆	:221 :45 1:113 1:454 3+	⑫Md Sp Wt	53 8 4 411 57½ 510 511½	Black C A	LB 118	4.80	67–11	Starstruck Miss114½ Powder Storm118½ Tooty Brush115⁸	No rally 8
15Sep93–4Dmr fst 6½f	:223 :461 1:114 1:182 3+	⑤Md 32000	58 8 5 63½ 42½ 41¾ 2hd	Stevens G L	LB 118	*2.10	79–13	Sew Attractive122hd Riviere Miss118² Other Wise122hd	Wide trip 10
1Sep93–4Dmr fst 6½f	:221 :454 1:12 1:182 3+	⑤Md 32000	60 9 6 76½ 62½ 41½ 21½	Pedroza M A	LB 118	4.10	78–16	Title In Doubt118¹½ Riviere Miss118¹½ Hillwalker118¹½	Good effort 9
12Aug93–4Dmr fst 6f	:223 :46 :582 1:111 3+	⑫Md 32000	65 4 4 54½ 45 33½ 31½	Black C A	LB 117	8.10	81–14	Ten Carat Diamond117¹ Treva117½ Riviere Miss117⁴	4 Wide stretch 9
24Jly93–6Hol fst 7f	:222 :451 1:102 1:232 3+	⑤Md 32000	49 6 9 93½ 84½ 67¾ 69	Black C A	LB 116	18.10	78–08	Always Glitter116½ Bold Dee Dee116¾ Talisman's Magic119½	11
Bumped start, wide trip									

WORKOUTS: Dec 22 SA 5f fst :59² H 6/64 Dec 14 SA 6f fst 1:13¹ H 7/26 Dec 6 SA 5f fst 1:01¹ H 15/29 Nov 27 SA 4f fst :47 H 2/37

Coral Isle

Own: Crockwell Craig D

B. m. 5
Sire: Mr Leader (Hail to Reason)
Dam: Grey Coral (Tudor Grey)
Br: Crockwell Craig D (Ky)
Tr: Hendricks Dan L (24 2 1 3 .08)

LOPEZ A D (45 3 1 3 .07) $35,000   L 118

handwritten: C 4|3

Lifetime Record:	7 M 0 2	$8,387			
1994	2 M 0 1	$2,862	Turf	0 0 0 0	
1993	3 M 0 0	$950	Wet	0 0 0 0	
SA	4 0 0 1	$3,312	Dist	1 0 0 0	$500

28Jan94–9SA fst 1	:223 :462 1:112 1:364	⑫Md 35000	71 1 2 2½ 11 11 31¼¾	Lopez A D	LB 118	9.40	84–15	Uncanny Ann118¼ Dances With Wolves113hd ᴰᴴCoral Isle118	Willingly 9
5Jan94–2SA fst 6f	:214 :451 :562 1:084	⑫Md 35000	55 4 7 1010 1012 815 513	Desormeaux K J	LB 118	7.60	80–09	Kalavrita119¹⁰ Smart Patch119²½ Time For A Storm117½	By tired ones 9
23Apr93–9Hol fst 1¹⁄₁₆	:231 :463 1:114 1:46 3+	⑫Md 35000	54 5 8 910 78½ 68½ 55½	Solis A	LB 121 f	15.40	65–20	Far Frau121½ Princess Waquoit115½ Northern Wool115²½	5 Wide stretch 10
5Mar93–2SA fst 6f	:213 :451 :581 1:111	⑫Md 35000	58 4 7 1114 810 77½ 67¾	Solis A	LB 118	10.90	73–15	SixQueens120⁵ WreakHavoc113½ PlyfulPosition120½	Improved position 12
3Feb93–6SA fst 6f	:214 :451 :574 1:11	⑫Md 40000	52 8 8 65 68 59 510½	Solis A	LB 119	2.50	72–14	Blurry119¹½ Precise Music117¾½ Carmine117²½	Rough start, wide 12
20Dec91–6Hol fst 6½f	:214 :451 1:102 1:171	⑫Md 35000	64 10 1 43 43¾ 32½ 34	Solis A	LB 115	6.10	82–13	Carefree Wonder118³½ Chile Missouri118½ Coral Isle115	Always close 10
6Dec91–6Hol fst 6½f	:22 :451 1:112 1:183	⑫Md 50000	58 1 4 33 33½ 25 42½	Solis A	LB 118	5.00	76–14	Clever Treasure118½ Flying Wish118nk Dublin R Delite118	No mishap 10

WORKOUTS: Jan 22 SA 4f fst :49 H 39/50 Jan 17 SA 4f fst :48 H 15/34 Dec 29 SA 5f fst 1:00⁴ H 23/52 Dec 23 SA 6f fst 1:11³ H 6/39 Dec 16 Hol 6f fst 1:16¹ Hg 12/18 Dec 11 Hol 5f fst 1:01² H 32/51

Alaskan Revelation

Own: Auerbach Ernest

Ch. m. 5
Sire: Yukon (Northern Dancer)
Dam: True Confession (Elocutionist)
Br: Jacqueline Getty Phillips, David Ri (Ky)
Tr: Robbins Jay M (11 0 0 3 .00)

PINCAY L JR (170 24 24 22 .14) $35,000   118

handwritten: C 6|3

Lifetime Record:	3 M 1 1	$7,650			
1994	1 M 0 1	$3,000	Turf	0 0 0 0	
1992	2 M 1 0	$4,650	Wet	1 0 1 0	$4,200
SA	3 0 1 1	$7,650	Dist	1 0 1 0	$4,200

17Jan94–9SA fst 6½f	:22 :451 1:101 1:163	⑫Md 35000	68 7 8 63 65½ 33 33	Stevens G L	B 118	5.60	86–09	BabyTkeTheGold119no Fcet117³ AlsknReveltion118²½	Bumped hard start 12
6Feb92–4SA sly 1¹⁄₁₆	:23 :462 1:102 1:431	⑫Md 35000	69 1 7 66 54½ 32½ 1nk	alvarado F T	B 115 b	7.40	87–09	ᴰAlaskan Revelation115nk Reservation Won117nk Perky Wonder117	9
Disqualified and placed second; Came out bumped 1/16									
16Jan92–2SA fst 6f	:212 :444 :58 1:122	⑫Md 35000	48 1 12 1011 715 811 55½	Castanon A L	B 117 b	93.50	69–19	Miss Juliet110¹½ Hail Lady King117¾ Joey's Love117	Broke slowly 12

WORKOUTS: Feb 5 SA 6f gd 1:16⁴ H 22/30 Jan 10 SA 6f fst 1:13² H 3/22 Jan 4 SA 5f fst :59⁴ H 10/45 Dec 29 SA 5f fst 1:01² H 32/52 ●Dec 21 SA 4f fst :46² H 1/34 Dec 11 SLR tr.t 5f fst 1:01⁴ B 5/9

Raise The Runway

Own: Jackson Bruce L

B. f. 4
Sire: Raise a Man (Raise a Native)
Dam: Runaway Lady (Caucasus)
Br: Vanier Mr & Mrs H & Varney Mr & Mrs L (Ky)
Tr: Jackson Bruce L (21 0 6 4 .00)

PEDROZA M A (110 8 16 9 .07) $35,000   L 117

handwritten: C 6|4

Lifetime Record:	15 M 4 5	$30,862			
1994	2 M 1 1	$6,562	Turf	0 0 0 0	
1993	7 M 2 2	$13,800	Wet	1 0 0 0	
SA	8 0 2 3	$17,512	Dist	5 0 2 1	$11,650

28Jan94–9SA fst 1	:223 :462 1:112 1:364	⑫Md 35000	71 3 5 63½ 52½ 43½ 31¼¾	Pedroza M A	LB 117	*1.50	84–15	UncnnyAnn118¼ DncesWithWolves113hd ᴰᴴCorlIsl118	Lacked room 3/8 9
17Jan94–4SA fst 1¹⁄₁₆	:234 :481 1:131 1:441	⑫Md 35000	71 4 4 53 2hd 2hd 21¾	Pedroza M A	LB 117	3.90	73–18	Smart Patch119¹½ Raise The Runway117⁴ Perilous Flyer117³	Bid, hung 7
31Dec93–9SA fst 1	:223 :462 1:112 1:37 3+	⑫Md 35000	71 7 7 73½ 51¾ 2hd 2½	Pedroza M A	LB 116	17.20	84–07	Noddy'sHalo121½ RaiseTheRunwy116² UncnnyAnn119nk	4 wide 2nd turn 9
4Dec93–6Hol fst 1¹⁄₁₆	:23 :462 1:112 1:451 3+	⑫Md 28000	54 6 6 67½ 45¼ 46½ 36½	Pedroza M A	LB 116	3.60	68–15	Yesternight118⁶½ RaiseTheRunway116³½ SweetLadyCzr118½	Second best 8
3Nov93–4SA fst 1¹⁄₁₆	:224 :464 1:121 1:381 3+	⑫Md 28000	65 1 4 31½ 2hd 21½ 35½	Perret C	LB 117	12.00	73–20	Chrlierusse117³½ LrkInTheMedow118½ RiseThRunwy117³½	Bid, weakened 10
21Oct93–6SA fst 1¹⁄₁₆	:23 :47 1:121 1:442 3+	⑫Md 28000	73 1 7 66½ 63½ 43 33½	Pedroza M A	LB 117	35.90	79–20	Hillwalker117²½ Uncanny Ann119½ Raise The Runway115³	Outfinished 7
70ct93–5SA fst 1¹⁄₁₆	:23 :47 1:113 1:441 3+	⑫Md 28000	44 3 4 52 51½ 812 819½	Atkinson P	LB 117	24.30	64–16	Ms. Jiles117¹½ Uncanny Ann119½ Charlierusse117²½	Steadied hard 5/16 11
18Sep93–6Fpx fst 6½f	:214 :461 1:114 1:184 3+	⑫Md 32000	44 1 3 43 4½ 812 812½	Atkinson P	LB 117	4.10	77–09	Mezquita116³½ Fifty Six Fiddles115nk Bold Dee Dee116no	Saved ground 10
7Jan93–4SA sly 1	:222 :461 1:112 1:383	⑫Md 32000	40 3 4 46 59½ 79 710½	Walls M K	LB 115	12.10	67–18	Becky Who117nk Cheta117½ Fundamental Rights117¹	4 Wide stretch 10
12Nov92–6Hol fst 7½f	:222 :454 1:112 1:302	⑫Md 32000	43 1 10 96¾ 75½ 77 714	Sorenson D	LB 118	10.00	69–12	Hail A Promise116⁵½ Creme Dela Talc116 Grand It Is118	Saved ground 10

WORKOUTS: Jan 9 SA 3f fst :36⁴ H 7/13 Dec 26 SA 4f fst :48³ H 22/46 Dec 19 SA 5f fst 1:01¹ H 10/16 Dec 13 SA 3f fst :35³ H 2/15 Nov 28 SA 6f fst 1:12¹ H 2/20 Nov 21 SA 5f fst 1:00³ H 9/39

© 1994 DRF, Inc.

The second race at Santa Anita on February 11th is an example of a low-priced overlay, dropping from the Maiden Special Weight level, that strongly figured. Looking over the field, the first thing to notice is that all but two of the horses have already lost for a $40,000 tag. Of these two, Bel's Sue is a 5-year-old, debut runner trying to win a route in her first start and at 128-1 is an automatic throwout. This leaves Anelda as the only horse dropping from the Maiden Special Weight level. Furthermore, she showed the requisite speed against better and is conditioned by Rodney Rash who excels at bringing horses back off a layoff.

While her Hibbert owner/breeder connections were initially high enough on Anelda to protect her, her class 6 rating suggests she is better placed at the Maiden Claiming $40,000 level. If she wins and gets claimed in this spot she will earn over $50,000 which will more than cover Pancho Villa's $10,000 stud fee and the other costs of getting her to the races.

Still, Anelda is coming back off a six month layoff and what if the filly is being dropped because she is damaged goods? This assumption, however, is undermined by her steady series of works, including a bullet blowout of :34.1 and a mile work for stamina. Horses with serious physical problems are seldom capable of withstanding this kind of rigorous training. The icing on the cake, however, is Anelda's 7-5 post-time odds.

So how to play? Given her speed against better, breeding, workouts, connections, and the fact that the other contenders have lost at today's level, Anelda should be even-money in this spot. At 7-5 she is a low-priced overlay and can either be bet on the Win or part-wheeled on top in the exacta to everyone except Bel's Sue, Fairy Flash and Riviere Miss.

Results: Anelda broke alertly from her inside post and wired the field. She paid $4.80 for the Win while Alaskan Revelation sat off the pace and closed for the Place with the $2 exacta returning $17. In this instance a $5 part wheel of Anelda on top would have cost $25 and returned $43, while the same amount would have returned $60 to Win. Difficult as it sometimes is to bet low-priced overlays, it is important to reiterate that they represent solid value and have the psychologically important function of limiting losing streaks and bolstering self-confidence.

Santa Anita Park

4

6½ Furlongs. (1:14) **MAIDEN CLAIMING. Purse $18,000. 3-year-olds, bred in California. Weight, 118 lbs. Claiming price $32,000; if for $28,000, allowed 2 lbs.**

START / 6½ FURLONGS / FINISH

Kleven's Best — Own: Alexander & Carrillo & Glassman
SOLIS A (193 30 31 31 .16) — $32,000
Ch. g. 3 (Jan) — Bt 7/4
Sire: Kleven (Alydar)
Dam: Ima Early Blossom (Imacornishprince)
Br: Martinez Silverio (Cal)
Tr: Spawr Bill (64 13 6 15 .20)
L 118

Lifetime Record :	2 M 0 0		$0
1994	2 M 0 0	Turf	0 0 0 0
1993	0 M 0 0	Wet	0 0 0 0
SA	2 0 0 0	Dist	0 0 0 0

30Jan94-6SA fst 7f :22^1 :44^3 1:09^4 1:23 [S]Md Sp Wt −0 6 2 1^{hd} 3½ 8^{15} 832½ Nakatani C S LB 118 3.70 54-07 College Town118^{2¾} Alybally118^6 Freeze Police118^1 8
Off bit awkwardly, not urged late
2Jan94-5SA fst 6f :21^2 :44^2 :56^3 1:09^2 [S]Md Sp Wt 68 4 2 1^{hd} 1^{hd} 31½ 8^5 Gonzalez S5 B 113 14.40 85-07 Superfluously118^{1½} Phone Prince118^{hd} Muirfield Village118^{hd} Inside duel 11
WORKOUTS: Feb 12 SA 4f fst :49^3 H *33/49* Jan 27 SA 4f fst :47^2 H *4/28* Jan 22 SA 6f fst 1:16^3 H *38/39* Jan 17 SA 4f fst :52 H *33/34* Dec 28 SA 7f fst 1:26 H *4/14* ●Dec 22 SA 4f fst :46^3 Hg *1/49*

Judy's Ultimo — Own: Kling Barbara
ATKINSON P (72 8 5 5 .11) — $32,000
B. g. 3 (Mar) — D 8/5
Sire: In Tissar (Roberto)
Dam: Judy's Pet (Petrone)
Br: Haug Fritz (Cal)
Tr: Sticka Ron (1 1 0 0 1.00)
118

Lifetime Record :	0 M 0 0		$0
1994	0 M 0 0	Turf	0 0 0 0
1993	0 M 0 0	Wet	0 0 0 0
SA	0 0 0 0	Dist	0 0 0 0

WORKOUTS: Feb 9 SLR tr.t 5f fst 1:00^3 Hg *1/1* Feb 3 SLR tr.t 4f fst :47^3 H *1/3* Jan 29 SLR tr.t 6f fst 1:14^1 H *2/8* Jan 22 SLR tr.t 6f fst 1:15 H *3/7* Jan 17 SLR tr.t 5f fst 1:01 H *1/3* Dec 24 SLR tr.t 5f fst 1:02 H *11/20* Dec 17 SLR tr.t 5f fst 1:02^3 Hg *8/10* Nov 27 Sac 3f fst :38^2 H *2/4* Nov 24 BM 5f fst 1:04^4 Hg *44/46* Nov 20 Sac 3f fst :39^3 Hg *3/3* Nov 6 Sac 5f fst 1:05^1 H *8/10* Oct 30 Sac 4f fst :53^4 H *2/3*

Air Apparently — Own: Anderson E J
GRYDER A T (123 5 9 17 .04) — $32,000
Ch. c. 3 (Mar) — Bt 6/4
Sire: Al Mamoon (Believe It)
Dam: Wild Drive (Iverness Drive)
Br: Edmund Gann (Cal)
Tr: Williams George L (—)
L 118

Lifetime Record :	2 M 0 0		$1,275
1993	2 M 0 0 $1,275	Turf	0 0 0 0
1992	0 M 0 0	Wet	0 0 0 0
SA	1 0 0 0 $1,275	Dist	0 0 0 0

10Dec93-4Hol fst 6f :22 :45^2 :57^3 1:10^1 Md 32000 5 8 5 73½ 9^{11} 11^{20} 12^{29} Delahoussaye E LB 119 f *2.80 60-12 A Treek For Roses119^6 Star Style119^{1½} Hunters Way119^1 Wide into lane 12
11Nov93-2SA gd 6f :22^1 :46^1 :59^1 1:12^4 Md 32000 36 8 6 6^4 67½ 55½ 48½ Delahoussaye E L 118 f 4.40 64-21 Toknight118^{6½} Gray Jove118^{nk} Don Icecapade118^{1¾} Broke in, bumped 8
WORKOUTS: Feb 13 Hol 5f fst 1:01 Hg *3/14* Feb 6 Hol 6f fst 1:14^2 H *7/20* Jan 28 Hol 6f fst 1:13^2 H *2/11* Jan 21 Hol 5f fst 1:02 H *12/22* Jan 14 Hol 4f fst :48 H *3/18* Dec 6 Hol 5f fst 1:02^4 H *17/18*

Pennyinthewater — Own: Rickerd Paige
ATHERTON J E (42 2 1 4 .05) — $32,000
Dk. b or br g. 3 (Mar) — D 7/4
Sire: Crystal Water (Windy Sands)
Dam: L'Penny (L'Natural)
Br: Warwick Family Trust (Cal)
Tr: Rickerd Paige (—)
113^5

Lifetime Record :	0 M 0 0		$0
1994	0 M 0 0	Turf	0 0 0 0
1993	0 M 0 0	Wet	0 0 0 0
SA	0 0 0 0	Dist	0 0 0 0

WORKOUTS: ●Feb 14 SA 3f fst :34 H *1/14* Feb 9 SA 4f gd :48^3 Hg *2/24* Feb 3 SA 6f fst 1:16^4 H *26/35* Jan 27 SA tr.t 6f gd 1:19^3 H *3/3* Jan 20 SA 6f fst 1:15^1 H *25/31* Jan 13 SA 5f fst 1:01^2 H *23/46* Jan 7 SA 6f fst 1:14^3 H *17/30* Dec 31 SA 5f fst 1:02^1 Hg *51/66* Dec 24 SA 5f fst 1:03^2 H *111/114* Dec 18 SA 4f fst :49^2 H *33/49* Dec 11 SA 4f fst :49^3 H *34/44* Dec 4 SA 3f fst :35^3 Hg *6/38*

Fleet Landing — Own: Busby & Elder & Epperson
STEVENS G L (191 23 42 30 .12) — $32,000
Ch. c. 3 (Mar) — C 9/5
Sire: Stanstead (Gummo)
Dam: Natural Wave (L'Natural)
Br: Warwick Family Trust (Cal)
Tr: Mitchell Mike (51 8 10 8 .16)
118

Lifetime Record :	0 M 0 0		$0
1994	0 M 0 0	Turf	0 0 0 0
1993	0 M 0 0	Wet	0 0 0 0
SA	0 0 0 0	Dist	0 0 0 0

WORKOUTS: Feb 10 Hol 5f fst 1:01^1 H *13/33* Feb 3 Hol 5f fst 1:01^4 H *15/35* Jan 27 Hol 5f fst 1:00^3 H *11/24* Jan 12 Hol 3f fst :37^2 H *4/11* Jan 6 Hol 3f fst :38 H *9/9*

Pingon — Own: Moreno Lisa M & Robert B
CEDENO E A (4 0 1 0 .00) — $28,000
Dk. b or br g. 3 (Feb) — F 5/3
Sire: Vigors (Grey Dawn II)
Dam: Consuming Passion (Jackie Fires)
Br: Mr. & Mrs. M. Malmuth (Cal)
Tr: Feld Jude T (24 0 0 3 .00)
116

Lifetime Record :	0 M 0 0		$0
1994	0 M 0 0	Turf	0 0 0 0
1993	0 M 0 0	Wet	0 0 0 0
SA	0 0 0 0	Dist	0 0 0 0

WORKOUTS: Feb 10 SA 5f fst 1:02 Hg *36/70* Feb 2 SA 4f fst :49^2 H *32/43* Jan 28 SA 7f fst 1:28 H *7/8* Jan 22 SA 7f fst 1:29^2 H *11/12* Jan 16 SA 6f fst 1:14^3 H *18/29* Jan 10 SA 6f fst 1:16^2 H *17/22* Jan 3 SA 5f fst 1:00^2 H *18/60* Dec 28 SA 5f fst 1:02 H *45/55* Dec 22 SA 4f fst :47^4 H *15/49* Dec 16 SA 4f fst :49^2 H *34/46* Dec 10 SA 5f fst 1:03^3 H *62/66* Nov 28 SA 4f fst :50^3 H *38/46*

Prince Quick — Own: Bettwy & Myers & Petruccione
DELAHOUSSAYE E (176 28 26 24 .16) — $32,000
Ch. c. 3 (Mar) — M 8/4
Sire: Fast Forward (Pleasant Colony)
Dam: Silver Princess (Prize Silver)
Br: Petruccione Joyce & Ronald (Cal)
Tr: Bacorn Herbert L (12 0 0 1 .00)
L 118

Lifetime Record :	3 M 0 0		$0
1994	2 M 0 0	Turf	0 0 0 0
1993	1 M 0 0	Wet	0 0 0 0
SA	2 0 0 0	Dist	0 0 0 0

30Jan94-6SA fst 7f :22^1 :44^3 1:09^4 1:23 [S]Md Sp Wt 29 7 4 2^{hd} 5^2 69½ 719½ Atherton J E5 L 113 b 18.60 67-07 College Town118^{2¾} Alybally118^6 Freeze Police118^1 8
4 wide, broke stride near 5/16
2Jan94-5SA fst 6f :21^2 :44^2 :56^3 1:09^2 [S]Md Sp Wt 67 11 6 73¼ 97¼ 8^6 95½ Atherton J E5 113 80.50 85-07 Suprfluously118^{1¼} PhonPrinc118^{hd} MuirfildVillg118^{hd} Wide, forced out 1/4 11
16Dec93-2Hol fst 6f :22 :45^3 :57^4 1:10^1 [S]Md Sp Wt 62 4 7 7^6 87¾ 7^8 812½ Black C A 118 58.40 77-15 Halloween Treat118^4 Camera Ready118^{hd} Superfluously118^3 No rally 8
WORKOUTS: Feb 11 SA 4f fst :50 H *47/57* Jan 22 SA 4f fst :48^1 H *29/50* Jan 12 SA 5f fst 1:00^1 H *8/46* Dec 26 SA 4f fst :48^3 H *22/46* Dec 9 Hol 6f fst 1:14^3 Hg *9/13* Dec 1 Hol 6f fst 1:14^4 H *21/25*

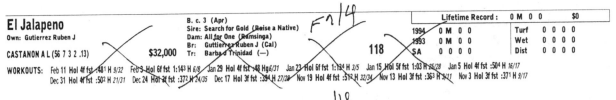

El Jalapeno — F7/14
Own: Gutierrez Ruben J
CASTANON A L (56 7 3 2 .13) — $32,000
B. c. 3 (Apr)
Sire: Search for Gold (Raise a Native)
Dam: All for One (Ramsinga)
Br: Guttierrez Ruben J (Cal)
Tr: Barba J Trinidad (—)
118

Lifetime Record :	0 M 0 0		$0
1994	0 M 0 0	Turf	0 0 0 0
1993	0 M 0 0	Wet	0 0 0 0
SA	0 0 0 0	Dist	0 0 0 0

WORKOUTS: Feb 11 Hol 4f fst :481 H 9/32 Feb 3 Hol 6f fst 1:143 H 6/8 Jan 29 Hol 4f fst :48 Hg6/31 Jan 23 Hol 6f fst 1:134 H 3/5 Jan 15 Hol 5f fst 1:03 H 25/28 Jan 5 Hol 4f fst :504 H 16/17
Dec 31 Hol 4f fst :502 H 21/31 Dec 24 Hol 3f fst :372 H 24/35 Dec 17 Hol 3f fst :394 H 27/28 Nov 19 Hol 4f fst :512 H 32/34 Nov 13 Hol 3f fst :363 H 3/11 Nov 3 Hol 3f fst :371 H 9/17

Drianas Gold — D8/14
Own: Elson Paul A
LOPEZ A D (52 3 1 3 .06) — $32,000
B. c. 3 (May)
Sire: Dance in Time (Northern Dancer)
Dam: Lady Driana (Imacornishprince)
Br: Paul A. Elson (Cal)
Tr: Truman Eddie (5 2 0 0 .40)
118

Lifetime Record :	0 M 0 0		$0
1994	0 M 0 0	Turf	0 0 0 0
1993	0 M 0 0	Wet	0 0 0 0
SA	0 0 0 0	Dist	0 0 0 0

WORKOUTS: Feb 14 SA 3f fst :382 H 14/14 – Feb 9 SA 6f gd 1:184 H 7/8 Feb 3 SA 6f fst 1:164 H 26/35 Jan 27 SA tr.t 6f gd 1:191 H 2/3 Jan 21 SA 5f fst 1:013 H 21/48 Jan 15 SA 4f fst :512 H 45/46
Jan 10 SA 5f fst 1:042 H 41/41

Flying Mate — B6/14
Own: Valpredo John
GONZALEZ S JR (180 23 15 22 .13) — $28,000
Ch. g. 3 (Apr)
Sire: Flying Victor (Flying Paster)
Dam: Air Watch (Dimaggio)
Br: John Valpredo (Cal)
Tr: Caganich Barbara (7 1 0 1 .14)
1115

Lifetime Record :	0 M 0 0		$0
1994	0 M 0 0	Turf	0 0 0 0
1993	0 M 0 0	Wet	0 0 0 0
SA	0 0 0 0	Dist	0 0 0 0

WORKOUTS: Feb 9 Hol 5f fst 1:003 H 3/7 Jan 31 Hol 6f fst 1:134 H 4/11 Jan 24 Hol 6f fst 1:134 H 5/8 Jan 18 Hol 5f fst 1:03 H 16/20 ●Jan 11 Hol 5f fst :594 H 1/9 Jan 5 Hol 4f fst :492 Hg9/17
Dec 28 Hol 5f fst 1:011 H 9/31 Dec 21 Hol 5f fst 1:011 H 11/17 Dec 14 Hol 5f fst 1:032 H 39/53 Dec 7 Hol 4f fst :494 H 12/20 Dec 2 Hol 4f fst :493 H 13/18 Nov 26 Hol 4f fst :491 H 21/35

Christopher Pete — D7/14
Own: Oakcrest Stable
DESORMEAUX K J (147 33 24 18 .22) — $32,000
Ch. g. 3 (Mar)
Sire: Torsion (Never Bend)
Dam: Miss Tres (Secretariat)
Br: Oakcrest Stable (Cal)
Tr: Sadler John W (38 7 5 5 .18)
118

Lifetime Record :	0 M 0 0		$0
1994	0 M 0 0	Turf	0 0 0 0
1993	0 M 0 0	Wet	0 0 0 0
SA	0 0 0 0	Dist	0 0 0 0

WORKOUTS: Feb 12 SA 5f fst 1:004 H 36/86 Feb 6 SA 6f fst 1:143 H 20/49 Jan 29 SA 6f fst 1:141 H 24/31 Jan 23 SA 6f fst 1:154 H 15/25 Jan 17 SA 5f fst 1:022 H 26/31 Jan 11 SA 5f fst 1:00 H 12/36
Jan 4 SLR tr.t 6f fst 1:15 H 3/4 ●Dec 28 SLR tr.t 5f fst 1:00 H 1/4 Dec 7 SLR tr.t 4f fst :473 H 2/7 Nov 24 SLR tr.t 3f fst :38 H 12/18 Nov 19 SLR tr.t 3f fst :363 H 4/5 Aug 17 SLR tr.t 3f fst :373 H 2/7

Fujiyama — M9/14
Own: Oda Harry
PEDROZA M A (123 8 18 10 .07) — $28,000
B. g. 3 (Feb)
Sire: Reconnaissance (Mr. Prospector)
Dam: Princess Baldski (Baldski)
Br: Harry Oda (Cal)
Tr: Palma Hector O (13 0 3 2 .00)
L 116

Lifetime Record :	3 M 0 0		$790		
1994	1 M 0 0	$450	Turf	0 0 0 0	
1993	2 M 0 0	$340	Wet	0 0 0 0	
SA	2 0 0 0	$450	Dist	0 0 0 0	

7Jan94-4SA	fst	6f	:213 :443 :571 1:10	[S]Md 28000	54	9 4 2½ 4¹¼ 6⁴½ 5⁸½	Pedroza M A	LB 116	37.70	78–08	Lords Irish Angel118¾ Capo1182¼ Victory Sun1183½	Weakened 12
15Oct93-4SA	fst	6f	:214 :444 :574 1:104	[S]Md 32000	–0	1 12 84 911 1218 1231	Castanon A L	118	115.90	52–12	ToolittleToolate118hd MountainPssge1185 BobsBikini1183	Off step slow 12
23Sep93-9Fpx	fst	6f	:214 :454 :582 1:111	[S]Md 32000	47	1 4 56 55 58 610½	Castanon J L	B 118	97.50	80–11	Alottanaught1185 Bobs Bikini1182 Dash Of Vanilla118nk	Wide into lane 10

WORKOUTS: Feb 12 SA 5f fst :594 H 8/86 Feb 6 SA 5f fst 1:004 H 13/108 Jan 30 SA 5f fst 1:01 H 37/58 Dec 31 SA 5f fst 1:001 H 18/66 Dec 24 SA 5f fst 1:014 Hg75/114 Dec 18 SA 5f fst 1:01 H 21/61

The fourth Race at Santa Anita on February 16th is an example of a well-meant, cheaply-bred maiden being dropped into a spot where it is more likely to win. The first thing to notice about this $32,000 Maiden Claiming race is that it is restricted to Cal-breds and that Kleven's Best has the most precocious (B+) rating and showed speed at the Maiden Special Weight level. With a class 7 rating and $2,500 stud fee, Kleven's Best was probably in over his head in his debut at the Maiden Special Weight level where horses with class ratings between 1-5 normally prevail. Looking at the gelding's workout pattern, his :46.3 gate work of December 22nd stands out and probably convinced Spawr and his recreational owner connections to protect the horse until they saw what they had.

In any event, having failed twice against better horses, Kleven's Best is now being dropped to a level where his breeding and his speed says he should be competitive. The public also agrees and has made him the 5-2 favorite.

84   Kleven's Best isn't however, the only class dropper, as Prince Quick, who showed speed last time out and finished right behind him to Superfluously, is also in the race. Furthermore, his owner/breeder Petruccione connections have replaced Atherton with Eddie D for this race and Prince Quick is being bet at 3-1.

Of the first-time starters, Flying Mate with his B6 rating also looks like a live horse for owner/breeder John Valpredo and is being moderately backed at 8-1. None of the other first-time starters are precociously bred. All of them, with the exception of Christopher Pete, look like they can be thrown out. Christopher Pete merits further respect, however, because John Sadler is good with debut runners, Kent Desormeux is the leading rider, and his Oakcrest Stable owner/breeder connections have spotted him at a level where his class 7 rating says he can win. Even more important, despite Torsion's poor debut rating as a sire, Christopher Pete has been bet down to 3-1.

So how to bet? With several possibilities, it's best to go light in this race. A Win bet on Kleven's Best plus a part backwheel exacta under Christopher Pete, Prince Quick and Flying Mate seems in order.

FOURTH RACE 6½ FURLONGS. (1.14) MAIDEN CLAIMING. Purse $18,000. 3-year-olds, bred in California. Weight, 118 lbs. Claiming price $32,000; if for $28,000, allowed 2 lbs.

Santa Anita
FEBRUARY 16, 1994

Value of Race: $18,000 Winner $9,900; second $3,600; third $2,700; fourth $1,350; fifth $450. Mutuel Pool $264,737.00 Exacta Pool $248,855.00 Quinella Pool $49,464.00

Last Raced	Horse	M/Eqt. A.Wt PP St	¼	½	Str	Fin	Jockey	Cl'g Pr	Odds $1
	Christopher Pete	B 3 118 10 8	9¹	6²	4²	1hd	Desormeaux K J	32000	2.90
	Fleet Landing	LB 3 118 5 4	3hd	44½	31½	22¾	Stevens G L	32000	6.30
	Flying Mate	B 3 111 9 9	8½	7²	6²	3hd	Gonzalez S Jr⁵	28000	8.70
30Jan94 6SA8	Kleven's Best	LBb 3 118 1 3	1hd	1hd	2¹	4no	Solis A	32000	2.50
30Jan94 6SA7	Prince Quick	Lb 3 118 7 2	2²	2½½	1hd	5²¾	Delahoussaye E	32000	2.90
7Jan94 4SA5	Fujiyama	LBb 3 116 11 1	42½	3hd	52½	6³	Pedroza M A	28000	13.20
	Pennyinthewater	Bb 3 113 4 7	102	101	8hd	7¾	Atherton J E⁵	32000	34.00
	Pingon	Bb 3 116 6 11	11	9½	9½	81½	Cedeno E A	28000	111.40
	Judy's Ultimo	L 3 118 2 10	7hd	82½	105	98	Atkinson P	32000	57.40
10Dec93 4Hol12	Air Apparently	LBf 3 118 3 5	5²	5½	7½	107½	Gryder A T	32000	40.30
	Drianas Gold	B 3 118 8 6	6¹	11	11	11	Lopez A D	32000	95.10

OFF AT 2:02 Start Good. Won driving. Time, :21³, :44⁴, 1:10³, 1:17¹ Track fast.

$2 Mutuel Prices:

11–CHRISTOPHER PETE	7.80	4.60	3.20
5–FLEET LANDING		5.60	4.40
10–FLYING MATE			5.00

$2 EXACTA 11–5 PAID $46.00 $2 QUINELLA 5–11 PAID $32.80

Ch. g, (Mar), by Torsion–Miss Tres, by Secretariat. Trainer Sadler John W. Bred by Oakcrest Stable (Cal).

© 1994 DRF, Inc.

Retrospectively, with four contenders, I shouldn't have bet the race the way I did. I got so carried away with the Maiden drop-down angle that I forced the issue and tried to split hairs rather than going with the value. Of my four contenders, three were under 3-1 and one (Flying Mate) was 8-1. Lapsing into a selection-oriented mode, I bet Kleven's Best rather than Flying Mate. But alas, this is one case where the "should have" bet actually lost, too, with Flying Mate finishing third, a head in front of Kleven's Best.

This race is an example of a dilemma that frequently confronts handicappers when choosing between Maiden Claiming first-time starters: For the action horse Christopher Pete had the right trainer but was not bred to win his debut, while Flying Mate had higher odds, but was indeed bred to fire first-time out. Generally speaking, Maiden Claiming races with this much ambiguity should be passed.

Santa Anita Park

3

6½ Furlongs. (1:14) MAIDEN CLAIMING. Purse $22,000. Fillies, 3-year-olds. Weight, 117 lbs. Claiming price $50,000; if for $45,000, allowed 2 lbs.

START ▾ 6½ FURLONGS ▴ FINISH

Sophie's Trick
Own: Fernandez & Granja Mexico
FLORES D R (94 4 13 17 .04) — $45,000
Dk. b or br f. 3 (Apr)
Sire: Clever Trick (Icecapade)
Dam: So Sophisticated (Quack)
Br: Richard Dick (Ky)
Tr: Palma Hector O (23 2 6 4 .09)
L 115

Lifetime Record :	2 M 0 1	$3,300		
1994	2 M 0 1	$3,300	Turf	0 0 0 0
1993	0 M 0 0		Wet	1 0 0 0
SA	2 0 0 1	$3,300	Dist	0 0 0 0

4Mar94-4SA fst 6f :214 :451 :58 1:113 ⒻMd 45000 56 4 4 33 33 34½ 32½ Cedeno E A LB 115 b 47.60 77-13 PromisingDawn117hd TequilRose1162½ Sophie'sTrick115no Drifted out late 9
4Feb94-9SA my 6f :22 :451 :572 1:094 ⒻMd 45000 32 8 4 74 76½ 813 923¼ Gonzalez S5 LB 110 b 15.20 65-16 Linda Lou B.1176 Queen Gen1174½ Regal Gentry1174 Wide 9
WORKOUTS: Mar 13 SA 4f fst :471 H 7/56 Feb 27 SA 4f fst :464 H 6/51 Feb 20 SA tr.t 5f sly 1:061 H 1/1 Feb 13 SA 5f fst 1:004 H 26/60 Jan 28 SA 6f fst 1:133 H 16/48 Jan 21 SA 6f fst 1:142 Hg 14/24

Phone The Bride
Own: Jones Brenda
NAKATANI C S (293 43 43 45 .15) — $50,000
Ch. f. 3 (Jun)
Sire: Phone Trick (Clever Trick)
Dam: June Bride (Riverman)
Br: Mrs. Dianne Charter (Ky)
Tr: Mitchell Mike (76 14 14 12 .18)
117

Lifetime Record :	0 M 0 0	$0		
1994	0 M 0 0		Turf	0 0 0 0
1993	0 M 0 0		Wet	0 0 0 0
SA	0 0 0 0		Dist	0 0 0 0

WORKOUTS: Mar 14 Hol 3f fst :363 H 7/17 Mar 9 Hol 4f fst :473 H 2/23 Mar 3 Hol 5f fst 1:02 Hg 12/16 Feb 19 Hol 6f fst 1:154 Hg 14/15 Feb 10 Hol 5f fst 1:004 H 7/33 Feb 3 Hol 5f fst 1:013 H 12/35
Jan 27 Hol 5f fst 1:01 H 14/24 Jan 13 Hol 5f fst 1:011 H 10/32 Jan 7 Hol 5f fst 1:021 H 20/27 Dec 31 Hol 5f fst 1:024 H 43/57 Dec 24 Hol 4f fst :491 H 18/37 Dec 4 Hol 3f fst :36 H 8/20

Gilded Lace
Own: Wygod Mr & Mrs Martin J
DESORMEAUX K J (257 58 44 30 .23) — $50,000
Dk. b or br f. 3 (Mar)
Sire: Pirate's Bounty (Hoist the Flag)
Dam: Lace Knickers (Reflected Glory)
Br: River Edge Farm, Inc. & Windy Hill T. B. A. (Cal)
Tr: Hendricks Dan L (43 4 4 4 .09)
L 117

Lifetime Record :	2 M 1 0	$9,625		
1994	2 M 1 0	$9,625	Turf	0 0 0 0
1993	0 M 0 0		Wet	0 0 0 0
SA	2 0 1 0	$9,625	Dist	0 0 0 0

26Feb94-6SA fst 6f :213 :444 :57 1:094 Ⓕ[S]Md Sp Wt 72 3 7 84½ 76 44½ 45¼ Desormeaux K J LB 117 *2.40 82-11 Serena's World1182 Masterful Dawn1173½ LaFrontera117no Checked start 9
12Feb94-3SA fst 6f :22 :452 :573 1:093 Ⓕ[S]Md Sp Wt 80 5 3 2hd 2hd 21 24½ Desormeaux K J LB 117 *.40e 84-11 Pirate's Revenge1174½ Gilded Lace1175 China Sky1173½ 2nd best 8
WORKOUTS: Mar 16 SA 3f fst :353 Hg 6/22 Mar 9 SA 4f fst :481 H 16/57 Feb 23 SA 4f fst :491 H 32/49 Feb 10 SA 3f fst :351 Hg 2/21 Jan 28 SA 6f fst 1:142 H 28/48 Jan 22 SA 6f fst 1:152 H 33/39

Streaking V.
Own: Naify Valerie
VALENZUELA P A (224 24 23 38 .11) — $50,000
B. f. 3 (May)
Sire: Risen Star (Secretariat)
Dam: Lt. Snoopy (Lt. Stevens)
Br: Robert C. Keller, J. W. Backer, et al. (Ky)
Tr: Griffiths Riley B (5 0 0 0 .00)
L 117

Lifetime Record :	1 M 0 0	$800		
1994	1 M 0 0	$800	Turf	0 0 0 0
1993	0 M 0 0		Wet	0 0 0 0
SA	1 0 0 0	$800	Dist	0 0 0 0

13Feb94-3SA fst 6f :214 :452 :572 1:10 ⒻMd Sp Wt 60 5 4 55 57½ 57¼ 511 Valenzuela P A LB 117 14.00 76-13 Prairie Pearl117hd Neeran117nk Prematurely Gray1175½ No factor 6
WORKOUTS: ●Mar 15 Hol 3f fst :344 H 1/9 Mar 3 Hol 6f fst 1:143 H 3/10 Feb 24 Hol 5f fst 1:01 H 10/29 Feb 5 Hol 4f gd :48 Hg 2/9 Jan 30 Hol 5f fst 1:003 H 14/31 Jan 23 Hol 4f fst :473 H 2/15

Mask De Naskra
Own: Ailshie & Shields & Shields
VALENZUELA F H (301 31 18 27 .10) — $50,000
B. f. 3 (Apr)
Sire: Star de Naskra (Naskra)
Dam: Graceful Creek (Darby Creek Road)
Br: Sultan Bruce & Bernstein David (Ky)
Tr: Bernstein David (36 5 6 5 .14)
L 117

Lifetime Record :	1 M 0 0	$550		
1994	1 M 0 0	$550	Turf	0 0 0 0
1993	0 M 0 0		Wet	0 0 0 0
SA	1 0 0 0	$550	Dist	0 0 0 0

4Mar94-4SA fst 6f :214 :451 :58 1:113 ⒻMd 50000 55 8 9 66 66½ 68 52¾ Valenzuela F H L 117 9.00 76-13 Promising Dawn117hd Tequila Rose1162½ Sophie's Trick115no Late bid 9
WORKOUTS: Mar 13 SA 4f fst :472 H 12/56 Feb 24 SA 5f fst 1:001 H 9/31 Feb 16 SA 6f fst 1:141 H 28/57 Feb 10 SA 5f fst 1:022 Hg 42/70 Feb 2 SA 5f fst 1:021 Hg 33/42 Jan 26 SA 4f gd :502 H 36/44

Queen Gen
Own: Philip & Sophie Hersh Trust
ANTLEY C W (278 39 28 16 .14) — $45,000
Ch. f. 3 (Jan)
Sire: General Assembly (Secretariat)
Dam: Queen Breeze (Breezing On)
Br: Pillar Stud Inc (Ky)
Tr: Bernstein David (36 5 6 5 .14)
115

Lifetime Record :	4 M 2 1	$12,100			
1994	3 M 2 1	$12,100	Turf	0 0 0 0	
1993	1 M 0 0		Wet	1 0 1 0	$4,400
SA	4 0 2 1	$12,100	Dist	2 0 1 1	$7,700

18Feb94-4SA fst 6½f :22 :452 1:104 1:172 ⒻMd 50000 74 5 1 1hd 11 11 2hd Antley C W B 117 2.60 85-11 Melrose Wine117hd Queen Gen1172½ Mostly Overcast1172 Game try 8
4Feb94-9SA my 6f :22 :451 :572 1:094 ⒻMd 50000 77 7 3 2½ 2½ 22½ 26 Desormeaux K J B 117 *2.30 82-16 Linda Lou B.1176 Queen Gen1174½ Regal Gentry1174 2nd best 9
21Jan94-3SA fst 6½f :214 :444 1:102 1:172 ⒻMd 50000 61 5 3 2hd 1hd 2hd 32 Gryder A T B 117 2.10 83-12 Antifreeze1171 Regal Gentry1151 Queen Gen117¾ Outfinished 8
26Dec93-2SA fst 6f :214 :444 :57 1:093 ⒻMd Sp Wt 65 3 12 83 74½ 77½ 99¼ Gryder A T B 117 b 93.10 80-09 Choice Claim1172 Meadow Moon117½ Whatawoman1173 Steadied start 12
WORKOUTS: Mar 10 SA 4f fst :473 H 7/47 Feb 28 SA 3f fst :361 H 7/13 Feb 15 SA 3f fst :363 H 10/28 ●Jan 17 SA 3f fst :341 Hg 1/22 Jan 10 SA 7f fst 1:264 H 5/5 Jan 3 SA 5f fst 1:004 Hg 34/60

Novelton
Own: Whittingham Charles
MCCARRON C J (169 30 28 26 .18) — $50,000
Ch. f. 3 (Feb)
Sire: Greinton (Green Dancer)
Dam: Novel (Plum Bold)
Br: Bradley, Chandler & Whittingham (Ky)
Tr: Whittingham Charles (82 7 11 10 .09)
117

Lifetime Record :	2 M 0 0	$1,650		
1994	2 M 0 0	$1,650	Turf	0 0 0 0
1993	0 M 0 0		Wet	0 0 0 0
SA	2 0 0 0	$1,650	Dist	0 0 0 0

4Mar94-4SA fst 6f :214 :451 :58 1:113 ⒻMd 50000 56 6 7 78 78¾ 78¼ 42½ Anderson K B 117 28.50 77-13 PromisingDawn117hd TequilRose1162½ Sophie'sTrick115no 4 wide into lane 9
7Jan94-9SA fst 6f :212 :441 :564 1:094 ⒻMd 50000 — 2 11 — — — — Iammarino M P B 117 43.80 — 08 Sophi'sFvorite1171½ FlyingRoylty1174 LindLouB.1171¼ Bobbled, lost rider 12
WORKOUTS: Mar 15 SA 5f fst 1:002 H 19/59 Mar 10 SA 3f fst :364 H 12/32 Feb 27 SA 6f fst 1:14 H 29/46 Feb 22 SA 6f fst 1:131 H 10/44 Feb 15 SA 5f fst 1:01 H 18/33 Feb 10 SA 5f fst 1:05 H 69/70

Debut Winners at the Maiden Claiming Level

Of course, not all Maiden Claiming races are won by class droppers. There are a number of mid-range sires (such as Al Mamoon, Copelan, Carr de Naskra, Fortunate Prospect, Allen's Prospect, Irish Tower, Known Fact, Pancho Villa, Phone Trick and Saratoga Six) whose first-time starters regularly win at the high Maiden Claiming level. The third race at Santa Anita on March 18th is one example. A seven-horse claiming affair, it consists of two class droppers, four horses that have already lost for $50,000 and one precociously-bred, first-time starter.

The heavy favorite at 3-5 is Gilded Lace who showed speed at the Maiden Special Weight level in her initial effort and was troubled her second time out. She has, however, been the beaten favorite twice in races restricted to State-breds and the drop in class today may be more apparent than real because she will be facing "open" competition from the progeny of much classier Kentucky sires.

Gilded Lace's drop into the claiming ranks, however, does make economic sense for her Wygod owner/breeder connections. Pirate's Bounty stands at their River's Edge Farm for $8,000 and, if Gilded Lace wins the race and is claimed, they will get over sixty grand and make a substantial profit. Still, the Wygod's and Dan Hendricks are astute judges of horseflesh and not likely to let a really good horse get away.

Streaking V is the other class dropper, but she didn't show any early speed first time out, nor did she make any move during the race. Droppers who made no move against better are unattractive because they never really competed at a higher level. Furthermore, Streaking V has modest connections, and isn't being backed on the board.

Of the horses that have already run for a $50,000 tag, Queen Gen looks the best. She has some speed, a competitive Beyer, and is the 5-2 second choice. If she can outbreak Gilded Lace, she might be able to steal the race. Still, since Queen Gen has already lost three times at this claiming level, her chances of winning seem considerably less than her odds. Mask De Naskra has a good B3 rating and might improve. But she ran a much lower Beyer than Queen Gen in her first start and at 22-1 is dead on the board.

The mystery horse in this race is Phone The Bride, a Mike Mitchell, first-time starter by Phone Trick. The filly has an enticing B4 rating and is out of a Riverman mare that, according to *Maiden Stats 1993*, has already thrown three winners from four starters. While Mitchell is primarily known as a shrewd claiming trainer, he is also competent with debut runners. On the downside, Phone The Bride sold at auction for only $21,000 which, given Phone Trick's $15,000 stud fee, suggests some conformation problems. On the other hand, the filly was a late June foal and may have been the runt of the auction. Her breeder Dianne Charter may have decided to unload her, get her stud fee back, and try again with another breeding. In the context of this race, however, Phone The Bride has been well placed by her current owner Brenda Jones. If the filly wins and is claimed, Jones will, after expenses, more than double her $21,000 investment. More worrisome, however, is the fact that Mitchell's live horses usually get bet and this filly is going off at 13-1. Still, Mitchell is a realistic trainer and must think Phone The Bride has a shot in this spot. Otherwise, he would have talked Jones into running her at the lower Maiden $32,000 level which, if the filly won and was claimed, would still yield her owner an attractive profit.

So what to do? Given her cheap auction price and light toteboard action, Phone The Bride can't be a prime play. Still, because of her precocious breeding and the possibility of a speed duel developing between Gilded lace and Queen Gen, and the fact that Gilded Lace has already been beaten twice as the favorite, Phone The Bride is worth a modest Win bet at 13-1 and should also be put behind the two favorites in the exacta as a Place bet.

THIRD RACE
Santa Anita
MARCH 18, 1994

6½ FURLONGS. (1.14) MAIDEN CLAIMING. Purse $22,000. Fillies, 3-year-olds. Weight, 117 lbs. Claiming price $50,000; if for $45,000, allowed 2 lbs.

Value of Race: $22,000 Winner $12,100; second $4,400; third $3,300; fourth $1,650; fifth $550. Mutuel Pool $275,781.00 Exacta Pool $298,653.00 Quinella Pool $45,049.00

Last Raced	Horse	M/Eql. A.Wt	PP	St	¼	½	Str	Fin	Jockey	Cl'g Pr	Odds $1
	Phone The Bride	LBb 3 117	2	7	7	5 1½	3 2	1 6	Nakatani C S	50000	13.80
26Feb94 6SA4	Gilded Lace	LBb 3 117	3	2	1½	2 3	1 1	2 3½	Desormeaux K J	50000	0.60
4Mar94 4SA3	Sophie's Trick	LBb 3 115	1	1	3 1	4 1½	5 4	3 1	Flores D R	45000	50.70
4Mar94 4SA4	Novelton	B 3 117	7	5	5 hd	3 2	4 1½	4 2	McCarron C J	50000	8.80
18Feb94 4SA2	Queen Gen	B 3 115	6	3	2 1	1 hd	2 ½	5 4½	Antley C W	45000	2.60
13Feb94 3SA5	Streaking V.	LB 3 117	4	6	6 5	6 ½	6 1½	6 5½	Valenzuela P A	50000	14.30
4Mar94 4SA5	Mask De Naskra	LB 3 117	5	4	4 hd	7	7	7	Valenzuela F H	50000	22.90

OFF AT 1:32 Start Good. Won driving. Time, :21³, :44³, 1:10⁴, 1:17¹ Track fast.

$2 Mutuel Prices:

2–PHONE THE BRIDE	29.60	6.60	5.40
3–GILDED LACE		2.60	2.60
1–SOPHIE'S TRICK			6.80

$2 EXACTA 2–3 PAID $67.00 $2 QUINELLA 2–3 PAID $20.00

Ch. f, (Jun), by Phone Trick–June Bride, by Riverman. Trainer Mitchell Mike. Bred by Mrs. Dianne Charter (Ky).

The ninth race on June lst at Hollywood Park is another example of a debut runner being properly spotted in the Maiden Claiming ranks by its owner/breeder connections. Looking over the field, most of the entrants are proven losers at this level and can be thrown out. This leaves Christina Maria, Allessea, Astrial, Comeback Kidder, Unpredictable Gal and the precociously-bred, debut runner Cabo Queen as the horses to consider.

Of the horses that have run, Allessea has already beaten Christina Maria and is the 3-2 favorite. She comes off a good initial effort and has the top Beyer. Furthermore, she is unlikely to bounce as she has been rested for three weeks and comes off a nice :35.3 work. Still, she will probably be tested once again by Christina Maria and may be vulnerable to Comeback Kidder who gets Gary Stevens in the irons and will likely sit just off the pace. While Comeback Kidder has lower Beyers, it is worth noting that she has already won at this level, only to be disqualified. Unpredictable Gal also has some

9

6 Furlongs. (1:08) MAIDEN CLAIMING. Purse $17,000. Fillies and mares, 3-year-olds and upward, bred in California. Weights: 3-year-olds, 115 lbs. Older, 122 lbs. Claiming price $32,000, if for $28,000, allowed 2 lbs. (Horses which have not started for a claiming price of $20,000 or less preferred.)

START ▼ 6 FURLONGS ▲ FINISH

Concorde's Music
Own: Minsky & Sloane
CASTANON E F (10 0 0 0 .00) $32,000
Dk. b or br f. 3 (May)
Sire: King Concorde (Super Concorde)
Dam: Sarah's Tune (Lord Stapleton)
Br: Minsky Bernard (Cal)
Tr: Maslonka Casey (10 0 1 0 .00)
L 1105

Lifetime Record :	1 M 0 0	$0		
1994	1 M 0 0		Turf	0 0 0 0
1993	0 M 0 0		Wet	0 0 0 0
Hol	1 0 0 0		Dist	1 0 0 0

5May94- 9Hol fst 6f :22 :452 :581 1:113 3↑ Ⓕ[S]Md 32000 17 11 3 106½ 76½ 1011 1014 Castanon A L B 115 b 36.10 68-08 B. Fancy1151 Unprectible Gal1157 Turf Princess122no Wide trip 12
WORKOUTS: May 25 Hol 3f fst :364 Hg 11/33 May 19 Hol 5f fst 1:02 H 25/30 May 12 Hol 4f fst :482 H 14/27 Apr 28 Hol 3f fst :353 Hg 3/30 Apr 22 SA 7f fst 1:31 H 10/11 Apr 16 SA 6f fst 1:15 Hg 13/24

Drona's Starlet
Own: Dwyer Austin
ATHERTON J E (75 7 5 5 .09) $32,000
B. f. 3 (Feb)
Sire: Nostalgia's Star (Nostalgia)
Dam: Fleet Drona (Drone)
Br: Austin Dwyer (Cal)
Tr: Bean Robert A (10 0 1 1 .00)
L 1105

Lifetime Record :	3 M 0 1	$3,825			
1994	3 M 0 1	$3,825	Turf	0 0 0 0	
1993	0 M 0 0		Wet	0 0 0 0	
Hol	2 0 0 1	$3,825	Dist	2 0 0 1	$3,825

14May94- 9Hol fst 6f :22 :452 :58 1:114 3↑ Ⓕ[S]Md 32000 64 10 1 73¾ 710 47½ 32 Long B LB 117 13.20 79-09 Copper Coinage1152 Accountable115hd Drona's Starlet1172¼ Far wide trip 11
28Apr94- 9Hol fst 6f :214 :45 :571 1:102 3↑ ⒻMd 32000 53 9 4 31½ 21½ 33½ 48¼ Long B LB 117 20.50 80-08 Queen Gen1161¾ Ruler Of Men1152 Our Cookie1154½ No late bid 9
15Apr94- 2SA fst 5½f :213 :45 :571 1:034 ⒻMd 28000 40 2 11 43 54 88½ 1014¾ Long B B 117 20.30 76-08 Kelly'sThing1172¼ TimeOfQueen112½ Snowy'sMrk1171¾ Off slow, bumped 12
WORKOUTS: May 25 Hol 4f fst :49 H 18/51 May 7 Hol 4f fst :481 H 19/44 Apr 24 Hol 5f fst 1:022 H 13/15 ●Apr 3 Hol 5f fst :592 H 1/23 Mar 7 Hol 5f fst :594 H 8/30 ●Mar 2 Hol 4f fst :471 Hg 1/18

Christina Maria
Own: Alvarez & Smith
NAKATANI C S (19 2 5 4 .11) $32,000
Ch. f. 3 (Mar)
Sire: Olympic Native (Raise a Native)
Dam: Black Jade (Torsion)
Br: Alvarez & Smith (Cal)
Tr: Sadler John W (27 4 3 1 .15)
L 115

Lifetime Record :	3 M 1 0	$5,325			
1994	3 M 1 0	$5,325	Turf	0 0 0 0	
1993	0 M 0 0		Wet	0 0 0 0	
Hol	1 0 0 0	$1,275	Dist	1 0 0 0	$1,275

8May94-10Hol fst 6f :22 :444 :57 1:093 3↑ Ⓕ[S]Md 32000 72 4 2 2½ 21½ 33 45¾ Antley C W LB 115 b *1.60 86-10 Sham Pain1154 We Got The Dinero115½ Allessea1221¼ Weakened 10
17Apr94- 2SA fst 5½f :22 :452 :574 1:042 Ⓕ[S]Md 32000 75 3 3 1hd 2hd 22 22¾ Gonzalez S5 LB 112 b 5.60 85-11 Star Sign1172¾ Christina Maria1122½ Pretty Kathleen1174 Inside duel 6
3Apr94- 9SA fst 5½f :214 :453 :583 1:052 ⒻMd 32000 52 6 7 74 86¾ 67 56 Solis A B 117 b 10.00 77-14 Lister Ridge1172¾ Our Cookie117½ Pala Canyon117¾ Improved position 12
WORKOUTS: May 27 SA 6f fst 1:153 H 5/9 May 21 SA 5f fst 1:033 H 23/30 May 6 SA 4f fst :472 H 4/28 Apr 30 Hol 5f fst 1:023 H 40/52 Apr 14 SA 4f fst :473 H 11/35 Mar 27 Hol 5f fst 1:012 Hg 11/21

Allessea
Own: Bradshaw Alicia J
MCCARRON C J (80 18 13 14 .23) $32,000
Ch. f. 4
Sire: Raise a Champion (Raise a Native)
Dam: Grenson (Grenfall)
Br: Alvarez & Smith (Cal)
Tr: Bradshaw Randy (11 1 3 2 .09)
122

Lifetime Record :	1 M 0 1	$2,550			
1994	1 M 0 1	$2,550	Turf	0 0 0 0	
1993	0 M 0 0		Wet	0 0 0 0	
Hol	1 0 0 1	$2,550	Dist	1 0 0 1	$2,550

8May94-10Hol fst 6f :22 :444 :57 1:093 3↑ Ⓕ[S]Md 32000 75 9 4 3½ 32 22½ 34½ Solis A B 122 4.10 87-10 Sham Pain1154 We Got The Dinero115½ Allessea1221¼ No late bid 10
WORKOUTS: May 29 Hol 3f fst :353 H 4/32 May 23 Hol 5f fst 1:014 H 29/47 Apr 29 Hol 6f fst 1:141 Hg 2/10 Apr 16 Hol 5f fst 1:03 H 20/26 Mar 27 Hol 6f fst 1:142 H 9/12 Mar 20 Hol 6f fst 1:132 H 5/9

Bolides Wish
Own: Carvajal & Dominguez & Everson
PEDROZA M A (114 12 19 16 .11) $28,000
Ch. f. 3 (Feb)
Sire: Bel Bolide (Bold Bidder)
Dam: Wishful Nickle (Plugged Nickle)
Br: Arnold Danny (Cal)
Tr: Dominguez Caesar F (29 3 5 2 .10)
L 113

Lifetime Record :	9 M 2 1	$5,600			
1994	7 M 2 1	$5,600	Turf	0 0 0 0	
1993	2 M 0 0		Wet	2 0 0 0	
Hol	4 0 1 0	$3,150	Dist	3 0 1 1	$2,450

13May94- 6Hol fst 6½f :221 :451 1:112 1:182 3↑ ⒻMd 22500 67 2 3 41¾ 21 1½ 2½ Pedroza M A LB 113 5.30 76-19 Maitlin115½ Bolides Wish113½ Fast Kicken Cara1156 Led, outfinished 8
30Apr94- 2Hol fst 7f :222 :454 1:11 1:24 3↑ ⒻMd 22500 39 2 4 52½ 2½ 24 512¾ Pedroza M A LB 113 24.20 71-10 LadyBirdJaklin1155½ SllsAlbmus1153½ StrikinglyVelvet1153 Bid, gave way 8
25Mar94- 2GG gd 6f :222 :462 1:001 1:142 ⒻMd 12500 52 5 2 31½ 21½ 2hd 22 Meza R Q L 117 6.60 65-22 J's Cinnamon Lady1172 Bolides Wish1172 Don't Spend It117no Game try 9
11Mar94- 2GG fst 6f :22 :45 :574 1:112 ⒻMd 12500 23 1 5 31 44¼ 57 612¼ Belvoir V T LB 117 *1.20 70-11 Innocent Cara1171¼ Classic Summer117½ K. L.'s Imagene1178 Tired 11
3Mar94- 5GG fst 6f :221 :46 :584 1:12 Ⓕ[S]Md 12500 51 3 1 21½ 32 31½ 33 Lopez A D LB 117 *1.90 76-17 Reb's Gal Sal1171½ Geri's Princess1171½ Bolides Wish1172½ Even late 8
20Feb94- 4SA wf 5½f :213 :443 :571 1:04 Ⓕ[S]Md 28000 46 8 3 73¾ 56¼ 66¼ 79½ Valenzuela F H LB 115 38.50 80-09 Missoula Lula1172 Costchoro1172 World Record Mama1172½ 4 wide turn 11
9Feb94- 4SA sl 6½f :213 :451 1:113 1:182 Ⓕ[S]Md 32000 -0 5 4 42½ 34 59 934¼ Valenzuela F H LB 117 b 12.30 46-21 Treasure's Pride117¾ Tia Veevee11712 Scoring Road1151¼ Tired 11
11Jun93- 2Hol fst 5f :221 :461 :59 ⒻMd 32000 22 4 4 41¾ 31½ 58 613 Atkinson P LB 118 b 10.10 76-09 Groovy Anni1183½ Fleta Kalem118¾ Trish's Trick1185 Faltered 9
13May93- 3Hol fst 4½f :214 :454 :521 ⒻMd 32000 39 6 3 31½ 33 68½ Nakatani C S B 118 14.40 86-11 DealEmDarling1183 SummertimeParty1183 GoldShivers118nk Weakened 9
WORKOUTS: Apr 23 Hol 5f fst 1:013 H 21/36 Apr 15 Hol 5f fst 1:04 H 29/30

---

speed, but she is parked on the outside and will have to be used early. Even though she has been bet down to less than 3-1, she has lower Beyers than Allessa and Christina Maria and figures to have little left in the stretch.

Youtellemshirley

Own: Kurth Ellen M

CASTANON A L (33 1 3 5 .03) — $28,000

B. f. 4
Sire: Walker's (Jaipur)
Dam: Miss Chatterly (Pass the Glass)
Br: Goemans Peg (Cal)
Tr: Perez Dagoberto L (4 0 0 0 .00)

L 120

Lifetime Record:	6 M 1 0	$6,400			
1994	6 M 1 0	$6,400	Turf	0 0 0 0	
1993	0 M 0 0		Wet	1 0 0 0	$2,400
Hol	2 0 0 0		Dist	4 0 1 0	$4,000

14May94-9Hol fst 6f :22 :45² :58 1:11⁴ 3↑ⒻⓈMd 28000 46 2 7 4¹½ 4⁶½ 7¹¹ 8⁹ Valenzuela P A LB 120 b 8.10 72-09 CopperCoinage115² Accountable115hd Drona'sStarlet117²¼ Saved ground 11
28Apr94-6Hol fst 6f :22 :45² :57¹ 1:09³ 3↑ⒻMd 32000 37 8 6 7⁵½ 8⁸½ 9¹⁵ 8¹⁹ Flores D R L 122 5.70 73-08 Time Of Queen1105½ Ruffmeup122½ Chelsea Castle120³ 5 wide turn 9
8Apr94-3SA fst 6f :22 :45¹ :57² 1:10 ⒻⓈMd 35000 71 4 3 4¹¼ 3¹ 3² 2³½ Flores D R LB 118 6.10 83-12 Royalty Doll120³½ Youtellemshirley118¾ Nancy'sLegend118⁶ Outfinished 7
25Mar94-2SA gd 1 1/16 :24³ :49¹ 1:14¹ 1:46¹ ⒻMd 35000 31 3 4 5⁴½ 7⁸ 7¹⁴ 7¹⁹½ Valenzuela F H LB 118 3.80 44-22 Tellalo120½ Dances With Wolves118⁵½ Sweet Lady Czar118³½ 7
Broke thru gate, pinched start, steadied 7/8
4Feb94-6SA my 7f :22⁴ :45⁴ 1:11¹ 1:25 ⒻMd Sp Wt 34 4 1 3¹ 4⁵ 4⁷ 4¹³ Martinez F F B 119 13.80 64-16 Cindazanno119³½ Ms. Ukulele119²½ On The Cheek119⁷ No factor 4
6Jan94-6SA fst 6f :21⁴ :44⁴ :56⁴ 1:09⁴ ⒻMd Sp Wt 50 2 8 5⁵ 8⁹½ 8¹⁴ 8¹³¾ Martinez F F B 119 44.40 74-11 Morning Showers119nk Suana119²½ Really Tops114³½ Brief speed 8

WORKOUTS: May 25 Hol 4f fst :48 H 2/51 May 11 Hol 4f fst :47 H 3/37 Apr 24 Hol 5f fst 1:00³ H 6/15 Apr 17 Hol 4f fst :49¹ H 12/24 ●Apr 3 Hol 4f fst :46² H 1/17 Mar 21 Hol 4f fst :48 H 2/19

Astrial

Own: Fanning Jerry

DELAHOUSSAYE E (58 12 11 13 .21) — $32,000

Ch. f. 3 (May)
Sire: Flying Paster (Gummo)
Dam: Astrious (Piaster)
Br: Cardiff Stud Farms (Cal)
Tr: Fanning Jerry (8 2 0 1 .25)

L 115

Lifetime Record:	1 M 0 0	$0			
1994	1 M 0 0		Turf	0 0 0 0	
1993	0 M 0 0		Wet	0 0 0 0	
Hol	1 0 0 0		Dist	1 0 0 0	

14May94-9Hol fst 6f :22 :45² :58 1:11⁴ 3↑ⒻⓈMd 32000 52 5 11 11¹⁵ 11¹⁹ 8¹⁵ 6⁶½ Valenzuela F H LB 115 b 9.50 74-09 Copper Coinage115² Accountable115hd Drona's Starlet117²¼ Off slow 11

WORKOUTS: May 27 SA 3f fst :35³ H 2/22 May 21 SA 4f fst :47³ H 7/33 May 8 SA 6f gd 1:15² Hg2/8 May 2 SA 6f fst 1:15¹ H 12/18 Apr 22 SA 5f fst :59³ Hg5/26 Apr 16 SA 5f fst 1:03 Hg41/51

Comeback Kidder

Own: Venneri & Welty

STEVENS G L (138 29 16 30 .21) — $32,000

Dk. b or br f. 3 (Mar)
Sire: Regalberto (Roberto)
Dam: One Lap to Go (Full of Drive)
Br: Halo Farms (Cal)
Tr: Sise Clifford Jr (23 3 3 7 .13)

L 115

Lifetime Record:	4 M 0 4	$9,570			
1994	1 M 0 1	$2,550	Turf	0 0 0 0	
1993	3 M 0 3	$7,020	Wet	0 0 0 0	
Hol	1 0 0 1	$2,550	Dist	4 0 0 4	$9,570

6May94-2Hol fst 6f :21³ :44¹ :56² 1:09³ 3↑ⒻMd 32000 65 9 2 2⁴ 2⁸ 3⁷ 3⁷½ Flores D R LB 115 b *1.80 84-06 Star Of Command115⁴ Rue D'alezan115³½ Comeback Kidder115½ No rally 10
28Sep93-6Fpx fst 6f :22¹ :46¹ :59³ 1:12³ ⒻMd 32000 58 7 5 2¹½ 1½ 1½ 1½ Flores D R LB 116 b *.80 84-09 DComebckKidder116½ TrckCheerledr116½ SingLrl116²½ Ducked out 3/16 10
Disqualified and placed third
10Sep93-9Dmr fst 6f :21⁴ :45¹ :57⁴ 1:11 ⒻⓈMd 32000 63 5 4 5¹¾ 2¹½ 2² 3²½ Desormeaux K J LB 118 b *1.90 81-13 NtiveLce118²¼ CusACommotion118nk CombckKiddr118⁵½ Not enough late 12
13Aug93-4Dmr fst 6f :22 :45² :57⁴ 1:11¹ ⒻⓈMd 32000 53 2 6 5²½ 4⁵½ 3⁶½ 3⁸ Stevens G L LB 117 *2.00 75-16 BrendsWildindin117⁴½ Gretchen'sAlm117³½ CombckKiddr117⁶½ No mishap 8

WORKOUTS: May 27 Hol 4f fst :49¹ H 22/40 May 22 Hol 5f fst 1:04⁴ H 41/43 May 16 Hol 4f fst :48⁴ H 17/42 Apr 24 Hol 7f fst 1:26⁴ H 2/2 Apr 18 Hol 6f fst 1:13 H 3/18 ●Apr 13 Hol 3f fst :35 H 1/18

Martial Artist

Own: Cronin J F & Sears W D

VALENZUELA F H (138 12 15 13 .09) — $32,000

B. f. 3 (May)
Sire: Martial Law (Mr. Leader)
Dam: Fillyraki (Faliraki)
Br: Sears William D & Cronin John (Cal)
Tr: Bacorn Herbert L (10 1 0 0 .10)

L 115

Lifetime Record:	1 M 0 0	$0			
1994	1 M 0 0		Turf	0 0 0 0	
1993	0 M 0 0		Wet	0 0 0 0	
Hol	1 0 0 0		Dist	1 0 0 0	

8May94-10Hol fst 6f :22 :44⁴ :57 1:09³ 3↑ⒻⓈMd 32000 53 8 5 4² 5⁹ 5¹³ 6¹³¼ Pedroza M A LB 117 35.40 79-10 Sham Pain115⁴ We Got The Dinero115½ Allessea122¹¼ Gave way 10

WORKOUTS: May 25 Hol 4f fst :48⁴ H 14/51 May 17 Hol 4f fst :50³ H 19/21 Apr 30 Hol 5f fst 1:02² H 39/52 Apr 21 SA 5f fst 1:00¹ Hg4/23 Apr 8 SA 5f fst 1:01 Hg16/36 Apr 1 SA 5f fst 1:01² Hg15/32

Buttercup On Rocks

Own: Molick & Quoq

ATKINSON P (58 5 10 4 .09) — $28,000

Dk. b or br f. 4
Sire: Aras an Uachtarain-Ir (Habitat)
Dam: Rochells Girl (Fluorescent Light)
Br: Richard Quog & Barbara Quog (Cal)
Tr: Bell Thomas R II (1 0 0 0 .00)

L 120

Lifetime Record:	4 M 1 0	$4,425			
1994	2 M 1 0	$4,425	Turf	0 0 0 0	
1993	2 M 0 0		Wet	0 0 0 0	
Hol	3 0 0 0	$425	Dist	3 0 0 0	$425

6May94-2Hol fst 6f :21³ :44¹ :56² 1:09³ 3↑ⒻMd 28000 52 5 4 4⁸ 4¹⁰ 5⁹ 5¹²½ Atherton J E⁵ LB 115 4.90 79-06 StrOfCommnd115⁴ RueD'lezn115³½ ComebckKidder115½ Came out start 10
14Apr94-2SA fst 5½f :21³ :45² :57⁴ 1:04³ ⒻMd 35000 58 5 5 6⁷¼ 3⁶ 3³ 2³ Atherton J E⁵ LB 113 19.50 84-09 LunchAPtriot118³ ButtrcupOnRocks113¹½ VlvtQust118hd Stumbled start 8
14May93-9Hol fst 6f :22¹ :45³ :58 1:10² 3↑ⒻMd 32000 38 3 10 5³½ 4²½ 4⁵½ 11¹⁶¾ Lopez A D B 115 b 35.40 71-10 Mood Elevator122³½ Irish Maccool116⁵ Icecapade Beauty115nk Gave way 12
24Apr93-1Hol fst 6f :22 :45¹ :57¹ 1:09² 3↑ⒻⓈMd Sp Wt 43 6 2 1hd 4½ 4⁷ 7¹⁸½ Nakatani C S B 115 b 24.10 75-07 Active122hd Malojen116⁷ Paster's Princess117⁷ Faltered 7

WORKOUTS: Apr 30 SA tr.t 4f hy :50⁴ H 9/16 Apr 6 SA 5f fst :59⁴ H 3/27 Mar 30 SA 5f fst 1:00¹ H 6/42 Mar 10 SA 3f fst :37² H 25/32

Cabo Queen

Own: Sahadi Scott & Stephen B

ANTLEY C W (112 26 22 9 .23) — $32,000

Ch. f. 3 (Mar)
Sire: Al Mamoon (Believe It)
Dam: Flying Hill (Flying Paster)
Br: Cardiff Stud Farm (Cal)
Tr: Jackson Bruce L (12 2 2 1 .17)

L 115

Lifetime Record:	0 M 0 0	$0			
1994	0 M 0 0		Turf	0 0 0 0	
1993	0 M 0 0		Wet	0 0 0 0	
Hol	0 0 0 0		Dist	0 0 0 0	

WORKOUTS: May 29 Hol 3f fst :36¹ H 11/32 May 20 Hol 4f fst :48¹ H 5/26 May 5 Hol 6f fst 1:13¹ Hg4/11 Apr 28 Hol 3f fst :37¹ H 16/30 Apr 20 SA 4f fst :49 H 21/32 Apr 13 SA 6f fst 1:14 Hg10/20 Apr 6 SA 4f fst :49¹ Hg25/31 Mar 30 SA 6f fst 1:14 H 9/22 Mar 23 SA 5f fst 1:00¹ H 14/112 Mar 13 SA 4f fst :49² H 40/56 Feb 26 SA 5f fst 1:01 H 20/40 Feb 19 SA 4f fst :48¹ H 23/61

Unprectible Gal

Own: Gallup Frank E

SOLIS A (153 29 15 22 .19) — $32,000

B. f. 3 (May)
Sire: Unpredictable (Tri Jet)
Dam: Genocide (Crozier)
Br: Frank E. Gallup (Cal)
Tr: Cenicola Lewis A (15 3 3 1 .20)

L 115

Lifetime Record:	2 M 2 0	$7,000			
1994	2 M 2 0	$7,000	Turf	0 0 0 0	
1993	0 M 0 0		Wet	0 0 0 0	
Hol	1 0 1 0	$3,400	Dist	1 0 1 0	$3,400

5May94-9Hol fst 6f :22 :45² :58¹ 1:11³ 3↑ⒻⓈMd 32000 49 2 7 3nk 1hd 1hd 2¹ Solis A B 115 *1.40 81-08 B. Fancy115¹ Unprectible Gal115⁷ Turf Princess122no Inside duel 12
6Apr94-4SA fst 6½f :22 :45¹ 1:10⁴ 1:17³ ⒻⓈMd 32000 55 6 1 4¾ 3nk 2hd 2⁴ Solis A B 117 11.90 80-12 Tia Veevee117⁴ Unprectible Gal117¹¾ Silent Draw117¾ Dueled between 8

WORKOUTS: May 28 Hol 4f fst :50² H 37/43 May 21 Hol 6f fst 1:15¹ H 14/25 May 15 Hol 4f fst :50 H 37/45 Apr 29 Hol 5f fst 1:01² H 5/25 Apr 21 SA 6f fst 1:17² H 11/11 Apr 15 SA 4f fst :48² H 12/29

Astrial is another possibility that can't be thrown out. She broke poorly in her initial start and might run to her precocious C+3 breeding second time out. She deserves another chance and is being bet some at 7-1.

But the really intriguing horse here is the first-time starter Cabo Queen who, with her B+6 rating, is ideally placed by her Sahadi owner/breeder connections to win this high Maiden Claiming event. Her sire Al Mamoon stands at Cardiff for some $7,500 and, if Cabo Queen wins this race and gets claimed, the Sahadi's will get back more than $40,000. Furthermore, Chris Antley is up, and the filly has been given a steady series of works.

Why then is Cabo Queen 19-1 on the board? One reason might be that Bruce Jackson isn't particularly noted for popping with first-time starters, another that this is Flying Hill's first foal to reach the races. But, perhaps more important is the fact that Allessea's 75 Beyer is above par for this level.

So how to bet? Since there are several contenders in the race, it can't be a prime bet. Nevertheless, Cabo Queen is definitely worth a moderate Win bet and should also be played behind Allessea, Astrial and possibly Comeback Kidder in the exacta in case she finishes second at those big odds.

NINTH RACE
Hollywood
JUNE 1, 1994

6 FURLONGS. (1.08) MAIDEN CLAIMING. Purse $17,000. Fillies and mares, 3-year-olds and upward, bred in California. Weights: 3-year-olds, 115 lbs. Older, 122 lbs. Claiming price $32,000, if for $28,000, allowed 2 lbs. (Horses which have not started for a claiming price of $20,000 or less preferred.)

Value of Race: $17,000 Winner $9,350; second $3,400; third $2,550; fourth $1,275; fifth $425. Mutuel Pool $220,833.00 Exacta Pool $198,372.00 Quinella Pool $30,777.00 Superfecta Pool $250,874.00

Last Raced	Horse	M/Eqt.	A.Wt	PP	St	1/4	1/2	Str	Fin	Jockey	Cl'g Pr	Odds $1
	Cabo Queen	LB	3 115	10	4	3¹	3¹½	1¹½	1⁵	Antley C W	32000	19.00
6May94 2Hol³	Comeback Kidder	LBb	3 115	8	2	5²	4½	4²	2⁴	Stevens G L	32000	6.90
8May94 10Hol⁴	Christina Maria	LBb	3 115	3	6	4ʰᵈ	5²½	5²	3³½	Nakatani C S	32000	8.30
5May94 9Hol²	Unprectible Gal	LB	3 115	11	1	2ʰᵈ	2ʰᵈ	2½	4ⁿᵒ	Solis A	32000	2.80
8May94 10Hol³	Allessea	B	4 122	4	8	1ʰᵈ	1ʰᵈ	3½	5¹½	McCarron C J	32000	1.40
14May94 9Hol⁶	Astrial	LBb	3 116	7	11	11	10²	8²	6¹½	Delahoussaye E	32000	7.20
13May94 6Hol²	Bolides Wish	LB	3 113	5	3	7¹½	6²	6³½	7⁵	Pedroza M A	28000	27.60
14May94 9Hol⁸	Youtellemshirley	LBb	4 120	6	5	9ʰᵈ	8½	9⁷	8½	Castanon A L	28000	60.00
14May94 9Hol⁶	Drona's Starlet	LB	3 110	2	9	8½	7³	7¹½	9¹¹	Atherton J E⁵	32000	20.60
8May94 10Hol⁶	Martial Artist	LB	3 115	9	7	10⁸	11	10¹	10⁴½	Valenzuela F H	32000	96.70
5May94 9Hol¹⁰	Concorde's Music	LBb	3 110	1	10	6ʰᵈ	9½	11	11	Castanon E F⁵	32000	118.60

OFF AT 5:16 Start Good. Won driving. Time, :21⁴, :45¹, :57¹, 1:09⁴ Track fast.

$2 Mutuel Prices:

11–CABO QUEEN	40.00	15.60	13.60
8–COMEBACK KIDDER		7.40	3.60
3–CHRISTINA MARIA			5.60

$2 EXACTA 11–8 PAID $326.80 $1 SUPERFECTA 11–8–3–12 PAID $4,449.90
$2 QUINELLA 8–11 PAID $151.00

Ch. f, (Mar), by Al Mamoon–Flying Hill, by Flying Paster. Trainer Jackson Bruce L. Bred by Cardiff Stud Farm (Cal).

Even though Cabo Queen went six-wide early, and continued to race wide into the stretch, she benefited from the early speed duel between Allessea and Unpredictable Gal and proved much the best. Besides demonstrating the power of the sire ratings, this race is another example that commercially-oriented owner/breeders run their debut horses where they belong and that those by precocious sires should never be underestimated. While in California, owner/breeders like Cardiff, Harris, River Edge and Golden Eagle Farms are particularly successful, it behooves handicappers to identify their counterparts in other regions of the country and to pay close attention whenever their precociously-bred horses debut in Maiden Claiming spots that make economic sense.

Not all Maiden Claiming races are won by horses with the highest sire ratings. The ninth race on June 26th at Hollywood Park is an example of a cheaper debut runner going off at a shorter price than his better-bred competition and prevailing. Looking at the most obvious contender, Star Doctor showed speed against better and must be respected on the class drop. But, he is coming off a two month layoff, with a big gap in his works,

Hollywood Park

9

6 Furlongs. (1:08) **MAIDEN CLAIMING. Purse $22,000. 3–year–olds and upward. Weights: 3–year–olds, 116 lbs. Older, 122 lbs. Claiming price $50,000; if for $45,000, allowed 2 lbs.**

START ▾ 6 FURLONGS ▴ FINISH

Sir Charles
Own: Southern Nevada Racing Stables Inc
CASTANON A L (61 1 5 7 .02) **$50,000**

Ch. c. 3 (May)
Sire: Ferdinand (Nijinsky II)
Dam: Nahema (Believe It)
Br: Waggener Ronald (Ky)
Tr: MacDonald Brad (8 0 1 1 .00)

L 116

Lifetime Record :	0 M 0 0		$0
1994	0 M 0 0	Turf	0 0 0 0
1993	0 M 0 0	Wet	0 0 0 0
Hol	0 0 0 0	Dist	0 0 0 0

WORKOUTS: Jun 24 SA 5f fst 1:01^1 Hg *16/23* Jun 20 SA 6f fst 1:13^2 H *5/13* Jun 15 SA 5f fst 1:02^4 H *23/27* Jun 10 SA 5f fst 1:02 H *16/28* Jun 4 SA 4f fst :49^3 H *11/34* May 30 SA 3f fst :37^4 H *12/19* May 2 SA 3f fst :36^3 H *7/37*

Final Orbit
Own: Putnam George
PEDROZA M A (211 21 37 34 .10) **$50,000**

Dk. b or br g. 4
Sire: Orbit Ruler (Our Rulla)
Dam: Boiled in Oil (Donut King)
Br: Putnam George F (Cal)
Tr: Hawthorne James W (4 0 0 0 .00)

122

Lifetime Record :	0 M 0 0		$0
1994	0 M 0 0	Turf	0 0 0 0
1993	0 M 0 0	Wet	0 0 0 0
Hol	0 0 0 0	Dist	0 0 0 0

WORKOUTS: ●Jun 15 Fpx 5f fst 1:01^1 Hg *1/4* ●Jun 2 Fpx 4f fst :49^2 H *1/4* May 26 Fpx 3f fst :38^1 H *5/5* Apr 22 Fpx 4f fst :49^2 H *5/15* Apr 15 Fpx 5f fst 1:03 H *6/8* Apr 9 Fpx 4f fst :49 H *8/12* Apr 2 Fpx 3f fst :36^1 H *6/21*

Call to the Hunt
Own: Crawford Margaret L
SCOTT J M (37 1 1 3 .03) **$45,000**

B. h. 6
Sire: Tally Ho the Fox (Never Bend)
Dam: Big Affair (Nanak)
Br: Crawford Margaret L (Cal)
Tr: Landers Dale (2 0 0 0 .00)

L 120

Lifetime Record :	26 M 3 0	$14,090			
1994	11 M 2 0	$10,500	Turf	0 0 0 0	
1993	12 M 1 0	$3,590	Wet	0 0 0 0	
Hol	10 0 1 0	$2,400	Dist	12 0 3 0	$12,190

12May94-6Hol	fst	6½f	:22^2	:45^1	1:09^3	1:16^1 3↑	[S]Md 28000	46	6	7	8^{12}	9^{12}	9^{14}	9$^{13\frac{1}{2}}$	Atherton J E	LB 120 b	21.00	74–09	[D]Pasmoso115^{nk} Santo Juliano117$^{1\frac{3}{4}}$ Boldtopic116$^{1\frac{1}{2}}$	No threat	10
30Apr94-6Hol	fst	1$\frac{1}{16}$	:23	:46^3	1:11^3	1:44^1 3↑	[S]Md 32000	36	8	10	10^{13}	11^{14}	9^{19}	10$^{22\frac{3}{4}}$	Long B	LB 123 b	15.90	57–17	Clever Ricky115^{no} Thirtyfive Black116$^{7\frac{1}{2}}$ However123^{hd}	Wide 7/8	12
14Apr94-5SA	fst	6f	:21^4	:44^1	:56^2	1:09^1	Md 35000	72	4	6	7$^{4\frac{1}{2}}$	7$^{5\frac{3}{4}}$	6$^{6\frac{3}{4}}$	5$^{7\frac{3}{4}}$	Long B	LB 118 b	42.10	83–09	Bolger's Lead120$^{\frac{1}{2}}$ Touch The Moon120$^{2\frac{1}{4}}$ Impressive Don118$^{\frac{1}{2}}$	No rally	9
7Apr94-6SA	fst	6f	:21^3	:45	:57	1:09^3	[S]Md 40000	66	11	8	8$^{7\frac{1}{2}}$	9$^{7\frac{3}{4}}$	8$^{6\frac{1}{2}}$	5^6	Long B	LB 120 b	49.90	83–11	Clackamas118$^{4\frac{1}{2}}$ Tabled Launch118^{no} Morrow's Comet118^1	Wide into lane	11
31Mar94-6SA	fst	6f	:22	:45	:57^2	1:10	Md 35000	58	7	8	8$^{8\frac{1}{4}}$	7^{12}	7^{10}	6$^{10\frac{1}{4}}$	Tejeira J	LB 118 b	21.40	77–13	Gallant Trejo120$^{\frac{1}{2}}$ Touch The Moon120^4 Good Presentation120$^{\frac{1}{2}}$	No rally	9
24Mar94-6SA	fst	6½f	:22^1	:45	1:09^4	1:16^1	[S]Md 35000	59	2	10	9$^{6\frac{1}{2}}$	8^8	8$^{7\frac{1}{4}}$	6$^{8\frac{1}{4}}$	Pincay L Jr	LB 118 b	12.80	83–09	Recommendtion120$^{1\frac{1}{4}}$ Bolger'sLed118$^{1\frac{1}{4}}$ HighrFlyr118^2	Improved position	10
10Mar94-6SA	fst	6f	:21^4	:44^4	:57^1	1:10	[S]Md 35000	70	6	9	8^6	7^6	5$^{7\frac{1}{2}}$	2^4	Castanon A L	LB 118 b	15.40	83–10	Title Defense118^4 Call To The Hunt118$^{\frac{3}{4}}$ Touch The Moon120$^{2\frac{1}{2}}$	Good try	11
24Feb94-2SA	fst	6½f	:22	:45^1	1:10^2	1:17	[S]Md 35000	47	1	6	4^3	5$^{5\frac{1}{2}}$	5^7	4$^{11\frac{1}{2}}$	Scott J M	LB 118 b	15.40	75–13	StolenProspct119$^{3\frac{1}{2}}$ TbldLunch117^1 Morrow'sComt118^7	Lugged out turn	8
3Feb94-4SA	fst	6f	:21^4	:44^4	:57	1:10	Md 35000	54	7	7	7^5	7$^{7\frac{1}{2}}$	8$^{8\frac{1}{2}}$	7$^{11\frac{3}{4}}$	Scott J M	L 118 b	16.00	75–09	The Informer119^5 Impressive Don118^2 Gus Gus120$^{\frac{1}{2}}$	Evenly	11
27Jan94-6SA	fst	6½f	:21^4	:44^2	1:09^4	1:16^3	[S]Md 35000	52	2	7	6$^{3\frac{1}{2}}$	7^9	6$^{7\frac{1}{2}}$	6$^{8\frac{3}{4}}$	Scott J M	L 118 b	8.60	80–16	Start'n Glo120$^{\frac{1}{2}}$ Touch The Moon119$^{1\frac{1}{4}}$ Stolen Prospect117^3	Weakened	11

WORKOUTS: ●Jun 23 SA 3f fst :34^3 H *1/14* Apr 11 SA 3f fst :39^4 H *20/20*

Quartz
Own: Tuck L L
ATKINSON P (106 9 13 9 .08) **$50,000**

B. g. 3 (Feb)
Sire: Nasty and Bold (Naskra)
Dam: The Best Years (Olden Times)
Br: Wild Plum Farm (Colo)
Tr: Tuck Mary Lou (3 0 1 0 .00)

116

Lifetime Record :	0 M 0 0		$0
1994	0 M 0 0	Turf	0 0 0 0
1993	0 M 0 0	Wet	0 0 0 0
Hol	0 0 0 0	Dist	0 0 0 0

WORKOUTS: Jun 23 Hol 3f fst :37 H *14/24* Jun 4 Hol 4f fst :49^2 H *31/48* May 29 Hol 4f fst :47 Hg *3/38* May 19 Hol 6f fst 1:15^4 H *18/19* May 13 Hol 6f fst 1:16^1 H *5/12* May 6 Hol 5f fst 1:01^4 H *31/35* Apr 28 Hol 4f fst :49^4 H *28/39* Apr 20 Hol 5f fst 1:01^4 H *10/20* Apr 15 Hol 3f fst :37^1 H *18/30* Mar 21 Hol 4f fst :48^4 H *5/19* Mar 14 Hol 4f fst :49^2 H *10/15* Mar 8 Hol 3f fst :36^4 H *5/15*

continued

Alongcameelliot

Own: Sandor Istvan J

Ch. g. 4
Sire: Highland Park (Raise a Native)
Dam: Mineola Lass (Swoon's Son)
Br: Mrs. Jack G. Jones, Sr. (Ky)
Tr: Cosme Ruben G (14 0 1 0 .00)

TRUJILLO V (62 2 3 1 .03) $50,000

[handwritten: C 8/4]

L 122

						Lifetime Record:	8 M 0 0	$7,000		
						1994	7 M 0 0	$7,000	Turf	0 0 0 0
						1993	1 M 0 0	$2,200	Wet	1 0 0 0 $2,400
						Hol	4 0 0 0	$4,050	Dist	4 0 0 0 $4,050

9Jun94-4Hol	fst 6½f	:22	:444 1:094 1:162	3+ Md 50000	78 5 5 2½ 42½ 41½ 53	Scott J M	LB 122 b 108.70	84-13	Air Corn116no Mr. Flower Power122½ Dazzled By Gold116½	Checked 3/8 10
27May94-2Hol	fst 6f	:222	:454 :574 1:102	3+ Md 50000	65 3 2 2½ 43½ 46 48¾	Atkinson P	L 122 fb 16.90	79-11	Walk Point115⁵ Respectable Rascal122nk Score O Gold116¾	Gave way 7
7May94-9Hol	fst 6f	:22	:45 :564 1:09	3+ Md Sp Wt	57 5 7 74¾ 108½ 912 913½	Trujillo V	LB 122 b 83.80	82-06	Bold Halo122¾ Slewpy Dewpy Doo115¼ Flying Marfa115³	No threat 10
18Mar94-6SA	fst 6½f	:221	:442 1:093 1:16	Md Sp Wt	65 7 4 66 76½ 77 610	Castanon A L	LB 120 b 33.80	82-10	Sikes120³½ Dion The Tyrant120hd J. R. Cigar120²	Wide into lane 7
4Mar94-6SA	fst 7f	:222	:452 1:094 1:22	Md Sp Wt	69 9 4 64 74½ 88½ 811½	Castanon A L	LB 120 b 79.60	81-13	Years Of Dreaming120½ Desperately1201 Cezind1203½	Gave way 11
4Feb94-3SA	my 6f	:214	:443 :571 1:102	Md Sp Wt	68 2 5 64½ 69½ 66½ 46	Castanon A L	LB 119 b 20.30	79-16	Gian's Big Break119½ Cezind119²½ Desert Spy119²	Closed belatedly 6
21Jan94-6SA	fst 6½f	:22	:444 1:094 1:161	Md Sp Wt	69 4 2 21 21 21 44	Castanon A L	LB 119 b 89.00	87-12	Greenspan119³ Juiceman119½ Desperately119½	Outfinished 10
9Nov93-3Hol	fst 6f	:22	:45 :564 1:092	3+ Md Sp Wt	27 6 4 55 612 623 625½	Castanon A L	B 115 38.00	68-11	Cigar117²½ Golden Slewpy116⁵½ Famous Fan115hd	Wide backstretch 6

WORKOUTS: Apr 24 SA 6f fst 1:13 H 2/12

Star Doctor

Own: Oldknow & Phipps

Ch. c. 3 (Apr)
Sire: Dr. Carter (Caro)
Dam: Retsina Star (Pia Star)
Br: William H. Oldknow & Robert W. Phipps (Ky)
Tr: Sadler John W (47 8 6 4 .17)

SOLIS A (283 55 27 44 .19) ✓ $50,000

[handwritten: D 6/3 early speed droppin]

L 116

						Lifetime Record:	2 M 0 0	$800		
						1994	2 M 0 0	$800	Turf	0 0 0 0
						1993	0 M 0 0		Wet	1 0 0 0
						Hol	0 0 0 0		Dist	0 0 0 0

15Apr94-6SA	fst 7f	:222	:451 1:094 1:222	3+ Md Sp Wt	65 4 2 2hd 33½ 51½ 510½	Flores D R	LB 114 24.50	80-08	Muirfield Village116hd Bold Halo118¹½ Explorateur116⁵	Inside duel 8
20Mar94-6SA	my 6½f	:22	:454 1:12 1:184	Md Sp Wt	46 2 8 813 710 77½ 616	Flores D R	B 118 23.80	62-19	Jade Master118⁷ Nucay118³ Le Dome118¾	Off bit slow 9

WORKOUTS: Jun 18 Hol 5f fst 1:01 H 13/32   Jun 12 Hol 4f fst :474 H 14/65   May 23 Hol 4f fst :474 H 10/46   ●May 3 Hol 7f fst 1:272 H 1/4   Apr 27 SA 5f fst 1:013 B 7/18   Apr 13 SA 4f fst :472 H 17/49

Rah Rah's Cheer

Own: Three Sisters Stable

B. c. 3 (Apr)
Sire: Rahy (Blushing Groom)
Dam: Trish Lum (Proud Clarion)
Br: Makk Laszlo (Ky)
Tr: Chambers Mike (6 1 1 1 .17)

DELAHOUSSAYE E (106 21 20 21 .20) $50,000

[handwritten: B+2/3 42K D 6/8]

116

						Lifetime Record:	0 M 0 0	$0		
						1994	0 M 0 0		Turf	0 0 0 0
						1993	0 M 0 0		Wet	0 0 0 0
						Hol	0 0 0 0		Dist	0 0 0 0

WORKOUTS: Jun 20 SA 5f fst :593 H 2/20   Jun 14 SA 5f fst 1:01 H 6/29   Jun 8 SA 5f fst 1:012 H 14/31   ●Jun 2 SA 6f fst 1:141 Hg 1/4   May 27 SA 7f fst 1:304 H 4/5   May 19 SA 6f gd 1:152 H 1/3   May 13 SA 6f fst 1:14 H 8/17   May 7 SA 6f gd 1:182 H 9/10   May 1 SA 5f sl 1:04 H 26/47   Apr 20 SA 5f fst 1:023 H 27/34   Apr 14 SLR tr.t 4f fst :473 H 1/8   Apr 7 SLR tr.t 4f fst :494 H 2/2

Cooper Indian

Own: Suffolk Racing Partnership Ltd

B. c. 3 (Apr)
Sire: Woodman (Mr. Prospector)
Dam: Believablee (Believe It)
Br: Nydrie Stud & Katalpa Farm (Ky)
Tr: Gregson Edwin (11 0 3 0 .00)

STEVENS G L (209 46 28 44 .22) $50,000

[handwritten: C 7.5/3]*

116

						Lifetime Record:	0 M 0 0	$0		
						1994	0 M 0 0		Turf	0 0 0 0
						1993	0 M 0 0		Wet	0 0 0 0
						Hol	0 0 0 0		Dist	0 0 0 0

WORKOUTS: Jun 23 Hol 4f fst :491 H 21/38   Jun 13 Hol 6f fst 1:142 H 7/13   Jun 7 Hol 6f fst 1:133 H 6/14   Jun 2 Hol 5f fst 1:004 H 18/42   May 28 Hol 5f fst 1:004 H 18/37   May 23 Hol 4f fst :493 H 31/46   May 18 Hol 4f fst :484 H 18/57   May 13 Hol 3f fst :363 H 3/18   May 8 Hol 3f fst :364 H 12/19

Dancing In Place

Own: Lewis Paul C

Ch. g. 3 (Feb)
Sire: Cliffs Place (Raised Socially)
Dam: Crozie Dancer (Somethingfabulous)
Br: Lewis Robert C (Nev)
Tr: Craigmyle Scott J (5 0 0 0 .00)

FLORES D R (130 14 8 11 .11) $45,000

[handwritten: U 10/5]

114

						Lifetime Record:	0 M 0 0	$0		
						1994	0 M 0 0		Turf	0 0 0 0
						1993	0 M 0 0		Wet	0 0 0 0
						Hol	0 0 0 0		Dist	0 0 0 0

WORKOUTS: Jun 19 Fpx 4f fst :484 H 2/5   Jun 12 Fpx 3f fst :354 H 1/2   Apr 19 TuP 5f fst :594 H 3/17   Apr 13 TuP 5f fst :592 H 2/15   Apr 10 TuP 3f fst :36 Hg 9/35   Mar 18 TuP 3f fst :39 B 38/44   Mar 14 TuP 5f fst 1:022 H 42/55

Yoda M D

Own: DeDomenico Mark

Dk. b or br g. 3 (Apr)
Sire: Knights Choice (Drum Fire)
Dam: Pamlisa's Delight (Drone)
Br: Northwest Farms (Wash)
Tr: Hofmans David (25 2 3 7 .08)

ANTLEY C W (226 49 43 25 .22) $50,000

[handwritten: B 6/5 2K D 6/7]

L 116

						Lifetime Record:	0 M 0 0	$0		
						1994	0 M 0 0		Turf	0 0 0 0
						1993	0 M 0 0		Wet	0 0 0 0
						Hol	0 0 0 0		Dist	0 0 0 0

WORKOUTS: Jun 22 Hol 5f fst 1:02 H 32/53   Jun 11 Hol 6f fst 1:134 H 4/19   Jun 3 Hol 6f fst 1:154 H 13/17   May 28 Hol 5f fst 1:013 H 25/37   May 22 Hol 5f fst 1:014 H 22/43   May 16 Hol 4f fst :491 H 26/42   Apr 18 YM 5f fst 1:001 H 4/10   ●Apr 7 YM 3f fst :35 Hg 1/23   Apr 3 YM 3f fst :37 B 7/19   Mar 26 WoC 4f fst :531 Bg 4/7   Mar 12 WoC 3f fst :412 B 7/8   Feb 26 WoC 4f sly :541 B 4/6

Respectable Rascal

Own: Horowitz Elliot J

Dk. b or br c. 4
Sire: Capote (Seattle Slew)
Dam: Merry Says So (Advocator)
Br: Calumet Farm & Glenmore Farm (Ky)
Tr: Smith Michael R (10 1 2 0 .10)

MCCARRON C J (128 22 23 21 .17) $50,000

[handwritten: D 5/4 Finished 2nd at 50,000]

L 122

						Lifetime Record:	8 M 4 1	$27,100		
						1994	1 M 1 0	$4,400	Turf	0 0 0 0
						1993	5 M 2 1	$17,100	Wet	1 0 0 1 $4,350
						Hol	2 0 2 0	$10,000	Dist	1 0 1 0 $4,400

27May94-2Hol	fst 6f	:222	:454 :574 1:102	3+ Md 50000	74 5 3 31½ 21½ 25	McCarron C J	LB 122 b *2.20	83-11	WlkPoint115⁵ RespectbleRscl122nk ScoreOGold116¾	Broke out, bumped 7
14Apr93-6SA	fst 1	:223	:462 1:104 1:362	3+ Md Sp Wt	76 1 2 21½ 22 46 68½	Valenzuela P A	LB 117 2.20	80-12	Harris County114¹½ Pleasant Tango118nk Outside The Line117³½	Faltered 8
14Mar93-4SA	fst 1	:23	:47 1:112 1:37	Md Sp Wt	65 1 3² 52½ 55½ 510	Stevens G L	LB 117 b 3.60	75-12	Altiplano117no Outside The Line117½ Mc Comas117³½	Brushed early 6
28Feb93-6SA	fst 1⅛	:233	:48 1:13 1:442	Md Sp Wt	80 1 3 31½ 2½ 22½ 22¾	Stevens G L	LB 117 b 3.30	79-17	Dinand117²¾ Respectable Rascal117⁵ Phoenician117⁸	Good effort 6
31Jan93-6SA	fst 7f	:232	:461 1:113 1:384	Md Sp Wt	66 8 1 1½ 1hd 21 23	Delahoussaye E L	LB 117 b 3.30	73-17	OnlyAlpha117³ RespctbleRscl117¹ OutsideTheLine117no	4 Wide 1st turn 9
9Jan93-6SA	my 7f	:232	:471 1:12 1:244	Md Sp Wt	69 4 6 53½ 33 31½ 33½	Pincay L Jr	B 118 b *1.60	75-15	MeadowBlaze118¹½ OnlyAlpha118² RespctbleRscl118⁵½	Not enough late 6
13Dec92-6Hol	fst 7½f	:223	:461 1:112 1:304	Md Sp Wt	68 7 3 52½ 51¾ 32½ 23	Pincay L Jr	B 119 b 8.70	80-13	BlazingAura119½ RespectableRascal119½ Jaltipan119⁴	Wide backstretch 6
17Oct92-6SA	fst 6½f	:213	:443 1:09 1:153	Md Sp Wt	37 6 4 52³ 89 814 720½	Atkinson P	B 117 b 32.30	71-12	Stuka117⁷ Tossofthecoin117 Cut To Run117	Brushed start 4

WORKOUTS: Jun 21 SA 6f fst 1:142 H 8/12   Jun 14 SA 5f fst 1:021 H 19/29   May 22 SA 4f fst :47 H 4/35   May 17 SA 7f fst 1:27 H 6/8   May 11 SA 7f fst 1:271 H 1/3   May 6 SA 6f fst 1:134 H 7/12

© 1994 DRF, Inc.

and his previous Beyers don't jump off the page. Even more important, despite the positive jockey switch to Solis, he is not the betting favorite when he should be.

Of the other experienced runners, Respectable Rascal is the second favorite at a little under 3-1. But he is parked on the outside, will have to use his speed early to keep up with Star Doctor, and may bounce in his second start back. Alongcameelliot has improving Beyers, but he has already lost twice at this level and along with Call To The Hunt, a chronic loser to lesser, looks like a throwout.

Of the six first-time starters, Sir Charles, Final Orbit and Dancing In Place are all by unprecocious sires and can be eliminated. Cooper Indian has a decent sire rating, but Woodman has a stamina index (3*) which suggests he will need more than six furlongs to prevail. This leaves us with Rah Rah's Cheer and Yoda M D to consider.

Since Rah Rah's Cheer's B+2 rating is both more precocious and classier than Yoda M D's B6, he seems more likely to upset Star Doctor. But that's not the way the tote sees it. Interestingly, the more cheaply-bred Yoda M D is 5-2 while Rah Rah's Cheer is 9-2. How could this be?

Checking our copy of *Maiden Stats 1993*, we discover that Rah Rah's Cheer sold for $52,000 at auction while Yoda M D went for $32,000. Why then is Rah Rah's Cheer being run for a $50,000 tag where Three Sisters Stable will obviously lose money on him? Yoda M D, on the other hand, is being run where he belongs according to his class 6 rating. If he wins and gets claimed out of this race, he will earn Mark DeDomenico a healthy profit. Given that David Hofmans is a competent debut trainer and Yoda M D is the actual betting favorite, we can assume that the Knights Choice gelding is well meant in this spot.

So how to bet? Given the tote action, a healthy Win bet on Yoda MD is in order along with keying him top and bottom in the exacta to both Rah Rah's Cheer and Star Doctor.

Results: Star Doctor and Rah Rah's Cheer dueled for the early lead with Yoda M D being reserved in third. Yoda M D caught Rah Rah's Cheer at the top of the stretch, and drew off by nearly three lengths, paying $6.60 to Win. The $2 exacta returned $29.80.

To Forgive is Divine

The fifth race at Golden Gate Fields on June 9th represents another kind of profitable Maiden Claiming play. This involves betting a horse dropping in class with a good debut rating that was shuffled back at the outset in its first start and wasn't really used thereafter. While the public might dismiss such a horse in its next start because it finished up the track, astute handicappers understand that maidens are an immature lot and some of them will inevitably have trouble breaking from or just out of the gate in their first starts. When this happens, especially with 2-year-old runners in short sprints, they are often eased by their jockeys who know that without a clean break they have lost all

Golden Gate Fields

5 *5 Furlongs.* (:56²) **MAIDEN CLAIMING.** Purse $13,000. 2-year-olds. Weight, 116 lbs. Claiming price $25,000; if for $22,500, allowed 2 lbs.

5 FURLONGS

Great Days Ahead
Own: Valpredo Donald J

MIRANDA V (234 17 16 24 .07) $25,000

B. c. 2 (Apr)
Sire: Somethingfabulous (Northern Dancer)
Dam: Cuci d'Or (Dimaggio)
Br: Donald John Valpredo (Cal)
Tr: Gilchrist Greg (65 16 23 4 .25)

C6/4 116

Lifetime Record :		0 M 0 0		$0
1994	0 M 0 0		Turf	0 0 0 0
1993	0 M 0 0		Wet	0 0 0 0
GG	0 0 0 0		Dist	0 0 0 0

WORKOUTS: Jun 3 GG 5f fst 1:03² H 26/33 May 27 GG 5f fst 1:02⁴ H 21/32 May 20 GG 4f fst :49¹ H 24/44 May 13 GG 4f fst :49⁴ H 20/39 May 6 GG 3f fst :36 H 4/16 Apr 29 GG 3f gd :37³ H 12/34 Apr 22 GG 3f fst :39 H 11/15 Apr 17 GG 2f fst :26¹ H 8/15

The Crowns Torch
Own: Irish Stable & Calvario & Partners

ESPINOZA V (405 46 57 58 .11) $25,000

Dk. b or br c. 2 (Mar)
Sire: Crowning (Raise a Native)
Dam: Lulutu (Envoy)
Br: Claudine D. Molick (Cal)
Tr: Calvario Manuel (37 8 6 7 .22)

U9/3 116

Lifetime Record :		0 M 0 0		$0
1994	0 M 0 0		Turf	0 0 0 0
1993	0 M 0 0		Wet	0 0 0 0
GG	0 0 0 0		Dist	0 0 0 0

WORKOUTS: May 29 BM 5f fst 1:00³ Hg 3/15 May 22 BM 4f fst :49³ H 7/14 May 15 BM 4f fst :51³ Hg 46/46 May 8 BM 4f fst :51¹ H 26/35 May 1 BM 3f fst :38³ Hg 9/12 Apr 23 BM 3f fst :37² H 3/19 Apr 16 BM 2f fst :26¹ H 3/4

Balcony's Surprise
Own: Roberts Gladys M

MERCADO P (313 19 31 44 .06) $25,000

Dk. b or br c. 2 (Jan)
Sire: Worthy Endevor (Ruken)
Dam: Balcony Kris (First Balcony)
Br: Gladys M. Roberts (Cal)
Tr: Henderson Frances (63 4 6 5 .06)

U9/5 116

Lifetime Record :		1 M 0 0		$0
1994	1 M 0 0		Turf	0 0 0 0
1993	0 M 0 0		Wet	0 0 0 0
GG	1 0 0 0		Dist	0 0 0 0

20May94-6GG fst 4½f :21⁴ :45⁴ :52 Md 25000 33 6 4 5⁶ 57¼ 6¹⁰ Chapman T M B 116 29.90 81-12 Agent Warrior116² I Promise Roses116½ Slow Cowboy115⁵ No threat 9
WORKOUTS: Jun 4 GG 5f fst 1:01⁴ H 18/36 May 15 GG 4f fst :49² H 26/57 May 9 GG 5f fst 1:05¹ H 45/46 Apr 29 GG 4f gd :50⁴ Hg 28/56 Apr 23 GG 4f fst :49¹ Hg 27/61 Apr 17 GG 3f fst :36² H 7/56

Panic Relief
Own: Lewis R M

HUMMEL C R (253 21 31 22 .08) $22,500

B. c. 2 (Mar)
Sire: Falstaff (Lyphard)
Dam: Hy Girls (Bargain Day)
Br: Richard J. Lewis & Rancho Yours Mine Ours (Cal)
Tr: Lewis Richard J (29 1 0 7 .03)

D7/5 114

Lifetime Record :		0 M 0 0		$0
1994	0 M 0 0		Turf	0 0 0 0
1993	0 M 0 0		Wet	0 0 0 0
GG	0 0 0 0		Dist	0 0 0 0

WORKOUTS: Jun 5 GG 4f fst :48⁴ H 30/69 May 29 GG 4f fst :48² Hg 11/45 May 22 GG 5f fst 1:03² H 56/65 May 15 GG 5f fst 1:02² H 28/55 May 4 GG 4f fst :48⁴ H 12/34 Apr 28 GG 3f gd :37 H 5/26 Apr 22 GG 3f fst :37² H 6/15 Apr 16 GG 3f fst :37 H 9/17 Apr 10 GG 2f fst :25 H 9/13

Glasslifter
Own: Lanning Curt & Lila

MEZA R Q (364 48 54 45 .13) *V* $25,000

Dk. b or br c. 2 (Apr)
Sire: Pass the Glass (Buckpasser)
Dam: Flying Gaelic (Gaelic Dancer)
Br: Curt Lanning & Lila Lanning (Cal)
Tr: Moger Ed Jr (94 19 8 13 .20)

B+6/4 116

Lifetime Record :		1 M 0 0		$0
1994	1 M 0 0		Turf	0 0 0 0
1993	0 M 0 0		Wet	0 0 0 0
GG	1 0 0 0		Dist	0 0 0 0

25May94-4GG fst 4½f :21² :45¹ :51³ Md Sp Wt 26 1 7 9¹⁷ 9¹⁹ 8¹⁶½ Lopez A D LB 117 19.40 76-15 Mocha Breeze116² Turko's Turn115¹½ Rocky T.116¹½ Shuffled back early 9
WORKOUTS: Jun 3 GG 5f fst 1:03³ H 29/33 May 19 GG 4f fst :50² Hg 22/27 May 7 GG 4f sly :55 H (d)8/9 ●Apr 28 GG 2f gd :25³ H 1/4

Tam's Chill
Own: Tam Richard

LOPEZ A D (425 72 63 61 .17) $25,000

Ch. c. 2 (Apr)
Sire: Big Chill (It's Freezing)
Dam: Princess Yoko (Raise a Native)
Br: Richard Tam (Cal)
Tr: Silva Jose L (166 34 28 17 .20)

M8/3 116

Lifetime Record :		0 M 0 0		$0
1994	0 M 0 0		Turf	0 0 0 0
1993	0 M 0 0		Wet	0 0 0 0
GG	0 0 0 0		Dist	0 0 0 0

WORKOUTS: Jun 3 GG 4f fst :49² H 19/36 May 28 GG 5f fst 1:03⁴ H 22/26 May 22 GG 4f fst :50³ Hg 56/71 May 14 GG 3f fst :37 H 3/14 May 6 GG 4f fst :49¹ H 19/31 Apr 29 GG 4f gd :52² H 50/56 Apr 16 GG 3f fst :36³ Hg 4/17 Apr 9 GG 3f gd :38⁴ Hg 9/13 Apr 2 GG 2f fst :24² H 1/2 Mar 26 GG 2f fst :27² H 4/4

Glorious Ghost
Own: Golden Eagle Farm

WARREN R J JR (419 68 51 49 .16) $25,000

Gr. c. 2 (Mar)
Sire: Bel Bolide (Bold Bidder)
Dam: Rare Gal (Caro)
Br: Mr. & Mrs. John C. Mabee (Cal)
Tr: Josephson Jedd B (66 11 9 11 .17)

C7/4 116

Lifetime Record :		0 M 0 0		$0
1994	0 M 0 0		Turf	0 0 0 0
1993	0 M 0 0		Wet	0 0 0 0
GG	0 0 0 0		Dist	0 0 0 0

WORKOUTS: Jun 4 GG 4f fst :48⁴ H 7/29 May 27 GG 5f fst 1:04 Hg 27/32 May 20 GG 5f fst 1:02¹ H 16/23 May 10 SLR tr.t 4f fst :49³ H 14/17 May 3 SLR tr.t 4f fst :49 H 4/11 Apr 19 SLR tr.t 4f fst :48⁴ H 8/20 Apr 13 SLR tr.t 3f fst :36¹ H 2/11

Grey Flyer
Own: Weigel M J

BAZE R A (581 153 109 95 .26) $25,000

Ro. g. 2 (Jan)
Sire: Polynesian Flyer (Flying Lark)
Dam: Definitely Sure (Pappagallo)
Br: James E. Lea & Dennis C. Lea (Wash)
Tr: Arterburn Tim (36 6 9 3 .17)

D10/4 116

Lifetime Record :		1 M 0 0		$300
1994	1 M 0 0	$300	Turf	0 0 0 0
1993	0 M 0 0		Wet	0 0 0 0
GG	1 0 0 0	$300	Dist	0 0 0 0

29Apr94-3GG gd 4½f :22 :46³ :53¹ Md c-25000 33 4 5 5¹¹ 56½ 58½ Boulanger G B 116 b *.70 76-20 Jupiter Nik116²½ I Promise Roses116½ Never Social116³½ Dull try 6
Claimed from Lea Dennis & James, Klokstad Bud Trainer
WORKOUTS: Jun 3 BM 5f fst 1:00⁴ Hg 3/8 May 27 BM 5f fst 1:01⁴ Hg 11/19 May 22 BM 5f fst 1:01³ H 3/14 May 13 GG 5f fst 1:02¹ H 10/26 May 8 BM 3f fst :38² H 10/17 Apr 27 BM 4f fst :48² H 3/28

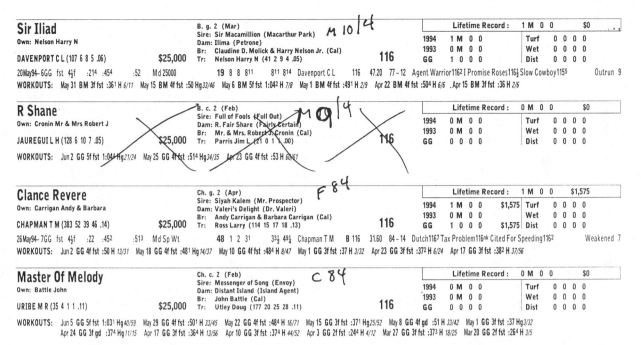

Sir Iliad
Own: Nelson Harry N
DAVENPORT C L (107 6 8 5 .06) $25,000
B. g. 2 (Mar)
Sire: Sir Macamillion (Macarthur Park)
Dam: Ilima (Petrone)
Br: Claudine D. Molick & Harry Nelson Jr. (Cal)
Tr: Nelson Harry N (41 2 9 4 .05)
116   *M 10/4*

	Lifetime Record :	1 M 0 0		$0	
1994	1 M 0 0		Turf	0 0 0 0	
1993	0 M 0 0		Wet	0 0 0 0	
GG	1 0 0 0		Dist	0 0 0 0	

20May94-6GG fst 4½f :214 :454 :52 Md 25000 19 8 8 8¹¹ 8¹¹ 8¹⁴ Davenport C L 116 47.20 77-12 Agent Warrior116² I Promise Roses116½ Slow Cowboy115⁵ Outrun 9
WORKOUTS: May 31 BM 3f fst :36¹ H 6/11 May 15 BM 4f fst :50 Hg33/46 May 6 BM 5f fst 1:04² H 7/8 May 1 BM 4f fst :49¹ H 2/9 Apr 22 BM 4f fst :50⁴ H 6/6 Apr 15 BM 3f fst :36 H 2/6

R Shane
Own: Cronin Mr & Mrs Robert J
JAUREGUI L H (128 6 10 7 .05) $25,000
B. c. 2 (Feb)
Sire: Full of Fools (Full Out)
Dam: R. Fair Share (Fairly Certain)
Br: Mr. & Mrs. Robert J. Cronin (Cal)
Tr: Parris Jim L (21 0 1 1 .00)
116   *M 9/4*

	Lifetime Record :	0 M 0 0		$0	
1994	0 M 0 0		Turf	0 0 0 0	
1993	0 M 0 0		Wet	0 0 0 0	
GG	0 0 0 0		Dist	0 0 0 0	

WORKOUTS: Jun 2 GG 5f fst 1:04¹ Hg21/24 May 25 GG 4f fst :51⁴ Hg34/35 Apr 23 GG 4f fst :53 H 60/61

Clance Revere
Own: Carrigan Andy & Barbara
CHAPMAN T M (383 52 39 46 .14) $25,000
Ch. g. 2 (Apr)
Sire: Siyah Kalem (Mr. Prospector)
Dam: Valeri's Delight (Dr. Valeri)
Br: Andy Carrigan & Barbara Carrigan (Cal)
Tr: Ross Larry (114 15 17 18 .13)
116   *F 84*

	Lifetime Record :	1 M 0 0		$1,575	
1994	1 M 0 0	$1,575	Turf	0 0 0 0	
1993	0 M 0 0		Wet	0 0 0 0	
GG	1 0 0 0	$1,575	Dist	0 0 0 0	

26May94-7GG fst 4½f :22 :45² :51³ Md Sp Wt 48 1 2 3¹ 3³½ 4⁹¼ Chapman T M B 116 31.60 84-14 Dutch116⁷ Tax Problem116nk Cited For Speeding116² Weakened 7
WORKOUTS: Jun 2 GG 4f fst :50 H 13/31 May 18 GG 4f fst :48¹ Hg14/37 May 10 GG 4f fst :48⁴ H 8/47 May 1 GG 3f fst :37 H 3/32 Apr 23 GG 3f fst :37² H 6/24 Apr 17 GG 3f fst :38² H 37/56

Master Of Melody
Own: Battle John
URIBE M R (35 4 1 1 .11) $25,000
Ch. c. 2 (Feb)
Sire: Messenger of Song (Envoy)
Dam: Distant Island (Island Agent)
Br: John Battle (Cal)
Tr: Utley Doug (177 20 25 28 .11)
116   *C 84*

	Lifetime Record :	0 M 0 0		$0	
1994	0 M 0 0		Turf	0 0 0 0	
1993	0 M 0 0		Wet	0 0 0 0	
GG	0 0 0 0		Dist	0 0 0 0	

WORKOUTS: Jun 5 GG 5f fst 1:03¹ Hg40/59 May 29 GG 4f fst :50¹ H 33/45 May 22 GG 4f fst :48⁴ H 16/71 May 15 GG 3f fst :37¹ Hg25/52 May 8 GG 4f gd :51 H 33/42 May 1 GG 3f fst :37 Hg3/32
Apr 24 GG 3f gd :37⁴ Hg11/15 Apr 17 GG 3f fst :36⁴ H 13/56 Apr 10 GG 3f fst :37⁴ H 44/52 Apr 3 GG 2f fst :24⁴ H 4/12 Mar 27 GG 3f fst :37³ H 18/25 Mar 20 GG 2f fst :26⁴ H 3/5

chance of winning. While some troubled maidens can be excellent bets in their second start, it is important to realize that others, especially those that have needed lots of gate works, simply may be chronic bad actors and poor betting risks.

Looking over this 5 furlong sprint, the first thing to note is that there are two class droppers in Clance Revere and Glasslifter and that none of the rest of the horses are by precocious sires. Grey Flyer, with his excellent Tim Arterburn/Russell Baze connections, is the heavy favorite at 8-5. Still, he looks vulnerable because after a six week break he is running back for the same price for which he was claimed. Furthermore, Grey Flyer didn't show any speed first-time out and his sire's class 10 rating suggests he may need to be dropped even further before being competitive.

Clance Revere, on the other hand, despite his sire's F8 rating, showed considerable ability in his first start against better and looks like a legitimate contender at 5-2 in this spot. In fact, given his competent connections, early speed, and the class drop, it's worth asking why Clance Revere isn't the favorite in this field.

With doubts about both Grey Flyer and Clance Revere, Glasslifter looms as an intriguing possibility, especially given his 15-1 odds. Since he was stuck on the rail and was shuffled back in his first start, Glasslifter never got a chance to run to his superior B+6 breeding. That race has to be thrown out. This time he's drawn a middle post and is running against cheaper which should give him a much better shot. Furthermore, the fact that Glasslifter wasn't used first time out may actually give him an advantage over Clance Revere. This is because the stress of being asked for too much speed in its debut can often leave a beaten 2-year-old feeling sore and "discouraged" for its second race. Clance Revere's work of :50 on June 2nd is slow compared to his previous works and suggests he might be a candidate to bounce his second time out.

So how to bet? Again the basic strategy is to bet the overlay contender Glasslifter to Win and place him behind the action horses Clance Revere and Grey Flyer in the exacta.

Results: The Crowns Torch outdueled Clance Revere for the lead to mid-stretch then was passed first by Master of Melody and then by the even faster closing Glasslifter. Glasslifter returned $32 to Win, $12.60 to Place and $6.20 to Show. With both Grey Flyer and Clance Revere finishing out of the money, it's tempting, retrospectively, to argue that I should have followed my intuition about Grey Flyer's vulnerability and bet Glasslifter to Win and Place rather than fooling around with the exotics. But numerous studies have shown that legitimate longshots finish second in the exacta to favorites much more often than they win, and that percentage wise it is costly to ignore this fact.

While class drops are important, they are not an "only factor" in handicapping Maiden Claiming races. The third race at Golden Gate Fields on April 7th is another example of a troubled debut horse running to its precocious breeding in its second start at the $25,000 Maiden Claiming level.

Looking over the field, the 4-5 favorite is Mira Santa, an early speed shipper from Santa Anita that is running for only a little less than the $32,000 tag she has already failed at eight previous times. Not only does she look like a vulnerable favorite, but at the price, she is definitely worth taking a stand against.

Golden Gate Fields

3

6 Furlongs. (1:073) **MAIDEN CLAIMING. Purse $12,000. Fillies, 3-year-olds. Weight 117 lbs. Claiming price $25,000; if for $22,500, allowed 2 lbs.**

Wild Royal
Own: F R L Thoroughbreds & Bonde Roselma
OLGUIN G L (106 9 7 18 .08) $25,000

Dk. b or br f. 3 (Mar)
Sire: Wild Again (Icecapade)
Dam: Royal Setting (Raise a Cup)
Br: Calumet Farm & Deters Charles H (Ky)
Tr: Bonde Jeff (53 8 10 6 .15)
117

Lifetime Record:	3 M 1 0	$3,800			
1994	3 M 1 0	$3,800	Turf	0 0 0 0	
1993	0 M 0 0		Wet	0 0 0 0	
GG	2 0 1 0	$3,800	Dist	3 0 1 0	$3,800

5Mar94-5GG fst 6f :214 :451 :58 1:11 ⒻMd Sp Wt 43 7 7 77¾ 66¼ 79 67½ Olguin G L LB 117 b 9.60 76-13 Key To V. A.1172½ Sea Heroine1171½ Deanna's Destiny117hd Raced far wide 10
13Feb94-6GG fst 6f :22 :45 :571 1:094 ⒻMd Sp Wt 48 8 6 44 33½ 37 210 Olguin G L LB 117 b 8.40 80-08 Gaelic Dowry11710 Wild Royal117nk Newcastle Blues117½ Second best 10
7Jan94-9SA fst 6f :212 :441 :564 1:094 ⒻMd 50000 50 12 7 910 1013 1012 712¼ Desormeaux K J B 117 fb 14.00 76-08 Sophi's Favorite1171¾ Flying Royalty1174 Linda Lou B.1171¼ 12
Carried wide by loose horse

WORKOUTS: Apr 5 GG 3f fst :391 H *5/5* Mar 30 GG 3f fst :371 H *7/15* Mar 24 GG 5f fst 1:003 H *5/26* Mar 18 GG 3f fst :394 H *15/15* Mar 2 GG 4f fst :513 H *42/55* Feb 24 GG 4f gd :51 H *40/65*

Wild Forever
Own: Beall Terry D
WARREN R J JR (183 28 15 27 .15) $25,000

Dk. b or br f. 3 (Feb)
Sire: Wild Again (Icecapade)
Dam: R. T.'s Flight (Grey Dawn II)
Br: Beall Terry D (Ky)
Tr: Caton Dent (12 2 0 2 .17)
117

Lifetime Record:	3 M 0 0	$0		
1994	2 M 0 0		Turf	0 0 0 0
1993	1 M 0 0		Wet	0 0 0 0
GG	0 0 0 0		Dist	1 0 0 0

17Mar94-4SA fst 1 1/16 :223 :46 1:104 1:442 ⒻMd 28000 21 4 3 58½ 714 718 728½ Pedroza M A LB 115 22.50 44-20 Our Passion1172½ Powerful Lad1172¼ Bundy's Sugarbabe1172 Done early 7
21Jan94-3SA fst 6½f :214 :444 1:102 1:172 ⒻMd 50000 32 4 2 33½ 66¾ 811 814¾ Gonzalez S5 B 112 25.40 70-12 Antifreeze1171 Regal Gentry1151 Queen Gen117¾ Done early 8
26Dec93-2SA fst 6f :214 :444 :57 1:093 ⒻMd Sp Wt 48 10 9 95 129¾ 1114 1115¾ Gonzalez S5 B 112 61.40 73-09 Choice Claim1172 Meadow Moon117½ Whatawoman1173 Wide trip 12

WORKOUTS: ●Mar 30 GG 5f fst 1:00 Hg *1/44* Mar 10 SA 5f fst 1:004 H *14/46* Mar 3 SA 3f fst :373 H *10/17* Feb 25 SA 5f fst :594 H *9/49* Feb 18 SA 6f gd 1:134 H *4/27* Feb 12 SA 4f fst :473 H *12/49*

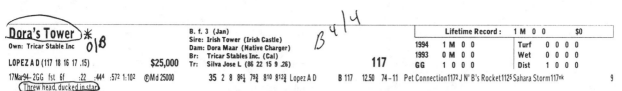

Dora's Tower
Own: Tricar Stable Inc
LOPEZ A D (117 18 16 17 .15) $25,000

B. f. 3 (Jan)
Sire: Irish Tower (Irish Castle)
Dam: Dora Maar (Native Charger)
Br: Tricar Stables Inc. (Cal)
Tr: Silva Jose L (86 22 15 9 .26)
117

Lifetime Record:	1 M 0 0	$0		
1994	1 M 0 0		Turf	0 0 0 0
1993	0 M 0 0		Wet	0 0 0 0
GG	1 0 0 0		Dist	1 0 0 0

17Mar94-2GG fst 6f :22 :444 :572 1:102 ⒻMd 25000 35 2 8 86¼ 79¾ 810 812¾ Lopez A D B 117 12.50 74-11 Pet Connection1172 J N' B's Rocket1125 Sahara Storm117nk 9
Threw head, ducked in start
WORKOUTS: Apr 2 GG 5f fst 1:012 H *20/45* Mar 26 GG 5f fst 1:042 Hg*44/51* Mar 7 GG 5f fst 1:034 H *22/25* Feb 28 GG 5f fst 1:01 H *2/12* Feb 19 GG 5f my 1:062 H (d)*3/8* Feb 10 GG 5f gd 1:024 Hg*26/40*

Distinctive Touch
Own: Lanning Curt & Lila
JUDICE J C (50 4 10 7 .08) $25,000

Ch. f. 3 (Apr)
Sire: Bet Big (Distinctive)
Dam: Brazen Maiven (Regal and Royal)
Br: Charlotte Gershaw & Marsha Mash (Fla)
Tr: Moger Ed Jr (43 9 3 6 .21)
117

Lifetime Record:	2 M 0 0	$0		
1994	2 M 0 0		Turf	0 0 0 0
1993	0 M 0 0		Wet	0 0 0 0
GG	2 0 0 0		Dist	2 0 0 0

24Mar94-2GG fst 6f :214 :45 :574 1:113 ⒻMd Sp Wt 47 1 5 31½ 42 55 68 Meza R Q LB 117 b 19.80 73-15 Sea Heroine1172 Deanna's Destiny1173 Castanera1171½ Gave way 8
5Mar94-5GG fst 6f :214 :451 :58 1:11 ⒻMd Sp Wt 42 3 10 87¾ 88½ 89 77¾ Meza R Q LB 117 b 21.80 76-13 Key To V. A.1172½ Sea Heroine1171½ Deanna's Destiny117hd Off slowly 10
WORKOUTS: Mar 18 GG 5f fst 1:004 Hg*8/34* Feb 25 GG 5f gd 1:013 H *21/77* Feb 16 GG 6f fst 1:151 Hg*7/10* Feb 9 GG 5f gd 1:02 H *8/30* Feb 3 GG 5f fst 1:01 Hg*6/19* Jan 28 BM 4f my :504 H *3/14*

Moscow Mystery
Own: Boatman Bill J
NOGUEZ A M (65 5 4 4 .08) $25,000

B. f. 3 (Apr)
Sire: Moscow Ballet (Nijinsky II)
Dam: In Prime Time (Boldnesian)
Br: Harris Farms Inc & Valpredo Don (Cal)
Tr: Cervantes Juan (32 1 1 2 .03)
117

Lifetime Record:	1 M 0 1	$1,050			
1994	1 M 0 1	$1,050	Turf	0 0 0 0	
1993	0 M 0 0		Wet	0 0 0 0	
GG	1 0 0 1	$1,050	Dist	1 0 0 1	$1,050

16Mar94-2GG fst 6f :214 :45 :58 1:11 ⒻMd c-12500 52 4 11 87 76½ 34½ 34½ Baze R A B 117 4.10 79-15 BergRosa1123½ GuarnteedToWin1171 MoscowMystery1175 Mild late bid 12
Claimed from Harris Farms Inc & Valpredo Donald, Gilchrist Greg Trainer
WORKOUTS: Mar 27 GG 5f fst 1:022 H *42/74* Mar 13 GG 4f fst :473 Hg*7/45* Mar 6 GG 5f fst 1:01 H *6/34* Feb 25 GG 5f gd 1:04 H *61/77* Feb 12 GG 5f fst 1:023 H *50/62* Feb 5 GG 5f fst 1:021 H *39/75*

Mira Santa
Own: Walter Mr & Mrs Robert H
MEZA R Q (172 22 28 15 .13) $25,000

B. f. 3 (Apr)
Sire: Batonnier (His Majesty)
Dam: La Beata (Tilt Up)
Br: Walter Robert H (Cal)
Tr: Pike Jelina (4 1 0 1 .25)
117

Lifetime Record:	8 M 2 1	$11,650			
1994	5 M 2 1	$9,900	Turf	0 0 0 0	
1993	3 M 0 0	$1,750	Wet	1 0 0 0	
GG	0 0 0 0		Dist	6 0 2 1	$11,250

17Mar94-9SA fst 6f :214 :444 :572 1:103 ⒻⓈMd 32000 49 11 1 1½ 2hd 21½ 97¾ Solis A B 117 2.90 76-11 Scoring Road1153½ Tia Veevee117nk Her Thoughts1171¼ Dueled, gave way 11
2Mar94-4SA fst 6f :212 :442 :57 1:104 ⒻⓈMd 32000 60 6 2 2hd 2hd 2hd 23½ Solis A B 117 11.30 79-13 Neli My Dear1173½ Mira Santa117½ Belocco117¾ Came in start 11
17Feb94-4SA wf 5½f :212 :443 :57 1:033 ⒻMd 32000 41 1 6 41½ 43 45½ 711 Solis A B 117 3.50 81-07 Paddy's Pick1174½ Dee Fox1151 Purely Political117½ Steadied into lane 12
27Jan94-1SA fst 6f :212 :443 :573 1:111 ⒻMd 32000 41 6 3 1hd 12 1hd 37 Valenzuela P A B 117 *1.20 74-16 Coco Luv1174½ Little Beau's Ms.1172½ Mira Santa1174½ Shortened stride 8
13Jan94-9SA fst 6f :214 :45 :58 1:12 ⒻⓈMd 32000 54 2 5 1hd 13½ 12½ 2hd Solis A B 117 *2.60 77-15 Latin Affair115hd Mira Santa117hd Guaranteed To Win1171¾ Game inside 11
28Dec93-4SA fst 6f :214 :442 :571 1:094 ⒻMd 32000 55 10 1 11½ 13½ 2hd 48 Atkinson P B 117 5.20 80-05 Serendipity1172 Saratoga Hope1174½ Quiet Pride1171½ Weakened 12
8Dec93-4Hol fst 6½f :22 :444 1:11 1:171 ⒻⓈMd 32000 31 3 6 11 12 21 514 Solis A B 119 33.60 69-08 DesertButterfly1175½ PapagoPeak1172½ Treasure'sPride1195 Speed, tired 12
3Nov93-6SA fst 6f :213 :443 :573 1:113 ⒻⓈMd Sp Wt -0 6 3 22 78¾ 917 933¾ Solis A 117 15.00 45-14 Concorde's Message117nk Linda LouB.117¾ FlyingProper1172 Done early 9
WORKOUTS: Mar 13 SA 4f fst :464 H *3/56* Mar 8 SA 4f fst :48 H *4/42* ●Feb 24 SA 4f fst :463 H *1/37* Feb 15 SA 4f fst :483 H *14/28* Feb 9 SA 5f gd 1:043 H *18/22* ●Feb 2 SA 4f fst :464 H *1/43*

Newcastle Blues
Own: Williamson Warren
BELVOIR V T (301 65 52 43 .22) $25,000

Dk. b or br f. 3 (Apr)
Sire: Apalachee (Round Table)
Dam: Money Player (Marshua's Dancer)
Br: Elia Mr & Mrs Chris (Ky)
Tr: Gaines Carla (16 2 3 1 .13)
117

Lifetime Record:	3 M 0 2	$5,325			
1994	3 M 0 2	$5,325	Turf	0 0 0 0	
1993	0 M 0 0		Wet	1 0 0 1	$2,475
GG	2 0 0 1	$2,850	Dist	3 0 0 2	$5,325

5Mar94-5GG fst 6f :214 :451 :58 1:11 ⒻMd Sp Wt 9 1 3 42 78¼ 1015 1020¾ Belvoir V T LB 117 b 6.60 63-13 Key To V. A.1172½ Sea Heroine1171½ Deanna's Destiny117hd Stopped 10
13Feb94-6GG fst 6f :22 :45 :571 1:094 ⒻMd Sp Wt 48 3 4 22 21½ 25 310¼ Belvoir V T L 117 *2.50 80-08 Gaelic Dowry11710 Wild Royal117nk Newcastle Blues117½ Pressed pace 10
23Jan94-6BM my 6f :221 :46 :592 1:131 ⒻMd Sp Wt 48 3 5 2½ 1½ 11 31¾ Meza R Q L 117 4.40 73-22 Joying117¾ Crusader Gal1171 Newcastle Blues1172 Weakened late 6
WORKOUTS: Apr 1 BM 4f fst :493 H *3/4* Mar 26 BM 7f fst 1:274 H *1/1* Mar 19 BM 5f fst 1:021 H *7/12* ●Mar 1 BM 4f fst :48 H *1/10* Feb 23 BM 5f gd 1:012 H *5/29* Feb 5 BM 5f fst 1:004 H *4/15*

But who can beat her? Wild Royal, Distinctive Touch and Newcastle Blues are all dropping from the Maiden Special Weight level and must first be considered. Since Newcastle Blues seems to want the lead and won't get it from Mira Santa, she looks vulnerable. Furthermore, she has already been beaten handily by Wild Royal and Distinctive Touch in the Key To V. A. race. Of the latter two, the cheaply-bred Distinctive Touch showed early speed her second time out, while finishing only a half length behind Wild Royal, and seems the more likely to further improve. At 13-1 Distinctive Touch is not being heavily backed, but seems a better bet than the stodgier Wild Royal who is going off at 7-1. While neither of these horses are worth a prime bet, since they have yet to lose at the Maiden Claiming level, they cannot be eliminated.

The only other horse that might improve is Dora's Tower. She broke poorly in her first start and, given the presence of a false favorite, deserves another chance to run to her sire's B4 rating. Viewed as a "debut" runner, Dora's Tower not only has the highest combined precocity and class rating in this field, but her trainer Jose Silva is excellent with young horses. Furthermore, the economics look realistic for her owner/breeder connections. While Tricar Stables has invested $7,500 in Irish Tower's stud fee, plus maybe another $10,000 in getting Dora's Tower to the races, they stand to earn $6500 in purse money on the Win and another $25,000 if she gets claimed in this spot. Still, even though Dora's Tower looks well meant here, her 13-1 odds and unexceptional works are grounds for concern.

So how to bet? Given that Mira Santa is a vulnerable favorite and the exacta unpredictable, the best strategy here is to play the most likely upsetters to Win and eschew the exotics. This means dutching Dora's Tower, Distinctive Touch and Wild Royal, to Win and hoping for the longest price.

THIRD RACE
Golden Gate

6 FURLONGS. (1.073) MAIDEN CLAIMING. Purse $12,000. Fillies, 3–year–olds. Weight 117 lbs. Claiming price $25,000; if for $22,500, allowed 2 lbs.

APRIL 7, 1994

Value of Race: $12,000 Winner $6,600; second $2,400; third $1,800; fourth $900; fifth $300. Mutuel Pool $87,150.00 Exacta Pool $96,400.00 Quinella Pool $19,978.00

Last Raced	Horse	M/Eqt. A.Wt	PP	St	¼	½	Str	Fin	Jockey	Cl'g Pr	Odds $1
17Mar94 2GG8	Dora's Tower	LB	3 117	3	3	1½	1½	14	19	Lopez A D	25000	13.10
24Mar94 2GG6	Distinctive Touch	LBb	3 117	4	4	3hd	34	2½	24½	Judice J C	25000	13.30
5Mar94 5GG6	Wild Royal	Lb	3 117	1	7	61	61½	52	3no	Olguin G L	25000	6.80
16Mar94 2GG3	Moscow Mystery	B	3 117	5	6	5hd	4hd	42	41	Noguez A M	25000	7.20
17Mar94 9SA9	Mira Santa	LB	3 117	6	1	22	22	34	52	Meza R Q	25000	0.70
17Mar94 4SA7	Wild Forever	LB	3 117	2	5	7	7	7	64½	Warren R J Jr	25000	13.80
5Mar94 5GG10	Newcastle Blues	LBb	3 117	7	2	43	52	6hd	7	Belvoir V T	25000	5.30

OFF AT 2:18 Start Good. Won handily. Time, :214, :444, :572, 1:101 Track fast.

$2 Mutuel Prices:

3-DORA'S TOWER	28.20	11.60	5.80
4-DISTINCTIVE TOUCH		10.80	10.00
1-WILD ROYAL			4.40

$5 EXACTA 3–4 PAID $618.50 $4 QUINELLA 3–4 PAID $222.80

B. f, (Jan), by Irish Tower–Dora Maar, by Native Charger. Trainer Silva Jose L. Bred by Tricar Stables Inc. (Cal).

Like Glasslifter, Dora's Tower ran to her precocious breeding at a big price when given a second chance, with the two class droppers, Distinctive Touch and Wild Royal, finishing second and third. The faint-hearted Mira Santa never got the lead and faded to fifth. For exotics players it is worth noting that had this been a trifecta race, playing the most precociously-bred horse and the two class droppers would have resulted in a huge score.

Class droppers with early speed are dangerous at all levels of the Maiden Claiming ladder. The fourth race on April 14th at Golden Gate Fields is an interesting example. A bottom-level affair, the field consists of six horses that have already lost for this tag, three class droppers and two non-precociously bred, first-time starters.

Of the horses that have already lost at this level, the lightly-raced Miss Fall and Reason To Prospect might further improve, but they will be facing class droppers Siete Legues, Sheshootshescores and Queen Linear, all of whom have shown speed against better. While Miss Fall and Reason To Prospect might win, as a rule, class droppers generally prevail in this kind of situation.

Golden Gate Fields

4

5½ Furlongs. (1:02) MAIDEN CLAIMING. Purse $7,000. Fillies, 3-year-olds. Weight, 117 lbs. Claiming price $12,500.

Miss Fall

Own: Marshall Mr & Mrs Frank

LOPEZ A D (151 26 20 24 .17) — $12,500

Ch. f. 3 (Feb)
Sire: Bold T. Jay (Island Agent)
Dam: Vista Fall (Grenfall)
Br: Marshall Mr & Mrs Frank (Cal)
Tr: Benedict Jim (42 5 6 8 .12)

117

Lifetime Record:	1 M 0 0	$525			
1994	1 M 0 0	$525	Turf	0 0 0 0	
1993	0 M 0 0		Wet	0 0 0 0	
GG	1 0 0 0	$525	Dist	1 0 0 0	$525

1Apr94-9GG fst 5½f :211 :443 :572 1:041	ⒻMd 12500	51 7 1 44 35 36½ 46½ Olguin G L	LB 117	44.20	82-12	Katelyn Alexis1175 Kimmie S.117½ Reason To Prospect1171	Weakened 11

WORKOUTS: Apr 10 GG 3f fst :374 H 44/52 Mar 27 GG 5f fst 1:034 H 61/74 Mar 19 GG 6f fst 1:162 Hg 14/17 Mar 13 GG 5f fst 1:012 Hg 21/55 Mar 6 GG 4f fst :502 H 21/40

Talinear

Own: Liebau F Jack & Sedlachek Harriet

MERCADO P (176 13 15 24 .07) — $12,500

B. f. 3 (Feb)
Sire: Talinum (Alydar)
Dam: Near Miss (Soy Numero Uno)
Br: Liebau F Jack (Cal)
Tr: Jimenez Luis A (17 1 2 3 .06)

117

Lifetime Record:	8 M 1 2	$4,050			
1994	4 M 1 1	$2,625	Turf	0 0 0 0	
1993	4 M 0 1	$1,425	Wet	1 0 0 1	$1,050
GG	3 0 1 1	$2,625	Dist	0 0 0 0	

16Mar94-2GG fst 6f :214 :45 :58 1:11	ⒻMd 12500	39 2 6 54 55½ 67½ 59½ Mercado P	LB 117 b	8.20	74-15	BergRos1123½ GurnteedToWin1171 MoscowMystery1175	Broke out start 12
6Mar94-2GG fst 6f :221 :46 :59 1:124	ⒻMd 12500	40 8 1 42½ 31½ 21½ 2no Mercado P	LB 117 b	3.90	75-12	Henderson's Gold117no Talinear1171 Lady Lahab117½	Rallied wide 8
18Feb94-9GG my 6f :221 :46 :584 1:121	ⒻⓈMd 12500	51 2 9 86 54 33 32 Mercado P	LB 117 b	21.60	76-16	Daring Dame1171½ Fabulent117½ Talinear1173	Wide late bid 12
12Jan94-2BM fst 6f :23 :461 :584 1:114	ⒻⓈMd 12500	35 8 7 62¾ 63½ 66 911 Mercado P	LB 117 b	3.20	71-16	Sarah's Boop1173 To Caro With Love1184 Thrillofawin117¾	Dull try 12
26Dec93-2BM fst 6f :23 :463 :592 1:122	ⒻⓈMd 12500	49 5 3 2hd 3nk 3½ 31 Mercado P	LB 117 b	9.00	78-17	Silent Daughter1171 Cup And A Half117no Talinear1175	Game try 12
5Dec93-6BM fst 6f :223 :461 :582 1:112	ⒻMd Sp Wt	23 3 3 3½ 43 713 814 Kaenel J L	LB 118 b	53.80	65-18	Woven Gold1176 Lovely Krissy117hd Sonata Sky1173½	Outrun 9
7Nov93-1BM fst 6f :222 :454 :584 1:122	ⒻMd Sp Wt	35 2 4 34 44½ 57 59¾ Tohill K S	LB 117 b	32.10	64-18	Little Brass1173 Day Rate117nk Boz Connection117½	Even try 6
28Oct93-6BM fst 6f :221 :46 :582 1:11	ⒻMd Sp Wt	35 1 5 31 712 713 713 Meza R Q	LB 117 b	20.00	68-17	Catnip1172 Sonata Sky1172 Valnesian1183½	Bumped, taken up 3/8 7

WORKOUTS: Apr 10 GG 7f fst 1:283 H 4/9 Apr 4 GG 7f fst 1:28 H 1/1 Mar 4 GG 5f fst 1:013 H 20/52 ●Feb 15 GG 3f fst :354 H 1/20 Feb 10 GG 5f gd 1:02 H 18/40

Reason To Prospect

Own: White T E & Vesteen

DAVENPORT C L (51 3 3 4 .06) — $12,500

Ch. f. 3 (Mar)
Sire: Prospective Star (Mr. Prospector)
Dam: Lovely Reason (Limit to Reason)
Br: Stabbe Mike & Pamela (Cal)
Tr: Nelson Harry N (16 0 2 2 .00)

117

Lifetime Record:	2 M 0 1	$1,575			
1994	2 M 0 1	$1,575	Turf	0 0 0 0	
1993	0 M 0 0		Wet	0 0 0 0	
GG	2 0 0 1	$1,575	Dist	1 0 0 1	$1,050

1Apr94-9GG fst 5½f :211 :443 :572 1:041	ⒻMd 12500	54 5 4 33½ 24 25 35½ Tohill K S	B 117 b	8.00	83-12	Katelyn Alexis1175 Kimmie S.117½ Reason To Prospect1171	Even late 11
25Mar94-2GG gd 6f :222 :462 1:001 1:142	ⒻMd 12500	47 1 7 2½ 33½ 64½ 44 Tohill K S	B 117 b	9.80	63-22	J's Cinnamon Lady1172 Bolides Wish1172 Don't Spend It117no	Even late 9

WORKOUTS: ●Mar 18 BM 5f fst 1:01 Hg 1/11 ●Mar 9 BM 3f fst :35 Hg 1/12 Mar 4 BM 4f fst :50 H 11/13 Feb 24 BM 3f fst :371 H 4/14 Feb 15 BM 4f fst :481 H 10/24

Siete Legues

Own: Rose Van

MEZA R Q (192 22 32 15 .11) — $12,500

Ch. f. 3 (Apr)
Sire: Pancho Villa (Secretariat)
Dam: All for Glory (Halo)
Br: Calcanadian Farms (Ky)
Tr: Caton Dent (14 2 0 2 .14)

117

Lifetime Record:	3 M 0 0	$600			
1994	3 M 0 0	$600	Turf	0 0 0 0	
1993	0 M 0 0		Wet	1 0 0 0	$300
GG	3 0 0 0	$600	Dist	0 0 0 0	

17Mar94-2GG fst 6f :22 :444 :572 1:102	ⒻMd 22500	48 7 6 52½ 53¾ 44½ 58 Meza R Q	L 115 b	22.00	79-11	Pet Connection1172 J N' B's Rocket1125 Sahara Storm117nk	No rally 9
27Feb94-5GG gd 6f :214 :451 :573 1:104	ⒻMd 20000	37 3 10 75¾ 45½ 410 814¾ Romero J A	L 117 b	36.30	70-15	Anna Grace1178 Sahara Storm117nk Flying Presence117½	12
Hopped, bobbled start							
6Feb94-6GG sly 6f :221 :451 :574 1:114	ⒻMd 25000	48 1 6 21½ 33 46½ 56¾ Meza R Q	L 117 b	8.00	73-17	Shiny Slew117¾ Day Rate117½ Summertime Party1173½	Weakened 9

WORKOUTS: Apr 10 GG 4f fst :471 H 2/67 Mar 30 GG 4f fst :501 H 15/18 Mar 10 GG 4f fst :49 H 12/36 Feb 1 GG 5f fst 1:032 H 29/32 ●Jan 26 GG 4f my :48 Hg 1/13 Jan 19 GG 5f fst 1:023 H 7/14

Sheshootshescores

Own: Klingler Dale H

CHAPMAN T M (197 32 22 22 .16) — $12,500

Ch. f. 3 (Mar)
Sire: Interco (Intrepid Hero)
Dam: Pearl Blue*Ir (Deep Blue Sea)
Br: Rodriguez Suzanne (Cal)
Tr: Hess R B (43 10 5 6 .23)

117

Lifetime Record:	3 M 0 0	$0		
1994	3 M 0 0		Turf	0 0 0 0
1993	0 M 0 0		Wet	0 0 0 0
GG	0 0 0 0		Dist	0 0 0 0

17Mar94-9SA fst 6f :214 :444 :572 1:103	ⒻⓈMd 28000	49 2 10 64½ 77½ 68 87¾ Rojas D S5	LB 110	30.40	76-11	Scoring Road1153½ Tia Veevee117nk Her Thoughts1171½	Off slowly 11
3Feb94-9SA fst 6f :213 :442 :564 1:092	ⒻⓈMd 32000	49 9 2 31 42 67½ 817 Desormeaux K J	LB 117	8.70	73-09	Siyah Relief1178 Zippen Miss117¾ Paddy's Pick117nk	Lugged out turn 12
2Jan94-9SA fst 6f :212 :443 :571 1:10	ⒻⓈMd c-32000	30 8 3 2hd 2hd 78 817½ Valenzuela F H	LB 117	5.80	70-07	Ed's Design1173 Paddy's Pick117½ Saros Lane1172½	Gave way 11

Claimed from All The Kngs Men& Cldwl& Crlm& Fell, Mayberry Brian A Trainer

WORKOUTS: Apr 11 GG 3f fst :374 H 4/6 Apr 4 GG 4f fst :473 H 3/20 Mar 10 SA 4f fst :502 H 38/47 Mar 4 SA 4f fst :473 H 5/39 Feb 25 SA 5f fst 1:02 H 34/49 Feb 18 SA 5f gd 1:032 H 23/29

continued

Dr J's Girl
Own: Abdun Nur J

HUMMEL C R (102 11 14 5 .11) $12,500

Lifetime Record :	1 M 0 0	$0

Dk. b or br f. 3 (Mar)
Sire: Bolger (Damascus)
Dam: Miss Uragold (El Rayo)
Br: Abel Thoroughbreds (Cal)
Tr: Monroe Sherrie (10 0 4 1 .00)   117

C 6/5

1994	1 M 0 0	Turf	0 0 0 0
1993	0 M 0 0	Wet	0 0 0 0
GG	1 0 0 0	Dist	1 0 0 0

1Apr94–9GG fst 5½f :211 :443 :572 1:041 ⓕMd 12500   34 11 9 77½ 6¹⁰ 6¹⁰ 712½ Hummel C R   LB 118   5.20 76–12 Katelyn Alexis117⁵ Kimmie S.117½ Reason To Prospect117¹   Raced wide 11
WORKOUTS: Apr 10 GG 4f fst :47² Hg5/67 ●Mar 29 GG 3f fst :35¹ H 1/9 Mar 23 GG 6f fst 1:17¹ H 5/7 Mar 16 GG 6f fst :15² Hg10/15 Mar 9 GG 5f fst :59⁴ H 5/31 Mar 3 GG 5f fst 1:03 H 19/46

Tyloch
Own: Cheek Sharon

ROMERO J A (56 3 5 9 .05) $12,500

Lifetime Record :	9 M 2 2	$5,062

B. f. 3 (Jun)
Sire: To B. Or Not (Don B.)
Dam: Bobby Socks (Revel)
Br: Neumann Duane 0 (Cal)
Tr: Hawthorne James W (14 0 1 2 .00)   117

D 9/4

1994	6 M 2 2	$4,900	Turf	0 0 0 0
1993	3 M 0 0	$162	Wet	0 0 0 0
GG	4 0 0 2	$2,100	Dist	6 0 2 2	$5,062

23Mar94–2GG fst 5½f :221 :46 :58⁴ 1:05³ ⓕMd 12500   44 1 6 3¹ 2½ 2⁵ 3⁸ Davenport C L   B 117 b 11.50 74–18 Magic Fable117³½ P. A. System117⁴½ Tyloch117ⁿᵒ   Weakened 10
11Mar94–2GG fst 6f :22 :45 :57⁴ 1:11² ⓕMd 12500   7 9 4 4² 3⁴ 711¹⁰18½ Meza R Q   B 117 fb 8.30 63–11 Innocent Cara117¹½ Classic Summer117½ K. L.'s Imagene117⁸   Stopped 11
23Feb94–4GG gd 5½f :21⁴ :45¹ :571 1:03⁴ ⓕMd 12500   49 2 3 2½ 2¹ 2⁶ 3¹⁵ Miranda V   B 117 b 4.50 76–15 Baby Shea117¹³ Stephanie's World117² Tyloch117¹   Weakened 6
10Feb94–5GG gd 6f :22³ :46¹ :59¹ 1:12² ⓕMd 12500   22 11 2 4¹½ 3² 3⁵ 11¹⁴¾ Gonzalez R M   B 117 b *2.40 62–19 Itsagenderthing112½ Daring Dame117⁶ Slew Run117²   Wide, stopped 12
20Jan94–2BM fst 5½f :22³ :46⁴ :59² 1:05⁴ ⓕⓈMd 12500   47 4 1 1¹ 1² 1ʰᵈ 2ʰᵈ Meza R Q   B 117 b *.80 82–12 Tie And Dye117ʰᵈ Tyloch117¹¹ T. Jay's Heart117³   Just missed 7
7Jan94–5BM fst 5½f :21⁴ :45³ :58⁴ 1:051 ⓕMd 12500   59 4 3 1½ 1ʰᵈ 2ʰᵈ 2½ Meza R Q   B 117 b 12.80 81–16 Maple Valley117½ Tyloch117⁴½ Desert Mystique117¹½   Held gamely 12
16Dec93–5BM gd 5½f :22 :46½ :59² 1:06² ⓕMd 12500   32 7 2 1¹½ 1¹½ 2¹ 5⁷ Gonzalez R M   B 117 b 11.80 72–21 Night Jazz Dancer117²½ Maple Valley117² Thrillofawin117¹   Tired 10
24Nov93–3BM fst 6f :22³ :46¹ :59² 1:13 ⓕⓈMd 12500   23 4 3 1ʰᵈ 1½ 2¹½ 9⁶½ Mercado P   B 117 40.00 65–20 UniversalAffair117ⁿᵒ ConstructATune117¹½ K.L.'sImgene117¹½   Weakened 11
14Nov93–2BM fst 5½f :21⁴ :45³ :59 1:05³ ⓕMd 12500   35 8 4 2⁴ 2⁵ 46½ 87½ Harvey B   B 117 22.00 76–19 Exclusive Root117¹ Blooming Star117²½ Expensive Star117¹   Faltered 12
WORKOUTS: Apr 10 BM 4f fst :48⁴ H 4/10 Mar 6 BM 3f fst :37⁴ H 2/4 ●Feb 4 BM 4f fst :48 H 1/10

Occident Me
Own: Negri Joseph F Jr

GONZALEZ R M (190 17 24 23 .09) $12,500

Lifetime Record :	0 M 0 0	$0

Ro. f. 3 (Feb)
Sire: Kala Native (Exclusive Native)
Dam: Best in Occidental (Truxton King)
Br: Joe Negri (Cal)
Tr: Brook Joseph (9 1 1 3 .11)   117

D 8/4

1994	0 M 0 0	Turf	0 0 0 0
1993	0 M 0 0	Wet	0 0 0 0
GG	0 0 0 0	Dist	0 0 0 0

WORKOUTS: Apr 9 BM 5f my 1:03⁴ H 3/5 Apr 3 BM 6f fst 1:15¹ Hg2/4 Mar 27 BM 5f fst 1:03⁴ H 18/19 Mar 20 BM 5f fst 1:03¹ Hg11/11 Mar 13 BM 5f fst 1:03 Hg15/20 Mar 6 BM 4f fst :52 H 8/10
Feb 22 BM 3f my :38¹ H 8/8 Feb 16 BM 3f fst :37 H 5/19 Feb 10 BM 3f fst :39¹ H 14/18

Dayes Of A Miracle
Own: Vistas Stables & Gordy

BAZE R A (318 87 60 39 .27) $12,500

Lifetime Record :	0 M 0 0	$0

Dk. b or br f. 3 (Apr)
Sire: Huddle Up (Sir Ivor)
Dam: Ayleen (Coursing)
Br: Fuqua Gwen G (Cal)
Tr: Shoemaker Leonard (56 12 9 5 .21)   117

n 6/5

1994	0 M 0 0	Turf	0 0 0 0
1993	0 M 0 0	Wet	0 0 0 0
GG	0 0 0 0	Dist	0 0 0 0

WORKOUTS: Apr 10 GG 4f fst :49³ H 40/67 Apr 3 GG 6f fst 1:15¹ H 5/10 Mar 27 GG 5f fst 1:01² H 25/74 Mar 20 GG 5f fst 1:00⁴ Hg19/60 Mar 14 GG 5f fst 1:01² H 5/13 Mar 8 GG 5f fst 1:01⁴ H 7/21
Mar 2 GG 4f fst :50³ Hg29/55 Feb 24 GG 4f gd :49⁴ H 22/65

Kimmie S.
Own: Swenson Vernon L & Taylor Faith

NOGUEZ A M (72 5 4 5 .07) $12,500

Lifetime Record :	1 M 1 0	$1,400

Ch. f. 3 (Apr)
Sire: Dunant (Sallust)
Dam: Sis Q. Annie (Lime)
Br: Swenson Mr & Mrs Vernon L (Cal)
Tr: Taylor Faith (6 0 1 0 .00)   117

F 7/5

1994	1 M 1 0	$1,400	Turf	0 0 0 0
1993	0 M 0 0		Wet	0 0 0 0
GG	1 0 1 0	$1,400	Dist	1 0 1 0	$1,400

1Apr94–9GG fst 5½f :211 :443 :572 1:041 ⓕMd 12500   55 9 6 5⁶ 57½ 47½ 2⁵ Noguez A M   B 117   16.50 84–12 Katelyn Alexis117⁵ Kimmie S.117½ ReasonToProspect117¹   Wide late run 11
WORKOUTS: Apr 12 BM 4f fst :53⁴ H 9/9 Apr 8 BM 4f fst :49⁴ H 8/11 ●Mar 29 BM 3f fst :36³ H 1/6 Mar 23 BM 5f fst 1:01⁴ Hg1/3 ●Mar 19 BM 6f fst 1:15 H 1/5 Mar 13 BM 5f fst 1:01¹ Hg5/20

Queen Linear
Own: Sprint Stable Inc

ESPINOZA V (253 32 36 30 .13)   ↓ $12,500

Lifetime Record :	2 M 0 0	$675

B. f. 3 (Feb)
Sire: Polish Navy (Danzig)
Dam: Colinear (Cohoes)
Br: Stiles Martin (Ky)
Tr: Coffey Junior (13 3 0 2 .23)   117

C 6/1   39k

1993	2 M 0 0	$675	Turf	0 0 0 0
1992	0 M 0 0		Wet	0 0 0 0
GG	0 0 0 0		Dist	1 0 0 0

18Oct93–4BM fst 6f :221 :45 :57² 1:10³ ⓕMd 20000   41 1 5 3¹½ 3¹½ 45½ 46½ Boulanger G   B 117 fb 4.30 76–13 Greenbank Lass117² Double Jewel117²½ La Bonne D'oyly117²   Weakened 8
6Sep93–6BM fst 5½f :22 :45⁴ :58² 1:04⁴ ⓕMd Sp Wt   27 4 8 6²½ 52½ 77½ 8¹³ Hansen R D   B 117 fb 7.30 74–14 Vail Link117½ Eishin Ohio117⁴ Bel Native117¾   Faltered 9
WORKOUTS: Apr 8 GG 6f fst 1:17⁴ H 13/17 Mar 30 GG 5f fst 1:03 B 37/44 Mar 23 GG 4f fst :51⁴ H 21/25 Mar 16 GG 4f fst :51¹ H 30/33 Mar 10 GG 3f fst :36⁴ H 6/12

On the surface, Sheshootshescores, with her B6 rating, looks like a legitimate favorite at 8-5. She's dropping in class, has speed and has been shipped north by R.B. Hess who is adept at this maneuver. On the downside, Sheshootshescores has tired more quickly in each of her three races and Hess, who claimed her for $32,000 in January, is now willing to lose her for only $12,500. So, the suspicion lingers that she may be damaged goods.

Given that possibility, the other class droppers Queen Linear and Siete Legues merit closer consideration, especially since they are the only Kentucky-breds in the race. Of the two, Queen Linear seems a less likely upsetter as she is coming off a seven-month layup and didn't fire in her first start. Still, Junior Coffey is good with layoff horses and has been known to pop at a price.

Given legitimate doubts about the other two class droppers, Siete Legues now looms as an intriguing possibility. She is by Pancho Villa who, with his B6 rating, is probably the best sire in the race. Although Siete Legues has lost three times at a higher Maiden Claiming level, two of those efforts were on off-tracks, a surface the Pancho Villas generally hate. Even more interesting, she's been rested a month and has a near bullet work of :47.1 on April 10th which suggests she's ready to run big in this spot at an attractive 7-1.

So how to bet? Because this is not a race with a single possibility, there are several ways to play it. One option is to assume the other class droppers will run well and bet Siete Legues to Win and second in the exacta to Sheshootshescores and Queen Linear. The other option is to play Siete Legues across the board and hope the vulnerable favorite finishes out of the money.

FOURTH RACE — 5½ FURLONGS. (1.02) MAIDEN CLAIMING. Purse $7,000. Fillies, 3-year-olds. Weight, 117 lbs. Claiming price $12,500.

Golden Gate

APRIL 14, 1994

Value of Race: $7,000 Winner $3,850; second $1,400; third $1,050; fourth $525; fifth $175. Mutuel Pool $91,530.00 Exacta Pool $95,734.00 Quinella Pool $21,645.00

Last Raced	Horse	M/Eqt.	A.Wt	PP	St	¼	¾	Str	Fin	Jockey	Cl'g Pr	Odds $1
17Mar94 2GG5	Siete Legues	Lb	3 117	4	3	1hd	1½	11½	14	Meza R Q	12500	6.70
1Apr94 9GG3	Reason To Prospect	LBb	3 117	3	2	62	71½	41	2hd	Davenport C L	12500	12.70
1Apr94 9GG4	Miss Fall	LB	3 117	1	7	8½	91	72	31½	Lopez A D	12500	15.50
23Mar94 2GG3	Tyloch	Bb	3 117	7	6	41	22	22	4no	Vergara O	12500	38.90
16Mar94 2GG5	Talinear	LB	3 117	2	8	7hd	6hd	6½	5no	Mercado P	12500	28.90
	Dayes Of A Miracle	LBbf	3 117	9	4	51	31	33	63	Baze R A	12500	3.60
18Oct93 4BM4	Queen Linear	LBb	3 117	11	1	3½	53	5hd	71	Espinoza V	12500	7.80
1Apr94 9GG2	Kimmie S.	B	3 117	10	11	9½	8½	9	8no	Noguez A M	12500	7.60
17Mar94 9SA8	Sheshootshescores	LB	3 117	5	10	104	106	81	9	Chapman T M	12500	1.80
1Apr94 9GG7	Dr J's Girl	LBb	3 117	6	5	2½	4½	—	—	Hummel C R	12500	20.40
	Occident Me	LB	3 117	8	9	11	11	—	—	Gonzalez R M	12500	55.70

Dr J's Girl:Eased; Occident Me:Eased

OFF AT 2:47 Start Good. Won driving. Time, :212, :452, :582, 1:051 Track fast.

$2 Mutuel Prices:

4–SIETE LEGUES	15.40	7.40	5.40
3–REASON TO PROSPECT		10.60	5.20
1–MISS FALL			8.60

$5 EXACTA 4–3 PAID $300.50 $4 QUINELLA 3–4 PAID $141.60

Ch. f, (Apr), by Pancho Villa–All for Glory, by Halo. Trainer Caton Dent. Bred by Calcanadian Farms (Ky).

Mom & Pop Breeders

While thus far we have concentrated on the classier sires, there are a number of lesser known regional stallions in the class 8-10 range (such as Slewdledo in Washington, Irish Conn in New Mexico, Hula Blaze in Texas, Black Mackee in Montana, Sea Aglo in Oregon, Feel The Power in West Virginia, Gaelic Christian in Oklahoma and Shanekite in California) whose progeny are well meant at the lower Maiden Claiming levels. Handicappers can profit handsomely by knowing who the cheaper precocious sires are, especially when they are shipped in from a neighboring circuit to win at a big price.

Stockton

3

5½ Furlongs. (1:02²) MAIDEN SPECIAL WEIGHT. Purse $8,500. 2-year-olds. Weight, 118 lbs. Claiming price $16,000. (California breds preferred.)

Major Mooch
O(B
Own: Capehart Howard & Liebau

HANNA M A (33 4 5 10 .12) $16,000

Dk. b or br c. 2 (Jan)   *D8*
Sire: Unpredictable (Tri Jet)
Dam: I'm Smoochin (L'Natural)
Br: Jack Liebau, Leigh Ann Howard, et al. (Cal)
Tr: Jimenez Luis A (2 0 0 2 .00)   **118**

	Lifetime Record :	2 M 0 0		$650					
1994	2 M 0 0		$650	Turf	0 0 0 0	
1993	0 M 0 0			Wet	1 0 0 0	$650
Stk	0 0 0 0			Dist	0 0 0 0	

26May94-7GG fst 4½f :22 :45² :51³ Md Sp Wt   34 3 3 4   58½ 613½ Espinoza V   LB 116 b 30.20   80-14 Dutch116⁷ Tax Problem116ⁿᵏ Cited For Speeding116²   Through early 7
6May94-4GG sly 4½f :21² :44⁴ :51¹ Md Sp Wt   46 4 8 7¹⁰   6¹⁰ 512½ Belvoir V T   LB 116 b 16.50   83-16 Charlie Galea116¹½ Dutch116³½ Mocha Breeze116½   Showed little 8

WORKOUTS: Jun 12 GG 3f fst :37¹ H 9/27   May 22 GG 3f fst :36 Hg4/28   May 15 GG 3f fst :36 H 3/52   Apr 29 GG 3f gd :37¹ Hg9/34   Apr 23 SLR tr.t 4f fst :48⁴ H 7/13   Apr 15 SLR tr.t 4f fst :53² H 4/4

Be Regal
D9
Own: Harralson Daniel

CABALLERO R (—) $16,000

Dk. b or br c. 2 (May)
Sire: Regalberto (Roberto)
Dam: Skipper Bea (Azirae)
Br: Myron F. Johnson (Cal)
Tr: Castellanos Rito (2 0 0 1 .00)   **118**

	Lifetime Record :	0 M 0 0		$0					
1994	0 M 0 0			Turf	0 0 0 0	
1993	0 M 0 0			Wet	0 0 0 0	
Stk	0 0 0 0			Dist	0 0 0 0	

WORKOUTS: May 27 Sac 4f fst :50³ H 18/27

Ryan's Folly
U10
Own: Ronin Mr & Mrs Robert J

BRIDGES K (23 1 1 2 .04) $16,000

Dk. b or br c. 2 (Jan)
Sire: Full of Fools (Full Out)
Dam: Sanghe (Blade)
Br: Mr. & Mrs. Robert J. Cronin (Cal)
Tr: Parris Jim L (4 0 0 1 .00)   **118**

	Lifetime Record :	1 M 0 0		$75					
1994	1 M 0 0		$75	Turf	0 0 0 0	
1993	0 M 0 0			Wet	0 0 0 0	
Stk	1 0 0 0		$75	Dist	1 0 0 0	$75

19Jun94-6Stk fst 5f :22¹ :45⁴ :58² Md 12500   14 2 8 86½ 77¼ 79½ 711½ Garrido O L   B 118 *1.70e 79-06 Corey Dean118² Deals Sweet Deal118³ Road Hog118²½   Broke slow 8

WORKOUTS: Jun 2 GG 5f fst 1:04⁴ Hg21/24   May 25 GG 4f fst :51⁴ Hg34/35   Apr 23 GG 4f fst :53 H 60/61

Jukabil
M6
Own: A & T Stock Farm

MIRANDA V (55 11 11 6 .20) $16,000

Dk. b or br c. 2 (Apr)
Sire: Present Value (Halo)
Dam: Scenic Springs (Power Ruler)
Br: David E McGlothlin & Harris Farms, Inc. (Cal)
Tr: Burch Eldwin (6 1 0 0 .17)   **118**

	Lifetime Record :	0 M 0 0		$0					
1994	0 M 0 0			Turf	0 0 0 0	
1993	0 M 0 0			Wet	0 0 0 0	
Stk	0 0 0 0			Dist	0 0 0 0	

WORKOUTS: Jun 19 Stk 5f fst 1:02¹ Hg5/5   Jun 14 Stk 4f fst :52³ H 16/18   Jun 6 Stk 4f fst :51³ H 1/2

Sir Iliad
M10
Own: Nelson Harry N

DEAN W K (1 0 0 0 .00) $16,000

B. g. 2 (Mar)
Sire: Sir Macamillion (Macarthur Park)
Dam: Ilima (Petrone)
Br: Claudine D. Molick & Harry Nelson Jr. (Cal)
Tr: Nelson Harry N (4 0 1 1 .00)   **118**

	Lifetime Record :	3 M 0 0		$75					
1994	3 M 0 0		$75	Turf	0 0 0 0	
1993	0 M 0 0			Wet	0 0 0 0	
Stk	1 0 0 0		$75	Dist	2 0 0 0	$75

19Jun94-6Stk fst 5f :22¹ :45⁴ :58² Md 12500   24 5 5 54½ 55½ 56½ 58 Dean W   B 118 22.40 83-06 Corey Dean118² Deals Sweet Deal118³ Road Hog118²½   Lacked response 8
9Jun94-5GG fst 5f :21⁴ :46³ 1:00 Md 25000   7 9 8 99½ 10¹³ 10¹⁴ 10¹¹½ Davenport C L   B 116 91.00 71-15 Glsslftr116½ MstrOfMlody116²½ ThCrwnsTrch116¹   Fractious post parade 11
20May94-6GG fst 4½f :21⁴ :45⁴ :52 Md 25000   19 8 8 8¹¹   8¹¹ 8¹⁴ Davenport C L   116 47.20 77-12 Agent Warrior116² I Promise Roses116½ Slow Cowboy115⁵   Outrun 9

WORKOUTS: May 31 BM 3f fst :36¹ H 6/11   May 15 BM 4f fst :50 Hg33/46   May 6 BM 5f fst 1:04² H 7/8   May 1 BM 4f fst :49¹ H 2/9   Apr 22 BM 4f fst :50⁴ H 6/6   Apr 15 BM 3f fst :36 H 2/6

Duplicator
U10
Own: McElhinney Lloyd C

MERCADO P (49 9 7 7 .18) $16,000

B. c. 2 (Feb)
Sire: Ventriloquist (Lyphard)
Dam: Heart n Class (Selectus)
Br: Lloyd C. McElhinney (Cal)
Tr: Sherman Art (7 2 0 0 .29)   **118**

	Lifetime Record :	0 M :0 0		$0					
1994	0 M 0 0			Turf	0 0 0 0	
1993	0 M 0 0			Wet	0 0 0 0	
Stk	0 0 0 0			Dist	0 0 0 0	

WORKOUTS: Jun 17 BM 4f fst :49³ Hg14/17   Jun 10 BM 4f fst :50¹ Hg21/25   Jun 1 BM 4f fst :49⁴ Hg5/9   May 27 Fpx 4f fst :50¹ H 13/13   May 12 Fpx 3f fst :38² H 6/7   May 5 Fpx 4f fst :53¹ H 6/6
   Apr 30 Fpx 3f fst :39¹ H 14/16   Apr 12 Fpx 3f fst :39⁴ H 7/8

American Ruler
AOTWN
Own: Golden State Stables

LOZOYA D A (8 2 0 2 .25) $16,000

Ch. c. 2 (Apr)   *D8*
Sire: Grand Ruler (Mr. Prospector)
Dam: So Intricate (Staff Writer)
Br: Golden State Stables Inc. (Cal)
Tr: White Dan (—)   **118**

	Lifetime Record :	0 M 0 0		$0					
1994	0 M 0 0			Turf	0 0 0 0	
1993	0 M 0 0			Wet	0 0 0 0	
Stk	0 0 0 0			Dist	0 0 0 0	

WORKOUTS: ●Jun 23 GG 3f fst :35³ Hg1/9   Jun 16 GG 5f fst 1:01 Hg8/23   Jun 11 GG 5f fst 1:03 H 39/44   Jun 4 GG 4f fst :51³ H 28/29   May 26 GG 3f fst :39² H 19/21   Apr 11 TuP 4f fst :50⁴ H 16/22

Rozzy Ridge ⟨circled⟩ G|B   A8/4   0|B
Own: Cameron Mary Ellen
B. c. 2 (Mar)
Sire: Slewdledo (Seattle Slew)
Dam: Compagnie (Canadian Gil)
Br: Paulson Bros. & Bob Cameron (Wash)
Tr: Fergason Rolland (3 0 1 0 .00)
DELGADILLO A (13 0 1 3 .00) $16,000
118

Lifetime Record :	0 M 0 0		$0	
1994	0 M 0 0		Turf	0 0 0 0
1993	0 M 0 0		Wet	0 0 0 0
Stk	0 0 0 0		Dist	0 0 0 0

WORKOUTS: Jun 19 Stk 4f fst :48³ H 3/5   Jun 14 Stk 4f fst :48¹ Hg3/18   ●Jun 3 PM 3f fst :36³ H 1/6   May 27 PM 3f fst :39² Bg5/6   May 20 PM 3f fst :37⁴ B 7/9   May 13 PM 3f fst :39² B 1/3

Never Dunn
Own: Calvin J & Dunn Rhea
Dk. b or br g. 2 (Feb)
Sire: Never Tabled (Never Bend)
Dam: Upward (Golden Eagle II)
Br: Peg Goemans (Cal)
Tr: James Greg (2 0 2 0 .00)
DREXLER H A (27 10 4 3 .37) $16,000
118

Lifetime Record :	1 M 0 0		$0	
1994	1 M 0 0		Turf	0 0 0 0
1993	0 M 0 0		Wet	0 0 0 0
Stk	0 0 0 0		Dist	1 0 0 0

17Jun94–2GG fst 5f :21² :45² :58² Md 14000 20 7 2 43½ 59 5¹² 7¹⁶½ Tohill K S LB 114 42.20 74–13 Clever Talk116⁵ Fiscal Dancer116⁴ Tam's Chilipie116² Wide, stopped 7
WORKOUTS: Jun 9 GG 5f fst 1:02³ H 26/30   Jun 3 GG 4f fst :51¹ H 30/36   May 28 GG 4f fst :51² H 48/51   May 9 GG 5f fst 1:05² H 46/46   Apr 30 GG 4f gd :49 Hg24/56   Apr 23 GG 4f fst :50 Hg38/61

Noble Groom
Own: Bringhurst Owen & Wilkinson Willard
Ch. g. 2 (Jan)
Sire: In the Swing (Blushing Groom)
Dam: Royal Scarlet (Crimson Satan)
Br: James J. Lindsey (Cal)
Tr: Bringhurst J Owen (—)
URIBE M R (4 0 1 1 .00) $16,000
118

Lifetime Record :	2 M 0 0	$1,950	Turf	0 0 0 0	
1994	2 M 0 0	$1,950	Turf	0 0 0 0	
1993	0 M 0 0		Wet	1 0 0 0	$1,950
Stk	0 0 0 0		Dist	1 0 0 0	

8Jun94–3GG fst 5f :21¹ :45¹ :57³ Md Sp Wt 42 1 2 1hd 1hd 42¼ 79½ Uribe M R B 116 6.00 85–09 Rocky T.116²½ Tax Problem116²½ Briareus116² Gave way 10
6May94–4GG sly 4½f :21² :44⁴ :51¹ Md Sp Wt 67 7 4 34 34 45¼ Chapman T M B 116 10.70 89–16 Charlie Galea116⁴½ Dutch116⁵½ Mocha Breeze116½ Wide trip 8
WORKOUTS: Jun 4 BM 3f fst :37¹ H 8/12   ●May 24 BM 4f fst :47³ H 4/7   Apr 29 BM 4f fst :50 Hg10/12   Apr 10 BM 4f fst :48³ H 2/10   Mar 28 BM 3f fst :37² H 4/9

Also Eligible (Not in Post Position Order):

Sea A Go ⟨circled⟩ B+9/4   0|B
Own: Hewes Edith M
Dk. b or br g. 2 (Feb)
Sire: Sea Aglo (Sea–Bird)
Dam: Bingo Day (Bargain Day)
Br: Calvin H. Hewes (Ore)
Tr: Hewes Calvin (4 0 1 0 .00)
HANNA M A (33 4 5 10 .12) BURNS $16,000
118

Lifetime Record :	0 M 0 0		$0	
1994	0 M 0 0		Turf	0 0 0 0
1993	0 M 0 0		Wet	0 0 0 0
Stk	0 0 0 0		Dist	0 0 0 0

WORKOUTS: Jun 23 Stk 3f fst :38 H 4/5   Jun 14 Stk 3f fst :35³ Hg2/4   Jun 3 GrP 4f fst :51 B 1/2   May 27 GrP 3f fst :38² Bg3/6   May 22 GrP 3f fst :39 B 5/8

The third race at Stockton on June 25, 1994 is a Maiden Claiming example which involves two debut shippers by cheaper sires that have been well-spotted by their "Mom & Pop" Northwest owner/breeder connections.

Looking over the nine juveniles entered into this $16,000 Maiden Claiming affair, Rossy Ridge and Sea A Go are the only first-time starters with an A or B rating. None of the other debut horses have even a C rating! Their prime competition looks to be American Ruler, the non-precociously bred favorite with that :35.1 gate work on June 23rd, and Major Mooch who is dropping in class at 5-2.

Rozzy Ridge with his A8 sire rating is the pedigree standout of the field. He also has decent works, including one bullet. Nevertheless, he appears to be lukewarm on the board. But, that could be deceptive because Rozzy Ridge has been shipped in from Portland Meadows, a cheaper track, and Slewdledo is a Washington sire who is probably unfamiliar to local Stockton Fair bettors. His 8-1 odds might actually signify heavy inside action, especially considering that Delgaldillo is a cold jockey and trainer Fergason has thus far had only modest success.

Sea A Go with his B+9 rating is only slightly less precocious and classy than Rozzy Ridge and has every right to be competitive at this $16,000 Maiden Claiming level. But

104   he has been shipped in from Grants Pass, an even cheaper Oregon track than Portland Meadows, and at 16-1 is lightly bet on the board.

Still, the favorites look vulnerable. Major Mooch at 5-2 is dropping in class, but he's only had one slow work in a month and has a low percentage trainer. American Ruler, on the other hand, cannot be easily dismissed. Despite his non-precocious D8 rating, he has a recent :35.3 bullet gate work and has been bet down to 8-5.

Of the other debut horses, Rozzy Ridge and Sea A Go merit a closer look. They may be just the kind of horses that low-profile connections can ship in to steal a nice pot. In particular, it's worth noting that Rozzy Ridge's owner/breeder connections thought enough of his chances to incur the additional expense of shipping him nearly a thousand miles to Stockton. If Rozzy Ridge wins he will earn 55 percent of the $8,500 purse for the Cameron family which will more than pay for his trip. And, if he gets claimed for another sixteen grand, that will add to the loot. The Camerons will then collect more than $20,000 for "losing" their horse which, after deducting Slewdledo's $1,300 stud fee, plus all the other costs of getting Rozzy Ridge to the races, should leave them with a more than ten grand profit.

But, even at this level, caution is in order and it is worth checking out the reproductive history of Rozzy Ridge's and Sea A Go's dams. Checking our copy of *Maiden Stats 1994* reveals that Compagnie (Rozzy Ridge's dam) is batting a hundred percent! She is 3-3 with 2-year-old winners, including Prince Charming who earned more than $129,000. Sea A Go, on the other hand, who is owned and bred by the Hewes family, is less promising because his dam Bingo Day has yet to throw a winner.

So how to play the race? Rozzy Ridge is a gift at 8-1. But Major Mooch and American Ruler are both getting action and must be respected. The best strategy here is to bet Rozzy Ridge to Win and key him top and bottom in the exacta with Major Mooch, American Ruler and Sea Ago. Since this is a trifecta race, one could also single Rozzy Ridge on top and play Major Mooch, American Ruler and Sea A Go in the second and third holes.

Results: Rozzy Ridge and Sea A Go (the two most precociously bred horses) dueled for the early lead. Rozzy Ridge then pulled away and paid $19.20 for the Win. American Ruler closed for the Place, while Sea A Go held on for the Show. The $2 exacta returned $57.20 and the trifecta a delightful $924.90.

Rozzy Ridge's win here illustrates the value of not only having information about a horse's sire and dam, but also of understanding how owner/ breeders think. There are a number of Mom & Pop operations in the horse business who make their money by breeding their mares to cheap, but precocious sires and then running the resulting get for a tag. Handicappers should always be alert, especially in cheaper Maiden Claiming races, to the possibility that a precociously-bred maiden has been shipped in by a cagey owner/breeder to steal a pot.

7

Handicapping for Stamina in Routes

For over 300 years breeders have been arguing over the relative merits of speed and stamina and this debate gets played out every day at the track, where racing secretaries try to accommodate both points of view by writing a variety of sprints and routes depending upon the kinds of horses they have on their backstretches.

As each racing season evolves and young horses mature, they are variously asked to increase the distance they must carry their speed from 4 1/2 furlongs up to 1 1/8 mile. While some of the precociously-bred horses that broke their maiden early in sprints have enough stamina in their pedigrees to successfully route, many do not. And conversely, a number of the more stoutly-bred maidens that were initially trounced in sprints now use the seasoning gained from those efforts to successfully stretch out.

From a pedigree handicapping point of view, this means that a young horse's precocity and speed ratings become much less important than its stamina index once it moves from a sprint to a route. Every day horses that won at first asking with high speed figures in sprints are defeated when they attempt to stretch out. Conversely, horses that earned low "figs" in sprints often show dramatic improvement when given the opportunity to run a further distance of ground.

Rather than castigating a sprinter for being a "quitter" in a route and a router for being a "plodder" in a sprint, it is better to appreciate that each of them are magnificent creatures in their own right and both blessed and limited by their genetic inheritance. The same quick-twitch muscle fibers that give sprinters their explosive turn of foot, also make them hit an invisible genetic wall once they try to carry their speed beyond six or seven furlongs. For sprinters still essentially hark back 300 years to those original Hobby "quarter horse" mares, while routers with their slower-twitched muscle fibers and metabolic rates have more of the long distance Oriental stallions in them.

The physiological explanation of these differing genetic inheritances revolves around the fact that a sprinter's explosive speed burns up a tremendous amount of oxygen. And that after six furlongs it can no longer aerobically replace the oxygen in its blood fast enough to prevent the anaerobic build up of lactic acid that causes muscle fatigue and loss of speed. The slower-twitched muscle fibers that a router possesses,

106   meanwhile, generally allows it to lope along and "ration" the amount of oxygen it uses, thus delaying the build up of lactic acid long enough so that it can carry its speed further than a sprinter can.

As mentioned earlier, the stamina indexes (SI's) assigned to each sire in this book have been derived from looking at the actual race records of their progeny. To recapitulate: horses by sires with SI's of 5 will generally be sprinters, those with 4's will get up to a mile, the 3's up to 1 1/16 mile, the 2's up to 1 1/8 mile, and finally those with SI's of 1 are likely to stay 1 1/4 mile and beyond.

Having outlined these parameters, however, it is important to emphasize that a stamina index, like every other racing statistic, is not an absolute indication of a horse's distance potential. A number of horses are neither pure sprinters nor pure routers and whether or not they can sprint or successfully stretch out often depends upon how they are trained and the pace of a race. Occasionally a well-spotted sprinter, if let loose on the lead, will steal a route, while an even-running plodder will "catch" the dying speed in a suicidal sprint.

But, most of the time, horses will run to either the sprint or route side of their breeding and their stamina indexes are strong indicators of how far they are likely to run. With that in mind, let us turn to some examples of Maiden, Allowance and Claiming events where appropriately-bred horses are stretching out for the first time.

Maiden Special Weight Routes

The first race at Hollywood Park on June 12th is a Maiden Special Weight route written by the Racing Secretary for those more stoutly-bred, yet talented, horses on the grounds who don't have enough early speed to win in sprints.

Surveying the field in this 1 1/16 mile event for Cal-Breds, the first thing to do is eliminate the inappropriately-bred horses and proven losers. Dabbler and Half Past Ten, with their SI's of 5, are bred to sprint and have already lost to Maiden Claimers. A La Plage and His Royal Grace both have decent Beyers and might improve. But, they have also been beaten by cheaper horses and the odds are against them winning when bumped up in class.

This leaves us with four horses to consider — Reality Changes, Having Our Say, Mill Sham and Persuasively — all of which are stretching out for the first time. Of these horses, Reality Changes, with his stamina index of 2 and Harris Farm owner/breeder connections, looks like a standout. He has already beaten Having Our Say and Persuasively and has a nice 7 furlong prep on June 4th for this stretch out. Furthermore, Reality Changes passed horses in his last race and his :22.3 early fractions look like the speed of this field. Last, but not least, this son of Hollywood Gold Cup winner Cutlass Reality has been bet down from 8-1 in his last race to 2-1 in this spot.

Hollywood Park

1

1 1/16 MILES. (1:40) **MAIDEN SPECIAL WEIGHT. Purse $37,000 (includes $3,000 from Cal-bred Race Fund).**
3-year-olds and upward, bred in California. Weights: 3-year-olds, 115 lbs. Older, 122 lbs.

1 1/16 MILES — FINISH ▲ ▲ START

Coupled – Dabbler and Half Past Ten

Dabbler
Own: Caiazzo Cynthia
SCOTT J M (25 1 1 2 .04)

Ch. c. 3 (Apr)
Sire: Fort Calgary (Bold Joey)
Dam: Honey Bronze (Tobin Bronze)
Br: Franklin Jim (Cal)
Tr: Anderson Arthur (3 0 0 0 .00)

115

Lifetime Record :	2 M 0 0	$0		
1994	2 M 0 0		Turf	0 0 0 0
1993	0 M 0 0		Wet	0 0 0 0
Hol	2 0 0 0		Dist	0 0 0 0

27May94- 2Hol fst 6f :22 2 :45 4 :57 4 1:10 2 3+ Md 45000 — 54 4 6 6 9½ 6 10 6 10 6 12¾ Scott J M 113 49.60 75-11 Walk Point115 5 Respectable Rascal122 nk Score O Gold116 3½ Outrun 7
13May94- 4Hol fst 6f :22 :45 2 :57 1 1:10 1 3+ [S]Md Sp Wt — -0 4 9 10 22 10 25 10 36 10 44¾ Castanon A L B 115 74.30 44-19 Touch The Moon122 2 Menacing Pirate115 3½ Reality Changes115 2 10
Awkward start, steadied, green

WORKOUTS: Jun 6 Fpx 4f fst :48 2 H 2/9 ●Jun 1 Fpx 3f fst :35 3 H 1/9 ●May 21 Fpx 6f fst 1:13 H 1/9 ●May 18 Fpx 5f fst 1:00 3 H 1/7 May 9 Fpx 5f fst 1:02 1 H 4/7 ●May 4 Fpx 5f fst 1:02 H 1/4

Reality Changes
Own: Harris Farms Inc
STEVENS G L (157 35 21 32 .22)

B. c. 3 (Feb)
Sire: Cutlass Reality (Cutlass)
Dam: Thaliard's Pearl (Thaliard)
Br: Harris Farms Inc (Cal)
Tr: Zucker Howard L (10 1 1 1 .10)

L 115

Lifetime Record :	2 M 0 2	$10,500		
1994	2 M 0 2	$10,500	Turf	0 0 0 0
1993	0 M 0 0		Wet	0 0 0 0
Hol	1 0 0 1	$5,250	Dist	0 0 0 0

13May94- 4Hol fst 6f :22 :45 2 :57 1 1:10 1 3+ [S]Md Sp Wt — 78 9 7 7 6½ 6 8½ 4 8½ 3 5½ Stevens G L LB 115 8.10 83-19 TouchTheMoon122 2 MenacingPirate115 3½ RealityChnges115 2 Wide trip 10
23Apr94- 4SA fst 6f :22 1 :45 1 :57 1 1:09 3 3+ [S]Md Sp Wt — 78 4 5 3 2 3 5 3 6 3 5½ Stevens G L LB 116 7.50 84-12 Phone Prince116 2¾ Shock Tippet116 2½ RealityChanges116 3½ Best of others 8

WORKOUTS: Jun 4 SA 7f fst 1:30 2 H 2/2 May 11 SA 4f fst :49 3 H 21/45 May 6 SA 4f fst :49 4 H 22/28 Apr 20 SA 4f fst :48 4 H 17/32 Apr 13 SA 6f fst 1:14 1 Hg 11/20 Apr 6 SA 4f fst :47 1 H 5/31

Having Our Say
Own: Von Flue Frank R & Tom C
VALENZUELA P A (134 10 18 16 .07)

B. g. 3 (Apr)
Sire: Caucasus (Nijinsky II)
Dam: Our Best Tell (Tell)
Br: Von Flue Frank R (Cal)
Tr: Davis Wes (2 0 0 0 .00)

115

Lifetime Record :	1 M 0 0	$875		
1994	1 M 0 0	$875	Turf	0 0 0 0
1993	0 M 0 0		Wet	0 0 0 0
Hol	0 0 0 0		Dist	0 0 0 0

23Apr94- 4SA fst 6f :22 1 :45 1 :57 1 1:09 3 3+ [S]Md Sp Wt — 64 1 6 7 5½ 6 8½ 6 10 5 10¾ Solis A 116 b 26.40 78-12 Phone Prince116 2¾ Shock Tippet116 2½ Reality Changes116 3½ Off step slow 8

WORKOUTS: Jun 8 SA 5f fst 1:04 3 H 29/31 May 29 SA 7f fst 1:29 H 6/6 May 17 SA 3f fst :37 2 H 8/15 May 9 SA 1 fst 1:42 H 1/2 May 2 SA 3f fst :38 3 H 33/37 Apr 22 SA 3f fst :34 3 H 3/25

Mill Sham
Own: Evergreen Farm
SOLIS A (213 43 22 31 .20)

Ro. g. 3 (Feb)
Sire: Mill Native (Exclusive Native)
Dam: Sham Say (Oh Say)
Br: Evergreen Farm (Cal)
Tr: Whittingham Charles (39 3 3 1 .08)

115

Lifetime Record :	3 M 0 1	$5,250		
1994	2 M 0 1	$5,250	Turf	0 0 0 0
1993	1 M 0 0		Wet	0 0 0 0
Hol	1 0 0 0		Dist	0 0 0 0

11May94- 4Hol fst 7f :22 2 :45 1:09 1:21 2 3+ Md Sp Wt — 73 4 5 5 4½ 6 6¼ 6 6¼ 6 10¼ Solis A B 115 5.80 87-10 Alphabet Soup115 hd Exalto116 2¼ Explorateur116 hd No rally 9
23Apr94- 6SA fst 6f :22 :45 :57 1 1:09 3 [S]Md Sp Wt — 83 4 7 8 13 8 9 6 8 3 3½ Solis A B 116 13.70 85-12 Nicolette's Pride116¾ Menacing Pirate116 2¾ Mill Sham116 no 8
Off step slow, greenly
12Sep93- 2Dmr fst 6f :21 4 :44 4 :57 1 1:10 1 [S]Md Sp Wt — 52 6 11 12 15 12 11 11 13 7 11½ Solis A B 118 83.80 77-10 WingedTip118 4½ HlloweenTret118 hd CollectbleWine118 3½ Off slowly, wide 12

WORKOUTS: Jun 8 Hol 5f fst 1:01 2 H 15/30 May 28 Hol 5f fst 1:01 4 H 27/37 May 23 Hol 5f fst 1:01 H 17/47 May 18 Hol 3f fst :36 2 H 9/26 May 7 Hol 4f fst :48 3 H 28/44 May 1 Hol 3f fst :37 1 H 18/26

Persuasively
Own: Figueroa Farms
MCCARRON C J (103 20 19 17 .19)

Ch. c. 3 (Mar)
Sire: Count Eric (Riverman)
Dam: Bold Juliet (Bold Tropic)
Br: Marvin E. Garrett (Cal)
Tr: Pierce Donald (2 0 0 1 .00)

L 115

Lifetime Record :	3 M 0 1	$7,875		
1994	3 M 0 1	$7,875	Turf	0 0 0 0
1993	0 M 0 0		Wet	0 0 0 0
Hol	2 0 0 1	$5,250	Dist	0 0 0 0

28May94- 2Hol fst 6f :22 :45 1 :57 2 1:10 2 3+ [S]Md Sp Wt — 75 2 6 4 3½ 3 3 3 1½ 3 2¼ McCarron C J LB 115 10.80 86-09 Menacing Pirate115 1¼ Capo116 1 Persuasively115 no Pinched start 6
13May94- 4Hol fst 6f :22 :45 2 :57 1 1:10 1 3+ [S]Md Sp Wt — 71 6 6 5 3 4 6 6 8¾ 6 8¾ Pincay L Jr B 117 5.10 81-19 Touch The Moon122 2 MenacingPirate115 3½ RealityChanges115 2 No rally 10
23Apr94- 6SA fst 6f :22 :45 :57 1 1:09 3 [S]Md Sp Wt — 83 1 8 4 2½ 4 3½ 4 4½ 4 3½ McCarron C J B 116 9.50 85-12 Nicolette's Pride116¾ Menacing Pirate116 2¾ Mill Sham116 no Off step slow 8

WORKOUTS: Jun 7 SA 5f fst 1:01 3 H 12/20 May 23 SA 4f fst :48 3 H 7/21 May 9 SA 5f fst 1:00 3 H 6/29 May 4 SA 4f fst :51 3 B 21/24 Apr 21 SA 4f fst :47 1 Hg 4/43 Apr 15 SA 6f fst 1:13 4 H 7/28

A La Plage
Own: Braly & Hecht & Redcher & Partners
VALENZUELA F H (174 15 19 18 .09)

Ro. g. 4
Sire: Caucasus (Nijinsky II)
Dam: Bangolure (The Tongan)
Br: Overton Timothy M Jr (Cal)
Tr: Bunn Thomas M Jr (8 0 1 2 .00)

L 122

Lifetime Record :	6 M 1 3	$15,475		
1994	3 M 0 1	$6,975	Turf	1 0 0 1 $3,150
1993	3 M 1 2	$8,500	Wet	0 0 0 0
Hol	3 0 1 1	$7,075	Dist	0 0 0 0

28May94- 5GG fm 1 1/16 [T] :22 3 :46 4 1:11 1 1:43 2 + 3+ Md Sp Wt — 82 7 5 4 3½ 4 1¼ 3 3½ 3 3½ Lopez A D LB 120 b 4.30 85-11 Trinity Pass116 1 Stampede Steve113 2½ A La Plage120 nk Mild late bid 10
6May94- 4Hol fst 7f :21 4 :44 2 1:09 3 1:22 4 3+ [S]Md 40000 — 68 2 8 6 8½ 6 9 5 5 4 3 Valenzuela F H LB 122 b 5.90 87-06 Victory Sun115½ Marinatedartichoke115¾ River Scout117 1¾ Late bid 10
18Mar94- 6SA fst 6½f :21 2 :44 2 1:09 3 1:16 Md Sp Wt — 75 2 7 7 6½ 6 6½ 6 5½ 4 5½ Black C A LB 120 b 9.70 86-10 Sikes120 3½ Dion The Tyrant120 hd J. R. Cigar120 2 Improved position 7
11Aug93- 9Dmr fst 1 :23 :47 1:11 4 1:37 3+ Md 40000 — 76 1 1 1 hd 4 1½ 3 1 3 2 Black C A LB 117 *2.00 80-17 I'm Ruined115 hd California Spirit117 2 A La Plage117 1¾ Weakened a bit 10
23Jly93- 9Hol fst 6½f :22 :44 4 1:10 2 1:17 3+ [S]Md c-32000 — 73 8 6 4¾ 4 2½ 2 1½ 2 1¾ Black C A LB 116 *1.30 82-13 Be Sum Interco109 1¾ A La Plage116 2½ Blind Touch114 5 Good effort 11
Claimed from Overton Jr & Sahadi, Sahadi Jenine Trainer
7Jly93- 9Hol fst 6½f :22 2 :45 1 1:10 3 1:17 3+ [S]Md 25000 — 76 9 9 10 7 8 3½ 6 2¾ 3½ Black C A 116 *2.10 84-09 Tony's Pirate116 hd Native Bearing116 nk A La Plage116½ 12
Troubled start, wide

WORKOUTS: May 22 Hol 6f fst 1:14 1 H 8/11 Apr 30 Hol 4f fst :52 3 H 40/41 Apr 20 Hol 7f fst 1:29 2 H 4/6 Apr 15 Hol 5f fst 1:02 2 H 24/30 Apr 10 Hol 3f fst :37 4 H 29/31 Mar 13 Hol 4f fst :47 3 H 2/14

continued

His Royal Grace
Own: Mevorach Samuel

ANTLEY C W (165 35 33 21 .21)

B. g. 3 (Apr)
Sire: Somethingfabulous (Northern Dancer)
Dam: Her Royal Grace (King of Kings)
Br: Old English Rancho (Cal)
Tr: Lewis Craig A (39 7 6 4 .18)

L 115

Lifetime Record :	3 M 0 2	$7,050		
1994	3 M 0 2	$7,050	Turf	0 0 0 0
1993	0 M 0 0		Wet	0 0 0 0
Hol	2 0 0 2	$5,700	Dist	0 0 0 0

27May94-6Hol fst 7f :224 :46 1:102 1:232 3↑ [S]Md c-40000 75 2 6 63½ 64¼ 53½ 33¾ McCarron C J B 115 3.30 83-11 Alybally1151 Diaghilev1152¾ His Royal Grace115hd Rail trip 7
Claimed from Johnston Betty & E W & Judy, Warren Donald Trainer
18May94-9Hol fst 7f :22 :45 1:094 1:223 3↑ Md 40000 82 4 5 64¾ 67½ 43½ 32¾ Solis A B 115 2.70 88-10 Geiger Time117½ DazzledByGold1152¼ HisRoyalGrace1151¼ Forced out 1/2 9
22Apr94-4SA fst 6f :22 :451 :572 1:10 Md 32000 78 6 6 64½ 66 63¾ 41½ Solis A B 118 18.30 86-11 Hot Papa Tom118¾ Hippiater118½ Scottish Slew118hd Came in start 11
WORKOUTS: Jun 6 Hol 5f fst 1:011 H 14/36 May 14 Hol 5f fst 1:011 H 14/46 ●May 9 Hol 5f fst :59 H 1/32 May 5 Hol 3f fst :372 H 20/23 Apr 18 SA 3f fst :362 Hg 12/22 Apr 10 SA 6f fst 1:134 Hg 10/33

Half Past Ten
Own: Caiazzo Cynthia

FERNANDEZ A L (5 0 0 0 .00)

Dk. b or br g. 4
Sire: Cad (Timeless Moment)
Dam: Love Amadeus (Pretense)
Br: Kerr Stock Farm Inc (Cal)
Tr: Anderson Arthur (3 0 0 0 .00)

122

Lifetime Record :	13 M 0 1	$4,170			
1994	2 M 0 0		Turf	1 0 0 0	
1993	3 M 0 0		Wet	0 0 0 0	
Hol	6 0 0 0		Dist	4 0 0 0	$900

1Jun94-4Hol fst 1$\frac{1}{16}$:231 :471 1:114 1:442 3↑ [S]Md 28000 45 7 10 910 910 710 717½ Scott J M 120 112.10 61-11 Never Topped1145½ Cinematic120½ Sun Discovery1205½ Pinched start 10
1May94-6Hol fst 7½f :222 :452 1:101 1:29 3↑ Md 35000 46 5 8 89¾ 87¾ 614 618½ Linares M G 120 38.00 71-08 White Statuary1152¾ Island Sport1178½ Essence Of Life120hd No factor 8
25Jly93-10Hol fst 7f :221 :441 1:091 1:222 3↑ Md Sp Wt 58 8 8 913 917 812 810½ Castaneda M 116 123.50 82-10 Mister Jolie1161¾ Family Firm116hd Top Senator121½ No factor 9
8Jly93-6Hol fm 1⅛ Ⓣ :483 1:123 1:363 1:484 3↑ Md Sp Wt 39 7 5 56 86¾ 915 920 Martinez F F 116 268.20 60-20 Wise Words1162 Daskiyawar1222¼ Jutland1222¼ Stopped 10
31Jan93-2SA fst 1 :223 :461 1:113 1:384 Md Sp Wt 55 4 8 910 89¾ 79 68½ Gulas L L5 LB 112 63.50 67-17 OnlyAlpha1173 RespectbleRscl1171 OutsideTheLine117no Carried out 7/8 9
30Dec92-4SA gd 1$\frac{1}{16}$:234 :483 1:133 1:45 Md 40000 48 2 4 44 65½ 717 517¾ Sanchez J M B 118 39.00 61-22 Hitumwhenyacan1183 Norm Breaker1181¾ Frozen River1189 No mishap 8
13Dec92-6Hol fst 7½f :223 :461 1:112 1:304 Md Sp Wt 23 3 6 84½ 76¼ 614 623¼ Berrio O A B 119 48.80 58-13 Blazing Aura119¾ Respectable Rascal1191 Jaltipan1194 Jostled start 9
28Nov92-4Hol fst 7f :22 :443 1:092 1:221 [S]Alw 31000N1X 51 1 9 911 912 910 714½ Castanon A L B 116 149.60 78-06 Offshore Pirate1172¾ Enlightened Energy119hd Silver Picea1191½ Outrun 9
20Nov92-3Hol fst 1$\frac{1}{16}$:232 :474 1:123 1:444 Md Sp Wt 39 5 5 42¾ 65¼ 611 615½ Castanon A L B 118 26.70 61-23 Tossofthecoin1184 Nautilus118½ Alycap118 Wide trip 6
23Oct92-2SA fst 7f :221 :444 1:103 1:241 Md 50000 60 1 8 812 814 89 35¾ alvarado F T B 118 84.00 76-14 Lucky'sFirstOne1183 I'mRuined1182¾ HalfPstTen118 Improved position 8
WORKOUTS: ●Jun 6 Fpx 5f fst 1:004 H 1/4 May 21 Fpx 6f fst 1:133 H 2/9 May 14 Fpx 6f fst 1:14 H 1/2 ●May 9 Fpx 5f fst 1:01 H 1/7 Apr 25 Fpx 6f fst 1:144 H 2/2 Apr 18 Fpx 4f fst :494 H 3/4

© 1994 DRF, Inc.

So how to bet? With all the positive factors going for him, Reality Changes should be even-money and is worth a prime Win bet at 2-1. While we could, alternatively, "spread" our win bet by wheeling Reality Changes on top in the exacta to everything except the Dabbler-Half Past Ten entry, this would only be more profitable if someone other than His Royal Grace finishes second.

FIRST RACE
Hollywood
JUNE 12, 1994

1$\frac{1}{16}$ MILES. (1.40) MAIDEN SPECIAL WEIGHT. Purse $37,000 (includes $3,000 from Cal-bred Race Fund). 3-year-olds and upward, bred in California. Weights: 3-year-olds, 115 lbs. Older, 122 lbs. (Day 36 of a 68 Day Meet. Clear. 74.)

Value of Race: $37,000 Winner $20,350; second $7,400; third $5,550; fourth $2,775; fifth $925. Mutuel Pool $252,593.00 Exacta Pool $212,313.00 Quinella Pool $30,180.00

Last Raced	Horse	M/Eqt.	A.Wt	PP	St	¼	½	¾	Str	Fin	Jockey	Odds $1
13May94 4Hol3	Reality Changes	LB	3 115	2	4	1½	1½	11½	11½	11¼	Stevens G L	2.20
27May94 6Hol3	His Royal Grace	LB	3 115	7	3	42½	41½	52½	21½	24½	Antley C W	2.80
11May94 4Hol6	Mill Sham	Bb	3 115	4	6	63	53½	41	32	34	Solis A	3.60
23Apr94 4SA5	Having Our Say		3 118	3	2	2hd	2hd	2hd	41½	41¾	Valenzuela P A	31.20
28May94 5GG3	A La Plage	LB	4 122	6	8	7hd	7½	62½	53½	52	Valenzuela F H	5.20
1Jun94 4Hol7	Half Past Ten		4 122	8	7	8	8	74	710	62¾	Fernandez A L	a-49.90
28May94 2Hol3	Persuasively	LB	3 115	5	1	32½	32	3hd	6½	716	McCarron C J	4.20
27May94 2Hol6	Dabbler		3 115	1	5	5½	6hd	8	8	8	Scott J M	a-49.90

a-Coupled: Half Past Ten and Dabbler.

OFF AT 1:00 Start Good For All But A LA PLAGE. Won driving. Time, :232, :47, 1:114, 1:373, 1:44 Track fast.

$2 Mutuel Prices:				
	2-REALITY CHANGES	6.40	3.20	2.40
	7-HIS ROYAL GRACE		3.40	2.60
	4-MILL SHAM			2.80

$2 EXACTA 2-7 PAID $25.00 $2 QUINELLA 2-7 PAID $12.40

© 1994 DRF, Inc.

The first race on March 4th at Golden Gate Fields is another example of a Maiden Special Weight route won by a horse stretching out for the first time. On the surface, the race looks like a lock for the 2-5 favorite Vanilla Wafer who finished third against winners in the Bay Meadows Oaks. But, Vanilla Wafer has already lost twice in routes at this level, at short prices, and comes out of her last race with declining Beyer

Golden Gate Fields

1

1 MILE. (1:33) **MAIDEN SPECIAL WEIGHT. Purse $19,000. Fillies, 3-year-olds, Weight, 117 lbs.**

1 MILE
START ▲ FINISH

Garbo N Charley
Own: Rubino Ronald C
HUMMEL C R (46 4 5 1 .09)

B. f. 3 (Apr)
Sire: Zaizoom (Al Nasr)
Dam: Dracena (Bold Tropic)
Br: Westfield Farms Inc. (Cal)
Tr: Maisenbach Warren (1 0 0 0 .00)

117

Lifetime Record : 1 M 0 0 $0

1994	1 M 0 0		Turf	0 0 0 0
1993	0 M 0 0		Wet	0 0 0 0
GG	1 0 0 0		Dist	0 0 0 0

13Feb94-6GG fst 6f :22 :45 :571 1:094 ⒻMd Sp Wt 21 6 10 1018 1018 1019 1020 Edgerly K L B 117 b 58.70 70-08 Gaelic Dowry11710 Wild Royal117nk Newcastle Blues117½ Outrun 10

WORKOUTS: Jan 31 Pln 6f fst 1:182 Hg 11/11 Jan 18 Pln 3f fst :374 H 7/8 Jan 7 Pln 1 fst 1:45 H 2/2 Dec 27 Pln 5f fst 1:03 H 14/22 Dec 20 Pln 4f fst :541 H 12/12 ●Dec 10 Pln 4f sly :512 H 1/5

Vanilla Wafer
Own: Jensen Luella M
BELVOIR V T (141 38 22 22 .27)

Dk. b or br f. 3 (Jan)
Sire: Grey Dawn II (Herbager)
Dam: Six n' Such (Hail the Pirates)
Br: Royster S Brooke (Ky)
Tr: Belvoir Howard (30 3 4 3 .10)

117

Lifetime Record : 8 M 3 4 $41,422

1994	2 M 0 2	$18,450	Turf	0 0 0 0	
1993	6 M 3 2	$22,972	Wet	0 0 0 0	
GG	1 0 0 1	$3,450	Dist	4 0 3 1	$21,609

24Feb94-8GG fst 1 1/16 :24 :481 1:132 1:464 ⒻAlw 23000N1X 55 5 5 57¼ 43 36 39¼ Belvoir V T LB 117 f 1.50 54-32 Not Too Proper117nk Saros Lane1179 Vanilla Wafer1174 Wide trip 5
17Jan94-4BM fst 1 1/16 :233 :473 1:122 1:434 ⒻB M Oaks-G3 72 5 5 55 55¼ 312 313¼ Belvoir V T L 116 f 12.40 70-25 Lindzell1121½ Work The Crowd11512 Vanilla Wafer1164 Even late 5
30Dec93-3BM fst 1 :222 :462 1:111 1:372 ⒻClmStk40000 72 3 7 713 77 43½ 3¾ Belvoir V T LB 120 12.70 82-17 DHMyBlueGenes116DHAvSingstheblus120¾VnillWfr120½ Wide late bid 7
14Nov93-3BM fst 1 :234 :482 1:133 1:392 ⒻMd Sp Wt 56 3 2 21 21½ 21½ 2nk Belvoir V T LB 117 *1.10 73-18 Lucky Jo B.117nk Vanilla Wafer1175 Sonata Sky11710 Slow late gain 5
17Oct93-1BM gd 1 :224 :474 1:133 1:401 ⒻMd Sp Wt 51 4 5 45 33½ 34 26 Belvoir V T LB 117 2.60 63-28 Jilly's Halo1176 Vanilla Wafer1171½ Day Rate1175 Wide late run 5
12Sep93-10YM fst 6f :22 :441 :563 1:084 ⒻMt Rnr Sprt35k 40 1 8 79 711 713 613¼ Malgarini T M 116 5.80 85-06 Happy La1123¼ Gap Gossip1133 Classy Rolls1122 No factor 8
28Aug93-10YM fst 1 :241 :482 1:131 1:391 ⒻEmrld Lassie55k 46 5 4 42½ 1½ 2hd 23 Malgarini T M 119 4.40 80-11 Sparkin With Jak1193 Vanilla Wafer1192 Classy Rolls1162½ 2nd best 5
7Aug93-3YM fst 6f :23 :454 :583 1:122 ⒻMd Sp Wt 43 2 8 88½ 78 44½ 32½ Malgarini T M 118 14.00 78-12 SparkinWithJak1161 MissCraftySlew1191½ VanillaWafer1181 Rallied late 9

WORKOUTS: Feb 16 GG 5f fst 1:011 H 9/37 Feb 9 GG 6f gd 1:17 H 17/23 ●Jan 31 GG 6f fst 1:142 H 1/4 Jan 14 BM 4f fst :493 H 20/40 Jan 8 BM 5f fst 1:011 H 21/51 Dec 28 BM 3f fst :374 H 17/28

Bel Native
Own: Evergreen Farm
BOULANGER G (130 12 32 23 .09)

Ro. f. 3 (Feb)
Sire: Mill Native (Exclusive Native)
Dam: Bel Real (In Reality)
Br: Evergreen Thoroughbreds Inc (Cal)
Tr: Bonde Jeff (25 4 6 4 .16)

117

Lifetime Record : 5 M 1 1 $6,300

1994	1 M 1 0	$2,400	Turf	0 0 0 0	
1993	4 M 0 1	$3,900	Wet	0 0 0 0	
GG	0 0 0 0		Dist	1 0 0 0	$375

8Jan94-1BM fst 6f :222 :453 :583 1:122 ⒻMd 32000 49 1 4 56 56½ 54½ 21 Baze R A LB 117 b 6.30 78-17 Irish Eagle1171 Bel Native117½ Glorius And Bold117hd 6
Broke out start, wide stretch
7Nov93-1BM fst 6f :222 :454 :584 1:122 ⒻMd Sp Wt 50 1 3 44½ 34½ 45 43¾ Boulanger G LB 117 b 9.60 70-18 Little Brass1173 Day Rate117nk Boz Connection117½ No rally 6
17Oct93-1BM gd 1 :224 :474 1:133 1:401 ⒻMd Sp Wt 29 2 2 33 58½ 512 518¼ Boulanger G LB 117 b 5.40 50-28 Jilly's Halo1176 Vanilla Wafer1171½ Day Rate1175 Gave way 5
6Sep93-6BM fst 5½f :22 :454 :582 1:044 ⒻMd Sp Wt 50 6 9 88½ 76½ 56 34½ Boulanger G LB 117 b 27.30 82-14 Vail Link117½ Eishin Ohio1174 Bel Native117¾ Mild late bid 9
4Aug93-6SR fst 5½f :23 :461 :592 1:053 ⒻMd Sp Wt 35 6 9 910 811 811 612¾ Hansen R D LB 117 b 13.50 72-12 Gentle Rainfall1171½ Valnesian11710 Athena's Elegance1171 Wide trip 9

WORKOUTS: ●Feb 27 GG 5f my 1:031 H (d) 1/15 Feb 21 GG 5f sly 1:07 H (d) 4/6 Feb 14 GG 5f fst 1:004 H 4/13 ●Feb 8 GG 4f my :514 H (d) 1/8 Feb 2 GG 4f fst :484 H 7/19 ●Jan 27 BM 5f sly 1:043 H (d) 1/10

Rustica Ridge
Own: Nor Fabio & Joanne H
WARREN R J JR (77 11 6 12 .14)

B. f. 3 (Feb)
Sire: Cox's Ridge (Best Turn)
Dam: Murcot (Royal And Regal)
Br: Nor Fabio & Joanne H (Ky)
Tr: Greenman Walter (19 1 3 4 .05)

117

Lifetime Record : 1 M 0 0 $0

1993	1 M 0 0		Turf	0 0 0 0
1992	0 M 0 0		Wet	0 0 0 0
GG	0 0 0 0		Dist	0 0 0 0

Entered 3Mar94- 9 SA
19Dec93-4Hol fst 5½f :221 :452 :574 1:04 ⒻMd Sp Wt 44 4 5 817 812 88¾ 610¼ Delahoussaye E B 118 13.70 81-13 Wild Impulse118½ Espadrille1182¾ Fresh Berries118¾ Outrun 8

WORKOUTS: Feb 26 Hol 5f fst 1:011 H 5/14 Feb 19 Hol 5f fst 1:014 H 20/37 Feb 11 Hol 4f fst :494 H 16/32 Feb 4 Hol 7f gd 1:28 H 1/1 Jan 28 Hol 1 fst 1:41 H 2/2 Jan 22 Hol 7f fst 1:314 H 5/5

Ali Capote
Own: Peter Redekop B C Ltd
CHAPMAN T M (71 9 8 4 .13)

B. f. 3 (May)
Sire: Capote (Seattle Slew)
Dam: Rendezvous (Meneval)
Br: Mickey Liber (Ky)
Tr: Hollendorfer Jerry (67 23 12 7 .34)

117

Lifetime Record : 3 M 0 1 $2,475

1994	2 M 0 1	$2,475	Turf	0 0 0 0	
1993	1 M 0 0		Wet	0 0 0 0	
GG	1 0 0 0		Dist	1 0 0 1	$2,475

13Feb94-6GG fst 6f :22 :45 :571 1:094 ⒻMd Sp Wt 34 5 3 66 89½ 914 715 Romero J A LB 117 b 7.70 75-08 Gaelic Dowry11710 Wild Royal117nk Newcastle Blues117½ Dull try 10
17Jan94-1BM fst 1 :23 :473 1:131 1:402 ⒻMd Sp Wt 52 2 1 1½ 1hd 31 35¼ Belvoir V T L 117 b 4.40 65-25 A Trifle To Spare1172½ Key To V. A.1123 Ali Capote1171½ Weakened 7
5Dec93-6BM fst 6f :223 :461 :582 1:112 ⒻMd Sp Wt 27 9 6 55½ 66½ 813 712¼ Belvoir V T B 117 5.30 67-18 Woven Gold1176 Lovely Krissy117hd Sonata Sky1173½ Wide trip 9

WORKOUTS: Feb 27 GG 4f my :512 H (d) 3/16 Feb 9 GG 5f gd 1:022 H 13/30 Feb 3 GG 4f fst :47 H 3/24 Jan 29 BM 5f gd 1:023 H 37/67 Jan 23 BM 4f sly :52 H (d) 8/18 Jan 9 BM 1 fst 1:423 H 3/3

Crusader Gal
Own: Aleo Harry J
BAZE R A (117 38 22 14 .32)

Ro. f. 3 (Feb)
Sire: Crusader Sword (Damascus)
Dam: Current Gal (Little Current)
Br: St. Francis Farm (Fla)
Tr: Gilchrist Greg (15 1 5 0 .07)

117

Lifetime Record : 2 M 1 0 $4,725

1994	2 M 1 0	$4,725	Turf	0 0 0 0	
1993	0 M 0 0		Wet	1 0 1 0	$3,300
GG	1 0 0 0	$1,425	Dist	0 0 0 0	

13Feb94-6GG fst 6f :22 :45 :571 1:094 ⒻMd Sp Wt 45 2 7 55½ 78½ 712 410¾ Baze R A B 117 b *2.50 79-08 Gaelic Dowry11710 Wild Royal117nk Newcastle Blues117½ Wide stretch 10
23Jan94-6BM my 6f :221 :46 :592 1:131 ⒻMd Sp Wt 50 1 6 64½ 53½ 31½ 2¾ Miranda V B 117 b 13.60 74-22 Joying117¾ Crusader Gal1171 Newcastle Blues1172 Closed wide 6

WORKOUTS: Feb 27 GG 5f my 1:042 H (d) 4/15 Feb 5 GG 5f fst 1:021 H 39/75 Jan 16 BM 5f fst 1:003 Hg 7/49 Jan 8 BM 5f fst 1:011 H 21/51 Dec 19 BM 5f fst 1:011 H 36/106 Dec 12 BM 3f my :381 Hg(d) 11/18

and *DRF* speed ratings. With three seconds and four thirds out of eight races, Vanilla Wafer may be the kind of horse that just doesn't want to win. Given her 2-5 price, she is worth trying to beat.

But who can beat her? Garbo N Charley is bred for the grass and at 42-1 can be thrown out. Bel Native has already lost to Vanilla Wafer by 12 lengths at this distance and can also be eliminated. Since there isn't much speed in the race, Ali Capote is an interesting possibility. She just might get brave on the front end. But, on the downside, she comes off a weak race and has already lost routing at this level while setting a slow pace.

This leaves Rustica Ridge and Crusader Gal who are both being stretching out for the first time and have stamina indexes which suggest they should improve with added distance. Crusader Gal retains leading rider Russell Baze, beat Garbo N Charley and Ali Capote in her last race, and is the second choice at 4-1.

Rustica Ridge is the mystery horse here and is by the classiest sire in the race. While she never contended in her first start, juvenile horses by Cox's Ridge almost never break their maidens in short sprints and Rustica Ridge did make up nearly seven lengths. Since then she has been given several months to further mature and brought up to this race with three long workouts plus a nice 5 furlong tightener on February 26th. It is also worth noting that she was scratched out of a race on March 3rd at Santa Anita by her owner/ breeder connections and shipped up for what they obviously think is an easier spot to break her maiden.

So How to bet? At 7-1 Rustica Ridge has been bet down to half of her debut odds and looks like a live horse. Everything considered, Rustica Ridge is definitely worth a Win bet and might also be boxed in the exacta with Crusader Gal in the hope that she will improve on the stretch out while Vanilla Wafer will continue to show declining form.

Results: Bel Native, Ali Capote and Crusader Gal dueled for the lead, but were caught by Rustica Ridge at the top of the stretch with Vanilla Wafer getting up for the Place. Rustic Ridge paid $14.80 on the Win and the $5 exacta returned $78.50. Though Crusader faded to fourth, she was still the value play in the exacta.

Maiden Claiming Routes

Not all horses with route breeding are fast enough to win in Maiden Special Weight company, so Racing Secretaries write races for them at the Maiden Claiming level. The second race at Hollywood Park on July 9th is an interesting example of a 3-5 favorite, with the highest speed figures in the race, who nevertheless is suspect because she has already lost nine times and is facing several horses that have never routed before.

Going over the field, the first thing to do is throw out all the proven losers at this distance and claiming level — which eliminates Vento E Fuoco, Lady Stonewalk, Turf Princess, That's Gin and Sailor's Grace, none of whom is getting bet. This leaves us with the stretchout horses Astrial, Her Thoughts, and Drona's Starlet as possible candidates to upset Silent Draw who has already lost twice routing at this level at low odds.

Hollywood Park

2

1 1/16 MILES. (1:40) **MAIDEN CLAIMING. Purse $18,000. Fillies and mares, 3-year-olds and upward, bred in California. Weights: 3-year-olds, 116 lbs. Older, 122 lbs. Claiming price $32,000; if for $28,000, allowed 2 lbs. (Horses which have not started for a claiming price of $20,000 or less preferred.)**

1 1/16 MILES
FINISH ▲ ▲START

Astrial
Own: Fanning Jerry
DELAHOUSSAYE E (145 27 25 27 .19) **$32,000**

Ch. f. 3 (May)
Sire: Flying Paster (Gummo)
Dam: Astrious (Piaster)
Br: Cardiff Stud Farms (Cal)
Tr: Fanning Jerry (14 2 3 1 .14)

L 116

Lifetime Record :	3 M 0 0		$425		
1994	3 M 0 0	$425	Turf	0 0 0 0	
1993	0 M 0 0		Wet	0 0 0 0	
Hol	3 0 0 0	$425	Dist	0 0 0 0	

23Jun94- 2Hol fst 7f :221 :451 1:104 1:25 3+ ⒻⓈMd 32000 **49** 8 5 69 78 69 55½ Delahoussaye E LB 116 b 5.80 74-14 Devil's Lass117½ StrikinglyVelvet116hd ComebackKidder1161¾ No late bid 10
1Jun94- 9Hol fst 6f :214 :451 :571 1:094 3+ ⒻⓈMd 32000 **58** 7 11 1113 1010 811 611½ Delahoussaye E LB 116 b 7.20 79-11 CboQun1155 CombckKiddr1154 ChristinMri115¾ Threw head start, wide 11
14May94- 9Hol fst 6f :22 :452 :58 1:114 3+ ⒻⓈMd 32000 **52** 5 11 1115 1119 815 66½ Valenzuela F H LB 115 b 9.50 74-09 Copper Coinage1152 Accountable115hd Drona's Starlet1172¼ Off slow 11

WORKOUTS: Jly 1 SA 4f fst :474 H *3/21* Jun 16 SA 6f fst 1:154 H *8/10* Jun 9 SA 5f fst 1:002 H *5/26* May 27 SA 3f fst :353 H *2/22* May 21 SA 4f fst :473 H *7/33* May 8 SA 6f gd 1:152 Hg*2/8*

Her Thoughts
Own: Jackson Bruce L
PEDROZA M A (256 25 41 35 .10) **$28,000**

Ch. f. 3 (Mar)
Sire: Samarid (Blushing Groom)
Dam: Little Bolder (Taufan)
Br: Jackson Bruce L (Cal)
Tr: Jackson Bruce L (26 4 3 3 .15)

L 114

Lifetime Record :	6 M 0 1		$3,750		
1994	6 M 0 1	$3,750	Turf	0 0 0 0	
1993	0 M 0 0		Wet	1 0 0 0	
Hol	2 0 0 0	$1,050	Dist	0 0 0 0	

16Jun94- 2Hol fst 6f :221 :46 :582 1:114 3+ ⒻMd 25000 **12** 11 2 72½ 86¾ 911 1022 Valenzuela F H LB 116 31.70 59-12 Alice Mine116½ Haloed Princess1204½ Miss Silver Sue1161½ Wide trip 11
13May94- 6Hol fst 6½f :221 :451 1:112 1:182 3+ ⒻMd 25000 **44** 8 4 52¼ 56 35½ 47 Valenzuela F H LB 115 8.30 70-19 Maitlin115½ Bolides Wish113½ Fast Kicken Cara1156 4 wide to turn 8
22Apr94- 9SA fst 5½f :214 :452 :572 1:033 ⒻⓈMd 28000 **33** 8 5 107¾ 109½ 711 917¾ Valenzuela F H LB 116 24.60 74-11 She's Enchanting1184 Accountable118½ Lady Kariba1186½ No factor 12
1Apr94- 4SA fst 6f :214 :444 :573 1:103 ⒻⓈMd 32000 **46** 8 2 23 35 35½ 1012½ Valenzuela P A LB 117 21.00 71-13 La Prieta1171 Copper Coinage1173 B. Fancy1172¼ Gave way 12
17Mar94- 9SA fst 6f :214 :444 :572 1:103 ⒻⓈMd 32000 **59** 6 8 74½ 54¾ 44 33¾ Valenzuela P A LB 117 70.60 80-11 Scoring Road1153½ Tia Veevee117nk Her Thoughts1171¼ Mild late bid 11
20Feb94- 4SA wf 5½f :213 :443 :571 1:04 ⒻⓈMd 28000 **24** 3 9 107¾ 1115 1113 1117½ Gonzalez S5 LB 110 39.50 72-09 MissoulLul1172 Costchoro1172 WorldRecordMm1172½ Steadied hard 1/2 11

WORKOUTS: Jly 2 Hol 4f fst :484 H *27/48* Jun 25 Hol 3f fst :352 H *3/32* Jun 9 Hol 6f fst 1:16 H *18/19* Jun 2 Hol 4f fst :483 H *17/41* May 25 Hol 3f fst :381 H *22/33* May 3 Hol 3f fst :381 H *19/23*

Vento E Fuoco
Own: Rappa Frank T & Margaret F
FERNANDEZ A L (14 0 0 0 .00) **$28,000**

Gr. f. 3 (May)
Sire: Incinderator (Northern Dancer)
Dam: First Advance (Advance Guard)
Br: Rappa Frank (Cal)
Tr: Craigmyle Scott J (10 0 0 0 .00)

L 114

Lifetime Record :	8 M 0 0		$1,775		
1994	3 M 0 0	$1,350	Turf	0 0 0 0	
1993	5 M 0 0	$425	Wet	0 0 0 0	
Hol	3 0 0 0	$1,775	Dist	2 0 0 0	$1,350

24Jun94- 4Hol fst 1 1/16 :231 :471 1:124 1:452 3+ ⒻⓈMd 32000 **41** 9 5 44½ 46 511 419 Scott J M LB 115 b 98.60 55-19 New Bounty1163½ Silent Draw11513 Turf Princess1202½ No late bid 10
5May94- 6Hol fst 6f :22 :451 :571 1:102 3+ ⒻⓈMd 32000 **13** 3 7 97¾ 1013 1018 922 Demesme E5 LB 113 b 138.00 66-08 Trish's Trick1155½ Earth Baby1144 Intergale1201 Outrun 10
6Apr94- 4SA fst 6½f :22 :451 1:104 1:173 ⒻⓈMd 28000 **—** 4 7 — — — — Demesme E5 LB 111 b 82.00 — 12 Tia Veevee1174 Unprectible Gal1171¾ Silent Draw117¾ 8
Stumbled, lost rider at start
22Sep93- 9Fpx fst 1 1/16 :224 :463 1:13 1:46 ⒻMd Sp Wt **35** 4 6 98¾ 913 924 921¾ Scott J M LB 116 109.40 56-21 Viz1161 Company Shot1163 Carmen Copy116nk Done early 10
30Aug93- 4Dmr fst 6f • :22 :452 :573 1:103 ⒻMd 32000 **26** 3 8 99¼ 912 917 822 Aragundi E L 117 99.60 64-14 Gretchen'sAlma1174 PrivateEquilizer117nk NativeLce1125 Showed little 9
20Aug93- 2Dmr fst 7f :22 :452 1:104 1:241 Md 32000 **14** 11 3 87 914 1019 1026½ Aragundi E B 114 156.10 53-14 Triple Thunder1171¾ Cologne1172 Set On Cruise1178½ Wide backstretch 11
4Aug93- 2Dmr fst 5f :222 :46 :583 ⒻMd 32000 **21** 10 10 129 127½ 911 815½ Black C A B 117 50.60 88-12 Gambling Mistress1174½ Drippingold117¾ Kelkyko1172½ Wide trip 12
22Jly93- 6Hol fst 5½f :222 :46 :583 1:05 ⒻMd 32000 **21** 1 4 711 712 714 515¾ Berrio O A 118 19.60 71-11 Superbe Slew1133 Falmora1188 Ashtabula118¾ 4 Wide stretch 8

WORKOUTS: Jly 6 Fpx 5f fst 1:021 H *5/9* Jun 18 Fpx 4f fst :484 H *3/8* ●Jun 12 Fpx 5f fst 1:01 H *1/4* May 28 Fpx 3f fst :362 H *8/15* ●May 22 Fpx 4f fst :473 H *1/6* May 1 Fpx 4f fst :483 Hg*3/10*

Drona's Starlet
Own: Dwyer Austin
ALMEIDA G F (59 3 8 8 .05) **$28,000**

B. f. 3 (Feb)
Sire: Nostalgia's Star (Nostalgia)
Dam: Fleet Drona (Drone)
Br: Austin Dwyer (Cal)
Tr: Bean Robert A (17 0 3 1 .00)

L 114

Lifetime Record :	5 M 0 1		$5,100		
1994	5 M 0 1	$5,100	Turf	0 0 0 0	
1993	0 M 0 0		Wet	0 0 0 0	
Hol	4 0 0 1	$5,100	Dist	0 0 0 0	

23Jun94- 2Hol fst 7f :221 :451 1:104 1:25 3+ ⒻⓈMd 28000 **55** 4 2 43½ 42½ 33 42½ Almeida G F LB 114 45.50 77-14 Devil'sLass117½ StrikinglyVelvet116hd ComebckKidder1161¾ Mild late bid 10
1Jun94- 9Hol fst 6f :214 :451 :571 1:094 3+ ⒻⓈMd 32000 **39** 2 9 85 76¾ 710 918¾ Atherton J E5 LB 110 20.60 72-11 Cabo Queen1155 Comeback Kidder1154 Christina Maria115¾ No threat 11
14May94- 9Hol fst 6f :22 :452 :58 1:114 3+ ⒻⓈMd 32000 **64** 10 1 73¾ 710 47½ 32 Long B LB 117 13.20 79-09 Copper Coinage1152 Accountable115hd Drona's Starlet1172¼ Far wide trip 11
28Apr94- 9Hol fst 6f :214 :45 :571 1:102 3+ ⒻMd 32000 **53** 9 4 31½ 21½ 33½ 48¼ Long B LB 117 20.50 80-08 Queen Gen1161¾ Ruler Of Men1152 Our Cookie1154½ No late bid 9
15Apr94- 2SA fst 5½f :213 :45 :571 1:034 ⒻMd 28000 **40** 2 11 43 54 88½ 1014¾ Long B B 117 20.30 76-08 Kelly'sThing1172¼ TimeOfQueen112½ Snowy'sMrk1171¾ Off slow, bumped 12

WORKOUTS: Jly 6 Hol 3f fst :37 H *13/17* Jun 14 Hol 4f fst :493 H *4/21* May 25 Hol 4f fst :49 H *18/51* May 7 Hol 4f fst :481 H *19/44* Apr 24 Hol 5f fst 1:022 H *13/15*

Lady Stonewalk
Own: Lindsey James J
ROMERO J A (3 0 0 0 .00) **$28,000**

B. f. 3 (May)
Sire: Stonewalk (Knightly Manner)
Dam: Green Signal (Roberto)
Br: Heidmann T & Lindsey James (Cal)
Tr: Stepp William T (17 0 0 3 .00)

L 114

Lifetime Record :	5 M 0 0		$0		
1994	4 M 0 0		Turf	0 0 0 0	
1993	1 M 0 0		Wet	0 0 0 0	
Hol	3 0 0 0		Dist	1 0 0 0	

24Jun94- 4Hol fst 1 1/16 :231 :471 1:124 1:452 3+ ⒻⓈMd 28000 **30** 8 10 1015 813 714 625¼ Skelly R V5 L 109 b 78.90 49-19 New Bounty1163½ Silent Draw11513 Turf Princess1202½ Off slowly 10
12Jun94-10Hol fst 6½f :22 :451 1:103 1:171 3+ ⒻMd 28000 **25** 9 10 1015 1019 921 923½ Trujillo V B 114 201.60 59-13 Explosive Illusion1162¼ Highly Flirtatious116hd Rue D'alezan116nk Outrun 10
27May94- 9Hol fst 6f :222 :453 :573 1:103 3+ ⒻMd 28000 **34** 1 9 815 811 919 920 Castanon E F5 B 109 79.70 67-11 Silent Assembly1223½ Rue D'alezan1151¼ Our Cookie1154 Off slowly 10
17Jan94- 1BM fst 1 :23 :473 1:131 1:402 ⒻMd Sp Wt **38** 3 7 712 710 69 613¼ Vergara O 117 7.10e 58-25 A Trifle To Spare1172½ Key To V. A.1123 Ali Capote1171½ Off slowly 7
28Dec93- 4SA fst 6f :214 :442 :571 1:094 ⒻMd 28000 **35** 11 9 1114 1116 813 615¾ Iammarino M P 115 103.60 72-05 Serendipity1172 Saratoga Hope1174½ Quiet Pride1171½ By tired ones 12

WORKOUTS: Jly 1 SA 5f fst 1:014 H *12/29* Jun 20 SA 5f fst 1:002 H *9/20* Jun 10 SA 3f fst :372 H *11/22* Jun 4 SA 4f fst :501 H *16/34* May 25 SA 3f gd :363 H *3/12* May 21 SA 4f fst :491 H *18/33*

continued

Turf Princess

Own: Allen Byron

VALENZUELA F H (259 20 27 29 .08) $28,000

Ch. f. 4
Sire: Coach's Call (Alydar) *SI 4*
Dam: Turf Skeets (Valiant Noble)
Br: Otto Bernier (Cal)
Tr: Allen Byron (10 0 0 4 .00)

L 120

Lifetime Record :	6 M 0 2	$5,250		
1994	6 M 0 2	$5,250	Turf	0 0 0 0
1993	0 M 0 0		Wet	0 0 0 0
Hol	3 0 0 2	$5,250	Dist	1 0 0 1	$2,700

24Jun94-4Hol	fst 1¹⁄₁₆	:231 :471 1:124 1:452	3+ ⒻⓈMd 28000	45	1	9	9¹²	7⁸¹⁄₂	4¹⁰	3¹⁶¹⁄₂	Valenzuela F H	LB 120	28.60	57 – 19	New Bounty116³¹⁄₂ Silent Draw115¹³ Turf Princess120²¹⁄₂	Wide into lane 10
20May94-2Hol	fst 7f	:224 :461 1:114 1:25²	3+ ⒻⓈMd 32000	10	5	9	10⁵	10⁷³⁄₄	8¹³	9²²¹⁄₂	Castanon A L	LB 122	13.80	54 – 14	Jill's Gift115³⁄₄ Riviere Miss122ⁿᵏ Four Step115¹	No threat 11
5May94-9Hol	fst 6f	:22 :45² :581 1:11³	3+ ⒻⓈMd 32000	32	9	11	12¹¹	10¹¹	7⁷³⁄₄	3⁸	Gonzalez S Jr	LB 122	88.60	74 – 08	B. Fancy115¹ Unprectible Gal115⁷ Turf Princess122ⁿᵒ	12
	Steadied start, lugged out, wide														
22Apr94-2SA	fst 7f	:224 :454 1:10⁴ 1:23³	ⒻⓈMd 40000	23	8	5	5²	5²¹⁄₂	7¹⁰	8²³	Gonzalez S⁵	B 115	23.50	61 – 11	Nncy'sLegend120⁶ BoomBoomBrbr118¹¹ WestCostBelle118ʰᵈ	4 wide turn 8
11Mar94-6SA	fst 6f	:221 :454 :58 1:10³	ⒻⓈMd 40000	19	9	5	5³¹⁄₂	5⁵¹⁄₂	6¹¹	8²²	Torres R V	120	73.10	62 – 14	ⒹNancy's Legend118¹ Fleet Sister118¹¹⁄₂ Launch A Patriot118³¹⁄₂	Gave way 9
26Feb94-4SA	fst 6f	:221 :45² :57³ 1:10²	ⒻⓈMd 35000	30	6	8	5²¹⁄₂	3³	6¹⁰	7¹⁸¹⁄₂	Torres R V	B 117	45.30	67 – 11	Wild Vickie119¹¹⁄₂ Fleet Sister117⁹ Boom Boom Barbara117³	8
	Took up start, wide, bumped 3 1/2														

WORKOUTS: Jly 3 SA 7f fst 1:30¹ H 2/2 • Jun 20 SA 7f fst 1:30 H 2/2 Jun 14 SA 1 fst 1:44¹ H 1/1 Jun 7 SA 4f fst :47³ H 3/34 May 31 SA 4f fst :50² H 13/22 May 14 SA 5f fst 1:00² H 10/27

That's Gin

Own: Cervantes & Garcia & Martinez

SHANNON E R (3 0 0 0 .00) $28,000

Ch. m. 5
Sire: Wajima (Bold Ruler) *SI 4*
Dam: Appalachian Dream (Apalachee) *SI-4*
Br: P. Valenti, J. Coelho, et al. (Cal)
Tr: Alvarez Fernando (2 0 0 0 .00)

120

Lifetime Record :	5 M 0 0	$0		
1994	2 M 0 0		Turf	0 0 0 0
1993	3 M 0 0		Wet	0 0 0 0
Hol	3 0 0 0		Dist	2 0 0 0

23Jun94-6Hol	fst 7f	:221 :45² 1:10³ 1:23⁴	3+ ⒻⓈMd 32000	32	1	6	9⁸¹⁄₂	10⁸¹⁄₂	8¹⁶	8¹⁹¹⁄₂	Shannon E R	B 120 b	182.60	65 – 14	LadyKariba117¹¹⁄₂ Cody'sShelby116¹¹⁄₂ WeGotTheDinero116²¹⁄₂	Saved ground 11
17Jun94-6Hol	fst 6f	:22 :45³ :581 1:11²	3+ ⒻⓈMd 32000	42	1	9	9⁸¹⁄₂	8¹¹	8¹¹	8¹⁵¹⁄₂	Shannon E R	B 122 b	88.90	67 – 15	Tracy's Paster115¹⁄₂ Christina Maria116² PattiO'pott117¹	Came out start 9
1Dec93-9Hol	fst 6f	:221 :451 :574 1:10⁴	3+ ⒻⓈMd 28000	51	8	3	9⁸	10¹⁰	9¹¹	8⁸	Martinez F F	B 119 b	128.60	78 – 13	Incndrll119ⁿᵏ KrstysLuckyLdy119ⁿᵏ DtWthRoylty119³⁄₄	Steadied hard 5 1/2 10
21Oct93-9SA	fst 1¹⁄₁₆	:23 :47 1:12¹ 1:44²	3+ ⒻMd 32000	45	8	8	8⁶¹⁄₂	9⁸	9¹¹	8¹⁹¹⁄₂	Martinez F F	B 121 b	105.50	62 – 20	Hillwalker117²³⁄₄ Uncanny Ann119¹⁄₂ Raise The Runway115³	Off slow, wide 12
7Oct93-9SA	fst 1¹⁄₁₆	:23 :46² 1:11³ 1:44¹	3+ ⒻMd 32000	36	1	10	10¹¹	10⁹¹⁄₂	10¹⁸	10²³¹⁄₂	Martinez F F	B 121	111.60	59 – 16	Ms. Jiles117¹¹⁄₂ Uncanny Ann119¹⁄₂ Charlierusse117²¹⁄₂	Outrun 11

WORKOUTS: Jly 6 SA 6f fst 1:14¹ Hg 3/13 Jun 12 SA 5f fst 1:01⁴ Hg 12/30 Jun 8 SA 5f fst 1:02² Hg 23/31 May 23 SA 5f fst 1:05 H 31/31 May 15 SA 4f fst :49³ Hg 18/21 Apr 23 Hol 4f fst :491 H 29/51

Silent Draw

Own: Fields Phillip W

ATKINSON P (129 11 16 12 .09) $32,000

B. f. 3 (Apr)
Sire: Martial Law (Mr. Leader) *SI 2*
Dam: Silent Garanda (Silent Screen) *SI 3*
Br: Fields Phillip W (Cal)
Tr: Stute Melvin F (74 9 15 7 .12)

116

Lifetime Record :	9 M 2 1	$15,900		
1994	9 M 2 1	$15,900	Turf	0 0 0 0
1993	0 M 0 0		Wet	0 0 0 0
Hol	4 0 2 0	$8,775	Dist	2 0 1 0	$4,950

24Jun94-4Hol	fst 1¹⁄₁₆	:231 :471 1:124 1:452	3+ ⒻⓈMd 32000	67	10	1	1¹	1ʰᵈ	2ʰᵈ	2³¹⁄₂	Atkinson P	B 115 b	2.10	70 – 19	New Bounty116³¹⁄₂ Silent Draw115¹³ Turf Princess120²¹⁄₂	5 wide 7 1/2 10
10Jun94-9Hol	fst 6¹⁄₂f	:213 :45 1:11 1:18	3+ ⒻⓈMd 32000	61	10	1	4¹¹⁄₂	6²¹⁄₂	4²¹⁄₂	3ⁿᵏ	Atkinson P	B 116 b	*2.10	75 – 13	La Dark114¹ Afleet Floozie118¹¹⁄₂ Sand The Coach116¹⁄₂	Wide trip 11
20May94-5Hol	fst 7f	:22 :45³ 1:104 1:241	3+ ⒻⓈMd 32000	69	6	3	2¹⁄₂	2ʰᵈ	2¹	2¹¹⁄₂	Atkinson P	B 115 b	4.60	82 – 14	Special Loot115¹¹⁄₂ Silent Draw115³ Gilded Lace117²³⁄₄	Game inside 9
27Apr94-3Hol	fst 1¹⁄₁₆	:22³ :46² 1:122 1:461	3+ ⒻⓈMd 32000	59	3	3	4⁴¹⁄₂	3¹¹⁄₂	4²	4³	Atkinson P	B 115 b	14.60	67 – 18	Judy's Joy115³⁄₄ New Bounty116²¹⁄₂ Impeccably Right115ʰᵈ	Checked 1/4 8
6Apr94-4SA	fst 6¹⁄₂f	:22 :451 1:104 1:171	ⒻⓈMd 32000	51	3	5	5⁴¹⁄₂	5³¹⁄₂	4⁴¹⁄₂	4⁹	Atkinson P	B 117 b	22.50	78 – 12	Tia Veevee117⁴ Unprectible Gal117¹¹⁄₂ Silent Draw117³	Mild late bid 9
23Mar94-9SA	fst 6f	:213 :443 :571 1:10³	ⒻⓈMd 32000	55	4	5	4³	3⁵¹⁄₂	4⁹¹⁄₂	5⁸¹⁄₂	Solis A	B 117 b	4.20	76 – 12	Amish Amanda117¹ Accountable117ⁿᵏ Star Signal124	Weakened 10
9Mar94-4SA	fst 6¹⁄₂f	:214 :452 1:111 1:174	ⒻⓈMd 32000	60	8	1	3ⁿᵏ	4¹¹⁄₂	3³¹⁄₂	4⁵	Flores D R	B 117 b	19.20	78 – 14	Watching Royalty117²¹⁄₂ Tropic Moon117¹⁄₂ Zippen Miss117²	Weakened 10
26Feb94-6SA	fst 6f	:213 :444 :57 1:094	ⒻⓈMd 32000	37	2	8	7⁴¹⁄₂	8⁸	9¹¹	9¹⁹	Solis A	B 117 b	35.90	69 – 11	Serena's World118² Masterful Dawn117³¹⁄₄ La Frontera117ⁿᵒ	No threat 9
12Feb94-3SA	fst 6f	:22 :45² :57³ 1:09³	ⒻⓈMd Sp Wt	58	1	7	5¹³⁄₄	6⁵¹⁄₄	4⁷	4¹³	Solis A	B 117 b	18.00	76 – 11	Pirate's Revenge117⁴¹⁄₂ Gilded Lace117⁵ China Sky117³¹⁄₂	Shuffled back 8

WORKOUTS: Jly 7 Hol 4f fst :47 H 3/33 Jly 1 Hol 4f fst :491 H 14/39 • Jun 23 Hol 3f fst :34⁴ H 1/24 Jun 17 Hol 4f fst :481 H 6/38 Jun 9 Hol 3f fst :36³ H 14/18 Jun 3 Hol 4f fst :48⁴ H 12/29

Sailor's Grace

Own: Galarneau Janice & Philip J

ATHERTON J E (148 8 12 9 .05) $28,000

B. f. 4
Sire: Fleet Twist (Fleet Nasrullah) *SI 4*
Dam: Jo Jo Dimaggio (Dimaggio) *SI-4*
Br: Galarneau Philip & Janice (Cal)
Tr: Chew Matthew (2 1 0 0 .50)

L 115⁵

Lifetime Record :	12 M 1 1	$1,704		
1994	8 M 1 1	$1,554	Turf	0 0 0 0
1993	4 M 0 0	$150	Wet	1 0 0 0	$150
Hol	1 0 0 0		Dist	3 0 0 1	$330

24Jun94-4Hol	fst 1¹⁄₁₆	:231 :471 1:124 1:452	3+ ⒻⓈMd 28000	16	6	7	8⁹	9¹⁵	9¹⁹	9³³¹⁄₂	Corral J R	120 b	42.50	41 – 19	New Bounty116³¹⁄₂ Silent Draw115¹³ Turf Princess120²¹⁄₂	Gave way 10
24Apr94-3TuP	fst 1	:234 :48 1:124 1:391	ⒻMd 5000	39	3	5	7⁵	7⁷	7⁸¹⁄₂	7¹⁰¹⁄₂	Stevens S A	120 fb	3.90	60 – 28	Glorious Hy Star120¹¹⁄₂ Miss Californier120ⁿᵒ Janessa120²	No bid 9
1Apr94-3TuP	fst 1	:23 :47 1:13 1:39⁴	ⒻMd 5000	40	5	9	9⁷¹⁄₂	6³³⁄₄	5⁴¹⁄₂	4⁷¹⁄₂	Stevens S A	120 fb	4.80	59 – 24	Go Kathy Go120³ Janessa120² Mermaids Singing120¹	Wide, no late bid 9
19Mar94-3TuP	fst 1¹⁄₁₆	:241 :48² 1:13 1:441	ⒻMd 5000	40	6	5	6⁵³⁄₄	4⁵	3³¹⁄₂	3⁸¹⁄₂	Stevens S A	120 b	2.70	76 – 13	Festive Vision120ʰᵈ Go Kathy Go120⁵¹⁄₂ Sailor's Grace120ⁿᵒ	Bid, tired drive 7
28Feb94-3TuP	fst 1	:231 :472 1:124 1:401	ⒻMd 5000	40	3	6	7⁸	6⁵	3²¹⁄₂	3⁴¹⁄₂	Stevens S A	120	10.90	62 – 25	Shewango120¹ Silor'sGrce120⁵ Arnie'sChoice115ⁿᵏ	6 Wide; closed well 7
18Feb94-4TuP	fst 1	:233 :471 1:114 1:38³	ⒻMd 5000	39	6	2	3¹	3²¹⁄₂	3³	4⁴¹⁄₂	Stevens S A	120	3.50	69 – 19	So Bubbly120² Go Kathy Go120ⁿᵏ Arnie's Choice115²	Tired 7
31Jan94-13TuP	fst 6¹⁄₂f	:222 :45³ 1:113 1:18²	ⒻMd 5000	37	2	7	5⁸	6⁷¹⁄₄	4³¹⁄₂	4³¹⁄₂	Stevens S A	120 f	4.90	74 – 16	Glowing Pleasure120¹ Retire Early120¹ So Bubbly120¹¹⁄₂	Not enough late 9
14Jan94-9TuP	fst 6¹⁄₂f	:22³ :45⁴ 1:133 1:19⁴	ⒻMd 5000	37	5	9	11⁷	12⁶	7³¹⁄₂	6⁵	Krasner S	120 f	24.20	74 – 16	Agitatin' Luck115ⁿᵒ Glowing Pleasure120² Kj's Reward120¹¹⁄₂	Late gain 12
25Jun93-3GG	fst 1	:231 :471 1:113 1:371	ⒻⓈMd 12500	34	1	10	9¹³	8¹⁸	7²⁰	6²³¹⁄₂	Warren R J Jr	B 117	71.30	60 – 20	Saved By The Moon117¹⁰ O'let It Snow117²¹⁄₂ Our Fast Ball117⁹	No threat 10
11Jun93-5GG	fst 1¹⁄₁₆	:22² :461 1:11 1:44²	ⒻMd 12500	37	5	10	12¹⁶	11¹⁴	11¹³	8¹⁴	Campbell B C	B 117	92.10	61 – 16	Hoedown's Gone117² O'let It Snow117³⁄₄ Our Fast Ball117¹	No threat 12

WORKOUTS: Jly 4 SA 4f fst :48⁴ H 9/23 Jun 22 SA 3f fst :38³ H 23/23 Jun 16 SA 1 fst 1:44³ H 1/1 Jun 10 SA 6f fst 1:17 H 13/14 Apr 20 TuP 5f fst 1:00 H 10/26 Apr 14 TuP 4f fst :51² H 19/22

Her Thoughts has some speed and might carry it further than her recent races indicate. On the downside, she is by an unknown sire with questionable stamina, comes off a miserable race and is dead on the board at 24-1. Furthermore, if she is sent, she will have to contend with Drona's Starlet and Silent Draw in the early going, which is likely to compromise her chances.

While Drona's Starlet comes off a much better race and has been bet down to 4-1, she too will probably vie for the early lead and, given her miler stamina indexes of 4, may come up short in the stretch.

This leaves us with Astrial, the 7-2 second choice, as the most probable winner. Despite her low Beyers, those were earned in sprints and she is definitely bred to get the 1 1/16 mile according to Flying Paster's stamina index of 3 on top, reinforced with Piaster's SI of 2 on the bottom. Furthermore, Astrial closed nicely at 7 furlongs which suggests that Eddie D can sit behind the likely speed duel and catch the leaders in the stretch.

So how to bet? Silent Draw has speed, ran an improved race last time, and has the stamina indexes to win at this distance. Still, she is grossly overbet at 3-5, probably because she has the highest known route Beyers in the race. But, since sprint Beyers and route Beyers can't be compared, the possibility exists that one of the stretchout horses will run faster routing and win the race. In this case, Astrial is worth backing on the Win end at 7-2 or, alternatively, can be put on top in the exacta to Silent Draw, who is extremely likely to finish second if she doesn't win.

SECOND RACE

Hollywood

JULY 9, 1994

1 1/16 MILES. (1.40) MAIDEN CLAIMING. Purse $18,000. Fillies and mares, 3–year–olds and upward, bred in California. Weights: 3–year–olds, 116 lbs. Older, 122 lbs. Claiming price $32,000; if for $28,000, allowed 2 lbs. (Horses which have not started for a claiming price of $20,000 or less preferred.)

Value of Race: $18,000 Winner $9,900; second $3,600; third $2,700; fourth $1,350; fifth $450. Mutuel Pool $227,064.00 Exacta Pool $175,759.00 Trifecta Pool $175,692.00 Quinella Pool $28,585.00

Last Raced	Horse	M/Eqt. A.Wt	PP	St	¼	½	¾	Str	Fin	Jockey	Cl'g Pr	Odds $1
23Jun94 2Hol5	Astrial	LBb 3 116	1	8	8½	8³½	3hd	2³	1½	Delahoussaye E	32000	3.40
24Jun94 4Hol2	Silent Draw	Bb 3 116	8	1	12½	11	1³	1⁷	2¹¹	Atkinson P	32000	0.60
24Jun94 4Hol3	Turf Princess	LB 4 120	6	5	6¹	6¹	5hd	4⁵	3⁷	Valenzuela F H	28000	13.40
23Jun94 2Hol4	Drona's Starlet	LB 3 114	4	3	2½	2hd	2⁸	3¹½	4²³	Almeida G F	28000	48.70
24Jun94 4Hol9	Sailor's Grace	LB 4 115	9	6	7⁶	7⁴	6¹½	5²½	5²½	Atherton J E5	28000	112.20
24Jun94 4Hol6	Lady Stonewalk	LBb 3 114	5	9	9	9	9	8¹½	6⁴½	Trujillo V	28000	86.80
24Jun94 4Hol4	Vento E. Fuoco	LBb 3 115	3	4	3¹½	4½	8⁴	7¹	7⁷½	Fernandez A L	28000	21.70
16Jun94 2Hol9	Her Thoughts	LB 3 114	2	7	5⁴	5²	4¹½	6hd	8²½	Pedroza M A	28000	87.10
23Jun94 6Hol8	That's Gin	Bb 5 122	7	2	4hd	3⁴	7¹	9	9	Shannon E R	28000	

OFF AT 1:36 Start Good. Won driving. Time, :23¹, :47¹, 1:11⁴, 1:38², 1:45¹ Track fast.

$2 Mutuel Prices:

1–ASTRIAL		8.80	2.80	2.20
8–SILENT DRAW			2.20	2.10
6–TURF PRINCESS				2.60

$2 EXACTA 1–8 PAID $19.20 $2 TRIFECTA 1–8–6 PAID $88.80 $2 QUINELLA 1–8 PAID $6.20

Sometimes a maiden with obvious route breeding is not initially stretched out by its trainer because there aren't any routes in the Condition Book or, more frequently, because it "looks" like a sprinter to him. Still, after enough failures sprinting, even the most adamant trainer will relent and take a shot on a horse's breeding. The tenth race at Pimlico on May 24th is an example of such a belated stretchout play at the lower Maiden Claiming level.

Looking over the field, Towering Oaks, Kryptonite and The Prints Match are all proven losers in routes and can be thrown out. Ditto with Cupid's Cannon whose stamina index of 5 indicates he is bred to sprint.

Pimlico

10

1 1/16 MILES. (1:40 4) **MAIDEN CLAIMING. Purse $7,000. 3-year-olds and upward. Weights: 3-year-olds, 112 lbs. Older, 124 lbs. Claiming price $8,500; for each $500 to $7,500, 2 lbs.**

1 1/16 MILES
START ▲ ▲ FINISH

Banker's Secret SI-2

Own: Silver Lake Stable

NO RIDER (—) $8,500

B. c. 3 (Feb)
Sire: Late Act (Stage Door Johnny)
Dam: Lila (Insubordination)
Br: HMS Farms (NY)
Tr: Lenzini John J (55 4 8 6 .07)

L 112

Lifetime Record:	13 M 0 0	$1,916			
1994	5 M 0 0	$806	Turf	0 0 0 0	
1993	8 M 0 0	$1,110	Wet	2 0 0 0	$200
Pim	4 0 0 0	$586	Dist	0 0 0 0	

26Apr94-5Pim	fst	6f	:233 :473 :593 1:13	3+ Md 7500	38 1 4 2½ 3½ 4½ 5½	Prado E S	L 112 b	11.10	72-12	Mixabatch1124½ Thewayitoughttobe1221½ Iron Aura1182½	Weakened 10
15Mar94-3Lrl	fst	7f	:234 :474 1:14 1:274	Md 11500	34 5 8 55 76¾ 910 79	Salazar A C	L 120 b	11.70	61-21	Loot 'n Burn1202½ Mixabatch1162½ Congressionalmedal1202	No factor 11
5Mar94-3Lrl	fst	6f	:231 :473 1:003 1:141	Md 9500	35 3 2 62¾ 53¾ 42½ 51½	Salazar A C	L 116 b	38.80	68-21	GryStrike120no LuckyChester120½ Sr'sGoldenBoy120¾	Lacked room 1/16 12
17Feb94-5Lrl	fst	5½f	:23 :481 1:011 1:074	Md 9500	16 8 1 66½ 911 77½ 59¾	Carle J D	L 116 b	73.40	72-22	Motion Offense1202½ A Regal Pole1204½ Hey Snipe120½	Bore in start 10
14Jan94-7Lrl	my	6f	:221 :471 1:011 1:153	Md 11500	-0 11 2 58½ 512 712 823¾	Reynolds L C	L 120 b	9.60	38-33	Illustrious Legacy1201¼ Lacekins1201¼ Ambit's Key1105½	Weakened 11
30Dec93-3Lrl	fst	7f	:244 :50 1:18 1:321	Md 9500	15 2 1 32 42½ 33 55	Reynolds L C	L 116 b	15.30	43-28	Ace's Token120hd Ambit's Key1104½ Terminatethegroom120nk	Weakened 12
17Dec93-5Lrl	fst	6f	:222 :464 1:004 1:133	Md 11500	20 9 3 25 26 109¾ 1110½	Skinner K	L 120 b	15.00	62-19	Beargrass Creek115½ A Regal Pole1162½ Bubba Jove1091	Gave way 14
24Nov93-3Lrl	fst	7f	:243 :491 1:152 1:28	Md 16500	28 1 3 1hd 1hd 21½ 89¼	Ayarza I	L 116 b	21.30e	60-21	Royal Accord120hd Foulis Joshua120nk Duke's Star1161	Tired 11
11Nov93-3Lrl	fst	6f	:23 :474 1:01 1:143	Md 11500	16 1 6 52¾ 910 912 78	Skinner K	L 120 b	37.40	59-14	Fighting Cat1201 Christopher Twist113¾ Red Genes1183	No factor 13
17Oct93-3Lrl	fst	6f	:223 :462 :592 1:123	Md 13500	25 8 11 1210 1216 1217 1015½	Elliott S	L 116	45.20	62-18	Monte Vallo1202½ Spring Grove120¾ All Mirth120½	Outrun 13

WORKOUTS: Apr 17 Lrl 6f fst 1:18 B *2/3*

Quick Comeback SI-3

Own: Kovin Bette

ROCCO J (109 14 13 15 .13) $8,500

Ch. g. 3 (May)
Sire: Raconteur (The Minstrel)
Dam: Cool Comment (Snow Knight)
Br: Kovin Mrs John F (Md)
Tr: Cartwright Ronald (42 7 9 4 .17)

L 112

Lifetime Record:	2 M 0 0	$0		
1994	2 M 0 0		Turf	0 0 0 0
1993	0 M 0 0		Wet	0 0 0 0
Pim	2 0 0 0		Dist	0 0 0 0

1May94-3Pim	fst	6f	:234 :48 1:003 1:133	3+ Md 14500	23 11 9 119¾ 1113 1119 1219¾	Rocco J	L 114	14.60	59-20	Klepto Kris1136 Congressionalmedal112¾ Work For Love1121¾	Wide 12
8Apr94-5Pim	fst	6f	:233 :473 1:003 1:134	3+ Md 14500	24 4 9 917 915 814 711	Rocco J	114	13.20	67-13	PenltyTime122¾ Sr'sGoldenBoy1121 CongrssionImdl1123¼	Steadied start 9

WORKOUTS: Apr 18 Lrl 4f fst :491 B *7/11* Mar 31 Lrl 5f fst 1:021 Hg *12/21* Mar 26 Lrl 5f fst 1:022 B *15/22* Mar 19 Lrl 5f fst 1:03 B *6/9* Mar 12 Lrl 4f fst :50 B *21/31* Mar 8 Lrl 5f fst 1:07 B *12/13*

Towering Oak

Own: Decantis Dominick P

KORTE K (21 0 2 3 .00) $8,500

Ch. g. 3 (Feb)
Sire: Oak Hill Judex (Mr. Judex)
Dam: Dreyfusard (Dreyfus II)
Br: DeCantis Dominick P (Va)
Tr: Decantis Dominick P (2 0 0 0 .00)

L 112

Lifetime Record:	24 M 1 0	$4,218			
1994	6 M 0 0	$636	Turf	0 0 0 0	
1993	18 M 1 0	$3,582	Wet	3 0 1 0	$1,116
Pim	4 0 0 0	$800	Dist	2 0 0 0	$400

20May94-3Pim	fst	6f	:233 :474 1:01 1:141	Md 9500	15 7 2 41¾ 96½ 97½ 810¼	Korte K	L 116	95.90	66-17	Promised Secret120½ Kipangani1201½ Saturday'sblast120nk	Weakened 11
30Apr94-4CT	fst	6½f	:234 :474 1:134 1:204	Md Sp Wt	6 6 5 64 66½ 711 619	Thornton J E	L 120	15.40	65-12	Snow Cap1205 Summery Encounter115½ Sir Ivor's View120no	Outrun 7
24Apr94-7CT	fst	7f	:251 :491 1:16 1:301	Md 5000	20 6 5 32½ 34 44 47½	Salvo P F	L 120	9.00	64-16	Half A Double120¾ Alcabir116¾ Adarleyforabuck1166	No mishap 7
6Mar94-5CT	gd	6½f	:234 :484 1:17 1:241	Md 10000	17 2 1 34½ 46 55 69¼	Hanna A T	L 120 b	6.70	58-23	Asby's Gap118½ Halo For Silky1162½ Major Momentom1163	No mishap 7
8Jan94-1CT	fst	6½f	:232 :46 1:142 1:212	Md 15000	23 2 1 65½ 67 611 68	Reeder G S	L 116	21.40	73-16	Hard Landing1161 Swelegant's Boy1161½ Kyle James1093	In tight 8
2Jan94-1Lrl	fst	1 1/16	:242 :481 1:131 1:444	Md Sp Wt	15 6 5 79½ 814 819 836	Douglas F G	L 120	114.30	50-17	P P Prospect1206¾ Electric Image1207½ Sidango120nk	No factor 9
26Dec93-7CT	fst	7f	:241 :482 1:152 1:30	Md 10000	20 5 2 54 54½ 84½ 55½	Graffagnini V M	L 120	8.30	67-18	Carpet Dance1162 Will Do Alright1161 Gettin' Rougher1201	No mishap 10
19Dec93-1Pen	fst	6f	:223 :472 1:01 1:152	Md 7500	23 8 2 58 44½ 45½ 43	Graffagnini V M	L 117	33.90	66-16	Space Dance114hd Will Do Alright114hd Gala's Derby Day1172¾	No rally 10
10Dec93-9Lrl	fst	1 1/16	:241 :48 1:14 1:484	Md 9500	6 3 8 915 915 920 722½	Kreidel K J	L 116	69.30	43-21	Two Plus1206 Jolly Memories1202 Broad Stitch1202	Outrun 10
24Nov93-2CT	fst	7f	:233 :481 1:144 1:283	Md Sp Wt	18 8 9 910 87¾ 610 418	Hanna A T	L 120	45.90	62-21	Hard Cold Cash1205 Swelegant's Boy12011 Perfect Year1202	No threat 10

WORKOUTS: Apr 20 CT 4f fst :491 H *5/19* Mar 26 CT 4f fst :492 H *5/32*

Kryptonite SI-4

Own: Davis Howard R

DOUGLAS F G (128 13 11 15 .10) $8,500

Dk. b or br c. 3 (Apr)
Sire: Illustrious Boat (Illustrious)
Dam: Jola (Dewan Keys)
Br: Miles Clyde L Jr (Va)
Tr: Dillon Jacob L (6 1 0 0 .17)

L 112

Lifetime Record:	3 M 0 0	$760			
1994	3 M 0 0	$760	Turf	0 0 0 0	
1993	0 M 0 0		Wet	1 0 0 0	$100
Pim	2 0 0 0	$660	Dist	2 0 0 0	$660

13May94-3Pim	fst	1 70	:241 :492 1:153 1:462	3+ Md 8500	37 5 4 32 32½ 45½ 410½	Douglas F G	L 112 b	7.80	68-21	Zulu Chief1177 Land Smartly1141½ Raggaad Valentine1122	Bumped early 7
9Apr94-5Pim	fst	1 1/16	:24 :482 1:141 1:493	Md 9500	26 5 9 912 89¾ 79½ 59½	Carle J D	L 116 b	38.20	49-26	Covert Ace1202½ Mo Valay1202½ Mixabatch1185	No threat 9
27Mar94-5Lrl	my	7f	:234 :482 1:143 1:282	3+ Md 7500	4 8 10 1019 1024 918 920¾	Douglas F G	113 b	23.70	46-19	Lacekins1143½ Pellston1121¾ Doc'sapeach1131½	Outrun 10

River Ply SI-4

Own: Nanez Danette E

PRADO E S (265 48 47 44 .18) $7,500

Ch. g. 4
Sire: Isella (Peace Corps)
Dam: Medango (Medaille d'Or)
Br: Corbett Farm (Md)
Tr: Nanez R Carlos (1 0 1 0 .00)

120

Lifetime Record:	1 M 1 0	$1,470			
1994	1 M 1 0	$1,470	Turf	0 0 0 0	
1993	0 M 0 0		Wet	0 0 0 0	
Pim	1 0 1 0	$1,470	Dist	1 0 1 0	$1,470

12May94-9Pim	fst	1 70	:221 :46 1:132 1:473	3+ Md 7500	32 5 8 817 715 56½ 24	Kreidel K J5	115 b	4.70	69-26	It's Only A Game1144 River Ply1151½ Aloma's Count110no	Rallied 9

WORKOUTS: May 10 Lrl 5f fst 1:022 B *6/15* ●Apr 22 Lrl 5f fst 1:02 H *1/5* Apr 17 Lrl 4f fst :513 B *5/8* Apr 8 Lrl 5f fst 1:053 B *10/10* Mar 26 Lrl 5f fst 1:02 Hg *5/22* Mar 19 Lrl 4f fst :484 H *9/25*

Aloma's Count SI-4

Own: The Chaney Ranch

CHAVEZ S N (50 5 4 4 .10) $8,000

B. g. 3 (Mar)
Sire: Aloma's Ruler (Iron Ruler)
Dam: Ringaround (Dancing Count)
Br: Charles R. Chaney (Md)
Tr: Calhoun Laurie A (8 0 1 2 .00)

L 110

Lifetime Record:	4 M 0 1	$1,370			
1994	2 M 0 1	$770	Turf	0 0 0 0	
1993	2 M 0 0	$600	Wet	0 0 0 0	
Pim	3 0 0 1	$1,070	Dist	1 0 0 1	$770

12May94-9Pim	fst	1 70	:221 :46 1:132 1:473	3+ Md 7500	30 3 1 15 13 11½ 35¼	Chavez S N	L 110 b	16.10	68-26	It's Only A Game1144 River Ply1151½ Aloma's Count110no	Tired 9
29Apr94-3Pim	fst	6f	:233 :471 1:001 1:133	Md 11500	10 5 3 44½ 57½ 713 919¼	Chavez S N	L 120 b	22.20	60-15	Jolly Memories1182¾ Leavinhome1181½ Little Sultan1202½	Gave way 10
7Aug93-3Pim	gd	5½f	:223 :462 :59 1:052	Md Sp Wt	8 5 4 1015 1018 1020 925¾	Salazar A C5	115 b	88.70	72-13	Blaise Winter1203½ Spartan Victory1205 Takeitlikeaman1202½	Outrun 10
2Jly93-5Lrl	fst	5f	:23 :46 :582	Md Sp Wt	4 8 7 1113 1017 1022 1025¼	Castillo F	120	36.30	76-05	Passing Reality1204 Blaise Winter1204½ Bald Greek120no	Outrun 11

WORKOUTS: May 10 Pim 3f fst :364 B *3/7* May 6 Pim 3f fst :364 B *3/10* Apr 26 Pim 5f fst 1:023 B *6/10* Apr 19 Pim 5f fst 1:013 Hg *2/12* Mar 31 Pim 4f fst :502 B *12/20* ●Mar 23 Pim 3f fst :37 B *1/7*

The Prints Match

Own: Torsney Jerome M

CHAVEZ S N (50 5 4 4 .10) **$8,500**

Ch. c. 3 (May)
Sire: Dactylographer (Secretariat)
Dam: The Neurologist (Ragstone)
Br: Dr. Regina Durkin (Ky)
Tr: Cremins Kevin J (4 0 0 0 .00)

112

Lifetime Record:	4 M 0 1	$1,590			
1994	1 M 0 0	$210	Turf	1 0 0 0	$300
1993	3 M 0 1	$1,380	Wet	0 0 0 0	
Pim	1 0 0 0	$210	Dist	2 0 0 1	$1,090

13May94-3Pim fst 170 :24^{1} :49^{2} 1:15^{3} 1:46^{2} 3↑ Md 7500	33 6 7 $7^{9\frac{1}{2}}$ 7^{10} 6^{11} $5^{12\frac{3}{4}}$ Chavez S N	110 b 23.30	66-21	Zulu Chief117^{7} Land Smartly$114^{1\frac{1}{2}}$ Raggaad Valentine112^{2}	Outrun 7
28Oct93-3Lrl fst $1\frac{1}{16}$:23^{4} :48^{2} 1:14 1:47^{1} Md 9500	–0 5 9 9^{20} 6^{21} 6^{22} $3^{32\frac{3}{4}}$ Hamilton S D	116 b 44.20	41-18	Dark Age Demon$120^{\frac{3}{4}}$ Sidango120^{32} The Prints Match116^{3}	Passed faders 9
5Oct93-4Lrl fst 6f :22^{3} :47 1:00^{1} 1:13 Md 20000	–0 7 10 10^{12} 10^{16} 10^{23} $10^{25\frac{1}{2}}$ Hamilton S D	116 b 69.70	49-17	Blood Bath$120^{4\frac{1}{4}}$ Distinct Effort115^{nk} True Freshman$113^{1\frac{1}{2}}$	Trailed 10
29Jly93-2Lrl fm 6f [T] :23^{1} :47^{2} :59^{1} 1:11^{1} Md Sp Wt	19 7 6 7^{11} 7^{13} 8^{16} $7^{20\frac{1}{2}}$ Saumell L	120 55.30	64-07	Gator Back120^{5} Lotsa Chile$120^{1\frac{3}{4}}$ Kaltondori$120^{1\frac{1}{2}}$	Outrun 9

WORKOUTS: Apr 28 Pim 4f fst :52 B *6/9*

Cupid's Cannon

Own: Hargett Farms

LOZANO W JR (63 13 9 8 .21) **$7,500**

B. g. 3 (Feb)
Sire: Cannon Shell (Torsion)
Dam: Two Skips (Bold Skipper)
Br: Cohn Seymour (NY)
Tr: Ferguson John W (13 1 2 2 .08)

L 108

Lifetime Record:	5 M 0 0	$200			
1994	5 M 0 0	$200	Turf	0 0 0 0	
1993	0 M 0 0		Wet	1 0 0 0	$100
Pim	1 0 0 0		Dist	0 0 0 0	

6May94-3Pim fst $5\frac{1}{2}$f :22^{3} :47^{2} 1:01^{1} 1:08 3↑ Md 8000	2 7 3 8^{10} 10^{14} 10^{15} $10^{17\frac{1}{4}}$ Stacy A T	L 110 b 71.30	65-23	Iron Aura$118^{\frac{3}{4}}$ Thewayitoughttobe122^{nk} Tough Terms112^{nk}	Outrun 12
27Mar94-5Lrl my 7f :23^{4} :48^{2} 1:14^{3} 1:28^{2} 3↑ Md 8500	— 2 6 1^{hd} 9^{23} — — Johnston M T	L 114 b 32.70	— 19	Lacekins$114^{3\frac{1}{4}}$ Pellston$112^{1\frac{3}{4}}$ Doc'sapeach$113^{1\frac{1}{4}}$	Bled 10
18Mar94-5Lrl fst 7f :24^{2} :48^{2} 1:14^{2} 1:27^{2} 3↑ Md 8500	7 9 4 $4^{3\frac{1}{2}}$ $7^{8\frac{1}{2}}$ 9^{18} 10^{21} Johnston M T	L 114 b 5.30	51-22	Beth'sLove118^{3} JollyMemories$112^{\frac{1}{2}}$ SunnyAndMllow$118^{\frac{1}{2}}$	Bumped start 11
20Jan94-5Aqu fst 6f [•] :23^{1} :47 :59^{3} 1:12^{4} Md 35000	8 8 7 10^{11} 11^{17} 10^{22} $10^{30\frac{1}{4}}$ Prado E S	122 89.70	49-23	Real Macher122^{10} Ibis Baba111^{6} [D]Haagbah118^{1}	No factor 12
9Jan94-7Aqu fst 6f [•] :23^{1} :47^{3} 1:00^{3} 1:13^{4} [S]Md Sp Wt	20 4 11 9^{8} $7^{8\frac{1}{2}}$ 8^{17} $10^{16\frac{1}{4}}$ Bisono C V	122 68.60	58-18	Hill Gallant122^{1} Gate Six$122^{2\frac{1}{2}}$ Arnoldovich122^{5}	Outrun 12

WORKOUTS: May 18 Lrl 5f fst 1:05^{2} B *6/6* May 3 Lrl 4f fst :50 B *13/24* Mar 6 Bow 5f fst 1:03^{2} B *4/4*

© 1994 DRF, Inc.

While River Ply and Aloma's Count also lost the first time they were stretched out, they ran creditably and might improve. Based on that assumption, the public has made them the 3-2 and 5-2 favorites. Still, neither of them seems like a lock. River Ply has no speed and Aloma's Count gave up an easy lead.

This leaves us with Banker's Secret and Quick Comeback the two stretchout horses to consider. Quick Comeback is dropping in class, has a competent trainer and is bred to get the distance. But, he has never been within nine lengths of the lead, and seems underlayed at 5-1. While Banker's Secret has lost thirteen times, he has never been beaten in a route and has the stamina index of 2 to get the distance. Furthermore, he has tactical speed which could allow him to sit behind Aloma's Count and come from off the pace as he did in his March 5th race. Also Banker's Secret gets the rail in a race that starts right before the turn. Given that both favorites look vulnerable, Banker's Secret at 5-1 is worth a Win and Place bet.

TENTH RACE **$1\frac{1}{16}$ MILES. (1.40^{4}) MAIDEN CLAIMING. Purse $7,000. 3-year-olds and upward. Weights: 3-year-olds, 112 lbs. Older, 124 lbs. Claiming price $8,500; for each $500 to $7,500, 2 lbs.**

Pimlico

MAY 24, 1994

Value of Race: $7,000 Winner $3,990; second $1,470; third $770; fourth $420; fifth $210; sixth $140. Mutuel Pool $21,083.00 Exacta Pool $38,394.00 Triple Pool $32,233.00

Last Raced	Horse	M/Eqt.	A.Wt	PP	St	¼	½	¾	Str	Fin	Jockey	Cl'g Pr	Odds $1
26Apr94 ^{5}Pim^{5}	Banker's Secret	L	3 114	1	1	2^{2}	$1^{1\frac{1}{2}}$	$1^{\frac{1}{2}}$	$1^{\frac{1}{2}}$	$1^{7\frac{3}{4}}$	Klinger C O	8500	5.00
13May94 ^{3}Pim^{4}	Kryptonite	Lb	3 113	4	7	$7^{3\frac{1}{2}}$	7^{3}	3^{3}	3^{8}	$2^{\frac{3}{4}}$	Delgado A	8500	6.40
12May94 ^{9}Pim^{2}	River Ply	b	4 120	5	3	3^{2}	3^{3}	2^{4}	2^{3}	3^{8}	Johnston M T	7500	1.40
1May94 ^{3}Pim^{12}	Quick Comeback	Lb	3 114	2	8	8	6^{1}	$6^{\frac{1}{2}}$	4^{2}	$4^{7\frac{1}{2}}$	Rocco J	8500	5.60
13May94 ^{3}Pim^{5}	The Prints Match	b	3 112	7	6	4^{2}	4^{2}	$5^{1\frac{1}{2}}$	5^{4}	$5^{4\frac{1}{2}}$	Chavez S N	8500	24.90
20May94 ^{3}Pim^{8}	Towering Oak	L	3 115	3	4	5^{hd}	5^{1}	7^{6}	6^{3}	$6^{\frac{1}{2}}$	Korte K	8500	84.80
6May94 ^{3}Pim^{10}	Cupid's Cannon	Lb	3 108	8	5	$6^{1\frac{1}{2}}$	8	8	8	7^{7}	Stacy A T	7500	82.50
12May94 ^{9}Pim^{3}	Aloma's Count	Lb	3 113	6	2	1^{2}	$2^{2\frac{1}{2}}$	$4^{\frac{1}{2}}$	7^{3}	8	Reynolds L C	8000	2.40

OFF AT 5:07 Start Good. Won driving. Time, :24, :49, 1:15, 1:42^{4}, 1:49^{4} Track fast.

$2 Mutuel Prices:			
1-BANKER'S SECRET	12.00	7.40	3.40
4-KRYPTONITE		7.00	2.80
5-RIVER PLY			2.20

$2 EXACTA 1-4 PAID $77.80 $2 TRIPLE 1-4-5 PAID $198.40

© 1994 DRF, Inc.

Golden Gate Fields

9

1 1/16 MILES. (1:39 2) **ALLOWANCE. Purse $23,000. 3-year-olds which have not won $3,000 other than maiden, claiming, starter or classified handicap. Weight, 120 lbs. Non-winners of a race other than claiming, starter, or classified handicap at one mile or over since December 15, allowed 3 lbs.**

1 1/16 MILES — START ▲ ▲ FINISH

Juan's Boy

Own: Clawson & Gellepes & Morrison

B. g. 3 (Mar)
Sire: Prospective Star (Mr. Prospector)
Dam: Sundiata (African Sky)
Br: Shoemaker Darlene (Cal)
Tr: Jenda Charles J (34 10 6 4 .29)

CHAPMAN T M (93 15 10 8 .16) — 117

Lifetime Record:	2 1 0 0	$10,450			
1994	2 1 0 0	$10,450	Turf	0 0 0 0	
1993	0 M 0 0		Wet	0 0 0 0	
GG	1 1 0 0	$10,450	Dist	0 0 0 0	

12Feb94-1GG fst 6f	:214 :441 :56 1:082	Md Sp Wt	95 6 1 33 2½ 21 1½	Chapman T M	LB 118	2.30	97-07	Juan's Boy118½ B. Charlie1187 Governor Elect1186	Closed gamely	6
22Jan94-5BM fst 6f	:221 :451 :573 1:103	ⓢMd Sp Wt	74 6 4 62¼ 72½ 41½ 3½	Chapman T M	LB 118	20.20	87-08	KingsburgGuy118nk B.Charlie118nk ⒹJun'sBoy1182	Drifted in midstretch	11

Disqualified and placed 7th

WORKOUTS: Mar 8 GG 4f fst :48 H 2/16 Mar 2 GG 1 fst 1:413 H 1/2 Feb 24 GG 4f gd :502 H 30/65 Feb 5 GG 6f fst 1:15 H 10/34 Jan 30 BM 4f fst :492 H 21/50 Jan 19 BM 4f fst :50 H 19/24

Suomi Power

Own: Biszantz Gary & Enberg Dick

Ch. g. 3 (Apr)
Sire: Explodent (Nearctic)
Dam: Yrmika Fr (Crystal Palace)
Br: Mandysland Farm & Enemy Stable (Ky)
Tr: Greenman Walter (27 5 4 5 .19)

BOULANGER G (170 20 35 28 .12) — 117

Lifetime Record:	12 1 4 2	$29,340			
1994	2 0 1 1	$8,725	Turf	1 0 0 1	$4,125
1993	10 1 3 1	$20,615	Wet	1 0 1 0	$4,600
GG	1 0 1 0	$4,600	Dist	1 0 1 0	$4,600

19Feb94-9GG sly 1 1/16	:232 :472 1:12 1:461	Alw 23000N1X	72 5 4 42½ 1hd 2hd 22	Grabowski J A	LB 117 b	5.80	64-29	Log Buster1202 Suomi Power1172½ Errante11716	Bumped stretch	5
22Jan94-1BM fm 7f Ⓣ	:23 :472 1:134 1:264	Crdf Stk F H27k	70 4 4 48½ 45 33½ 35½	Grabowski J A	LB 114 b	9.20	— —	Makinanhonestbuck1152 Flying Kajo1163½ Suomi Power11411	Even try	4
17Dec93-1BM gd 1	:23 :471 1:123 1:392	Md 20000	67 2 2 21 1hd 15 13½	Warren R J Jr	LB 118 b	*1.10	73-25	Suomi Power1183½ Healbequick1161 Cordial Hello1182	Steady drive	7
21Nov93-3BM fst 1	:223 :464 1:124 1:382	Md 20000	62 5 4 41½ 3½ 1½ 22	Boulanger G	LB 118 b	3.60	76-19	Barts Demon1182 Suomi Power1183 Notinthislifetime1162½	Held well	8
13Nov93-1BM fst 1	:23 :473 1:123 1:374	Md Sp Wt	53 2 3 32 32 26 310½	Warren R J Jr	LB 118 b	12.50	71-23	River Flyer11810 Catch The Native118nk Suomi Power1183	Even late	6
22Oct93-6SA fst 1	:224 :464 1:123 1:384	Md 32000	53 4 2 1hd 2hd 3½ 46½	Desormeaux K J	LB 117 b	4.00	70-21	Veracity117hd Prevasive Force1172¾ Sierra Real1123½	Inside duel	9
11Oct93-2SA fst 6f	:212 :443 :574 1:112	Md 32000	53 6 1 41 51¾ 41½ 42½	Solis A	LB 118 b	1.80	77-11	A Tribute To Max118hd Beau Finesse1181 Toknight1181½	4 Wide into lane	9
26Sep93-5Fpx fst 6f	:214 :464 :584 1:122	Md 32000	50 4 2 32½ 41¼ 41½ 21½	Nakatani C S	LB 118 b	*1.30	85-09	Mob Stage1181½ Suomi Power1181 Gary's Lucky One1181	Split rivals	10
6Sep93-2Dmr fst 6f	:222 :461 :584 1:111	Md 32000	62 9 2 31 1hd 2hd 21½	Desormeaux K J	LB 118 b	5.50	82-10	Cologne1181½ Suomi Power118nk Stone County Road1188½	Good effort	11
20Aug93-2Dmr fst 7f	:22 :452 1:104 1:241	Md 28000	44 10 2 21½ 23½ 47 412½	Desormeaux K J	LB 116 b	3.70	67-14	Triple Thunder1171½ Cologne1172 Set On Cruise1178½		11

Erratic lane, steadied 1/8

WORKOUTS: Mar 6 GG 5f fst 1:024 H 20/34 Feb 28 GG 3f fst :381 H 5/13 Feb 16 GG 3f fst :353 H 2/20 Feb 11 GG 5f fst 1:031 H 30/35 Feb 6 GG 4f sly :543 H (d) 10/13 Jan 31 GG 3f fst :37 H 2/3

Bai Brun

Own: Clark & Domush & Silver & Thiriot

B. c. 3 (Feb)
Sire: White Mischief (Roberto)
Dam: Royal Burgandy (Native Royalty)
Br: Azpurua Eduardo (Fla)
Tr: Sumja Brent (5 2 0 1 .40)

WARREN R J JR (105 18 8 17 .17) — 117

Lifetime Record:	4 1 0 1	$10,900			
1994	3 0 0 1	$4,100	Turf	0 0 0 0	
1993	1 1 0 0	$6,800	Wet	0 0 0 0	
GG	0 0 0 0		Dist	1 0 0 1	$3,120

3Feb94-8GP gd 1 1/16 ⊗	:233 :481 1:132 1:453	Alw 26000N2L	74 9 2 12½ 2½ 22½ 36¾	Douglas R R	L 119	25.10	75-20	Lahint1224¼ Bay Street Star1192½ Bai Brun1191¼	Weakened	9
16Jan94-7GP fst 7f	:222 :451 1:102 1:24	Alw 23000N1X	57 5 8 88 712 57¼ 56	Smith M E	117	*2.10	77-14	Gold Coin117hd Thrilla In Manila117hd Line Dance120¾		8

Eight wide top str, drifted out, mild bid

1Jan94-4Crc fst 7f	:221 :451 1:112 1:252	Alw 15000N1X	75 5 6 63¾ 41¼ 1hd 2¾	Jimenez E J	117	33.90	84-15	Fortunate Joe117¾ ⒹBai Brun117¾ Just A Roman1101½		7

Bid wide, bore in stretch Disqualified and placed 4th

30Nov93-6Crc fst 7f	:23 :464 1:13 1:27	Md 25000	57 5 4 44 2hd 13 14¼	Jimenez E J	119	11.80	80-11	Bai Brun1194¼ Adel Boy109no Moneyinthewallet1171½	Driving	6

WORKOUTS: Mar 9 GG 3f fst :37 H 10/17 ●Mar 2 GG 5f fst 1:00 H 1/29 Feb 24 GG 4f gd :502 H 30/65 Jan 29 Crc 5f gd 1:03 B 25/54 Jan 14 Crc 4f sly :483 H (d) 3/27 Jan 11 Crc 4f sly :482 H (d) 2/29

Rock Em Steady

Own: Serendipity Stable

Ch. c. 3 (Feb)
Sire: Swiss Trick (Damascus)
Dam: Time to Succeed (Time to Explode)
Br: Serendipity Stbls & Lightner J II (Cal)
Tr: Silva Jose L (47 11 10 5 .23)

LOPEZ A D (19 3 3 5 .16) — 117

Lifetime Record:	1 1 0 0	$3,850			
1994	1 1 0 0	$3,850	Turf	0 0 0 0	
1993	0 M 0 0		Wet	1 1 0 0	$3,850
GG	1 1 0 0	$3,850	Dist	0 0 0 0	

17Feb94-9GG sly 6f	:22 :452 :582 1:113	ⓢMd 12500	67 8 7 58 46 33 12	Belvoir V T	LB 118	5.20	81-16	Rock Em Steady1182 Navarro Natural1182½ Land O Jazz118¾	Rallied wide	10

WORKOUTS: Mar 10 GG 3f fst :393 H 12/12 Mar 2 GG 6f fst 1:163 H 4/5 Feb 12 GG 4f fst :50 H 58/78 Feb 5 GG 6f fst 1:15 Hg 10/34 Jan 29 BM 6f gd 1:163 H 3/8 Jan 23 BM 6f sly 1:193 H (d) 4/4

Warning Label

Own: Lanning Curt & Lila

Dk. b or br g. 3 (Apr)
Sire: Never Tabled (Never Bend)
Dam: T. V. Residual (Pirate's Bounty)
Br: Wygod Martin J & Pam (Cal)
Tr: Moger Ed Jr (23 6 1 4 .26)

MEZA R Q (92 17 16 7 .18) — 117

Lifetime Record:	8 1 1 4	$21,957			
1994	3 1 0 1	$9,612	Turf	0 0 0 0	
1993	5 M 1 3	$12,345	Wet	2 1 0 0	$7,725
GG	1 1 0 0	$6,600	Dist	0 0 0 0	

21Feb94-1GG sly 1	:224 :462 1:112 1:373	Md 25000	77 2 1 1½ 1½ 12 16	Belvoir V T	LB 118	*1.10	83-21	Warning Label1186 Clever Luck1187 Batrango118nk	Drew off	6
22Jan94-5BM fst 6f	:221 :451 :573 1:103	ⓢMd Sp Wt	66 1 5 41 41¼ 62¼ 63¼	Day P	LB 118	4.50	84-08	Kingsburg Guy118nk B. Charlie118nk ⒹJuan's Boy1182		11

Lacked room midstretch Placed 5th through disqualification.

14Jan94-3BM fst 1	:224 :471 1:122 1:38	Md Sp Wt	71 5 1 15 1½ 21½ 32¼	Belvoir V T	LB 118 b	*1.00	81-22	Log Buster1181½ Bensred118¾ Warning Label1186	Weakened	5
4Dec93-1BM fst 6f	:224 :454 :582 1:112	Md Sp Wt	69 5 4 2hd 21 21½ 33	Kaenel J L	LB 118 b	*2.70	76-16	Emperor Jared1181 Dinadidit1182 Warning Label1181	Weakened	8
18Nov93-3BM fst 6f	:223 :453 :58 1:103	Presidio25k	69 4 4 31½ 21 23 25	Kaenel J L	LB 117 b	9.50	83-21	Jillyball1155 Warning Label1172½ Raison Buisson1136	Second best	6
6Nov93-4BM fst 6f	:223 :453 :58 1:104	Md Sp Wt	61 6 5 52½ 54 46 37½	Kaenel J L	LB 118 b	6.90	74-21	SaratogaBandit1182½ Reality'sConquest1185 WarningLbel1181½	Even late	8
16Oct93-1BM sly 6f	:222 :453 :581 1:112	Md Sp Wt	63 1 8 32½ 21½ 22 44	Warren R J Jr	LB 118 b	6.20	75-14	Mari's Notes118½ Kingsburg Guy1182½ Saratoga Bandit1181	Off slowly	8
22Aug93-8Bmf fst 5½f	:212 :452 :582 1:05	Md 20000	50 8 2 32 22 31 32¾	Kaenel J L	LB 118	9.00	87-08	Frank C1181¾ Flash The Glass1181 Warning Label1181½	Pressed pace	10

WORKOUTS: Mar 7 GG 5f fst 1:011 H 4/25 ●Mar 1 GG 5f fst 1:001 H 1/9 Feb 15 GG 5f fst 1:02 H 30/41 Feb 8 GG 4f my :552 H (d) 7/8 Jan 9 BM 6f fst 1:144 H 4/9 Jan 2 BM 5f fst 1:02 H 24/40

Champagne Shane
Own: Halo Farms

BELVOIR V T (176 42 30 24 .24)

Ch. g. 3 (Feb)
Sire: Shanekite (Hoist Bar)
Dam: Jerell's Girl (Raise Your Glass)
Br: Halo Farms (Cal)
Tr: Hollendorfer Jerry (78 26 14 10 .33)

117

Lifetime Record:	5	1	1	0	$8,275					
1994	1	0	1	0	$3,000	Turf	0	0	0	0
1993	4	1	0	0	$5,275	Wet	0	0	0	0
GG	1	0	1	0	$3,000	Dist	0	0	0	0

2Feb94- 7GG	fst	1	:22^4	:46^3	1:10^4	1:36^4	Clm 16000	77	1	2	3^2	2^3	2^2	2^{nk}	Belvoir V T	117 b	3.20	87 – 17	He Wanted More117^{nk} Champagne Shane117^4 Taj Raj112^{nk}	Closed well	7
27Sep93- 3BM	fst	1	:22^4	:47	1:12^2	1:38^3	Md 12500	61	6	2	1^{hd}	1^2	1^5	1^{10}	Baze R A	B 118 b	*1.50	77 – 22	Champagne Shane118^{10} T. C. Blue118^8 Golden Art118^2	Ridden out	9
9Sep93- 6Dmr	fst	6f	:22	:45^1	:57	1:09^2	[S]Md 32000	55	5	8	7^3	6^5	4$^{7\frac{1}{2}}$	4$^{12\frac{1}{2}}$	Solis A	LB 118 b	14.60	79 – 09	DoublThWthr118$^{7\frac{1}{2}}$ StillSwngn'118no GoldnExplo113^5	Bumped start, wide	12
20Aug93- 2Dmr	fst	7f	:22	:45^2	1:10^4	1:24^1	Md 32000	36	1	10	9^8	8^9	5^{12}	5$^{15\frac{3}{4}}$	Solis A	LB 117 b	22.00	63 – 14	Triple Thunder117$^{1\frac{3}{4}}$ Cologne117^2 Set On Cruise117$^{8\frac{1}{2}}$	Off slowly, wide	11
1Aug93- 6Dmr	fst	5f	:22	:45^1		:57^4	[S]Md Sp Wt	26	8	5	5^6	6$^{8\frac{1}{2}}$	7^{15}	7$^{16\frac{3}{4}}$	Solis A	B 117 b	22.60	91 – 05	OneHppyFell117$^{1\frac{1}{2}}$ SmokInMotion117$^{1\frac{3}{4}}$ Don'sRlity117$^{1\frac{3}{4}}$	Wide backstretch	9

WORKOUTS: Mar 9 GG 5f fst 1:03^2 H *31/32* Mar 3 GG 5f fst 1:03^4 H *38/46* Feb 16 GG 4f fst :49^2 H *15/34* Feb 9 GG 5f gd 1:03^1 H *22/30* Jan 27 BM 5f sly 1:05 H (d)*2/10* Jan 22 BM 1 fst 1:41^4 H *1/2*

He Wanted More
Own: Lingsch Norman & Sherman Art

BAZE R A (153 44 30 18 .29)

B. c. 3 (Mar)
Sire: Singular (Nodouble)
Dam: Unreal Upstart (Unreal Zeal)
Br: Siegal Jan (Fla)
Tr: Sherman Art (23 4 2 3 .17)

117

Lifetime Record:	12	3	2	2	$33,348						
1994	3	2	0	0	$19,388	Turf	0	0	0	0	
1993	9	1	2	2	$13,960	Wet	2	0	0	1	$2,330
GG	2	2	0	0	$18,150	Dist	0	0	0	0	

23Feb94- 1GG	gd	1	:23^3	:47^4	1:12^2	1:39^1	Clm 20000	79	4	2	2$^{\frac{1}{2}}$	1^{hd}	1^2	1^2	Meza R Q	LB 117 b	*1.10	75 – 28	He Wanted More117^2 T. C. Blue119^4 DoubleTheBridge117$^{4\frac{1}{2}}$	Steady drive	5
2Feb94- 7GG	fst	1	:22^4	:46^3	1:10^4	1:36^4	Clm 16000	77	6	3	1^{hd}	1^3	1^2	1^{nk}	Meza R Q	LB 117 b	3.30	87 – 17	He Wanted More117^{nk} Champagne Shane117^4 Taj Raj112^{nk}	Held gamely	7
13Jan94- 6BM	fst	1	:22^1	:45^3	1:11^2	1:39^2	Clm 20000	51	1	2	2$^{1\frac{1}{2}}$	2$^{\frac{1}{2}}$	1^{hd}	4$^{5\frac{1}{2}}$	Mercado P	LB 117 b	3.30	70 – 22	T. C. Blue117$^{1\frac{1}{2}}$ In Focus118$^{3\frac{1}{2}}$ Taj Raj117$^{\frac{1}{2}}$	Drifted out stretch	6
12Dec93- 3BM	my	6f	:22^1	:45^1	:57^2	1:10^2	Clm 20000	68	6	7	6$^{4\frac{1}{4}}$	4^4	4^5	3$^{3\frac{3}{4}}$	Mercado P	LB 114	3.00e	80 – 20	Crafty Johnny115$^{\frac{3}{4}}$ Imua Keoki109^3 He Wanted More114^{hd}	Even late	8
14Nov93- 6BM	fst	6f	:22^2	:45^2	:58^1	1:11^2	Clm 16000	65	4	5	2$^{\frac{1}{2}}$	1^1	1$^{1\frac{1}{2}}$	2^{nk}	Warren R J Jr	LB 115 b	3.60	79 – 19	T. C. Blue115^{nk} He Wanted More115$^{\frac{1}{2}}$ Colorado Swinger115^2	Game try	8
29Oct93- 1BM	fst	1	:22^4	:47	1:12^2	1:38^4	Clm 25000	50	3	3	2^1	2$^{1\frac{1}{2}}$	3^3	3^7	Boulanger G	LB 114 b	2.50	69 – 21	Jeep Shot115^1 Bolger My Boy116^6 He Wanted More114^2	Weakened	7
9Oct93- 7BM	fst	6f	:22^1	:45^2	:58^3	1:11^3	Clm c-16000	64	7	7	5^5	4^4	3^3	2^2	Boulanger G	LB 114 b	10.40	76 – 23	TrdeSecrets114^2 HeWntedMore114^1 HlosPreferred115$^{1\frac{1}{2}}$	Raced far wide	9
Claimed from Bally Stable & Mountain High Stable, Webb Bryan Trainer																					
16Sep93- 2BM	fst	1	:22^4	:47^1	1:13^4	1:41^2	Md 12500	48	8	5	3^4	1^2	1^4	1$^{4\frac{1}{2}}$	Boulanger G	LB 118 b	*2.20	63 – 25	HeWntedMore118$^{4\frac{1}{2}}$ ClydeTheGlide118^6 Acutboverelity118^6	Rallied wide	8
6Sep93- 1BM	fst	5½f	:21^3	:45^2	:58^2	1:04^4	Md 20000	44	5	7	6$^{6\frac{3}{4}}$	6$^{5\frac{3}{4}}$	6$^{6\frac{1}{4}}$	5$^{6\frac{3}{4}}$	Boulanger G	LB 118 b	4.90	80 – 14	Bold Gold Hill118$^{2\frac{1}{2}}$ Greek Crown118^2 Imua Keoki118^2		7
Hesitated, ducked in start																					
21Jly93- 6Crc	sly	5f	:23	:47		:59^4	Md 25000	29	3	5	6$^{3\frac{1}{4}}$	5^5	5$^{6\frac{3}{4}}$	6^9	Coa E M^5	L 111 b	*2.20	88 – 04	Medical Pro109^3 Always There116^3 Twentyeight Paces116$^{2\frac{1}{2}}$	Weakened	12

WORKOUTS: Mar 7 GG 4f fst :48 H *6/17* Feb 12 GG 4f fst :47^2 H *10/78* Jan 27 BM 4f sly :52^3 H (d)*5/14* Jan 7 BM 5f fst 1:00^2 H *4/30* ●Dec 22 BM 4f fst :47 H *1/33*

NW1 Allowance Races

After breaking its maiden, the next challenge for a horse with promising ability is usually at the Allowance NW1 level. The ninth race on March 12th at Golden Gate Fields is an example of a horse that won sprinting in a quick time and is now being asked to carry his speed 1 1/16 mile. Impressed by his 95 Beyer sprint rating, the public has jumped on Juan's Boy and made him the even-money favorite. The short price seems further justified because his competition primarily consists of a bunch of horses that have already lost at the claiming level. Of these Warning Label, Champagne Shane and He Wanted More all won wire to wire at a mile, but are suspect because they are unlikely to get the easy lead they seem to need from Juan's Boy. Furthermore they all have stamina indexes of 4 and 5 which suggest they are likely to come up short going an extra sixteenth of a mile.

If Juan's Boy is going to get beat, it will probably be by a horse that is bred to get the 1 1/16 mile distance, as indicated by a SI of 3 or lower. Looking at the field Suomi Power, Bai Brun and Rock Em Steady all qualify. But Rock Em Steady just beat bottom-level maiden claimers and his 39-1 odds reflect the fact that he is statistically very unlikely to be competitive at this classier NW1 Allowance level. Suomi Power, on the other hand, comes off a good race at this level, but likes to finish second and is surprisingly dead on the board at 16-1. So, of our stretchout possibilities, only Bai Brun at 3-1 is getting bet, probably because he showed tactical closing speed in a 7 furlong event at Calder and comes off a bullet 5 furlong work.

So how to bet? Juan's Boy has speed, class, a mile work and a high-percentage trainer in Chuck Jenda. The only question mark is his breeding. Both his sire and broodmare sire have a stamina index of 4, indicating Juan's Boy may have trouble running further than a mile. Furthermore, his dam Sundiata, according to *Maiden Stats 1993,* has a Dosage of 5.67 which, since it is greater than 4, indicates a speed-oriented pedigree. Given legitimate doubts about Juan's Boy's stamina, he is worth trying to beat at even-money. Since Juan's Boy is a vulnerable favorite who may run off the board, the best way to play this race is to dutch Bai Brun and Suomi Power on the Win and box them in the exacta.

Results: Juan's Boy was challenged early by Bai Brun and Warning Label and relinquished the lead to those two after 6 furlongs. Bai Brun, the bet-down horse that was bred to get the distance, pulled away at the top of the stretch and returned $7.80 to Win and $4.80 to Place. Juan's Boy is a classic example of the dangers of jumping on a horse coming out of a sprint with high speed figures that isn't bred to go on. Pedigree handicappers have a distinct edge on the public by understanding that sprint figures and route figures are simply not comparable.

The eighth race at Gulfstream on January 22nd is another NW1 route example with an interesting "speed" stretchout play. Of the eleven horses entered, the two favorites, Golden Larch at 7-5 and Turkomatic at 5-2 look tough. Golden Larch has the highest sprint Beyers and is bred to stretch out. He is also a Bill Mott shipper coming off a layoff which Olmsted's *Trainer Guide* touts as a winning play. Turkomatic is dropping from two stakes races, has been bet down, and figures to dramatically improve in this spot.

Gulfstream Park

8

1 1/16 MILES. **(1:40¹) ALLOWANCE. Purse $24,000 (plus $2,400 FOA). 3-year-olds which have not won a race other than maiden, claiming or starter. Weight, 120 lbs. Non-winners of a race other than claiming at one mile or over since November 15, allowed 3 lbs.**

1 1/16 MILES START▲ ▲FINISH

One Small Step
Own: Wimpfheimer Jacques D
HERNANDEZ R (11 1 0 0 .09)

Ro. c. 3 (May)
Sire: Slewpy (Seattle Slew)
Dam: Bright Omen (Grey Dawn II)
Br: Wimpfheimer Jacques D (Fla)
Tr: Wimpfheimer Donald C (1 0 0 0 .00)

L 117

Lifetime Record:	8	1	1	1	$15,770						
1993	8	1	1	1	$15,770	Turf	1	0	0	0	$150
1992	0	M	0	0		Wet	1	0	0	0	$800
GP	0	0	0	0		Dist	1	1	0	0	$10,200

25Dec93-9Crc	sly	1	:23⁴	:48²	1:14⁴	1:42⁴	Alw 16000N1X	63	5	9	75¼	4³	31½	42½	Hernandez R	L 117 b	15.10	74-21	Ego Runner117¾ Atomic Power117½ Grey Chandon1171¼		9
Bobbled start, bid wide, hung																					
13Nov93-5Crc	sf	1 1/16 Ⓣ	:24	:48⁴	1:13⁴	1:46²	Alw 15000N1X	50	2	4	63¼	9¹¹	9¹²	9¹⁵	Hernandez R	L 120 b	27.20	53-35	City Forest120⁵ Danger Ranger1102¼ Fortunate Joe117hd	Gave way	11
30Oct93-7Crc	fst	1	:23³	:47¹	1:13²	1:40³	Alw 15000N1X	50	3	6	78½	6⁸	6⁸	715¾	Hernandez R	L 115 b	6.50	72-11	Eleven Eleven115¾ Fortunate Joe115⁵ Booms Prospect1151¼		7
Bobbled start, no factor																					
15Oct93-3Crc	fst	1 1/16	:24¹	:48³	1:13⁴	1:48²	Md Sp Wt	65	4	1	1²	12½	1²	1³	Hernandez R	L 117 b	*.80	79-18	One Small Step117³ N Dever117¾ Ensign Mickey B117¾	Driving	6
17Sep93-6Crc	fst	1	:24²	:48²	1:13²	1:40⁴	Md Sp Wt	70	3	1	13½	1½	1½	2½	Hernandez R	L 116 b	*1.60	86-17	GrndGlmmr116½ OnSmllStp116¹¹ EnsgnNoblty116²	Gave way grudgingly	10
4Sep93-6Crc	fst	7f	:22³	:46¹	1:12¹	1:25⁴	Md Sp Wt	61	7	1	1³	3½	51¾	3¾	Hernandez R	L 117 b	13.40	85-13	Bay Street Star117½ Ready To Cope117nk One Small Step1174¾		7
Speed early, eased back, six wide top str, late rally																					
24Jly93-6Crc	fst	6f	:22	:46¹	:59	1:12¹	Md Sp Wt	41	8	8	6⁵	66¾	5⁸	5¹¹	Hernandez R	L 116	8.10e	76-05	Gooseby1162½ Silver Music1165½ Royal Venture116¾		10
Lacked response, eight wide str																					
19Jun93-3Crc	fst	5f	:22⁴	:47		1:00¹	Md Sp Wt	28	4	10	95½	87½	78¾	68½	Hernandez R	117	15.90	87-01	Ego Runner1171½ Pad1171½ Rock Tock117hd	Mild bid	12

WORKOUTS: ● Jan 15 **Crc 4f fst :48² Hg** *1/47* Jan 9 **Crc 3f sly :38³ B (d)** *4/7* Dec 19 **Crc 4f fst :48³ H** *7/52* Dec 12 **Crc 4f fst :49² B** *3/36* Dec 5 **Crc 4f fst :51 B** *27/39* Nov 7 **Crc 4f fst :49³ B** *13/31*

Ensign Mickey B

Own: Barbere B
FELIX J E (6 1 0 0 .17)

Dk. b or br g. 3 (Apr)
Sire: Blue Ensign (Hoist the Flag)
Dam: Strapped (Ambernash)
Br: Chabboquasset Farm (Ky)
Tr: Mastronardi Nick Jr (6 0 0 1 .00)

120

Lifetime Record :	12 1 3 1	$19,160			
1994	2 0 0 0	$240	Turf	4 1 1 0	$11,580
1993	10 1 3 1	$18,920	Wet	1 0 0 0	$140
GP	1 0 0 0	$240	Dist	1 0 0 1	$1,820

15Jan94- 7GP fst 1⅛ :472 1:123 1:391 1:524 Alw 24000N1X 54 6 8 710 79 812 812 Rodriguez P A 120 f 12.70 61-19 Silver Profile120no Jade Fortune1171¾ Critical Mass1172 Showed little 9
2Jan94-10Crc fm *1⅛ (T) 1:464 Trp Pk Dby-G2 48 13 12 116 1311 1317 1318 Ramos W S 113 f 29.00f 67-17 Fabulous Frolic1121½ Wake Up Alarm1172 Gator Back119½ Never close 14
12Dec93- 5Crc fm *1 1/16 (T) 1:44 Alw 16000N1X 69 9 10 108½ 107¾ 46 23¾ Ramos W S 120 3.70 80-14 Silver Music1173¾ Ensign Mickey B1202½ Cielo De Oro1171½ Good effort 10
7Dec93- 6Crc fm *1 1/16 (T) 1:472 Md Sp Wt 67 1 8 86½ 78 53¼ 2¾ Rodriguez P A 119 3.10 66-35 [D]DncInThRing119¾ EnsignMickyB119no [DH]GottScor119 Bothered late 8
Placed first through disqualification.
27Nov93-10Crc gd 1 1/16 (T) :231 :482 1:132 1:444 Cty of Miami35k 64 5 8 88½ 76¼ 76¾ 65¼ Toribio A R 113 44.00 71-22 Gator Back113no Fly Cry1152 Star Of Manila120nk Wide 8
20Nov93- 6Crc fst 7f :231 :471 1:124 1:26 Md Sp Wt 66 12 6 62½ 42½ 33½ 23¼ Rodriguez P A 119 12.30 82-14 Red Tazz1193¼ Ensign MickeyB119no MyMagicTouch119¾ 4 Wide, bumped 12
9Nov93- 3Crc fst 1 :242 :49 1:142 1:42 Md Sp Wt 60 4 6 73¾ 66¾ 54¾ 56¾ Douglas R R 118 b 4.10 74-16 Federal Court118nk Mr. Music Man1182¼ Fleet Eagle1183½ Mild bid 7
15Oct93- 3Crc fst 1 1/16 :241 :483 1:134 1:482 Md Sp Wt 59 2 2 36 46½ 43½ 33¾ Douglas R R 117 f 4.70 75-18 One Small Step1173 N Dever117¾ Ensign Mickey B117¾ Rallied 6
2Oct93- 6Crc sly 1 :234 :474 1:141 1:411 Md Sp Wt 51 7 2 21 21 512 514¼ Valles E S 117 f 8.80 71-13 Blazing Affair1175 Fleet Eagle1175 Gem Digger1172 10
Drifted out leaving final turn, clipped heels, faded
18Sep93- 3Crc fst 1 :241 :483 1:134 1:394 Md Sp Wt 64 5 2 2hd 2hd 23 26 Russell W B5 111 f 47.70 86-07 FbulousFrolic1116 EnsignMickeyB1113 ForceOfNture1092 Second best 8

WORKOUTS: Dec 29 Crc (T) 4f fm :49 B (d) 2/2 Dec 24 Crc 5f fst 1:032 B 20/40 Nov 18 Crc 3f fst :364 B 8/31 Nov 5 Crc 4f fst :483 H 4/41 Oct 29 Crc 5f fst 1:014 B 3/21

Forward To Lead

Own: Pomerantz L J & Sullivan R J
DOUGLAS R R (88 12 14 12 .14)

Gr. g. 3 (May)
Sire: Stutz Blackhawk (Mr. Prospector)
Dam: Honeytab (Al Hattab)
Br: John Yeoman & Jean Yeoman (Fla)
Tr: Ziadie Ralph (14 3 2 0 .21)

L 117

Lifetime Record :	7 1 1 2	$57,950			
1993	7 1 1 2	$57,950	Turf	1 0 0 0	$150
1992	0 M 0 0		Wet	1 0 1 0	$2,520
GP	0 0 0 0		Dist	3 0 0 2	$46,080

12Dec93-10Crc fst 1 1/16 :234 :483 1:133 1:47 Wht Plsure-G3 38 7 7 78¾ 712 613 728 Douglas R R 113 12.50 56-13 Gator Back1173 [DH]Federal Court113 [DH]Ride The Rails1201 Never close 7
27Nov93- 4Crc fst 1 1/16 :234 :48 1:133 1:48 Alw 16000N1X 66 3 6 55½ 53½ 23 35 Douglas R R 117 fb *.60 76-19 FederalCourt1204¼ SilverMusic117¾ ForwrdToLed1174¼ Lugged in stretch 7
13Nov93- 5Crc sf 1 1/16 (T) :24 :484 1:134 1:462 Alw 15000N1X 42 10 9 74¼ 77 1013 1118½ Douglas R R 117 b 2.60 49-35 City Forest1205 Danger Ranger1102½ Fortunate Joe117hd No factor 11
23Oct93-11Crc fst 1 1/16 :23 :462 1:113 1:461 FS InReality400k 79 4 12 1114 710 57½ 38½ Douglas R R 120 f 38.60 81-12 Holy Bull1207½ Rustic Light1201 Forward To Lead1201½ Late rally 12
25Sep93-11Crc fst 7f :221 :451 1:104 1:243 [R]F S Affirmed125k 66 10 10 106½ 85½ 79¼ 811 Henry W T 118 f 9.30 81-11 Rustic Light1181½ Blazing Affair1181½ Mr. Meadow1182½ Mild bid 13
11Sep93-10Crc sly 7f :223 :461 1:121 1:251 Alw 14000N1X 77 4 10 95¾ 63¼ 2½ 22½ Henry W T 118 f 3.30 86-13 Ride The Rails1182½ Forward To Lead1183½ March Fifth1154½ 10
Six wide top str, best of rest
28Aug93- 6Crc fst 5½f :223 :462 :59 1:053 Md Sp Wt 64 5 6 53¾ 33½ 22 1nk Henry W T 116 f 5.90 99-09 Forward To Lead116nk Gold Alex1164¾ Sunshine Man116nk Fully extended 9

WORKOUTS: Jan 18 Crc 5f sly 1:044 B 14/16 ●Jan 12 Crc 5f sly 1:02 H (d) 1/7 Dec 20 Crc 4f fst :49 B 2/27 Nov 24 Crc 3f fst :37 B 3/18 Nov 9 Crc 4f sly :48 H (d) 2/22 Nov 6 Crc 3f fst :37 B 5/28

Lt. Hill

Own: White Oak Farm
BRAVO J (67 9 8 7 .13)

B. c. 3 (May)
Sire: Temperence Hill (Stop the Music)
Dam: Pretty Driver (Lt. Stevens)
Br: Victor DiVivo (Md)
Tr: Tagg Barclay (6 0 0 1 .00)

L 117

Lifetime Record :	8 1 1 1	$25,270			
1993	8 1 1 1	$25,270	Turf	1 0 0 0	$300
1992	0 M 0 0		Wet	0 0 0 0	
GP	0 0 0 0		Dist	2 1 0 1	$11,440

11Dec93-10Lrl fst 1⅛ :473 1:121 1:374 1:503 [S]Md Juvnile125k 75 10 6 55½ 56½ 610 611¾ Luzzi M J L 122 b 4.40 73-26 Run Alden122¾ Canton River1222 Don's Sho1224 No factor 10
16Nov93- 7Lrl fst 7f :233 :464 1:113 1:243 Rollicking60k 79 3 9 52¼ 43½ 33 2¾ Luzzi M J L 113 b 9.80 85-16 Canton River115¾ Lt. Hill1132 Concern1132¾ Closed 10
24Oct93- 7Lrl fst 1 1/16 :241 :484 1:132 1:46 Alw 18500N1X 70 9 3 2½ 2½ 1hd 31½ Luzzi M J L 120 b 2.20 78-17 Prince Of Royalty1151 Concern120½ Lt. Hill1202½ Wide early 10
21Sep93- 6Pim gd 170 :231 :481 1:132 1:432 Md Sp Wt 68 5 2 2½ 2hd 12½ 18 Luzzi M J L 120 b 4.00 — — Lt. Hill1208 Melissa's Quickie120¾ Blood Bath1202¾ Driving 8
9Sep93- 5Pim fm 1 (T) :233 :48 1:134 1:414 Md Sp Wt 51 10 10 108½ 88½ 79¼ 76½ Johnston M T L 120 b 6.70 53-38 Takeitlikeaman1201¼ Marvinalay120no Blood Bath1204 Outrun 11
26Aug93- 4Pim fst 6f :224 :46 :582 1:111 Md Sp Wt 64 1 2 21 22 23½ 48 Prado E S L 120 b 6.10 81-14 Mr. Meadow1207¾ Blue Salute120no Blood Bath120no Weakened 10
15Jly93- 3Lrl fst 5f :224 :463 :583 Md Sp Wt 27 6 3 31½ 52½ 69½ 716¼ Johnston M T 120 23.10 84-13 Texturizer1203 Bald Greek120½ Blood Bath1204½ Gave way 9
2Jly93- 5Lrl fst 5f :23 :46 :582 Md Sp Wt 36 9 11 88 913 916 715 Johnston M T 120 9.00 86-05 Passing Reality1204 Blaise Winter1204½ Bald Greek120no Outrun 11

WORKOUTS: Jan 20 GP 4f fst :49 B 19/67 Jan 15 GP 4f gd :512 B (d) 29/50 Jan 3 GP 3f gd :371 B 2/12 ●Dec 2 Lrl 5f fst 1:012 H 1/7 Nov 15 Lrl 3f fst :361 H 2/7 ●Nov 10 Lrl 4f fst :471 H 1/12

Montaya (GB)

Own: Drey A
CASTILLO H JR (36 3 2 5 .08)

B. c. 3 (Mar)
Sire: Taufan (Stop the Music)
Dam: Kellys Reef (Pitskelly)
Br: The Hall Stud Ltd (GB)
Tr: Ebert Dennis W (8 2 1 0 .25)

117

Lifetime Record :	6 1 2 0	$13,908			
1993	6 1 2 0	$13,908	Turf	6 1 2 0	$13,908
1992	0 M 0 0		Wet	0 0 0 0	
GP	0 0 0 0		Dist	0 0 0 0	

13Nov93- 4Haw gd 1 (T) :224 :462 1:134 1:42 Alw 15000N1X 59 3 4 34 2hd 21 23 Bourque C C 117 fb *.60e 56-41 Winter St. Blues1173 Montaya117nk Michaelsdiscretion1174 Led briefly 8
31Oct93- 9Haw fm 1 (T) :222 :451 1:112 1:372 Royal Glint40k 60 5 7 67½ 44½ 43 55¾ Sibille R 115 fb 12.20 76-18 Sir Court115½ Galaxy1172 Midwest Blues115½ Brief wide bid 9
18Jun93♦ Ascot(GB) sf 5f (T)Str 1:041 Alw 49000 46 Swinburn W R 125 20.00 Great Deeds1183 Bid For Blue125hd Imperial Bailiwick1203 13
Tr: J M Troy — Windsor Castle Conditions Stks — *Pressed pace, led 3f to 2f out, ridden and one-paced late*
31May93♦ Doncaster(GB) gd 6f (T)Str 1:171 Alw 7900 46½ Geran A 127 3.00 Ochos Rios1233½ Sporting Warrior1231 Brocturne Gold1272 9
Toyota Cross & Sons Cndtns Stk — *Well placed in 4th, shaker. up 2f out, no late response*
24May93♦ Sandown(GB) yl *5f (T)Str 1:041 Maiden 8600 1hd Geran A 126 5.00 Montaya126hd Silver Wedge1261½ Lively Stream1262½ 14
Pizza Hut Maiden Stakes — *Close up on uphill course, led halfway, ridden out*
15Apr93♦ Newmarket(GB) gd 5f (T)Str 1:004 Maiden 8700 22 Swinburn W R 126 *3.50 Wajiba Riva1262 Montaya1261½ Greatest1261½ 9
EBF Stuntney Maiden Stakes — *Pushed along to chase leaders, brief bid 1f out, hung*

WORKOUTS: Jan 20 GP 3f fst :37 B 14/39 Jan 6 GP 5f fst 1:02 Bg 18/44 Dec 24 GP 4f fst :491 B 24/71 Nov 24 Haw 5f fst 1:012 H 3/44

Turkomatic

Own: Eaglestone Farm
FERRER J C (57 2 4 7 .04)

B. c. 3 (May)
Sire: Turkoman (Alydar)
Dam: Matinee Mimic (Silent Screen)
Br: Eaglestone Kugler Partnership (Ky)
Tr: Tammaro John J (6 1 0 0 .17)

117

Lifetime Record :	8 1 1 2	$26,081			
1994	1 0 0 0	$2,240	Turf	0 0 0 0	
1993	7 1 1 2	$23,841	Wet	0 0 0 0	
GP	1 0 0 0	$2,240	Dist	1 0 0 0	

5Jan94- 8GP fst 6f :23 :462 :583 1:101 Spect Bid BC70k 79 4 7 76 54¼ 45½ 56¾ Velez J A Jr 113 36.90 83-14 Hlo'sImge1143¾ DistinctRelity119nk SenorConquistdor1131¼ Broke slowly 7
24Oct93- 8Kee fst 1 1/16 :232 :472 1:111 1:421 Brdrs Fty-G2 78 6 7 57½ 58 58 710¼ Sarvis D A L 121 10.10 83-17 Polar Expedition1215 Goodbye Doeny1212 Solly's Honor1211 Tired 8
10Oct93- 4Kee fst *7f :224 :453 1:113 1:261 Alw 18000N2L 72 6 8 76¾ 78¾ 46½ 31¾ Sarvis D A L 118 6.10 96-05 Goodbye Doeny1181¼ Ocean Crest121½ Turkomatic1185 10
Ducked in bumped, start, mild gain
26Sep93- 9TP gd 6f :223 :462 :591 1:12 Alw 12400N2L 70 5 6 88½ 55½ 42½ 2½ Sarvis D A L 121 13.20 82-25 Flight Forty Nine121½ Turkomatic1211 Tarzans Blade115½ Finished fast 8
10Sep93- 6TP fst 6f :223 :47 1:01 1:134 Md Sp Wt 59 5 6 46½ 44 1½ 13 Sarvis D A L 121 4.20 74-25 Turkomatic1213 Dr. Right1215 Milt's Overture121½ Drew clear, urging 7
4Jly93- 1CD fst 6f :213 :454 :582 1:113 Md Sp Wt 39 5 6 98½ 710 713 612¾ Arguello F A Jr L 118 45.00 75-07 Saddleridge118¾ Judge T C1181½ Look For Trouble1185 No factor 12
16Jun93- 1CD fst 5½f :23 :463 :593 1:053 Md Sp Wt 31 8 9 85½ 55¼ 410 415¼ Bartram B E L 119 9.20 79-23 Strawberry A. OK.1191¼ LookForTrouble11910 Busta1194 No late response 12
18Apr93- 4Kee fst 4½f :222 :454 :521 Md Sp Wt 50 7 6 53½ 34 34 Bailey J D L 120 4.50 94-09 Seattle Rob1201 Armband1203 Turkomatic1202 Mild bid 10

WORKOUTS: ●Jan 16 GP 6f fst 1:12 H 1/15 Dec 29 GP 7f fst 1:27 H 1/3 Dec 23 GP 6f fst 1:14 H 2/8 Dec 17 GP 5f fst 1:013 B 4/23 Dec 10 GP 4f fst :473 H 2/31 Dec 4 GP 4f fst :49 B 8/22

continued

Sultan Of Java

Own: Condren W J & Cornacchia J & Farish

BARTON D M (20 1 4 2 .05)

B. c. 3 (Feb)
Sire: Java Gold (Key to the Mint)
Dam: Dalila (Hold Your Peace)
Br: Farish W S (Ky)
Tr: Zito Nicholas P (13 0 1 2 .00)

SI-3
SI-4

117

Lifetime Record :	10 1 3 0	$35,290			
1994	1 0 0 0		Turf	5 1 2 0	$27,300
1993	9 1 3 0	$35,290	Wet	1 0 0 0	
GP	0 0 0 0		Dist	0 0 0 0	

2 Jan94-10Crc	fm 1⅛ ⊕		1:464	Trp Pk Dby-G2	63 5 9 93¼ 118½ 111² 1011	Bailey J D	112	4.70e	74-17	Fabulous Frolic112½ Wake Up Alarm117² Gator Back119½	No factor 14
22 Nov93-8Aqu	gd 1	:243	:483 1:133 1:372	Alw 30000N1x	75 7 3 42 32½ 32½ 23½	Smith M E	117	2.50e	86-11	PennineRidge1193½SultanOfJava1171½GrandContinent11191½	Up for place 8
17 Oct93-11Bel	sf 1¼ ⊕	:241	:474 1:124 1:454	Alw 30000N1x	72 3 5 72¼ 61¼ 5½ 2¾	Bailey J D	117	6.00	66-35	Tomorrow'sComet117¾ SultanOfJava117¾ SeattleRob1171	Finished well 7
2 Oct93-8Bel	yl 1⅛ ⊕	:223	:461 1:114 1:443	Pilgrim-G3	59 4 5 52½ 55½ 59½	Alvarado F T	114	*1.30e	63-28	DoveHunt1134½ Tomorrow'sComet1133½ DynmiteLugh113½	Saved ground 9
19 Sep93-7Bel	my 1	:223	:46 1:111 1:373	Alw 30000N1x	64 7 5 54 72½ 53¼ 57¼	Alvarado F T	119	10.00	73-16	Tabasco Cat117hd Amathos117¾ Linkatariat117¾	Wide turn 7
23 Aug93-3Sar	fm ⊕ 1	:243	:481 1:12½ 1:383	Md Sp Wt	73 6 2 29 28 23 1¾	Alvarado F T	118	20.00	84-16	Sultan of Java118⁴ Silver Target118hd Exclusive Casino1181½	Driving 10
14 Aug93-3Sar	fst 7f	:224	:462 1:111 1:234	Md Sp Wt	57 6 1 1½ 1hd 45 49¾	Antley C W	118	6.10	76-14	Amathos118nk Fleet Stalker118nk Colonel Slade1189½	Dueled, tired 8
31 Jly93-3Sar	fst 6f	:22	:46 :583 1:111	Md Sp Wt	51 8 2 1hd 3nk 62⅜ 68½	Alvarado F T	118	12.10	77-08	Gold Tower1181½ Colonel Slade1182 Retrospection118nk	Dueled, tired 8
23 Jun93-5Bel	fst 5½f	:222	:46 :59 1:054	Md Sp Wt	51 5 2 43½ 42 44¾	Perret C	118	*1.00	81-14	Linkatariat118nk Caherdaniel1182 Mcdee1182½	Flattened out 8
3 Jun93-3Bel	fst 5½f	:222	:461 :584 1:05¹	Md Sp Wt	70 4 2 1½ 1½ 1½ 2¾	Perret C	118	16.10	88-13	Slew Gin Fizz118¾ Sultan Of Java118¾ Peace Negotiations118²	Gamely 8

WORKOUTS: Jan 15 GP 5f gd 1:03² B (d) 16/26 • Jan 1 GP 3f fst :37² B 7/14 Dec 26 GP 5f fst 1:02 B 11/28 Dec 18 GP 5f fst 1:01² B 14/40 Dec 10 GP 4f fst :49 B 10/31 Nov 17 Bel 5f fst 1:01³ H 10/19

Baling Wire

Own: Backer William M

BAILEY J D (58 14 8 3 .24)

Ch. c. 3 (Feb)
Sire: Tejano (Caro)
Dam: Bailrullah (Bailjumper)
Br: Backer William H (Ky)
Tr: Tagg Barclay (6 0 0 1 .00)

SI-3
SI-4

117

Lifetime Record :	2 1 0 0	$8,985			
1994	1 0 0 0	$150	Turf	0 0 0 0	
1993	1 1 0 0	$8,835	Wet	0 0 0 0	
GP	0 0 0 0		Dist	0 0 0 0	

| 1 Jan94-4Crc | fst 6f | :221 | :451 1:112 1:252 | Alw 15000N1x | 66 7 2 42 52½ 55½ 55 | Bailey J D | 120 | 3.80 | 80-15 | Fortunate Joe117¾ Bai Brun117½ Just A Roman1101½ | Tired 7 |
| 20 Nov93-7Lrl | fst 6f | :222 | :464 1:00 1:134 | Md Sp Wt | 63 1 3 41 31½ 2½ 1no | Prado E S | 120 | 5.70 | 71-24 | Baling Wire120no Hood Island120½ Back Dated120½ | Driving 14 |

WORKOUTS: Jan 16 GP 5f fst 1:01³ B 24/72 Dec 30 GP 4f fst :483 Bg 12/44 Dec 26 GP 4f fst :493 B 8/20 Dec 20 GP 4f fst :51⁴ B 39/45 Dec 11 Lrl 4f fst :49 B 10/24 Nov 17 Lrl 3f fst :36² B 3/7

Golden Larch

Own: Darley Stud Management Inc

SMITH M E (69 9 11 8 .13)

Ch. c. 3 (Mar)
Sire: Slew o' Gold (Seattle Slew)
Dam: Golden Petal (Mr. Prospector)
Br: Firestone Mr & Mrs Bertram R (Va)
Tr: Mott William I (33 4 4 4 .17)

SI-2
SI-4*

L 117

Lifetime Record :	1 1 0 0	$12,558			
1993	1 1 0 0	$12,558	Turf	0 0 0 0	
1992	0 M 0 0		Wet	0 0 0 0	
GP	0 0 0 0		Dist	0 0 0 0	

| 26 Nov93-4CD | gd 7f | :231 | :464 1:12² 1:253 | Md Sp Wt | 84 1 8 53 52 12½ 13 | Davis R G | L 122 | 4.10 | 82-15 | Golden Larch122³ Andover Road122no Tidal Song122½ | Inside, driving 12 |

WORKOUTS: Jan 16 GP 6f fst 1:14 H 7/15 Jan 6 GP 6f fst 1:19 B 16/16 Dec 23 GP 5f fst 1:02² B 19/31 Dec 16 GP 4f fst :494 B 27/42 Nov 18 CD 3f my :36³ B 6/18 Nov 13 CD 4f my :51 Bg 16/26

Java Royal

Own: Murrell John R

SELLERS S J (41 8 5 6 .20)

B. c. 3 (Feb)
Sire: Java Gold (Key to the Mint)
Dam: Wakonda (Fappiano)
Br: Nerud John A Irrevocable Trust (Ky)
Tr: Bracken James E (10 2 0 1 .20)

SI-3
SI-4*

L 117

Lifetime Record :	9 1 0 1	$13,185			
1994	1 0 0 1	$2,860	Turf	0 0 0 0	
1993	8 1 0 0	$10,325	Wet	3 0 0 0	$420
GP	1 0 0 1	$2,860	Dist	0 0 0 0	

8 Jan94-10GP	fst 6f	:22	:45 :57² 1:10	Alw 22000N2L	74 2 7 52 42½ 33¾	Sellers S J	L 117 b	16.50	87-09	ArrivalTime114½ CodeHome114² JavaRoyl117½	Lacked needed response 7
25 Dec93-7Crc	sly 1	:234	:474 1:14 1:422	Alw 16000N1x	53 1 7 54½ 44½ 46½ 610½	Ramos W S	L 120 b	3.10	68-21	Mr. Angel1203½ Roman Seige1121¼ Change Of Venue117¾	10
										Bumped start, thru after 6 furlongs
11 Dec93-6Crc	fst 7f	:231	:47 1:113 1:254	Md Sp Wt	74 4 7 72¼ 2hd 13½ 11¾	Ramos W S	L 120 b	3.40	86-10	Java Royal120½ Fleet Eagle1206¾ Honor Colony120¾	Driving, clear 9
2 Dec93-6Crc	fst 7f	:241	:482 1:141 1:422	Md Sp Wt	58 2 8 87 77 54 41½	Castaneda M	L 119 b	5.60	78-20	Runaway Witness119nk Mr. Music Man1191 Fleet Eagle119hd	8
										Pinched back start, 5 wide rally
20 Nov93-6Crc	fst 7f	:231	:471 1:124 1:26	Md Sp Wt	65 5 10 118¾ 1010 56 44	Castaneda M	L 119 b	13.10	81-14	Red Tazz119³½ Ensign Mickey B119nk My Magic Touch119¾	Wide, gaining 9
23 Oct93-6Crc	sly 7f	:224	:46 1:114 1:25	Md Sp Wt	37 2 10 51¾ 42½ 57¾ 716½	Rodriguez P A	118 b	25.90	73-07	Trialist1182½ My Magic Touch1181 Federal Court1185	Gave way 8
9 Oct93-5Crc	sly 7f	:224	:46 1:123 1:261	Md Sp Wt	42 1 6 1hd 32 711 716¾	Coa E M5	113	10.70	67-13	Ali'Ibito'reality1186½ Sunshine Man1182 Sir H. C.118nk	Stopped 7
25 Sep93-6Crc	fst 6f	:214	:454 :592 1:124	Md Sp Wt	33 5 8 41½ 53 67¼ 611¼	Coa E M5	112	15.00	71-11	Fortunate Joe117¾ Crystal Pistol1174 Mighty Bowl117nk	Tired 10
14 Jly93-3Crc	sly 5f	:223	:462	:59	Md Sp Wt	–0 11 12 1010 1226 1223 1227¾	Douglas R R	117	25.60	74-08	Birdie King1177 Super Review117² Ming's Court112¾	Outrun 12

WORKOUTS: Jan 16 Crc 5f fst 1:02⁴ B 12/32 Jan 2 Crc 4f fst :49³ Bg 9/31 Dec 9 Crc 4f fst 1:04 B 18/26 Nov 29 Crc 5f fst 1:03² B 4/10 Nov 6 Crc 5f fst 1:03 B 11/47

Bermuda Cedar

Own: Burning Daylight Farms

SANTOS J A (51 8 10 7 .16)

Ch. c. 3 (Mar)
Sire: Talinum (Alydar)
Dam: Rue Reality (Fantasy 'n Reality)
Br: Backer J W & Baumohl A & Partners (Ky)
Tr: Carroll David (8 1 2 3 .13)

SI-2

117

Lifetime Record :	8 1 3 2	$56,766			
1994	1 0 0 1	$2,760	Turf	0 0 0 0	
1993	7 1 3 1	$54,006	Wet	3 0 3 0	$35,156
GP	1 0 0 1	$2,760	Dist	1 0 0 1	$2,760

4 Jan94-6GP	fst 1⅛	:232	:47 1:122 1:46	Alw 23000N2L	72 9 1 11 2hd 31½ 37	Smith M E	114 b	3.30	73-21	Warm Wayne114¾ Jade Fortune114¾ BermudaCedar1144	Svd grnd, tired 7
19 Dec93-4Aqu	fst 6f ⊡	:223	:46 :584 1:12	Safe Ground50k	71 4 8 65 65¼ 76¾ 57¾	McCauley W H	117 b	10.70	75-22	Mr. Shawklit117¾ Joe Casey1194 Hussonet119no	No rally 8
5 Dec93-2Aqu	sly 6f ⊡	:233	:481 1:002 1:123	Alw 28000N1x	71 4 2 31 21 31½ 31¾	McCauley W H	117 b	*1.70	77-18	Hussonet122¾ Bermuda Cedar117¾½ Fabulous Force117²½	Second best 5
29 Oct93-8Med	fst 6f	:222	:452 :574 1:103	Comet35k	69 5 1 2hd 3½ 33½ 37½	McCauley W H	113 b	1.80	80-18	Distinct Reality1205¾ Storm Street113¼ Bermuda Cedar1131	No late rally 7
20 Oct93-8Aqu	sly 7f	:222	:452 1:094 1:223	Cowdin-G2	88 4 5 3½ 1hd 21½ 25	McCauley W H	122 b	13.90	86-13	You And I122⁵ Bermuda Cedar12211 Gulliviegold122hd	Second best 7
30 Sep93-5Bel	fst 6f	:222	:46 :582 1:112	Md Sp Wt	87 4 1 2½ 1½ 1½ 1¾	Davis R G	118 b	3.60	83-22	Bermuda Cedar118¾ Prank Call118nk Plutonius1181¾	Driving 9
10 Sep93-2Bel	my 6f	:221	:453 :574 1:104	Md 75000	78 4 5 22 2½ 21½ 21¾	Davis R G	118 b	3.50	85-13	Mobile114¾ Bermuda Cedar118nk Jo Ran Express1183	Held place 8
28 Jly93-5Sar	fst 5f	:214	:453	:582	Md Sp Wt	47 2 7 76½ 73¾ 74½ 78	Davis R G	118	6.50	85-08	Wire Squire118½ Mcdee118½ Mobile118nk	Broke slowly, green 9

WORKOUTS: Jan 19 GP 4f gd :48 H 2/19 Dec 31 GP 5f fst 1:01 B 5/51 Dec 17 Bel tr.t 3f fst :374 B 12/26 Nov 28 Bel tr.t 4f gd :494 H 3/12 Nov 9 Bel tr.t 6f fst 1:15 B 1/2

© 1994 DRF, Inc.

All the other horses, except for Bailing Wire, have already lost at this level and can be thrown out. Interestingly, Bailing Wire's trainer Barclay Tagg also has Lt. Hill as part of an uncoupled entry running in this spot at 5-1. But, Lt. Hill has been off six weeks, has less speed than Bailing Wire and has an underlined broodmare sire stamina index of 5, which may be why he faded at Laurel in his October 24th route while tracking slow fractions.

Bailing Wire, despite his high 18-1 odds, is the more intriguing part of the "entry" because Jerry Bailey remains up, he is bred to stretch out and his :22.2/:44.3 early fractions figure to put him on or near the lead. Furthermore, Tagg gave him a useful 7 furlong prep at Calder on January lst. Apart from his lack of action, the only other grounds for concern is that Bailing Wire may lack stamina on his broodmare sire side. That doubt, however, is assuaged by discovering in *Maiden Stats 1993* that Bailrullah has a route-qualifying Dosage of under 4.

So how to bet? Given his lack of action and the credentials of the two favorites, Bailing Wire cannot be a prime Win bet. Still, speed stretching out is dangerous and Bailing Wire is a threat to steal the race. The best strategy here is a moderate Win bet as well as putting him behind Golden Larch and Turkomatic in the exacta.

Results: Bailing Wire led into deep stretch, only to get beat a neck and nipped a nose by the two favorites. Golden Larch paid $4.80 to Win, Turkomatic $3.20 to Place and Bailing Wire $6.20 to Show. Since Bailing Wire was covered on the Win end and for second in the exacta to the two favorites, one more sensible bet would have been to play him to hang on for third in the trifecta behind a Golden Larch and Turkomatic box. The $2 trifecta with this combination returned $204.

Marathon Routes

Of all the different kinds of routes, the stamina indexes in *Exploring Pedigree* are probably most determinant in "marathon" races of a 1 1/4 mile and beyond. The fifth race on May 15th at Pimlico is a 1 1/2 mile turf example which includes several "action" horses that are simply not bred to get the distance.

Looking at the stamina indexes for each of the horses, the first thing to notice is that Final Alliance, Machete Road, Gotcha Cornered, Mr. Mouse and Lunch Connection all come up short. So, even though Machete Road and Mr. Mouse come off good recent races and are being bet below their 8-1 and 4-1 morning lines, they should be thrown out.

This leaves us with Hamr It, Takonius, Wedges of Lemon and Take Heed as the remaining contenders. While Wedges of Lemon technically qualifies according to his stamina index, he has already faded three times at shorter distances and seems to need a lead which he is not likely to get in this spot. Since actual race experience supercedes pedigree analysis, Wedges of Lemon can also be thrown out.

Of the remaining horses, Take Heed has been running against better, had a prep race at Keeneland and looks like a legitimate favorite at 7-5. Hamr It is a bit of a mystery horse since we don't have a stamina rating for his English sire. But, since he ran creditably at 1 3/8 mile as a maiden and passed horses in his last two races at a mile, he can't be thrown out. Furthermore, despite his layoff, he is being bet at 6-1.

Pimlico

5

1½ MILES. (Turf). (2:272) **ALLOWANCE. Purse $18,500. 3-year-olds and upward which have not won a race other than maiden, claiming, starter or hunt meeting. Weights: 3-year-olds, 112 lbs. Older, 124 lbs. Non-winners of a race other than claiming at one mile or over since April 19, allowed 3 lbs. Such a race since March 19, 5 lbs.**

TURF COURSE START — 1½ MILES — FINISH

Final Alliance

Own: Kushner Arlene E

DELGADO A (137 25 19 18 .18)

Gr. c. 4 — SI 4
Sire: J O Tobin (Never Bend)
Dam: Charming Dawn (Vigors)
Br: Conway & Mactier & Malcho & Wiesemann (Ky)
Tr: Heil Nancy B (9 2 1 2 .22)

L 119

Lifetime Record :	12	1	0	0	$13,990						
1994	5	0	0	0	$1,765	Turf	2	0	0	0	$1,480
1993	7	1	0	0	$12,225	Wet	0	0	0	0	
Pim Ⓣ	1	0	0	0	$370	Dist Ⓣ	0	0	0	0	

1May94-11Pim	fm	1 Ⓣ	:231 :463 1:103 1:371	3↑ Alw 18500N1X	77	5 8 99¾ 912 97¾ 73	Delgado A	L 119 f	14.40	79-15	HoratiusSwinger1191 MacheteRoad110¾ ManassaSttion1191	Rank early 12
Placed 6th through disqualification.												
22Apr94-7Pim	fst	6f	:231 :46 :582 1:121	Alw 16500N1X	63	4 6 69 612 610 57½	Douglas F G	L 117 f	24.70	78-15	Johnny Eager117¾ Dixieland Music1171½ Harundale1173¾	Pinched back 7
24Mar94-8Lrl	fst	6f	:224 :462 :583 1:11	Alw 16800N1X	55	5 1 2hd 31½ 78½ 79½	Douglas F G	L 117 f	33.70	75-17	Fortuna Al Ponte1172¼ Algebar Henderson117¾ Harundale117¾	Gave way 7
12Mar94-9Lrl	fst	5½f	:222 :46 :581 1:042	Alw 18300N1X	57	8 9 1112 1114 1011 1112¼	Douglas F G	L 117 f	42.10	87-15	Red River Gorge1173 Seventh Summer1174½ Night Spirit117¾	Outrun 12
6Jan94-8Lrl	fst	6f	:231 :464 :59 1:112	Alw 17400N1X	33	6 5 32½ 46½ 915 916½	Delgado A	L 117 f	13.80	67-20	Roovan117no Stelios Art117nk Distinctive Dan1171¾	Gave way 9
27Aug93-6Pim	fst	6f	:224 :454 :581 1:11	3↑ Alw 17100N1X	56	7 3 21 42½ 69 711	Delgado A	L 113 fb	29.20	79-11	Ziggy's Dream1132¼ Danzig Diplomat1173 Kahli Kisu1173½	Gave way 8
7Aug93-9Pim	fst	6f	:231 :463 :591 1:123	3↑ Alw 17100N1X	61	7 2 31 2hd 33½ 56¾	Luzzi M J	L 114 fb	7.90	75-13	Brittany's Boy112nk Ziggy's Dream114no Kahli Kisu1173½	Gave way 8
30Jly93-6Lrl	fst	6f	:224 :463 :583 1:111	3↑ Alw 16800N1X	70	7 3 31 21½ 21¼ 44¾	Luzzi M J	L 112 fb	9.70	79-15	Longest Drive1121¼ Kahli Kisu117¾ Danzig Diplomat1172¾	Weakened 7
17Jly93-5Lrl	fm	1 Ⓣ	:231 :471 1:123 1:37	3↑ Alw 20000N1X	71	9 2 2hd 2hd 35 410½	Luzzi M J	L 112 f	16.50	75-15	No Delay112hd Gayquare1179 Agent Cooper1171¼	Weakened 11
10Jly93-8Lrl	fst	6f	:221 :454 :581 1:112	3↑ Alw 16800N1X	52	5 1 33 33½ 38 58½	Luzzi M J	L 115 f	2.80	75-17	Mur Wick1133½ No Delay112hd Line Pro1103	Wide 7

WORKOUTS: Apr 11 Bow 6f fst 1:132 H *1/1* ●Mar 5 Bow 6f fst 1:13 H *1/11* Feb 17 Bow 6f fst 1:17 B *1/1*

Machete Road

Own: R D M Racing Stable

STACY A T (109 11 17 20 .10)

B. g. 3 (Apr) — SI 2
Sire: Jungle Blade (Blade)
Dam: Royal Lorrium (Mandate)
Br: Conway James & Millard Abel (Ky)
Tr: Trombetta Michael J (13 0 6 0 .00)

L 109

Lifetime Record :	14	1	4	1	$29,030						
1994	8	1	3	1	$22,540	Turf	1	0	1	0	$3,885
1993	6	M	1	0	$6,490	Wet	2	1	1	0	$13,290
Pim Ⓣ	1	0	1	0	$3,885	Dist Ⓣ	0	0	0	0	

1May94-11Pim	fm	1 Ⓣ	:231 :463 1:103 1:371	3↑ Alw 18500N1X	82	9 7 43 44 43 21	Stacy A T	L 110 b	9.20	81-15	HoratiusSwinger1191 MacheteRoad110¾ ManassaSttion1191	Wide early 12
21Apr94-8Pim	fst	1 1/16	:231 :471 1:123 1:45	Alw 18500N1X	62	4 7 64½ 64 88½ 815	Wilson R	L 117 b	5.30	67-25	South Bend1152 Brass Scale1152¼ Owned By Us1151½	Wide, gave way 9
10Apr94-5Pim	my	1 1/16	:232 :464 1:122 1:464	Alw 18500N1X	80	1 5 33 35 23 22¾	Wilson R	L 117 fb	6.60	70-36	Lt. Hill1152¾ Machete Road1176 Alleged Impression1152½	Steadied 3/8 8
26Mar94-1Aqu	fst	7f	:231 :471 1:123 1:251	Clm 40000	61	2 7 42½ 53 65 69¼	Luzzi M J	119 b	35.50	69-22	Doctrinaire1172 Jacksome1171¼ Grateful Appeal1172	No threat 7
10Mar94-2Lrl	my	1 1/16	:242 :484 1:14 1:471	Md Sp Wt	77	1 1 1½ 1hd 11 13¼	Pino M G	L 120 fb	2.00	74-28	Machete Road1203¼ Electric Image1201½ Ⓓ Elberton Sheriff1208½	Driving 8
1Mar94-3Lrl	fst	6f	:23 :462 :59 1:113	Md Sp Wt	69	2 2 34½ 35 34 33¼	Johnston M T	L 120 b	6.90	79-24	Chisholm Trail120¾ Frozen Kable1102½ Machete Road1204½	Evenly 7
6Feb94-2Lrl	fst	1 1/16	:241 :484 1:134 1:452	Md 25000	64	2 3 31½ 21½ 22½ 22¾	Pino M G	L 120 b	3.70	80-14	Seventy Five North1202¾ MacheteRoad120nk AngleAllay1165½	Held second 8
1Feb94-5Pha	fst	170	:223 :463 1:131 1:434	Md Sp Wt	36	3 3 32½ 34 411 420½	Lloyd J S	L 122 b	*1.40	60-23	Devil'sFortress12211 MorningJrrod1174 WiseJudgement1225½	Tired badly 7
30Dec93-3Aqu	fst	1 1/16 ⊡	:233 :482 1:144 1:50	Md Sp Wt	60	4 4 52¼ 43 48 48	Prado E S	118 b	14.20	47-37	Concoctor1181 Final Clearance1182 Smartweed1185	Even trip 10
8Dec93-1Lrl	fst	1 1/16	:24 :483 1:141 1:473	Md 25000	58	6 6 65¼ 2½ 2½ 2no	Pino M G	L 120	2.50	72-22	Duke's Star120no Machete Road1202 Sidango120no	Drifted late 8

WORKOUTS: Feb 19 Lrl 5f fst 1:02 B *4/12*

Hamr It

Own: Graham Patrick J

KLINGER C O (124 21 18 16 .17)

Ch. c. 4 — ?
Sire: Soukab (Good Counsel)
Dam: I've Gotta Crow (Crow)
Br: Eichhorn Richard W (Pa)
Tr: Graham Patrick J (—)

119

Lifetime Record :	5	1	2	1	$16,100						
1993	5	1	2	1	$16,100	Turf	5	1	2	1	$16,100
1992	0	M	0	0		Wet	0	0	0	0	
Pim Ⓣ	1	0	0	0	$495	Dist Ⓣ	0	0	0	0	

17Jly93-7Lrl	fm	1 Ⓣ	:224 :464 1:122 1:372	3↑ Alw 20000N1X	76	2 8 79½ 98½ 53¾ 32	Hamilton S D	113	5.90	81-15	Esteban117no Val Gleam1172 Hamr It113¾	Closed 11
25Jun93-5Lrl	fm	1 Ⓣ	:232 :463 1:12 1:38	3↑ Md Sp Wt	77	2 3 37 25 23 1¾	Rocco J	113	3.60	80-18	Hamr It113¾ Roovan108nk Judge Connelly114nk	Driving 12
10Jun93-5Lrl	fm	1⅛ Ⓣ	:454 1:101 1:354 1:483	3↑ Md Sp Wt	81	6 5 56½ 45 24 23½	Rocco J	113	3.30	83-17	Geewhillikins1233½ Hamr It1134¾ Judge Connelly11417	Rallied 9
30May93-5Pim	fm	1⅜ Ⓣ	:51 1:184 1:45 2:223	3↑ Md Sp Wt	61	4 1 2hd 1½ 33½ 54¾	Prado E S	112	4.90	— —	Pre Med124¾ Broadford1122¼ Bucky's Boy1141½	Gave way 8
15May93-6Mgo	gd	*1⅛ Ⓣ	1:492	3↑ Md Sp Wt	–0	5 3 33½ 33½ 44 22½	Hendriks E5	137	-	— —	Sadie's Tornado1472½ Hamr It1372½ Takonius1503½	10

WORKOUTS: ●May 11 Fai 4f fst :49 B *1/6* Apr 20 Fai 4f fst :52 B *3/3*

Gotcha Cornered

Own: Lakeville Stables

JOHNSTON M T (178 35 21 24 .20)

Gr. g. 3 (May) — SI-4
Sire: Weshaam (Fappiano)
Dam: Shoot It Out (Wolfgang)
Br: Casey James W (WV)
Tr: Leatherbury King T (123 21 18 11 .17)

L 112

Lifetime Record :	12	2	2	2	$20,555						
1994	7	1	0	2	$15,100	Turf	0	0	0	0	
1993	5	1	2	0	$5,455	Wet	2	0	1	0	$2,050
Pim Ⓣ	0	0	0	0		Dist Ⓣ	0	0	0	0	

29Apr94-6Pim	fst	1 1/16	:231 :464 1:122 1:461	Clm 25000	66	3 4 47 45½ 54½ 37¼	Johnston M T	L 119	3.40	69-21	All Gale117nk Duke's Star1177 Gotcha Cornered1193	No factor 6
19Apr94-4Pim	fst	1 1/16	:241 :49 1:132 1:46	Clm 25000	77	1 2 2hd 2hd 11½ 1¾	Johnston M T	L 117	3.50	77-21	Gotcha Cornered117¾ All Gale117no Battleship Grey1174½	Driving 6
10Apr94-5Pim	my	1 1/16	:232 :464 1:122 1:464	Alw 18500N1X	66	7 6 69 59½ 49 411¼	Johnston M T	L 115	26.60	62-36	Lt. Hill1152¾ Machete Road1176 Alleged Impression1152½	No threat 8
26Mar94-2Lrl	fst	1⅛	:491 1:142 1:402 1:534	Clm c-18500	56	4 4 53 46 56 35½	Rocco J	L 117	2.80	63-19	Gator's Rocket1172¾ True Freshman1172¾ Gotcha Cornered117½	No threat 7
Claimed from Casey Eleanor M, Casey James M Trainer												
11Mar94-6Lrl	fst	1 1/16	:231 :47 1:13 1:461	Alw 18500N1X	65	3 5 67¼ 54½ 59½ 59½	Stacy A T	L 115	27.10	69-27	Mister Persister115hd P P Prospect1154½ Triform115nk	Weakened 6
27Feb94-7Lrl	fst	1 1/16	:234 :48 1:13 1:46	Alw 19100N1X	69	3 6 58½ 47½ 57¼ 49½	Wilson R	115	11.90	70-31	Forejove1151 P P Prospect1178 Hit The Green115½	No factor 8
5Feb94-6Lrl	fst	1 1/16	:242 :492 1:15 1:47	Clm 25000	40	7 7 31 31 78 713½	Pino M G	117	7.60	62-16	Mister Persister1171¾ Galactic Light1171 Take The Flight113hd	Tired 7
22Dec93-1Lrl	fst	7f	:231 :463 1:123 1:263	Clm 17500	51	5 6 65½ 78½ 79 55½	Kuykendall L M	115	16.10	70-20	Heavenwood1162 Bad Dude114nk Gala Performer1163¼	Outrun 7
4Dec93-10CT	sly	6½f	:233 :473 1:141 1:204	Alw 4700N4L	50	6 5 2½ 2½ 1hd 21	Dupuy L	116	6.50	83-16	Safeway Matt1141 Gotcha Cornered1166 Ara Too114½	Lead, tired 8
23Oct93-10CT	fst	7f	:23 :461 1:12 1:26	Ⓡ Tri St Fty57k	47	7 7 77 78¾ 711 713½	Dupuy L	117	3.00e	79-12	Whiz Pass117hd Run Alden1244 Again And Again1175	Outrun 8

WORKOUTS: ●Mar 6 Lrl 4f fst :48 H *1/15* Feb 18 Lrl 4f fst :512 B *7/8*

Takonius

Own: Chesterbrook

DOUGLAS F G (110 11 10 12 .10)

Ch. h. 5
Sire: Foretake (Forli)
Dam: Quidonia (Northern Jove)
Br: Paxson Mrs Henry D (NY)
Tr: Gilliam Jeremy J (—)

WON AT DISTANCE

119

Lifetime Record:	6 1 0 1	$9,890			
1993	5 1 0 1	$9,890	Turf	3 1 0 1	$9,790
1992	1 M 0 0		Wet	0 0 0 0	
Pim Ⓣ	1 1 0 0	$9,405	Dist Ⓣ	0 0 0 0	

2Nov93-7Lrl fst 1¾ ⊗ :534 2:13 2:374 3:024 3+ Fisherman28k 49 2 6 67½ 614 621 645 Douglas F G 115 24.00 — — Reggae1223½ Legal Choice117nk Asserche1222½ Outrun 7
19Aug93-5Pim fm 1⅜ Ⓣ :531 1:203 1:452 2:24 3+ Md Sp Wt 62 1 6 45 54 21 1¾ Douglas F G 122 39.60 — — Takonius122¾ Charlie Tango1221½ Strong Prospect1221½ Driving 10
20Jly93-9Lrl fst 1 1/16 :234 :484 1:143 1:471 3+ Md 8500 16 8 2 31 68 714 825¾ Castillo F 122 5.80 48–20 Truly Rowdy1039 Pete Radue114¾ South Of Java1131½ Faltered 8
15May93-6Mgo gd *1⅛ Ⓣ 1:492 3+ Md Sp Wt –0 10 5 57½ 57½ 54½ 35 Gillam J 150 - — — Sadie's Tornado1472½ Hamr It1372½ Takonius1503½ 10
20Mar93-1Aik fm *6½f Ⓣ 1:30 3+ Md Sp Wt –0 12 10 95½ 84 710 68¾ Marzullo V7 143 - 71 — Northern Pat1502½ T. V. Gold140no Roman Impulse1334 12
28Nov92-2Aik gd *2¼ Hurdles 4:32 3+ Md 12000 — 1 8 — — — — Walsh P B 132 - — — Keltie1521 Willowark1472½ Fill My Card1471 Lost rider 8

Wedges Of Lemon

Own: Daney Mrs Bernard J Jr

REYNOLDS L C (181 27 27 23 .15)

B. g. 4
Sire: Caveat (Cannonade)
Dam: Let Me Sleep (Cyane)
Br: Black Gates Nursery Trust (Md)
Tr: Boniface J William (25 4 2 4 .16)

SI-1

119

Lifetime Record:	8 1 0 0	$9,660			
1994	1 0 0 0		Turf	3 0 0 0	$770
1993	7 1 0 0	$9,660	Wet	0 0 0 0	
Pim Ⓣ	1 0 0 0		Dist Ⓣ	0 0 0 0	

1May94-11Pim fm 1 Ⓣ :231 :463 1:103 1:371 3+ Alw 18500N1X 64 3 4 2hd 21 22½ 109 Reynolds L C 119 b 4.40 73–15 HoratiusSwinger1191 MacheteRoad110¾ ManssSttion1191 Shut off 1/16 12
Placed 9th through disqualification.
22Nov93-10Pha fst 7f :222 :453 1:112 1:251 3+ Alw 15578N2L 45 2 10 76 75¾ 810 1010½ Umana J L 113 b 10.40 70–21 Absolutely Gold1151½ Biting Irish1191¾ Merlington119½ Off slow 10
9Nov93-8Pha fst 7f :221 :451 1:11 1:243 3+ Alw 15500N1X 54 10 8 911 97¾ 89¾ 68½ Umana J L 113 b 11.10 75–15 SherPrctice1166 JohnsI.D.116nk Fergi'sSpidr113no Came in,impeded rival 10
27Oct93-9Med fst 1 1/16 ⊗ :24 :474 1:131 1:453 3+ Alw 19000N1X 16 3 2 1hd 52¾ 818 832 Colton R E 114 b 8.80 48–18 Patty Mitts1152½ Super Arti1162½ Rokee To Cohee1162½ Tired 8
15Oct93-4Med yl 1 1/16 Ⓣ :234 :484 1:142 1:473 3+ Alw 17000N1X 73 8 1 1hd 1hd 4½ 54½ Colton R E 113 b 8.90 56–34 Land Axe111¾ Ojin111no Real Stutz1163 Weakened 11
2Oct93-5Med yl 170 Ⓣ :22 :464 1:114 1:43 Clm 35000 73 6 2 2½ 2½ 21 44 Molina V H 115 24.60 72–26 D J's Rainbow115½ Ojin1153½ Isobars115nk Needed more 10
13Sep93-8Pha fst 1 1/16 :233 :474 1:124 1:472 3+ Alw 16000N2L 33 1 1 32 58½ 616 625¾ Colton R E 112 f 2.20 46–28 Hard Hat1165½ I'm Ferdinand1132½ Gone Regal1115½ Stopped 6
12Apr93-3Pha fst 1 1/16 :232 :481 1:142 1:49 Md Sp Wt 69 5 2 2½ 1½ 13 16½ Colton R E 122 3.00 64–34 Wedges Of Lemon1226½ Caro's Swan Song122nk Mortal1226½ Drew off 7

Mr Mouse

Own: Campbell Alexander G

ROCCO J (85 12 10 13 .14)

Dk. b or br c. 4
Sire: Storm Bird (Northern Dancer)
Dam: Petite Danceuse (Little Current)
Br: Campbell Alex G Jr (Ky)
Tr: Fisher John R S (4 0 1 1 .00)

SI-4*

119

Lifetime Record:	7 1 0 3	$17,785			
1994	1 0 0 1	$3,480	Turf	5 1 0 2	$15,380
1993	6 1 0 2	$14,305	Wet	1 0 0 1	$2,035
Pim Ⓣ	0 0 0 0		Dist Ⓣ	1 0 0 0	

21Apr94-4Aqu yl 1 1/16 Ⓣ :242 :484 1:143 1:481 3+ Alw 29000N1X 72 9 10 816 510 38 34¾ Bravo J 119 10.00 63–28 Knight Course1191½ GrandContinental1123½ MrMouse1191½ Steadied early 10
18Nov93-7Lrl fst 1⅛ :47 1:121 1:394 1:532 3+ Alw 19400N1X 66 7 5 515 520 710 68½ Wilson R 114 6.00 63–26 San American1171½ Stelios Art1141½ Roanoke's Image1124¾ Steadied early 9
30Oct93-8Lrl sly 1¼ ⊗ :481 1:151 1:421 2:082 3+ Alw 19100N1X 79 1 1 1hd 2½ 31½ 32½ Wilson R 113 2.80 62–32 Bullseye Daredevil117½ Off The Cuff1172 Mr Mouse11310 Weakened 8
5Oct93-8Med yl 170 Ⓣ :23 :474 1:13 1:423 3+ Alw 17000N1X 76 7 6 63 63½ 53½ 42 Thomas D B 113 6.70 76–28 Electric Spark116½ Land Axe1111½ Clipped111hd Needed more 10
13Aug93-4Sar fm 1½ [T] :501 1:143 2:052 2:293 3+ Alw 28500N1X 67 9 6 716 56 68 611½ Santos J A 114 4.90 70–15 Groomed To Win1121½ Matchless Dancer1133 Glenbarra1125½ No factor 9
24Jly93-5Lrl fm 1⅛ Ⓣ :48 1:133 1:39 1:511 3+ Md Sp Wt 63 8 4 43 32 3½ 11¾ Rocco J 113 2.20 74–21 Mr Mouse1131¾ Pfoney Pfelix1081 Have At It113¾ Driving 9
4Jly93-5Lrl fm 1⅛ Ⓣ :471 1:13 1:391 1:521 3+ Md Sp Wt 65 3 9 818 817 411 33¾ Rocco J 113 6.80 65–29 Roovan1143¾ Half A Crown113no Mr Mouse1131¾ Closed 9

WORKOUTS: May 5 Fai 4f gd :53 B 2/4 Apr 7 Fai 4f gd :53 B 4/5 Mar 25 Fai 5f gd 1:044 B 1/1 ●Mar 20 Fai 4f fst :49 B 1/5

Take Heed

Own: Bayard Mrs James A

WILSON R (55 7 5 7 .13)

Dk. b or br c. 4
Sire: Caveat (Cannonade)
Dam: My Mafalda (Smarten)
Br: Bayard Mrs James A (Md)
Tr: Delp Grover G (16 3 1 5 .19)

SI-1

L 119

Lifetime Record:	8 1 0 2	$31,468			
1994	1 0 0 0	$1,100	Turf	6 1 0 1	$22,495
1993	7 1 0 2	$30,368	Wet	1 0 0 1	$5,973
Pim Ⓣ	2 0 0 0	$2,190	Dist Ⓣ	0 0 0 0	

28Apr94-4Kee fm 1 1/16 Ⓣ :233 :484 1:14 1:443 Alw 26750N2L 75 2 4 32 41½ 41½ 43¾ Wilson R L 114 b *1.90 79–17 Tidy Colony1133½ Early Thaw112no Southern Money115nk 9
In tight start no rally
13Nov93-10Lrl fst 1⅛ :472 1:121 1:37 1:492 [S]Nthn Dncr100k 86 8 6 612 68½ 69½ 511½ Wilson R L 115 b 11.40 79–20 Jest Punching1155½ Wa Bert1151½ Pescagani1151 Outrun 8
30Oct93-11Lrl sly 1¼ ⊗ :493 1:151 1:412 2:064 Japan Rcg As54k 77 6 3 32½ 21 35½ 312½ Johnston M T L 115 b 3.90 60–32 Pescagani1153½ Oscar Max1159 Take Heed1153 Weakened 6
9Oct93-7Lrl fm 1⅛ Ⓣ :471 1:121 1:371 1:493 Md Mln Turf95k 92 8 7 87 88 54 31 Johnston M T L 115 b 34.30 81–21 Awad122nk Dancing Douglas119¾ Take Heed1151¾ Bumped 1/8 pole 9
11Sep93-8Pim fm 1⅛ Ⓣ :482 1:122 1:373 1:50 3+ Alw 21500N2X 78 3 4 55 33 22 44½ Johnston M T L 114 b 18.10 80–11 Telegrapher119¾ Geewhillikins1193¾ Blue Highways108no Weakened 11
17Aug93-7Pim fm 5f Ⓣ :22 :452 :574 3+ Alw 17700N1X 63 4 10 109½ 810 76¾ 46½ Saumell L L 114 b 4.90 86–12 Bates Return1134½ Slickem117nk Dixieland Music1171¾ Mild rally 10
17Jly93-5Lrl fm 1 Ⓣ :231 :471 1:123 1:37 3+ Alw 20000N1X 71 6 5 64½ 96 68½ 610½ Johnston M T 114 7.30 74–15 No Delay112hd Gayquare1179 Agent Cooper1171½ No threat 11
8Jly93-3Lrl fm 5½f Ⓣ :224 :471 :591 1:052 3+ Md Sp Wt 62 7 4 53 54½ 33 1½ Johnston M T 115 3.50 94–06 Take Heed115½ Oliver Witha Twist1151¾ J. B. Chris1152½ Driving 8

WORKOUTS: May 10 Lrl Ⓣ 5f fm 1:013 H (d) 2/5 ●Apr 23 Lrl 5f fst 1:002 H 1/14 Apr 16 Lrl 6f gd 1:18 B 2/3 Apr 10 Lrl 6f fst 1:15 H 1/2 Mar 25 Lrl 5f my 1:023 B 4/8 Mar 19 Lrl 4f fst :49 B 10/25

Lunch Connection

Own: Hensley & Redmond & Moreland Stable

NO RIDER (—)

B. g. 4
Sire: Linkage (Hoist the Flag)
Dam: Parisian Honey (Honey Jay)
Br: Audley Farm Inc (Va)
Tr: Bailes W Robert (30 5 5 3 .17)

SI-4

L 119

Lifetime Record:	13 1 2 0	$20,329			
1994	5 0 1 0	$2,050	Turf	0 0 0 0	
1993	4 0 1 0	$7,179	Wet	0 0 0 0	
Pim Ⓣ	0 0 0 0		Dist Ⓣ	0 0 0 0	

7May94-8Del fst 1 :234 :474 1:133 1:394 3+ Clm 10000 75 1 3 41 2½ 1hd 21 Juarez C L 113 b 5.70 83–24 Memorable Sal1121 Lunch Connection1134¾ Moycullen106no Gamely 7
17Apr94-6Pha fst 1 1/16 :24 :481 1:134 1:474 3+ Alw 17479N2L 51 4 5 42 42¾ 57½ 614½ Ayarza I L 116 b 3.20 58–38 Mr. New Salem1162½ Gotchas Kid1066½ Loyal Opposition1161¾ Wide, tired 8
9Apr94-7Pim fst 1 1/16 :233 :47 1:12 1:462 Alw 18500N1X 62 4 8 711 76½ 66 75¾ Stacy A T L 117 b 4.00e 69–26 Bunny's Live Wire117¾ Busy Me1171¾ Solar Angle117¾ Wide 11
24Mar94-8Lrl fst 6f :224 :462 :583 1:11 Alw 16800N1X 59 1 7 710 77 68½ 68 Stacy A T L 117 27.30 77–17 Fortuna Al Ponte1172½ Algebar Henderson117¾ Harundale117¾ Outrun 7
12Mar94-9Lrl fst 5½f :222 :46 :581 1:042 Alw 18300N1X 62 2 12 1217 1219 1112 710½ Stacy A T L 117 b 48.30 88–15 Red River Gorge1173 Seventh Summer1174½ Night Spirit117¾ Belated bid 12
27Feb93-10Pha fst 170 :223 :472 1:133 1:443 Alw 15000N2L 54 9 4 34 52½ 66¾ 48½ Nicol P A Jr L 119 b *3.70 64–23 Falcon Delight122hd Lil Em'n Em1111 Another Huey1127½ Wide, tired 9
19Feb93-7Lrl fst 1 1/16 :241 :481 1:13 1:451 Alw 19100N1X 72 7 5 56½ 57 510 417½ Delgado A L 117 b 13.80 66–35 Bounding Daisy11712 Snooky's Taylor1131½ Art History1084 No threat 8
6Feb93-8Lrl fst 7f :233 :472 1:12 1:243 Dncng Cnt42k 52 3 5 55 56½ 511 520 Delgado A L 113 b 12.60 66–25 Wolf Prince119nk Asset Impression11311 Without Dissent1144¾ Outrun 5
17Jan93-8Lrl fst 1 1/16 :233 :47 1:131 1:464 Alw 18130N1X 58 1 2 42½ 32 31½ 2no Delgado A L 120 b 3.00 76–19 FiftyCentDollrs114no LunchConnection1201½ FrdisBoy112¾ Gamely inside 5
27Dec92-7Lrl fst 1 1/16 :233 :471 1:13 1:464 Md Sp Wt 63 4 4 33½ 31 1hd 1no Delgado A L 120 b 6.50 79–14 Lunch Connection120no Hope Of Peace1201½ King Kapalua1206 Driving 9

WORKOUTS: May 3 Bow 3f fst :362 B 3/10 Apr 3 Bow 4f fst :502 B 3/3 Mar 8 Bow 3f gd :37 Bg 5/9 Mar 1 Bow 6f fst 1:143 H 1/2 Feb 22 Bow 5f fst 1:032 B 4/6 Feb 16 Bow 4f fst :50 B 6/10

Takonius is also coming off a layoff and at 18-1 is way above his 5-1 morning line. Nevertheless, if only because he is the only horse to have won at this distance, he cannot be eliminated. Furthermore, trainer Jeremy Gilliam, according to Olmsted's *Trainer Guide*, wins races with layoff horses.

So how to bet? At 7-5, Take Heed is scary to go against with two layoff horses. The best strategy here is to box Takonius and Hamr It with the favorite and hope one of them either wins or comes in second to him. Alternatively, we could take a flyer and single Take Heed on top in the trifecta and box Takonius and Hamr It behind him in the second and third holes.

Results: Take Heed ran like a 7-5 favorite should and crushed the field by eleven lengths. Hamr It nosed out Takonius for the Place and the latter beat the rest of the field by four lengths. The $2 exacta paid $23.40 and the trifecta returned $140. The key handicapping question in this race was: is it harder to overcome a layoff (Takonius and Hamr) or more difficult to overcome a negative stamina index; both Machete Road and Mr. Mouse were legitimate contenders coming off good recent races but for the fact that they didn't have the stamina index of 1 necessary to get the marathon distance of 1 1/2 mile. In this case, the stamina indexes were the key to handicapping the race.

Stretchouts in Claiming Races

Horses that are stretched out for the first time in claiming races are often overlooked and can go off at surprisingly high mutuels. The ninth race on July 30th at Del Mar is one juicy example. Going over the field, Sal's Wish, Falcon Bid, Double Sec, Mark Ninety and Decidedly Friendly have all lost to cheaper and, despite the action on some of them, are not likely winners. Even though Falcon Bid and Mark Ninety are multiple winners, it is worth noting that they had to be dropped down to win and that they earned higher Beyers while contesting a slower pace than they are likely to get today. Furthermore, Mark Ninety is parked on the outside and will have to use his speed early, which is likely to hurt him late.

The lukewarm favorite at nearly 3-1 is the class dropper Veracity, who is reunited with Eddie D, who won with him for $50,000 on April 27th. Count Con, despite his 17-1 odds, also has to be considered because he ran right behind Veracity while coming off a layoff on the 27th and might do so again.

This leaves us with Bold Capital and Luckscaliber as the two stretchout horses who have never lost at this level to consider. Of the two horses, Bold Capital has more speed and is out of a Private Account mare which also adds stamina to his pedigree. Private Account is worth emphasizing as he is listed in bold as one of the superior broodmare sires. With his superior :21.4/:44.4 early fractions and route breeding, Bold Capital figures to have the lead over Mark Ninety, if he wants it, and be able to sustain his run. Furthermore, according to Olmsted's *Trainer Guide*, Ed Gregson wins races with

Del Mar

9

1 MILE. (1:33¹) **CLAIMING. Purse $34,000. 3-year-olds. Weight, 120 lbs. Non-winners of two races at one mile or over since June 1, allowed 3 lbs. Of such a race since July 1, 5 lbs. Claiming price $40,000; if for $35,000, allwoed 2 lbs. (Maiden or races when entered for $32,000 or less not considered.)**

1 MILE
START & FINISH

Sal's Wish

Own: Ringler Nancy & Reed
SOLIS A (4 0 0 1 .00) $40,000

B. c. 3 (Mar)
Sire: Lyphard's Wish*FR (Lyphard)
Dam: Sal's High (Poker)
Br: Ringler Nancy (Ky)
Tr: Cenicola Lewis A (—)

L 115

Lifetime Record :	12 1 3 0	$31,300			
1994	8 1 3 0	$27,700	Turf	2 0 0 0	$1,950
1993	4 M 0 0	$3,600	Wet	1 0 0 0	$1,350
Dmr	0 0 0 0		Dist	4 0 1 0	$6,450

4Jly94- 1Hol fst 1 1/16 :23¹ :46⁴ 1:11¹ 1:43 Clm 32000 85 7 6 4² 4¾ 2¹ 2¹¾ Solis A LB 115 5.10 84-14 A Treek For Roses1151¾ Sal's Wish1154 R. J.'s Raider1156½ 4 wide 2nd turn 7
5Jun94- 3GG fm 1 1/16 Ⓣ :23 :46² 1:10³ 1:44 + Clm 32000 74 3 4 48½ 44 43 43 Warren R J Jr LB 117 4.50 83-12 Dinner Affair1221½ Noble Danz117hd Robbie J.1171¼ Even late 7
8May94- 5Hol fm 1 1/16 Ⓣ :23 :46 1:10³ 1:42¹ Alw 40000N1X 70 1 9 11¹⁴ 107½ 107¾ 88¾ Solis A LB 115 28.10 76-12 Hot Number119½ Muirfield Village1171¼ Mint Green1151 No rally 11
21Apr94- 2SA fst 1 1/16 :23¹ :47¹ 1:12² 1:45 Md 32000 70 7 4 3² 2½ 1² 13½ Solis A LB 117 *.70 70-22 Sal's Wish1173½ Gray Jove1172½ Doc Kline1171¾ Ridden out 8
7Apr94- 9GG fst 1 1/16 :23 :47² 1:12¹ 1:43² Md Sp Wt 78 8 5 52¾ 42½ 42 43¼ Meza R Q LB 118 2.70 77-24 WesternTrder118¾ Asterisc'sPrince118½ CleverRmrk1182 Wide 2nd turn 9
10Mar94- 4SA fst 1 :22² :46¹ 1:11¹ 1:37² Md 40000 73 9 9 9⁷ 85½ 57½ 21¾ Desormeaux K J LB 117 7.10 80-14 Attleboro1171¾ Sal's Wish117½ Uronurown1174 Closed fastest 10
10Feb94- 1SA fst 1 1/16 :23 :47¹ 1:13¹ 1:46³ Md 32000 64 1 5 53½ 41¼ 1hd 2nk Solis A LB 117 2.80 62-27 Noble Danz117nk Sal's Wish1173½ Z. Leader1151¼ Game try 8
23Jan94- 5SA fst 1 :22⁴ :46² 1:10¹ 1:35¹ Md Sp Wt 71 5 6 8⁵ 77¾ 7¹² 8¹⁷ Solis A LB 117 27.80 76-13 DramaticGold117nk MuirfieldVillage1178½ Devil'sMirge1173½ Came in start 10
26Dec93- 9SA fst 1 :22¹ :45 1:10¹ 1:36³ Md Sp Wt 62 10 9 8¹³ 8⁶ 8⁹ 8¹⁴ Antley C W LB 117 18.90 73-10 Almaraz1171¾ DramaticGold1171¾ AdvantgeMiles1176 Steadied 1/4 & 3/16 10
12Dec93- 5Hol my 6f :22 :45² :58² 1:12¹ Md 50000 70 4 6 87½ 6¹¹ 49½ 44 Desormeaux K J LB 119 3.50 75-20 Mr. Cooperative1192 Mr Peter P.117hd Danny The Demon1192 8
Steadied 1/2, 4 wide into lane

WORKOUTS: Jly 23 **Hol 6f fst 1:13¹ H** *4/24* Jly 13 **Hol 4f fst :49¹ H** *29/51* Jun 29 **Hol 4f fst :48⁴ H** *21/57* Jun 22 **Hol 4f fst :48¹ H** *11/36* Jun 15 **Hol 5f fst 1:01² H** *13/45* May 31 **Hol 5f fst 1:00³ H** *5/24*

Veracity

Own: Dutton & Greene
DELAHOUSSAYE E (5 0 1 0 .00) $40,000

Gr. g. 3 (Apr)
Sire: Vigors (Grey Dawn II)
Dam: Summer Treasure (Summing)
Br: Grossman Jack (Ky)
Tr: Dutton Jerry (—)

L 115

Lifetime Record :	10 2 1 2	$51,100			
1994	6 1 1 1	$38,200	Turf	2 0 0 1	$6,750
1993	4 1 0 1	$12,900	Wet	1 0 0 1	$3,000
Dmr	0 0 0 0		Dist	5 1 1 1	$21,350

15Jun94- 5Hol fm 1 1/16 Ⓣ :24 :47⁴ 1:12 1:42⁴ Alw 40000N1X 70 2 9 107¾ 10⁷ 105½ 89½ Antley C W LB 115 b 34.00 72-14 Gracious Ghost117½ Saltgrass1192 Sharp Phase117hd Saved ground 10
18May94- 4Hol fm 1 Ⓣ :23⁴ :47² 1:12 1:37¹ Clm 62500 75 6 7 74¼ 75¼ 6³ 33¼ Antley C W LB 116 b 2.90 75-19 Set On Cruise1162¼ Baja Bill1171 Veracity116nk Wide rally 7
27Apr94- 3Hol fst 1 1/16 :24 :47⁴ 1:12¹ 1:44³ Clm 50000 79 4 5 5⁵ 43½ 31½ 1½ Delahoussaye E LB 116 b 3.50 78-18 Veracity116½ Count Con1164 Baja Bill1161½ Ducked in 1/8 5
1Apr94- 1SA fst 1 :23¹ :46³ 1:11³ 1:37³ Clm 32000 72 4 7 7¹⁰ 76½ 3⁵ 24½ Delahoussaye E LB 116 b *1.80 76-15 Gary's Lucky One1154½ Veracity1165½ Chief Brody1184½ Off step slow 7
3Mar94- 6SA fst 1 :22³ :46² 1:11³ 1:38¹ Clm 50000 77 6 7 67¾ 43¾ 42¾ 42¾ Valenzuela F H LB 115 b 4.90 75-21 Welcome Discovery1162¼ Iron Groove115½ Sport's Staff115no Late bid 7
30Jan94- 3SA fst 1 1/16 :23⁴ :47⁴ 1:11⁴ 1:41⁴ Alw 40000N1X 74 3 5 5³ 55½ 5⁸ 5¹² Valenzuela F H LB 115 19.70 74-12 Pollock'sLuck115no Numrous1166½ CrownngDcson1183½ Off bit awkwardly 5
28Dec93- 2SA fst 1 :22¹ :45³ 1:10³ 1:36¹ Clm c-50000 70 1 6 68½ 73¾ 44½ 76¾ Flores D R LB 115 4.80 82-12 Arezzo1152 He's Fabulous1152 Smoke In Motion1152 Wide trip 10
Claimed from Annuzzi & Knight, Knight Terry Trainer
8Dec93- 8BM my 1 :23¹ :46⁴ 1:11² 1:37¹ Alw 20000N1X 71 7 7 7⁹ 66½ 63¾ 32¼ Baze R A B 115 *2.30 82-19 Saratoga Bandit1142 Night Letter119nk Veracity115½ Far wide stretch 7
22Oct93- 6SA fst 1 :22⁴ :46⁴ 1:12³ 1:38⁴ Md 32000 65 9 6 66¾ 64½ 51¾ 1hd Flores D R B 117 3.40 76-21 Veracity117hd Prevasive Force1172¾ Sierra Real1123½ Wide rally 9
10Oct93- 9SA fst 6f :21⁴ :44³ :57 1:09⁴ Md Sp Wt 55 7 9 11¹² 10¹³ 9¹¹ 6¹¹ Flores D R B 117 37.70 77-16 Sky Kid117hd Gracious Ghost1175 Al's River Cat1171¼ By tired ones 11

WORKOUTS: Jly 21 **Dmr 7f fst 1:29² H** *1/1* Jly 13 **SA 4f fst :47³ H** *2/30* Jly 7 **SA 4f fst :47² H** *5/20* Jun 30 **SA 5f fst 1:01⁴ H** *10/20* Jun 22 **SA 3f fst :37 H** *17/23* Jun 8 **SA 5f fst 1:00⁴ H** *10/31*

Bold Capital

Own: Conejo Ranch & Pendleton
GOMEZ G K (3 0 1 0 .00) $40,000

Dk. b or br g. 3 (Mar)
Sire: Bold Ruckus (Boldnesian)
Dam: Star Account (Private Account)
Br: Windfields Farm (Ont-C)
Tr: Gregson Edwin (1 0 0 0 .00)

115

Lifetime Record :	2 1 0 0	$10,625		
1994	2 1 0 0	$10,625	Turf	0 0 0 0
1993	0 M 0 0		Wet	0 0 0 0
Dmr	0 0 0 0		Dist	0 0 0 0

25Jun94- 6Hol fst 6½f :21⁴ :44⁴ 1:10¹ 1:16⁴ 3↑ Md 32000 81 5 2 1hd 2hd 1¹ 1nk Solis A B 116 b 2.60 85-10 Bold Capital116nk Top Gear1165½ Low Level Attack1161 Gamely 10
29May94-10Hol fst 6½f :21³ :44² 1:10¹ 1:17 3↑ Md 32000 64 10 4 3¹ 2hd 2¹ 43¼ Solis A B 115 b 16.30 81-09 Skim The Gravy116½ Top Gear1152½ Sack Lucifer113nk Weakened 11

WORKOUTS: Jly 28 **Dmr 4f fst :49 H** *32/53* Jly 21 **SA 6f fst 1:15² H** *5/11* Jly 15 **SA 6f fst 1:15¹ H** *8/15* Jly 9 **SA 5f fst 1:01³ H** *26/41* Jly 4 **SA 4f fst :50 H** *14/23* Jun 17 **SA 5f fst 1:01³ H** *7/21*

Falcon Bid

Own: Adams & St Cyr
BLACK C A (3 0 0 0 .00) $40,000

Gr. c. 3 (May)
Sire: Imperial Falcon (Northern Dancer)
Dam: Qui Bid (Spectacular Bid)
Br: Hedgestone Management (Ont-C)
Tr: St Cyr Robert (—)

L 115

Lifetime Record :	8 2 1 0	$31,925		
1994	8 2 1 0	$31,925	Turf	0 0 0 0
1993	0 M 0 0		Wet	0 0 0 0
Dmr	0 0 0 0		Dist	0 0 0 0

Entered 29Jly94- 5 DMR
16Jly94- 5Hol fst 1 1/16 :23 :46² 1:11 1:42⁴ Clm c-22500N2L 92 6 2 2¹ 2½ 12½ 12½ Black C A LB 116 5.10 87-12 Falcon Bid1162½ Island Sport1165½ Major Ruler1141¾ Driving 10
Claimed from Graham Decourcy W, Hendricks Dan L Trainer
11Jun94- 1Hol fst 1 1/16 :22⁴ :45³ 1:10² 1:43² Clm 28000 44 2 2 2½ 33½ 5¹¹ 520¾ Black C A LB 115 8.80 63-16 Tulwar1181¾ Decidedly Friendly1185½ Lucks Mine1153½ Bobbled start 6
4May94- 6Hol fst 1 1/16 :23³ :47 1:11³ 1:43⁴ 3↑ Md c-32000 79 2 2 1½ 1¹ 1½ 1hd Black C A LB 115 *1.60 82-18 FlconBid115hd Momonymomonymomony1211¼ ShinngColors1211½ Gamely 10
Claimed from Philip & Sophie Hersh Trust, Bernstein David Trainer
21Apr94- 4SA fst 1 1/16 :22⁴ :47 1:12¹ 1:44² Md 32000 76 2 4 22½ 1hd 1hd 2nk Black C A LB 117 5.20 73-22 Al Monsoon117nk Falcon Bid1174 Pacific Edition1171¼ Game try 8
6Apr94- 2SA fst 6f :21³ :44⁴ :57² 1:10 Md 50000 65 4 5 42½ 3¹ 52¼ 5⁶ Pedroza M A LB 118 8.90 81-12 Raised State1182½ Capo1181¼ Granby118nk Weakened 10
6Mar94- 2SA fst 6f :22 :45¹ :57² 1:10 Md Sp Wt 59 4 2 63¾ 4⁴ 4⁹ 4¹³ Pedroza M A LB 118 15.10 74-09 De Vito1184 Minero1186½ Hello Chicago1182½ Broke in, bumped 11
5Feb94- 4SA fst 6½f :22¹ :45 1:09² 1:15⁴ Md Sp Wt — 6 2 3½ — — — McCarron C J LB 118 4.80 — 11 Irgun1184 Scenic Route118nk Devon Dancer11813 6
Lost his action, fell 3/8
22Jan94- 4SA fst 6f :21² :44¹ :56 1:08² Md Sp Wt 71 7 4 66½ 65½ 57½ 511¾ Pedroza M A LB 118 14.40 83-09 Fly'n J. Bryan1184½ Laabity1183½ Paster's Caper1182 Wide into lane 7

WORKOUTS: Jly 25 **SLR tr.t 4f fst :47⁴ H** *5/8* Jly 11 **Hol 6f fst 1:17 H** *13/13* Jly 5 **Hol 5f fst :59³ H** *3/35* Jun 28 **Hol 4f fst :49¹ H** *13/27* Jun 22 **Hol 4f fst :49 H** *18/36* Jun 8 **Hol 4f fst :48 H** *5/56*

continued

Double Sec

Own: Budann Stable

NAKATANI C S (6 1 0 1 .17) $40,000

Ch. c. 3 (Apr)
Sire: Desert Wine (Damascus)
Dam: Ameridouble (Nodouble)
Br: Cardiff Stud Farm (Cal)
Tr: Truman Eddie (2 0 1 1 .00)

115

Lifetime Record :	7 1 2 1	$27,700			
1994	7 1 2 1	$27,700	Turf	2 0 0 1	$3,750
1993	0 M 0 0		Wet	1 0 0 0	$3,000
Dmr	0 0 0 0		Dist	2 1 0 0	$11,550

18Jun94- 4GG fm 1 1/16 Ⓣ :223 :464 1:112 1:433+ Alw 25000N1X 64 6 8 9¾ 117 1111 912¾ Lopez A D B 117 b 5.90 75-12 Attleboro1176 Just For Dino117hd Dancethetides117hd Bumped break 11
28May94- 7Hol fst 1 1/16 :231 :464 1:112 1:442 Clm 32000 77 2 4 42 42½ 41½ 2hd Pincay L Jr B 117 b 4.30 79-21 Water Test116hd Double Sec1171¾ Al Monsoon119nk 7
Lugged in, steadied 1/8, lugged in late
14May94- 6GG fm 1⅛ Ⓣ :493 1:134 1:384 1:513 Alw 25000N1X 76 6 6 86½ 78½ 54¾ 33 Boulanger G B 117 b 7.00 80-07 T. V. Producer1171 Asterisca's Prince1202 Double Sec117nk Mild late bid 8
2Apr94- 2SA fst 1 :224 :462 1:112 1:372 Clm 50000 68 8 8 67½ 65½ 76½ 77½ Black C A B 115 b 17.40 75-15 Set On Cruise115½ Eagle's Deed1151 Tulwar115½ 4 wide trip 8
19Mar94- 3SA sly 1 1/16 :24 :48 1:131 1:483 Alw 39000N1X 62 4 4 412 49½ 48½ 412½ Pincay L Jr B 117 2.50 40-41 Heartless Wager117¾ Almaraz1204 Margie's Boy1177½ Outrun 4
25Feb94- 6SA fst 1 :224 :47 1:12 1:38 Ⓢ Md c-40000 73 6 8 95¾ 73¾ 43 1½ Pincay L Jr B 117 *1.00 79-25 Double Sec117½ Decidedly Friendly1151¾ Mulvihill115½ Wide rally 9
Claimed from Selznick Jeffrey, Stute Warren Trainer
28Jan94- 2SA fst 6f :213 :444 :571 1:101 Md 32000 68 9 7 910 109½ 610 24½ Pincay L Jr B 118 17.00 81-12 Heartless Wager1184½ Double Sec1181¾ Bobs Bikini118hd 11
Awkward start, green early

WORKOUTS: Jly 25 Dmr 7f fst 1:26 H 1/2 Jly 18 SA 6f fst 1:133 H 5/14 Jly 12 SA 4f fst :484 H 10/26 Jly 6 SA 6f fst 1:144 H 5/13 Jun 29 SA 3f fst :371 H 7/12 Jun 15 SA 5f fst 1:031 H 25/27

Decidedly Friendly

Own: Friendly Ed & Natalie

STEVENS G L (8 3 0 3 .38) $40,000

B. c. 3 (Feb)
Sire: Literati (Nureyev)
Dam: Natalie Knows (Decidedly)
Br: Friendly Ed (Cal)
Tr: Sadler John W (1 0 0 0 .00)

L 115

Lifetime Record :	10 1 4 0	$29,225			
1994	7 1 3 0	$25,825	Turf	2 0 0 0	
1993	3 M 1 0	$3,400	Wet	0 0 0 0	
Dmr	0 0 0 0		Dist	1 0 1 0	$4,200

24Jun94- 6Hol fm 1 1/16 Ⓣ :232 :47 1:112 1:413 Clm 65000 69 8 4 42½ 43 58½ 611¼ Stevens G L LB 115 11.60 77-13 Jabbawat116no Raised State1163½ Waltzing Pleasant116½ Gave way 8
11Jun94- 1Hol fst 1 1/16 :224 :453 1:102 1:432 Clm 32000 77 3 5 510 46 33 21¾ Solis A LB 118 5.60 82-16 Tulwar1181¾ Decidedly Friendly1185½ Lucks Mine1153½ Second best 6
15Apr94- 9SA fst 1 1/16 :23 :473 1:124 1:442 Ⓢ Md 32000 77 3 1 11½ 12 16 15 McCarron C J LB 117 *.80 73-18 Decidedly Friendly1175 Clever Ricky1172 Pennyinthewater1122¾ Handily 12
11Mar94- 9SA fst 1 1/16 :23 :471 1:122 1:434 Ⓢ Md 32000 80 8 3 31½ 3½ 2hd 21¼ Solis A LB 117 3.10 75-22 Sir Litchi1171¼ Decidedly Friendly1171 Apurado1172¾ Bid, outfinished 11
25Feb94- 6SA fst 1 :224 :47 1:12 1:38 Ⓢ Md 35000 72 3 2 2hd 1hd 12 2½ Nakatani C S LB 115 5.90 78-25 Double Sec117½ Decidedly Friendly1151¾ Mulvihill115½ Drifted out late 9
2Feb94- 4SA fst 7f :223 :452 1:102 1:233 Ⓢ Md 32000 55 9 8 85½ 75¾ 67 45¼ Nakatani C S LB 118 11.70 79-13 OneRichRunnr1181¾ KismtHookUs116¾ Uronurown1182¾ Widest into lane 9
16Jan94- 4SA fst 6½f :22 :444 1:102 1:17 Ⓢ Md 32000 59 3 7 64¼ 67½ 77 57¼ Gryder A T LB 118 7.50 80-09 Heezfor Gramps1183 Mulvihill1181¾ Lightning N'ice118¾ No rally 9
3Dec93- 6Hol fst 6f :22 :452 :581 1:112 Ⓢ Md 32000 49 8 3 83 56¼ 68½ 67¼ Gryder A T LB 119 *1.90 76-15 Go To Bat119no Tam Rise1191¼ Win The Case1171¾ Wide, bumped 3/8 9
17Nov93- 4Hol fm 1 Ⓣ :463 1:111 1:361 Md Sp Wt 28 6 6 69 75½ 913 922¾ Desormeaux K J B 118 3.50 61-14 Valiant Nature1183½ Saltgrass1183½ Unrivaled1181 Saved ground 9
Quarter time unavailable
30Oct93- 2SA fst 6f :211 :444 :581 1:112 Md 32000 66 2 12 1113 912 73¾ 2no Desormeaux K J B 118 17.30 80-12 Water Garden118no Decidedly Friendly1182¾ Cajero1182½ 12
Off slow, wide, jumped mirror image

WORKOUTS: Jly 23 Hol 7f fst 1:271 H 4/8 Jly 17 Hol 4f fst :493 H 33/47 Jly 11 Hol 5f fst 1:013 H 25/43 Jly 5 Hol 4f fst :483 H 10/35 Jun 20 Hol 4f fst :481 H 6/42 Jun 6 Hol 5f fst 1:011 H 14/37

Luckscaliber

Own: Lee Newton Jr & Wrayanna T

DESORMEAUX K J (—) $40,000

B. g. 3 (Apr)
Sire: Cutlass Reality (Cutlass)
Dam: Ready for Luck (Run of Luck)
Br: Walsh Adele & Blankman Sydney (Cal)
Tr: Hess R B Jr (3 0 0 0 .00)

L 115

Lifetime Record :	2 1 1 0	$12,750		
1994	2 1 1 0	$12,750	Turf	0 0 0 0
1993	0 M 0 0		Wet	0 0 0 0
Dmr	0 0 0 0		Dist	0 0 0 0

22Jun94- 9Hol fst 6½f :221 :452 1:102 1:164 3↑ Ⓢ Md 32000 81 10 3 73¾ 63½ 1½ 1hd Desormeaux K J LB 118 *1.00 85-12 Luckscaliber118hd Bolgability1164 Reflux1224½ Wide, driving 10
26May94- 9Hol fst 6f :22 :451 :572 1:102 3↑ Ⓢ Md 32000 80 5 2 21 21½ 11½ 21 Desormeaux K J LB 116 3.40 87-13 Flight Of Majesty1221 Luckscaliber1164½ Chained1143½ Overtaken late 11

WORKOUTS: Jly 25 Dmr 4f fst :501 H 18/22 Jly 18 Hol 7f fst 1:292 H 6/6 Jly 12 Hol 4f fst :501 H 24/31 Jly 7 Hol 3f fst :38 H 18/23 Jun 10 Hol 4f fst :491 H 7/24 May 21 Hol 4f fst :49 H 26/42

Count Con

Own: Hawn W R

ATKINSON P (1 0 0 0 .00) $40,000

Dk. b or br g. 3 (May)
Sire: Fast Account (Private Account)
Dam: Con's Sister (Summer Time Guy)
Br: Hawn W R (Cal)
Tr: MacDonald Mark (—)

L 115

Lifetime Record :	7 1 1 1	$22,650			
1994	5 1 1 1	$22,650	Turf	1 0 0 0	
1993	2 M 0 0		Wet	0 0 0 0	
Dmr	0 0 0 0		Dist	1 0 0 1	$4,200

22May94- 4Hol fm 1 Ⓣ :23 :462 1:104 1:35 Alw 40000N2L 40 1 8 810 89¾ 715 723¼ Atkinson P LB 115 36.60 66-10 Vaudeville117½ Saltgrass1192 Timbalier1193½ 8
Lugged in early, steadied twice
27Apr94- 3Hol fst 1 1/16 :24 :474 1:121 1:443 Clm 50000 78 5 4 42 2hd 1hd 2½ Atkinson P LB 116 9.70 77-18 Veracity116½ Count Con1164 Baja Bill1161½ 4 wide 2nd turn 5
27Feb94- 3SA fst 1 :222 :463 1:112 1:373 Clm 32000 63 1 7 64½ 52½ 54 36½ Atkinson P LB 115 6.60 74-18 Set On Cruise1151½ Gary's Lucky One1145 Count Con115nk 7
Lacked room 5/16, altered path
10Feb94- 4SA fst 1 1/16 :231 :473 1:123 1:451 Md 32000 72 8 9 97¾ 75½ 32½ 11½ Atkinson P LB 117 74.00 69-27 Count Con1171½ Apurado1151 Pacific Edition1172 9
Broke very slow, bumped hard, circled field, lugged in 1/16
28Jan94- 2SA fst 6f :213 :444 :571 1:101 Md 32000 57 4 11 1113 1112 913 79 Esposito M LB 119 95.40 77-12 Heartless Wager1184½ Double Sec1181¾ Bobs Bikini118hd 11
Bore in, steadied start
4Dec93- 4Hol fst 1 1/16 :233 :472 1:121 1:45 Md 32000 33 8 3 31 42½ 710 720 Gryder A T LB 119 b 17.60 56-15 Sierra Real1141 My Ideal1192 Stage Door Tyler1191 Wide both turns 8
15Nov93- 6SA fst 7f :232 :47 1:114 1:243 Ⓢ Md 32000 23 2 10 1010 1010 917 818½ Antley C W B 117 15.60 61-15 He's Fabulous1171¾ Win The Case115hd Al's Sunshine1173½ Off slow, green 10

WORKOUTS: Jly 27 Dmr 5f fst 1:013 H 27/53 Jly 20 SA 1 fst 1:433 H 2/2 Jly 13 SA 6f fst 1:134 H 2/9 Jly 7 SA 4f fst :484 H 10/20 Jun 4 SA 4f fst :494 H 13/34 May 19 SA 4f gd :483 H 4/25

Mark Ninety

Own: Manzani Ronald

PINCAY L JR (4 0 0 1 .00) $40,000

Gr. g. 3 (Mar)
Sire: Verbatim (Speak John)
Dam: Grey Prism (Grey Dawn II)
Br: Elmendorf Farm Inc (Ky)
Tr: Spawr Bill (1 0 0 0 .00)

L 115

Lifetime Record :	12 2 2 1	$39,399		
1994	5 1 0 1	$15,550	Turf	1 0 0 0
1993	7 1 2 0	$23,849	Wet	0 0 0 0
Dmr	2 1 1 0	$14,950	Dist	0 0 0 0

3Jly94- 3Hol fst 1 1/16 :231 :461 1:102 1:432 Clm 16000 81 5 1 11 11½ 12 13½ Pincay L Jr LB 117 b 3.40 84-19 Mark Ninety1173½ Kismet Hook Us1165½ Brand New Dance1162¾ Driving 7
4May94- 3Hol fst 6f :214 :453 :58 1:111 Clm c-20000 27 5 1 1hd 2hd 47 415½ McCarron C J LB 116 b 2.60 68-09 Smooth Shot116½ Tajo116hd Heezfor Gramps11615 5
Veered out start, drifted out lane Claimed from Cooke Jack Kent, Smithwick Daniel M Jr Trainer
20Apr94- 5SA fst 6f :213 :45 :573 1:10 Clm 40000 52 6 4 73½ 96½ 109½ 915 McCarron C J LB 115 b 21.10 72-17 Going To The Fair1165 Mamystere116hd Smoke In Motion115½ No rally 10
7Apr94- 5SA fst 6f :211 :44 :562 1:093 Clm 50000 67 5 6 610 67½ 67 75¼ Gonzalez S5 LB 110 b 22.80 83-11 Cold Cool N'bold1151¾ Mr Peter P.115hd Sixat The Bay1152¾ Wide into lane 9
27Feb94- 6SA fst 6f :212 :442 :564 1:094 Clm 45000 74 4 3 2hd 1hd 1½ 33¾ Gonzalez S5 LB 109 b 24.20 84-10 Smoke InMotion1152 ColdCoolN'bold1161¾ MarkNinety109¾ Outfinished 7
10Dec93- 2Hol fm 1 1/16 Ⓣ :234 :481 1:133 1:444 Alw 32000N1X — 5 1 2½ 78 723 — Black C A LB 116 17.90 — 27 Duca1191½ Daggett Peak1194 Sundays Cream1171 Stopped, eased 7
21Oct93- 7SA fst 6f :213 :45 :572 1:094 Alw 32000N1X 56 6 7 31 3½ 75 912½ McCarron C J LB 115 b 14.20 75-14 Blumin Affair1182½ Bullamatic1152 Just A Roman118¾ 4 Wide into lane 10
4Oct93-12Fpx fst 6f :214 :452 :583 1:123 Fairplex BrC36k 66 4 6 55 44½ 44½ 21½ Warren R J Jr L 120 b 5.10 82-10 Top Success1201½ Mark Ninety120hd Mi Profe120nk Good effort 6
27Aug93- 6Dmr fst 5½f :22 :452 :573 1:041 Md 45000 77 4 1 2hd 1½ 11 12 McCarron C J LB 115 b *1.60 90-11 Mark Ninety1152 Prevasive Force1171½ Ninety Knots1171 Driving 8
6Aug93- 6Dmr fst 6f :221 :452 :58 1:112 Md 32000 69 2 2 11 14 12 2½ Warren R J Jr B 117 b 8.50 81-15 Baja Bill117½ Mark Ninety117½ Cologne1173½ Set pace 10

WORKOUTS: Jly 11 Hol 4f fst :491 H 25/47 ● Jun 27 Hol 7f fst 1:252 H 1/4 Jun 20 Hol 7f fst 1:271 H 5/8 Jun 13 Hol 6f fst 1:154 H 12/13 Jun 6 Hol 5f fst 1:011 H 14/37 May 30 Hol 5f fst 1:01 H 20/56

stretchouts. On the downside, Bold Capital loses Alex Solis, is dead on the board at 26-1 and comes directly from a Maiden Claiming victory.

On the surface Luckscaliber looks more promising. He retains Kent Desormeux and is getting some action at 8-1. Still, while he closed at 6 1/2 furlongs on June 25th, he did give up a length-and-half lead the time before and might not be able to keep up with Bold Capital's faster fractions. Also Run of Luck is not nearly as stout a broodmare sire as Private Account. Checking further in *Maiden Stats 1993*, we discover that Luckscaliber's dam has a sprint-oriented dosage of 5.0 which might prevent him from getting the mile. Still, because of Cutlass Reality's SI of 2, Luckscaliber can't be confidently thrown out.

So how to bet? While it's never wise to put too much money on maiden claimers facing winners for the first time, both Bold Capital and Luckscaliber have competitive Beyers, route breeding and the speed to make them dangerous. They are worth betting to Win and putting behind Veracity in the exacta.

NINTH RACE

Del Mar

JULY 30, 1994

1 MILE. (1.33¹) CLAIMING. Purse $34,000. 3-year-olds. Weight, 120 lbs. Non-winners of two races at one mile or over since June 1, allowed 3 lbs. Of such a race since July 1, 5 lbs. Claiming price $40,000; if for $35,000, allwoed 2 lbs. (Maiden or races when entered for $32,000 or less not considered.)

Value of Race: $34,000. Winner $18,700; second $6,800; third $5,100; fourth $2,550; fifth $850. Mutuel Pool $355,928.00 Exacta Pool $288,807.00 Trifecta Pool $333,518.00 Quinella Pool $46,925.00

Last Raced	Horse	M/Eqt.	A.Wt	PP	St	¼	½	¾	Str	Fin	Jockey	Cl'g Pr	Odds $1
25Jun94 6Hol¹	Bold Capital	Bb	3 115	3	2	2¹	4¹	5²	4½	1½	Gomez G K	40000	26.20
15Jun94 5Hol⁸	Veracity	LBb	3 117	2	8	9	8½	6²	51½	21½	Delahoussaye E	40000	2.70
3Jly94 3Hol¹	Mark Ninety	LBb	3 117	9	5	4¹	31½	3¹	1hd	31½	Pincay L Jr	40000	4.70
16Jly94 5Hol¹	Falcon Bid	LB	3 115	4	3	5½	5hd	4½	3¹	4nk	Black C A	40000	3.90
4Jly94 1Hol²	Sal's Wish	LB	3 115	1	1	1½	2hd	1hd	2½	5⁴	Nakatani C S	40000	4.90
24Jun94 6Hol⁶	Decidedly Friendly	LB	3 115	6	9	8¹	9	71½	78	68	Stevens G L	40000	8.50
22Jun94 9Hol¹	Luckscaliber	LB	3 115	7	4	31½	1½	2hd	6½	75½	Desormeaux K J	40000	7.90
22May94 4Hol⁷	Count Con	LB	3 115	8	7	6hd	61½	8³	8³	84½	Atkinson P	40000	17.60
18Jun94 4GG⁹	Double Sec	Bb	3 115	5	6	7⁷	7³	9	9	9	Antley C W	40000	14.40

OFF AT 6:18 Start Good. Won driving. Time, :22², :46², 1:11, 1:23⁴, 1:36² Track fast.

$2 Mutuel Prices:

3-BOLD CAPITAL	54.40	19.60	10.20
2-VERACITY		4.80	3.00
9-MARK NINETY			4.20

$2 EXACTA 3-2 PAID $316.20 $2 TRIFECTA 3-2-9 PAID $2,200.00 $2 QUINELLA 2-3 PAID $115.40

Dk. b. or br. g, (Mar), by Bold Ruckus–Star Account, by Private Account. Trainer Gregson Edwin. Bred by Windfields Farm (Ont-C).

© 1994 DRF, Inc.

Bold Capital's $54.00 mutuel is a poignant illustration of the value of using a horse's stamina indexes, along with pace analysis, in order to cash a big winning ticket. More often than not, horses with tactical speed that are bred to stretch out will be undervalued compared to horses that have already routed and have higher known route Beyers. One of the beauties of stretchout plays is that, in contrast with maiden debut runners, lack of toteboard action doesn't necessarily lessen a horse's chance of winning. While the crowd will typically bet down the horses with the highest route speed figures, pedigree handicappers can profitably bet the 'unknown" horse whose stamina indexes suggest he will love the added distance.

8
Handicapping for the Turf

Horses that are bred for the turf and trying it for the first or second time, can be among the most profitable plays at North American tracks. Though one would think that all horses should have the breeding to run well on the grass, for cultural and historical reasons, this simply is not so.

Compared to the rest of the world, American sires have been primarily selected for their ability to get horses that can win on the dirt. To appreciate how important this fact is one has only to see that the best English and French horses are bred to win such races as the Epsom Derby and the Arc de Triomphe on the turf, while in America the Kentucky Derby and the Breeders' Cup Classic are the most coveted races.

An important consequence of this historical development is that, apart from Northern Dancer's influence, relatively few North American sires transmit turf ability to their get. This gives handicappers, familiar with who the good turf sires are, a distinct edge over the rest of the wagering public. This is especially true in Maiden, NW1 Allowance and some claiming turf races where none of the horses have as yet won on the grass. In this circumstance, horses bred for the turf and running on it for the first time often go off at attractive prices because the public generally prefers to bet the horse with the best dirt form, or the one that has the fastest *losing* time on the turf, than on a promising horse "debuting" on the grass.

An even greater source of pari-mutuel value is found in horses with good turf breeding that have been forced, by lack of opportunity, to initially labor on the dirt where they cannot excel. But, once switched to the grass, they often show a dramatic improvement which the public does not anticipate because it assumes they will continue to replicate their dismal dirt form. And the worse their dirt form looks, the higher the odds at which they will go off!

Important as the turf ratings are for evaluating the chances of horses debuting on the grass, they cannot be used without also looking closely at their trainers, workout patterns, stamina indexes and how far they are being asked to run. Additionally, in ambiguous situations, referring to *Maiden Stats* to see if a dam has produced any turf winners or referring to Tomlinson's excellent *Mudders/Turfers* can also be quite helpful. Interestingly, toteboard action is relatively unimportant with horses debuting on the grass, as long as they are bred for the surface, because high odds probably reflect previous poor dirt form. With this as background let us now turn our attention to a variety of races where horses bred for the turf are trying it for the first or second time.

Maiden Special Weight Turf Routes

The fourth race at Hollywood Park on June 8th is an example of a Maiden Special Weight event specifically written by the Racing Secretary to give horses with grass breeding an opportunity to break their maidens. Going over the field, the first thing to notice is that Fair Marryann, On The Cheek, Virgo Rising, Erin Sweeney and Golden Wreath have all tried the turf against other maidens and failed while posting modest Beyers. This gang of losers makes the race ripe for a first-time-on-the-turf horse, bred for the surface, to prevail.

Hollywood Park

4 — **1 1/16 MILES.** (Turf). (1:38⁴) MAIDEN SPECIAL WEIGHT. Purse $34,000. Fillies and mares, 3–year–olds and upward. Weights: 3–year–olds, 115 lbs. Older, 122 lbs. (Horses which have started for $32,000 or less in their last three starts have second preference.)

TURF COURSE 1 1/16 MILES START FINISH

Fair Marryan
Own: Walter Barbara A
DELAHOUSSAYE E (80 14 17 17 .18)

Ro. f. 4
Sire: Runaway Groom (Blushing Groom–Fr)
Dam: Home From the Fair (Northern Dancer)
Br: Walter Barbara A (Cal)
Tr: Stute Melvin F (46 7 7 3 .15) 122

Handwritten: E6/2 All14 NO speed

	Lifetime Record:	3 M 0 1	$5,950	
1994	2 M 0 1	$5,950	Turf	2 0 0 1	$5,950
1993	1 M 0 0		Wet	0 0 0 0	
Hol ⑦	0 0 0 0		Dist ⑦	0 0 0 0	

17Apr94–5SA fm 1 ⑦ :22² :46³ 1:11² 1:37 3+ⒻMd Sp Wt 71 8 8 7¹⁹ 79 66¾ 62½ Delahoussaye E B 123 *2.20 77–14 Crimson Steel123½ On The Cheek123² Fair Marryan123²½ Late bid 8
3Apr94–5SA fm 1 ⑦ :22⁴ :47 1:12¹ 1:37 ⒻMd Sp Wt 76 3 7 8¹⁰ 86½ 64½ 51 Solis A B 120 25.70 79–21 Sabedoria120½ Crimson Steel120ⁿᵏ Chasse120ʰᵈ Far wide into lane 8
28Mar93–6GG gd 6f :22¹ :45² :58 1:10⁴ ⒻMd Sp Wt 41 2 9 9¹⁴ 8¹⁴ 8¹³ 8¹³ Castaneda M B 117 22.30 75–13 Vedra117½ Rushin Rocket117¾ O'let It Snow117ⁿᵏ Off slowly 9
WORKOUTS: Jun 3 Hol 4f fst :48³ H 10/29 May 28 Hol 7f fst 1:30⁴ H 4/4 May 16 Hol 5f fst 1:03¹ H 40/42 May 10 Hol 4f fst :50³ H 24/34 May 4 Hol 7f fst 1:30¹ H 3/4 Apr 28 Hol 5f fst 1:01² H 34/63

Fast A Foot
Own: Kirby & Schow
MCCARRON C J (95 19 17 15 .20)

Ch. f. 3 (Apr)
Sire: Afleet (Mr. Prospector)
Dam: Reassert Yourself (Caucasus)
Br: Bud Boschert's Stables Inc (Ky)
Tr: Jackson Declan A (6 1 0 3 .17) L 115

Handwritten: C5/9 B8/3

	Lifetime Record:	3 M 0 2	$10,200	
1994	3 M 0 2	$10,200	Turf	0 0 0 0	
1993	0 M 0 0		Wet	0 0 0 0	
Hol	0 0 0 0		Dist ⑦	0 0 0 0	

21May94–3Hol fst 1⅛ :23² :46⁴ 1:11⁴ 1:44² 3+ⒻMd Sp Wt 68 4 8 75 74½ 54 35 Atherton J E5 B 110 31.60 74–18 Neeran115½ Flying Royalty116⁴ Fast A Foot110¾ 4 wide 2nd turn 8
7May94–4Hol fst 1⅛ :22⁴ :46 1:10³ 1:43¹ 3+ⒻMd Sp Wt 59 5 7 8¹⁰ 7¹³ 5¹¹ 3¹² Atherton J E5 B 110 54.00 73–09 La Frontera115⁶ Turkomine115⁶ Fast A Foot110³ Best of others 10
9Apr94–2SA fst 6f :21⁴ :45 :57² 1:09⁴ ⒻMd Sp Wt 58 8 10 11¹⁴ 10¹⁰ 10¹² 8¹² Black C A B 117 116.10 76–10 Accountable Lady117¹½ Top Rung117⁴½ Locate117²¾ Off bit slow 11
WORKOUTS: Jun 5 SA 4f fst :51² H 39/39 May 30 SA 5f fst 1:02² H 10/15 May 14 SA 4f fst :51³ H 38/38 May 4 SA 4f fst :50³ H 20/24 Apr 29 SA tr.t 3f gd :38² H 8/13 Apr 23 SA 6f fst 1:12¹ H 2/17

Pharma
Own: Paulson Allen E
BLACK C A (155 17 21 20 .11)

Handwritten: M.L. 6/8

B. f. 3 (May)
Sire: Theatrical (Nureyev)
Dam: Committed (Hagley)
Br: Allen E. Paulson (Ky)
Tr: Hassinger Alex L Jr (17 4 1 2 .24) L 115

Handwritten: A4/14 F4/14 DRM 3B ON TURF

	Lifetime Record:	3 M 0 0	$1,500	
1994	1 M 0 0	$800	Turf	0 0 0 0	
1993	2 M 0 0	$700	Wet	0 0 0 0	
Hol ⑦	0 0 0 0		Dist ⑦	0 0 0 0	

22May94–6Hol fst 6f :21⁴ :44² :56³ 1:09⁴ 3+ⒻMd Sp Wt 80 4 4 42½ 45 45½ 52½ Black C A LB 115 53.30 88–08 Redress115ⁿᵒ Track Gal115½ Strong Colors115½ Mild late bid 9
31Oct93–4SA fst 6f :21³ :44⁴ :57² 1:10 ⒻMd Sp Wt 60 6 6 52½ 52½ 52½ 6¹0½ Black C A L 117 33.20 77–12 Princess Mitterand117ʰᵈ Whatawoman117⁴ Gliding Lark117⁴½ No rally 9
17Oct93–4SA fst 6½f :22 :45¹ 1:09⁴ 1:16¹ ⒻMd Sp Wt 63 4 6 3½ 3½ 3² 58½ Valenzuela P A LB 117 9.60 82–09 Dancing Mirage112¾ Whatawoman117¾½ Devoted Danzig117½ Weakened 8
WORKOUTS: May 19 Hol 4f fst :47³ H 12/55 May 14 Hol 6f fst 1:15⁴ H 14/18 May 9 Hol 6f fst 1:12⁴ H 3/19 May 4 Hol 5f fst 1:02³ B 25/39 Apr 28 Hol 5f fst 1:01³ H 41/63 Apr 23 Hol 5f fst 1:02⁴ H 30/36

Golden Wreath
Own: Oldknow & Phipps
FLORES D R (105 9 6 10 .09)

Ch. f. 4
Sire: Majestic Light (Majestic Prince)
Dam: Golden Shore (Windy Sands)
Br: Oldknow William H & Phip Robert W (Ky)
Tr: Whittingham Charles (37 3 3 1 .08) 122

Handwritten: C4/14 C5/14

	Lifetime Record:	2 M 0 0	$2,550	
1994	2 M 0 0	$2,550	Turf	2 0 0 0	$2,550
1993	0 M 0 0		Wet	0 0 0 0	
Hol ⑦	0 0 0 0		Dist ⑦	0 0 0 0	

17Apr94–5SA fm 1 ⑦ :22² :46³ 1:11² 1:37 3+ⒻMd Sp Wt 66 1 6 67 54 35 45 Black C A B 123 3.90 75–14 Crimson Steel123½ On The Cheek123² Fair Marryan123²½ No late bid 8
3Apr94–5SA fm 1 ⑦ :22⁴ :47 1:12¹ 1:37 ⒻMd Sp Wt 69 5 8 76½ 73¾ 54½ 64 Black C A B 120 12.80 76–21 Sabedoria120½ Crimson Steel120ⁿᵏ Chasse120ʰᵈ 6 wide into lane 8
WORKOUTS: Jun 5 Hol 4f fst :48 H 15/60 May 30 Hol 5f fst 1:01⁴ H 38/56 May 25 Hol 7f fst 1:28² H 4/7 May 18 Hol 6f fst 1:14⁴ H 10/22 May 11 Hol 6f fst 1:15⁴ H 31/32 May 6 Hol 5f fst 1:02 H 32/35

continued

On The Cheek
Own: West Gary L & Mary E

PINCAY L JR (128 17 17 15 .13)

Dk. b or br f. 4
Sire: The Minstrel (Northern Dancer)
Dam: Likeashot (Gun Shot)
Br: Ranch K D (Ky)
Tr: Bradshaw Randy (14 1 3 3 .07)

L 122

C 5/3
makes Beyers

	Lifetime Record :	6 M 1 1	$15,850
1994	6 M 1 1	$15,850	Turf 2 0 1 0 $9,350
1993	0 M 0 0		Wet 1 0 0 1 $4,800
Hol 0 0 0 0			Dist 0 0 0 0

7May94–4Hol fst 1 1/16 :224 :46 1:103 1:431 3+ (F)Md Sp Wt 49 8 2 3 31 310 518 Valenzuela P A LB 123 b *2.90 67–09 La Frontera1156 Turkomine1156 Fast A Foot1103 Dueled, gave way 10
17Apr94–5SA fm 1 (T):222 :463 1:112 1:37 3+ (F)Md Sp Wt 76 6 2 21 11 2hd 21 McCarron C J LB 123 b 3.00 79–14 Crimson Steel1231 On The Cheek1232 Fair Marryan1232 Game try 8
3Apr94–5SA fm 1 (T):224 :47 1:121 1:37 (F)Md Sp Wt 76 1 1 11 11 1hd 41 Stevens G L LB 120 b 11.10 79–21 Sabedoria1201 Crimson Steel120nk Chasse120hd Game try 8
4Mar94–3SA fst 1 :223 :463 1:114 1:374 (F)Md Sp Wt 31 4 3 411 56 510 5242 Nakatani C S LB 121 6.60 55–22 Madrona121nk Sabedoria12110 Tropez12131 Forced wide 1st turn 5
4Feb94–6SA my 7f :22 :454 1:111 1:25 (F)Md Sp Wt 49 2 2 2hd 31 32 36 Nakatani C S B 119 3.30 71–16 Cindazanno11931 Ms. Ukulele11921 On The Cheek1197 Pressed pace 6
6Jan94–6SA fst 6f :214 :444 :564 1:094 (F)Md Sp Wt 54 6 4 651 66 711 712 Stevens G L B 119 16.90 76–11 Morning Showers119nk Suana11921 Really Tops1143 Stumbled start 8

WORKOUTS: ●May 31 Hol 3f fst :35 H 1/18 May 25 Hol 6f fst 1:143 H 12/26 May 19 Hol 5f fst 1:013 H 21/30 May 1 Hol 4f fst :48 H 13/57 Mar 30 Hol 4f fst :482 H 2/23 Mar 24 Hol 5f fst :592 H 3/22

Starlit Bronze
Own: Cannata Mr & Mrs Carl

ANTLEY C W (152 32 28 19 .21)

B. f. 3 (Feb)
Sire: General Assembly (Secretariat)
Dam: Five Star Kelly (Northern Prospect)
Br: Mr. & Mrs. Carl Cannata (Ky)
Tr: Spawr Bill (29 4 4 3 .14)

L 115

C 6/3
D 6/4
pgm 016

	Lifetime Record :	6 M 2 2	$23,300
1994	3 M 2 0	$14,450	Turf 0 0 0 0
1993	3 M 0 2	$8,850	Wet 1 0 1 0 $6,800
Hol 0 0 0 0			Dist 0 0 0 0

6Apr94–6SA fst 1 :224 :462 1:112 1:361 (F)Md Sp Wt 71 6 6 551 54 55 591 Pincay L Jr B 117 b 3.20 78–17 Devoted Danzig1173 Neeran1741 November Morn1173 4 wide 2nd turn 7
20Mar94–3SA my 1 1/16 :232 31 1:474 (F)Md Sp Wt 75 3 4 431 321 31 21 Pincay L Jr B 117 b *1.50 75–33 Taxquenya1171 Starlit Bronze1173 Devoted Danzig11715 Inside bid 6
19Feb94–9SA fst 1 :231 :464 1:113 1:431 (F)Md Sp Wt 83 5 5 661 66 54 22 Pincay L Jr B 117 5.70 77–15 Fair Kris1172 Starlit Bronze117nk Devoted Danzig1171 4 wide into lane 9
11Oct93–6SA fst 1 :222 :453 1:102 1:363 (F)Md Sp Wt 74 2 9 912 87 46 36 Solis A B 116 7.10 81–07 Twice The Vice1161 Besar11651 Starlit Bronze1164 Best of rest 9
11Sep93–4Dmr fst 1 :223 :444 :57 1:10 (F)Md Sp Wt 68 7 10 1014 910 58 341 Solis A B 118 83.20 84–12 Malibu Light118hd Palestrina1184 Starlit Bronze1182 10
Off slowly, inside rally
28Aug93–5Dmr fst 6f :22 :453 :58 1:102 (F)Md Sp Wt 60 11 10 107 661 56 610 Flores D R B 117 22.00 77–12 Cee's Maryanne1171 Besar11741 Mostly Overcast11711 Wide trip 11

WORKOUTS: Jun 5 Hol 5f fst 1:024 H 43/47 Jun 1 Hol 3f fst :373 H 21/35 May 27 Hol 7f fst 1:302 H 7/9 May 12 Hol 6f fst 1:152 H 24/25 Apr 25 SA 4f fst :494 H 27/37 Apr 18 SA 4f fst :514 H 30/32

Virgo's Rising
Own: Lee & Scholz

VALENZUELA P A (129 9 17 16 .07)

Ch. f. 3 (Apr)
Sire: Alysheba (Alydar)
Dam: Main Prospect (Mr. Prospector)
Br: Singer Craig (Ky)
Tr: Van Berg Jack C (94 6 5 13 .06)

L 115

C 4/1
O 1/4

	Lifetime Record :	8 M 1 2	$17,025
1994	5 M 1 1	$9,750	Turf 2 0 0 0 $2,175
1993	3 M 0 1	$7,275	Wet 1 0 0 0 $2,550
Hol 2 0 0 0	$2,175		Dist 1 0 0 0 $2,175

26May94–1Hol fst 1 1/16 :232 :472 1:121 1:46 3+ (F)Md 40000 61 7 7 621 621 42 311 Nakatani C S LB 115 3.90 69–29 Top Clearance1171 New Bounty1151 Virgo's Rising11511 Wide trip 9
7May94–4Hol fst 1 1/16 :224 :46 1:103 1:431 3+ (F)Md Sp Wt 43 4 4 571 613 715 621 Valenzuela F H LB 115 10.20 64–09 La Frontera1156 Turkomine1156 Fast A Foot1103 Saved ground 10
20Mar94–3SA my 1 1/16 :232 :474 1:14 1:474 (F)Md Sp Wt 49 1 5 511 561 411 4161 Stevens G L LB 117 10.20 40–33 Taxquenya1171 Starlit Bronze1173 Devoted Danzig11715 No threat 6
3Mar94–9SA fst 1 :33 :472 1:13 1:382 (F)Md 40000 72 7 4 31 2hd 211 22 Stevens G L LB 117 4.20 75–21 Danza Regio1122 Virgo's Rising11711 Our Passion1173 Outfinished 9
22Jan94–6SA fst 1 :224 :461 1:104 1:361 (F)Md Sp Wt 70 5 5 641 75 77 893 Delahoussaye E LB 117 2.70 78–12 Contraindication1172 Devoted Danzig11721 Fair Kris1173 Gave way 9
29Dec93–6SA fst 1 :221 :461 1:104 1:361 (F)Md Sp Wt 74 8 6 631 421 341 341 Nakatani C S B 117 9.80 85–15 Tulgey Wood1172 Krispy1172 Virgo's Rising1173 No late gain 9
10Dec93–6Hol fm 1 1/16 (T):232 :47 1:123 1:46 (F)Md Sp Wt 63 9 4 46 321 311 431 Solis A B 119 37.40 62–27 Girls Who Care1192 Mostly Overcast11911 Jet Gray119hd Mild late bid 9
18Nov93–6Hol fm 1 (T):24 :48 1:122 1:372 (F)Md Sp Wt 50 4 10 1093 971 983 9101 Nakatani C S B 118 8.90 67–16 Lady Danz118hd Whatawoman118no Carmen Copy1183 Rough start 10

WORKOUTS: May 17 Hol 5f fst 1:011 H 10/25 Apr 25 Hol 6f fst 1:153 H 10/15 Apr 19 Hol 5f fst 1:02 H 8/20 Apr 2 Hol 4f fst :481 H 7/31 Mar 12 Hol 4f fst :474 H 5/29

⟶ ∗

Turkomine
Own: McCaffery & Toffan

GONZALEZ S JR (110 8 9 6 .07)

Ch. f. 3 (May)
Sire: Turkoman (Alydar)
Dam: Let's Get Pinned (Dewan)
Br: Brereton C. Jones (Ky)
Tr: Gonzalez J Paco (14 3 3 1 .21)

115

B 4/1
B 4/3
D AM 1/2

	Lifetime Record :	3 M 1 0	$9,200
1994	3 M 1 0	$9,200	Turf 0 0 0 0
1993	0 M 0 0		Wet 1 0 0 0 $2,400
Hol 0 0 0 0			Dist 0 0 0 0

7May94–4Hol fst 1 1/16 :224 :46 1:103 1:431 3+ (F)Md Sp Wt 70 7 1 11 111 22 211 Gonzalez S Jr B 115 6.20 79–09 La Frontera1156 Turkomine1156 Fast A Foot1103 Reluctant to load 10
9Apr94–2SA fst 6f :214 :45 :572 1:094 (F)Md Sp Wt 50 6 3 31 851 791 1015 Gonzalez S5 B 112 7.50 73–10 Accountable Lady11711 Top Rung11741 Locate11721 11
In tight 1/2, greenly
19Mar94–4SA sly 61/2f :221 :454 1:124 1:201 (F)Md Sp Wt 43 5 6 531 341 36 49 Gonzalez S5 B 112 b *1.40 62–21 Capote Peak1176 Gondwanaland1171 She's A D. A.11721 Broke in air 6

WORKOUTS: Jun 5 Hol 4f fst :492 H 40/60 May 30 Hol 5f fst 1:011 H 24/56 May 24 Hol 5f fst 1:03 H 12/19 May 1 Hol 5f fst 1:014 H 35/46 Apr 25 SA 6f fst 1:13 H 3/19 Apr 19 SA 4f fst :484 H 24/37

Erin Sweeney
Own: Stathatos Dan S

ALMEIDA G F (17 1 2 4 .06)

B. f. 4
Sire: Erins Isle (Busted)
Dam: Tara Sweeney (Tisab)
Br: Stathatos Dan (Ky)
Tr: Speckert Christopher (17 1 3 2 .06)

L 122

B 7/2
F 10/5

	Lifetime Record :	5 M 1 1	$10,650
1993	5 M 1 1	$10,650	Turf 4 0 1 1 $10,650
1992	0 M 0 0		Wet 0 0 0 0
Hol 2 0 1 0	$6,000		Dist 4 0 1 1 $10,650

25Nov93–4Hol fm 1 1/16 (T):233 :473 1:122 1:441 3+ (F)Md Sp Wt 65 4 3 31 21 22 32 McCarron C J L 118 5.10 73–25 Yerna1182 Erin Sweeney11821 Dancing Quoit1185 Forced out 7/8 8
11Nov93–6SA gd 1 1/16 (T):233 :48 1:124 1:443 (F)Md Sp Wt 54 5 6 631 681 691 691 McCarron C J LB 118 34.60 62–24 Devil's Beware1181 Dolce Amore11841 Sky Safari1202 4 Wide trip 7
4Sep93–2Dmr fm 1 1/16 (T):234 :49 1:132 1:442 (F)Md Sp Wt 65 1 3 421 32 341 341 Black C A LB 118 47.30 79–12 Portugee Starlet1181 Chasse1184 Erin Sweeney1181 Not enough late 8
21Jly93–5Hol fm 1 1/16 (T):242 :483 1:131 1:442 3+ (F)Md Sp Wt 48 4 7 841 851 883 8121 Sorenson D LB 116 139.30 61–26 WingsOfDreams1172 Noddy'sHlo122nk StrletMinister11651 4 Wide stretch 10
28May93–6GG gd 1 1/16 (T):232 :473 1:124 1:45 + 3+ (F)Md Sp Wt 44 4 6 673 761 78 711 Baze R A LB 116 b 13.70 70–14 Miz Interco1141 Go Ali117hd Sew Original11431 No threat 10

WORKOUTS: Jun 4 SA 5f fst 1:031 H 24/34 May 29 SA 6f fst 1:153 H 7/14 May 24 SA 7f fst 1:264 H 1/3 May 19 SA 7f gd 1:301 H 3/4 May 13 SA 7f fst 1:292 H 2/2 May 3 SA 6f fst 1:141 H 5/9

Blue Tess
Own: Pulliam C N

VALENZUELA F H (163 14 17 16 .09)

B. f. 3 (Mar)
Sire: Guadalupe Peak (Mr. Prospector)
Dam: Quilting Party (Confederate)
Br: C. Norman Pulliam (Cal)
Tr: Pulliam Vivian M (3 0 0 0 .00)

L 115

M 8/4

	Lifetime Record :	2 M 0 0	$0
1994	2 M 0 0		Turf 0 0 0 0
1993	0 M 0 0		Wet 0 0 0 0
Hol 0 0 0 0			Dist 0 0 0 0

14May94–6Hol fst 6f :213 :443 :564 1:091 3+ (F)SMd Sp Wt 64 11 6 783 78 691 Valenzuela F H LB 115 73.70 84–09 Snowy's Mark1151 Klassy Kim1165 Plum Flo1151 Wide to turn 11
24Apr94–9SA fst 6f :214 :444 :57 1:094 3+ (F)SMd Sp Wt 53 9 7 910 713 713 7141 McCarron C J LB 116 18.60 73–13 Two Steppin' Gal1163 Klassy Kim11621 Flying Royalty11123 No threat 11

WORKOUTS: Jun 4 SA 3f fst :351 H 3/36 May 28 SA 7f fst 1:27 H 2/4 May 22 SA 3f fst :36 H 6/21 May 10 SA 4f fst :482 H 4/22 May 4 SA 3f fst :353 B 4/23 Apr 17 SA 5f fst 1:001 H 9/43

© 1994 DRF, Inc.

Of the horses debuting on the turf, Fast A Foot, Starlit Bronze and Blue Tess can be thrown out because of their sire's mediocre ratings. This leaves us with Pharma and Turkomine, with their A and B turf ratings, as the debut horses most likely to upset the more experienced horses.

Besides Theatrical's outstanding turf rating, Pharma has several other things going for her. With her :22.1/:45.2 early fractions, she is clearly the speed of this field and, if allowed to set a slow pace, is a definite threat to steal the race for her excellent Paulson/Hassinger connections. While toteboard is less important on the turf, it is also worth noting that Pharma is the actual betting favorite at 2.8-1, which underscores the fact that the horses that have already run are vulnerable to a "first-time-on-the-turf" horse.

Pharma's primary negatives are that she is being asked to do two new things at the same time, stretch out and try the turf, and is out of a mare whose sire has an F turf rating. Looking at Theatrical's SI of 1, Pharma should have no trouble getting the distance and we already know she is bred for the surface. But Hagley as a broodmare sire is a potential problem and the best thing to do is to check *Maiden Stats 1993* to see how Pharma's dam Committed has done with her other turf runners. Doing so we discover Committed is 3-3 with her progeny on the turf which eliminates her breeding as a potential problem and, if anything, makes her as strong on the dam side as she is with her sire.

Turkomine with her B rating and 6-1 odds, however, must also be considered. While she has a lower sire rating than Pharma she does have the benefit of already having run a route and improving Beyers. But, on the downside, Turkomine has never passed another horse, tired in all her previous races, and is unlikely (with her :22.4/:46 early fractions) to get the lead from Pharma. Furthermore, trainer Paco Gonzales isn't noted for his prowess on the turf.

So how to bet? While Pharma was 53-1 in her last race, today she has been bet down to less than 3-1. Given all of her positives, she should be 8-5 and is worth a prime bet. If Turkomine is not for real, second place becomes a crapshoot, so the exacta is out. Best thing to do is play Pharma to Win and Place and box her with Turkomine as a secondary bet.

Results: Allowed to set leisurely early splits of :23/:47 Pharma easily wired the field and paid $7.60 to Win and $4.40 to Place. Starlit Bronze closed from off the pace for second with the $2 exacta returning $38.80. Long shot Blue Tess completed the trifecta which paid $678.60. Pharma's victory in this spot shows how powerful the combination of good breeding and early speed can be. Turkomine the other horse with good turf breeding made a bold move but then tired and finished eighth. Since some turf horses need a start on the surface before winning, Turkomine is worth serious consideration her next time out.

Hollywood Park

1

1 1/16 MILES. **(Turf, chute). (1:38⁴) MAIDEN SPECIAL WEIGHT. Purse $34,000. Fillies and mares, 3–year–olds and upward. Weights: 3–year–olds, 116 lbs. Older, 122 lbs. (Horses than have started for $32,000 or less in their last three starters will have second preference.)**

TURF COURSE — 1 1/16 MILES — START — FINISH

On The Cheek

Own: West Gary L & Mary E

FLORES D R (178 16 15 14 .09)

Dk. b or br f. 4
Sire: The Minstrel (Northern Dancer)
Dam: Likeashot (Gun Shot)
Br: Ranch K D (Ky)
Tr: Bradshaw Randy (32 3 3 5 .09)

L 122

Lifetime Record :	7 M 1 1	$15,850			
1994	7 M 1 1	$15,850	Turf	3 0 1 0	$9,350
1993	0 M 0 0		Wet	1 0 0 1	$4,800
Hol Ⓣ	1 0 0 0		Dist Ⓣ	1 0 0 0	

8Jun94-4Hol	fm	1 1/16 Ⓣ	:23	:47¹	1:11³	1:42¹	3+ ⒻMd Sp Wt	61	3 2	2¹	2²	3⁴½	713½	Pincay L Jr	LB 122 b	3.90	71–13	Pharma1164½ Starlit Bronze1154½ Blue Tess1151		Weakened	9
7May94-4Hol	fst	1 1/16	:22⁴	:46	1:10³	1:43¹	3+ ⒻMd Sp Wt	49	8 2	2½	3¹½	3¹⁰	5¹⁸	Valenzuela P A	LB 123 b	*2.90	67–09	La Frontera1156 Turkomine1156 Fast A Foot1103		Dueled, gave way	10
17Apr94-5SA	fm	1 Ⓣ	:22²	:46³	1:11²	1:37	3+ ⒻMd Sp Wt	76	6 2	2½	1¹½	2hd	2½	McCarron C J	LB 123 b	3.00	79–14	Crimson Steel123½ On The Cheek1232 Fair Marryan1232½		Game try	8
3Apr94-5SA	fm	1 Ⓣ	:22⁴	:47	1:12¹	1:37	ⒻMd Sp Wt	76	1 1	1½	1¹½	1hd	4¹	Stevens G L	LB 120 b	11.10	79–21	Sabedoria120½ Crimson Steel120nk Chasse120hd		Game try	8
4Mar94-3SA	fst	1	:22³	:46³	1:11⁴	1:37⁴	ⒻMd Sp Wt	31	4 3	4¹½	5⁶	5¹⁰	524¾	Nakatani C S	LB 121	6.60	55–22	Madrona121nk Sabedoria12110 Tropez1213½		Forced wide 1st turn	5
4Feb94-6SA	my	7f	:22⁴	:45⁴	1:11¹	1:25	ⒻMd Sp Wt	49	2 2	2hd	3¹	3²	3⁶	Nakatani C S	B 119	3.30	71–16	Cindazanno1193½ Ms. Ukulele1192½ On The Cheek1197		Pressed pace	4
6Jan94-6SA	fst	6f	:21⁴	:44⁴	:56⁴	1:09⁴	ⒻMd Sp Wt	54	6 4	6⁵½	6⁶	7¹¹	7¹²	Stevens G L	B 119	16.90	76–11	Morning Showers119nk Suana1192½ Really Tops1143½		Stumbled start	8

WORKOUTS: Jly 13 Hol 4f fst :47¹ H *2/51* Jly 7 Hol 6f fst 1:14¹ H *13/25* Jun 30 Hol 5f fst 1:01¹ H *12/41* Jun 21 Hol 4f fst :50³ H *24/25* ●May 31 Hol 3f fst :35 H *1/18* May 25 Hol 6f fst 1:14³ H *12/26*

It's Ben Freezing

Own: Tpl Dot Dsh Stbl Trst & Vandervoort

DESORMEAUX K J (227 52 40 36 .23)

Ch. f. 3 (May)
Sire: It's Freezing (T. V. Commercial)
Dam: Benzina (Jaazeiro)
Br: Joyce Vandervoort (Ky)
Tr: Vienna Darrell (49 6 5 9 .12)

116

Lifetime Record :	3 M 0 0	$2,575			
1994	1 M 0 0	$550	Turf	0 0 0 0	
1993	2 M 0 0	$2,025	Wet	0 0 0 0	
Hol Ⓣ	0 0 0 0		Dist Ⓣ	0 0 0 0	

28May94-10Hol	fst	6f	:22	:45²	:57⁴	1:10²	3+ ⒻMd 50000	62	2 4	2½	1hd	3¹½	5⁷½	Valenzuela P A	B 117	18.40	81–09	Flying Ellie116½ Gliding Lark1154½ Tequila Rose1151½	Inside duel	9
19Dec93-4Hol	fst	5½f	:22¹	:45²	:57⁴	1:04	ⒻMd Sp Wt	63	8 4	4⁵	4⁵½	4⁴¼	4⁴	Stevens G L	B 118	11.30	88–13	Wild Impulse118½ Espadrille1182¾ Fresh Berries118¾	Mild bid	8
24Oct93-4SA	fst	6f	:21⁴	:45	:57¹	1:09¹	ⒻMd Sp Wt	48	1 6	5¹¾	7⁵¾	7⁸¼	7¹⁷	McCarron C J	B 117	35.70	74–12	Lakeway117nk Madder ThanMad1172 DesertStormette1177½	Checked 5/16	7

WORKOUTS: Jly 11 SA 5f fst 1:03¹ H *21/26* Jly 6 SA 6f fst 1:16³ H *9/13* Jly 1 SA 5f fst 1:01¹ H *10/29* Jun 25 SA 5f fst 1:02² H *18/24* Jun 19 SA 6f fst 1:14² H *13/16* Jun 13 SA 5f fst 1:03 H *15/23*

Erin Sweeney

Own: Stathatos Dan S

PEDROZA M A (280 26 46 37 .09)

B. f. 4
Sire: Erins Isle (Busted)
Dam: Tara Sweeney (Tisab)
Br: Stathatos Dan (Ky)
Tr: Speckert Christopher (27 1 4 2 .04)

L 122

Lifetime Record :	6 M 1 1	$10,650			
1994	1 M 0 0		Turf	5 0 1 1	$10,650
1993	5 M 1 1	$10,650	Wet	0 0 0 0	
Hol Ⓣ	3 0 1 0	$6,000	Dist Ⓣ	5 0 1 1	$10,650

8Jun94-4Hol	fm	1 1/16 Ⓣ	:23	:47¹	1:11³	1:42¹	3+ ⒻMd Sp Wt	62	7 4	5³½	6⁴½	6⁷	6¹³¼	Almeida G F	LB 122	30.90	72–13	Pharma1164½ Starlit Bronze1154½ Blue Tess1151	Weakened	9
25Nov93-4Hol	fm	1 1/16 Ⓣ	:23³	:47³	1:12²	1:44¹	3+ ⒻMd Sp Wt	65	4 3	3¹	2¹	2¹	2²	McCarron C J	L 118	5.10	73–25	Yerna1182 Erin Sweeney1182½ Dancing Quoit1185	Forced out 7/8	8
11Nov93-6SA	gd	1 1/16 Ⓣ	:23³	:48	1:12⁴	1:44³	3+ ⒻMd Sp Wt	54	6 5	5²	6³½	6⁸½	6¹⁹¼	McCarron C J	LB 118	34.60	62–24	Devil's Beware1187 Dolce Amore1184½ Sky Safari1202	4 Wide trip	7
4Sep93-2Dmr	fm	1 1/16 Ⓣ	:23⁴	:49	1:13²	1:44²	3+ ⒻMd Sp Wt	65	1 3	4²½	3²	3⁴½	3⁴½	Black C A	LB 118	47.30	79–12	Portugese Starlet118½ Chasse1184 Erin Sweeney118½	Not enough late	8
21Jly93-5Hol	fm	1 1/16 Ⓣ	:24²	:48³	1:13¹	1:44²	3+ ⒻMd Sp Wt	48	4 7	8⁴½	8⁵½	8⁸¾	8¹²½	Sorenson D	LB 116	139.30	62–26	WingsOfDreams117½ Noddy'sHlo122nk StrletMinister1161¾	4 Wide stretch	10
28May93-6GG	gd	1 1/16 Ⓣ	:23²	:47³	1:12⁴	1:45	+ 3+ ⒻMd Sp Wt	44	4 6	6⁷¾	7⁶½	7⁸	7¹¹	Baze R A	LB 116 b	13.70	70–14	Miz Interco114¾ Go Ali117hd Sew Original1143½	No threat	10

WORKOUTS: Jly 14 SA 3f fst :39¹ H *30/30* Jly 10 SA 5f fst 1:01³ H *16/33* Jly 4 SA 1 fst 1:42³ H *2/2* Jun 27 SA 6f fst 1:16³ H *3/4* Jun 16 SA 4f fst :52 H *24/24* Jun 4 SA 5f fst 1:03¹ H *24/34*

Medinilla

Own: Paulson Allen E

BLACK C A (267 28 32 34 .10)

B. f. 3 (Apr)
Sire: Seattle Slew (Bold Reasoning)
Dam: Eastland (Exceller)
Br: Paulson Allen E (Ky)
Tr: Hassinger Alex L Jr (34 10 5 4 .29)

L 116

Lifetime Record :	0 M 0 0	$0			
1994	0 M 0 0		Turf	0 0 0 0	
1993	0 M 0 0		Wet	0 0 0 0	
Hol Ⓣ	0 0 0 0		Dist Ⓣ	0 0 0 0	

WORKOUTS: Jly 14 Hol 3f fst :36 Hg*6/23* Jly 9 Hol 7f fst 1:27³ H *4/5* Jly 4 Hol 6f fst 1:15⁴ H *11/14* Jun 22 Hol 6f fst 1:14³ H *7/21* Jun 16 Hol 6f fst 1:14 H *7/20* Jun 11 Hol 3f fst :36³ Hg *13/27* Jun 6 Hol 6f fst 1:16³ H *15/16* Jun 1 Hol 4f fst :48⁴ H *18/39* May 23 Hol 5f fst 1:03³ H *42/47* May 19 Hol 4f fst :50² H *45/55* May 9 Hol 6f fst 1:15⁴ H *17/19* May 4 Hol 6f fst 1:15³ H *21/26*

Crissy Aya

Own: Doi Mutsuaki

NAKATANI C S (137 24 28 24 .18)

Dk. b or br f. 3 (May)
Sire: Saros (Sassafras)
Dam: Iza Valentine (Bicker)
Br: Green Thumb Farm (Cal)
Tr: Sahadi Jenine (38 6 5 9 .16)

L 116

Lifetime Record :	2 M 0 1	$6,125			
1994	2 M 0 1	$6,125	Turf	0 0 0 0	
1993	0 M 0 0		Wet	0 0 0 0	
Hol Ⓣ	0 0 0 0		Dist Ⓣ	0 0 0 0	

18Jun94-3Hol	fst	6½f	:21⁴	:44¹	1:09	1:15⁴	3+ ⒻⓈMd Sp Wt	52	6 1	2½	4¹½	5¹³	5²¹¾	Pedroza M A	B 116	4.90	68–12	Persistant Sal1162¾ You'renotlistening1164½ Kelkyko11614	Gave way	8
29Apr94-1Hol	fst	6f	:22³	:46¹	:58¹	1:10⁴	3+ ⒻⓈMd Sp Wt	58	2 5	3²	3³	3⁸	3⁹½	Black C A	B 115	*1.10	76–13	Masterful Dawn1154½ Gilded Lace1155 Crissy Aya1151	Weakened	5

WORKOUTS: Jly 14 Hol 3f fst :38² H *18/23* Jly 8 Hol 5f fst 1:01⁴ H *13/40* Jly 2 Hol 5f fst :59⁴ H *5/40* Jun 26 Hol 3f fst :37² H *31/46* Jun 8 Hol 5f fst 1:01 H *11/30* Jun 2 Hol 7f fst 1:27¹ H *6/11*

Lady Lela

Own: Moss Mr & Mrs J S

STEVENS G L (301 58 38 60 .19)

Dk. b or br f. 3 (Feb)
Sire: Silver Hawk (Roberto)
Dam: Indian Maiden (Chieftain)
Br: Moss Mr & Mrs J S (Ky)
Tr: Mandella Richard (66 11 11 13 .17)

116

Lifetime Record :	2 M 0 0	$2,550			
1994	2 M 0 0	$2,550	Turf	0 0 0 0	
1993	0 M 0 0		Wet	0 0 0 0	
Hol Ⓣ	0 0 0 0		Dist Ⓣ	0 0 0 0	

18Jun94-9Hol	fst	1 1/16	:22³	:45²	1:10²	1:45¹	3+ ⒻMd Sp Wt	77	3 7	8¹²	9⁹¾	8⁷½	4¹³½	Stevens G L	115	23.00	73–18	Et Voila1151½ Step In Toe115½ Katy's Lady115no	Came in late	9
4Jun94-1Hol	fst	6½f	:21⁴	:44²	1:09¹	1:15³	3+ ⒻMd Sp Wt	57	6 7	8¹⁰	8¹³	8¹⁶	7¹⁶	Atkinson P	116	11.50e	75–07	Cathy's Dynasty1162¾ Desert Stormette1163 Ladies Cruise1164½	Outrun	8

WORKOUTS: Jly 11 Hol 5f fst 1:01¹ H *21/43* Jly 6 Hol 6f fst 1:17² H *20/21* Jun 14 Hol 5f fst 1:03⁴ H *26/27* May 29 Hol 5f fst 1:01² Hg*31/46* May 23 Hol 6f fst 1:13 H *6/24* May 18 Hol 6f fst 1:14³ H *7/22*

Gliding Lark

Own: Roncari & Team Valor & Tsujimoto

SOLIS A (379 74 44 56 .20)

Ch. f. 3 (Mar)
Sire: Woodman (Mr. Prospector)
Dam: Eighty Lady (Flying Lark)
Br: Robert S West Jr & London Throughbrd Services Ltd. (Ky)
Tr: Hennig Mark (20 1 3 5 .05)

L 116

Lifetime Record :	8 M 1 2	$17,100		
1994	4 M 1 1	$9,000	Turf	0 0 0 0
1993	4 M 0 1	$8,100	Wet	0 0 0 0
Hol Ⓣ	0 0 0 0		Dist Ⓣ	0 0 0 0

28May94-10Hol	fst	6f	:22 :452 :574 1:102	3↑ ⒻMd 50000	80	1	8	811	75¾	41¾	2½	Stevens G L	LB 115	4.70	87–09	Flying Ellie116½ Gliding Lark1154½ Tequila Rose1151½	Closed gamely	9
1May94-2Hol	fst	7f	:221 :452 1:101 1:223	3↑ ⒻMd 62500	74	5	6	711	76	32½	34	Stevens G L	LB 115	4.40	87–08	Music St. Melody1153½ Pala Canyon115½ Gliding Lark1151¾	5 wide turn	8
19Feb94-6SA	fst	1 1/16	:231 :464 1:113 1:431	ⒻMd Sp Wt	65	9	6	55½	55	67	712½	Stevens G L	LB 117	9.80	67–15	Fair Kris1172 Starlit Bronze117nk Devoted Danzig1171½	5 wide into lane	9
22Jan94-6SA	fst	1	:224 :461 1:104 1:361	ⒻMd Sp Wt	76	9	4	31½	3½	43	56½	Black C A	LB 117	7.30	82–12	Contraindication117¾ Devoted Danzig1172½ Fair Kris1173	Weakened	9
26Dec93-2SA	fst	6f	:214 :444 :57 1:093	ⒻMd Sp Wt	75	2	8	106	85¾	66	45½	Stevens G L	LB 117	10.00	83–09	Choice Claim1172 Meadow Moon117½ Whatawoman1173	Late bid	12
26Nov93-9Hol	fst	6½f	:213 :442 1:093 1:16	ⒻMd Sp Wt	70	2	9	79½	67½	64½	66	Valenzuela F H	B 118	5.60	83–09	Jacodra's Devil1132¾ Sliced Twice1181¼ DevotedDanzig118no	Saved ground	10
31Oct93-4SA	fst	6f	:213 :444 :572 1:10	ⒻMd Sp Wt	76	1	8	42½	41¾	42	34	Valenzuela F H	B 117	9.30	83–12	Princess Mitterand117hd Whatawoman1174 Gliding Lark1174½	No late bid	9
10Oct93-5Bel	fst	6f	:223 :463 :591 1:121	ⒻMd Sp Wt	53	7	8	3½	3nk	43	46	Davis R G	117	3.50	73–20	Aly's Conquest117nk Annie Bonnie1174½ Vibelle1171½	Dueled, tired	10

WORKOUTS: Jly 12 Hol 4f fst :502 H 27/31 Jly 6 Hol 5f fst 1:032 H 42/44 Jun 30 Hol 5f fst 1:011 H 12/41 Jun 24 Hol 4f fst :494 H 11/32 Jun 18 Hol 4f fst :491 H 27/41 Jun 11 Hol 4f fst :484 H 31/60

Nijmis

Own: Cooke Jack Kent

ARIAS J C (10 0 0 0 .00)

B. f. 3 (Feb)
Sire: Nijinsky II (Northern Dancer)
Dam: Foreign Missile (Damascus)
Br: Brushwood Stable (Pa)
Tr: Smithwick Daniel M Jr (8 0 1 1 .00)

L 116

Lifetime Record :	2 M 0 0	$0		
1993	2 M 0 0		Turf	1 0 0 0
1992	0 M 0 0		Wet	0 0 0 0
Hol Ⓣ	1 0 0 0		Dist Ⓣ	0 0 0 0

18Nov93-6Hol	fm	1 Ⓣ	:24 :48 1:122 1:372	ⒻMd Sp Wt	55	10	6	63½	52¾	86½	88	Almeida G F	LB 118	57.50	70–16	Lady Danz118hd Whatawoman118no Carmen Copy1183	Drifted wide 1/4	10
17Oct93-4SA	fst	6½f	:22 :451 1:094 1:161	ⒻMd Sp Wt	44	6	8	88½	88¾	77¾	716¼	Black C A	B 117	40.50	74–09	Dancing Mirage1123 Whatawoman1173½ Devoted Danzig117½	Outrun	8

WORKOUTS: Jly 9 SA 7f fst 1:26 H 2/4 Jly 2 SA 7f fst 1:29 H 7/10 Jun 25 SA 7f fst 1:283 H 2/2 Jun 4 SA 5f fst 1:02 H 12/34 May 29 SA 4f fst :49 H 16/44 May 23 SA 4f fst :504 H 19/21

Castle Gardens (Ire)

Own: Bloomer Robert L

MCCARRON C J (189 36 34 31 .19)

B. f. 3 (Mar)
Sire: Common Grounds (Kris)
Dam: Very Sophisticated (Affirmed)
Br: Leslie Mellon (Ire)
Tr: Cross Richard J (15 1 2 5 .07)

116

Lifetime Record :	1 M 1 0	$1,490			
1993	1 M 1 0	$1,490	Turf	1 0 1 0	$1,490
1992	0 M 0 0		Wet	0 0 0 0	
Hol Ⓣ	0 0 0 0		Dist Ⓣ	0 0 0 0	

16Oct93♦ Naas(Ire)	yl	6f Ⓣ LH	1:134	ⒻTifrums EBF Maiden	2½	Hogan D	122	10.00	Market Slide122½ Castle Gardens122hd Masawa122hd		21
Tr: Michael Grassick				Maiden 9300					Close up, dueled for lead 2f out, headed 1f out, gamely		

WORKOUTS: Jly 7 SA 6f fst 1:172 H 10/11 Jly 2 SA 7f fst 1:271 H 3/10 Jun 26 SA 7f fst 1:27 H 4/8 Jun 20 SA 6f fst 1:13 H 2/13 Jun 14 SA 5f fst 1:02 H 16/29 ●Jun 8 SA 5f fst 1:001 H 1/31

Starlit Bronze

Own: Cannata Mr & Mrs Carl

*PINCAY L JR (247 33 30 37 .13)

B. f. 3 (Feb)
Sire: General Assembly (Secretariat)
Dam: Five Star Kelly (Northern Prospect)
Br: Mr. & Mrs. Carl Cannata (Ky)
Tr: Spawr Bill (65 12 11 9 .18)

L 116

Lifetime Record :	7 M 3 2	$30,100			
1994	4 M 3 0	$21,250	Turf	1 0 1 0	$6,800
1993	3 M 0 2	$8,850	Wet	1 0 1 0	$6,800
Hol Ⓣ	1 0 1 0	$6,800	Dist Ⓣ	1 0 1 0	$6,800

8Jun94-4Hol	fm	1 1/16 Ⓣ	:23 :471 1:113 1:421	3↑ ⒻMd Sp Wt	81	4	7	64½	54½	22½	24½	Antley C W	LB 115 b	4.80	80–13	Pharma1164½ Starlit Bronze1154½ Blue Tess1151	Second best	9
6Apr94-6SA	fst	1	:224 :462 1:112 1:361	ⒻMd Sp Wt	71	6	6	55½	54	55	59½	Pincay L Jr	B 117 b	3.20	78–17	Devoted Danzig1173½ Neeran1174½ November Morn117¾	4 wide 2nd turn	7
20Mar94-3SA	my	1 1/16	:232 :474 1:14 1:474	ⒻMd Sp Wt	75	3	4	43½	32½	31	2½	Pincay L Jr	B 117 b	*1.50	55–33	Taxquenya117½ Starlit Bronze117¾ Devoted Danzig11715	Inside bid	6
19Feb94-6SA	fst	1 1/16	:231 :464 1:113 1:431	ⒻMd Sp Wt	83	5	5	66½	66	54	22	Pincay L Jr	B 117	5.70	77–15	Fair Kris1172 Starlit Bronze117nk Devoted Danzig1171½	4 wide into lane	9
11Oct93-6SA	fst	1	:222 :453 1:102 1:363	ⒻMd Sp Wt	74	2	9	912	87	46	36	Solis A	B 116	7.10	81–07	Twice The Vice116½ Besar1165½ Starlit Bronze1164½	Best of rest	9
11Sep93-4Dmr	fst	6f	:213 :444 :57 1:10	ⒻMd Sp Wt	68	7	10	1014	910	58	34½	Solis A	B 118	83.20	84–12	Malibu Light118hd Palestrina1184½ Starlit Bronze1182		10
Off slowly, inside rally																		
28Aug93-5Dmr	fst	6f	:22 :453 :58 1:102	ⒻMd Sp Wt	60	11	10	107	66½	56	610	Flores D R	B 117	22.00	77–12	Cee's Maryanne117½ Besar1174½ Mostly Overcast1171¼	Wide trip	12

WORKOUTS: Jly 14 Hol 4f fst :512 B 54/61 Jly 10 Hol 7f fst 1:313 H 5/5 Jun 30 Hol 5f fst 1:011 H 12/41 Jun 26 Hol 4f fst :512 H 42/49 Jun 5 Hol 5f fst 1:024 H 43/47 Jun 1 Hol 3f fst :373 H 21/35

Cloudy Line

Own: Pine Creek Ranch

ATKINSON P (144 11 18 13 .08)

Gr. f. 3 (Apr)
Sire: Skywalker (Relaunch)
Dam: Baby Duck (Quack)
Br: Cardiff Stud Farm (Cal)
Tr: Stute Melvin F (86 12 16 9 .14)

116

Lifetime Record :	4 M 0 1	$2,850		
1993	4 M 0 1	$2,850	Turf	0 0 0 0
1992	0 M 0 0		Wet	0 0 0 0
Hol Ⓣ	0 0 0 0		Dist Ⓣ	0 0 0 0

2Sep93-2Dmr	fst	1	:222 :463 1:12 1:38	ⒻMd 40000	50	7	8	86¾	66½	55½	38½	Atkinson P	B 118	12.40	71–18	BurningDesire1163 AvSingstheblues1135½ CloudyLin118no	5 Wide stretch	8
8Aug93-6Dmr	fst	6½f	:223 :46 1:103 1:17	ⒻMd Sp Wt	51	5	7	72¾	62½	96½	811¼	Solis A	B 117	87.50	75–11	Top Of The Sky117nk Tin117¾ Malibu Light117no	Outrun	11
11Jly93-6Hol	fst	5½f	:22 :451 :573 1:041	ⒻⓈMd Sp Wt	36	9	10	1113	1114	813	717	Black C A	B 117	33.90	74–12	Choobloo1178½ Top Of The Sky117no Masterful Dawn1171¾	Wide, greenly	12
12Jun93-4Hol	fst	5f	:214 :453 :59	ⒻⓈMd Sp Wt	29	5	5	67	810	812	611¼	Torres H	B 117	8.50e	78–15	SpiritualPath117½ SeToShiningSe1172 FlyingPresence1121¼	Took up 1/4	8

WORKOUTS: Jly 14 Hol 4f fst :471 H 5/61 Jly 8 Hol 7f fst 1:30 H 10/11 Jly 2 Hol 4f fst :47 H 3/48 Jun 27 Hol 1 fst 1:40 H 1/1 Jun 21 Hol 7f fst 1:291 H 4/5 Jun 15 Hol 5f fst 1:021 Hg27/45

Designatoree

Own: McCaffery & Toffan

VALENZUELA F H (282 21 30 33 .07)

B. f. 3 (Apr)
Sire: Alysheba (Alydar)
Dam: Never Scheme (Never Bend)
Br: Hugh G. King & Dorothy Scharbauer (Ky)
Tr: Gonzalez J Paco (25 4 5 3 .16)

116

Lifetime Record :	6 M 0 1	$7,650		
1994	5 M 0 1	$7,650	Turf	0 0 0 0
1993	1 M 0 0		Wet	0 0 0 0
Hol Ⓣ	0 0 0 0		Dist Ⓣ	0 0 0 0

18Jun94-9Hol	fst	1 1/16	:223 :452 1:102 1:451	3↑ ⒻMd Sp Wt	68	9	6	66½	65¾	56½	77	Pincay L Jr	B 117	9.20	68–18	Et Voila1151½ Step In Toe115½ Katy's Lady115no	Drifted in lane	9
21May94-3Hol	fst	1 1/16	:232 :464 1:114 1:442	3↑ ⒻMd Sp Wt	67	6	6	52½	42	43½	45¾	Gonzalez S Jr	B 115	13.70	73–18	Neeran115½ Flying Royalty1164½ Fast A Foot110¾	Bid, weakened	8
6Apr94-6SA	fst	1	:224 :462 1:112 1:361	ⒻMd Sp Wt	53	1	3	34	64½	613	619½	Gonzalez S5	B 112	6.30	68–17	Devoted Danzig1173½ Neeran1174½ November Morn117¾	Rail trip	7
6Mar94-6SA	fst	1	:223 :463 1:121 1:382	ⒻMd Sp Wt	82	2	1	1½	1hd	2½	31	Gonzalez S5	B 112	30.60	76–18	Espadrille117hd Neeran1171 Designatoree1122	Inside trip	7
22Jan94-6SA	fst	1	:224 :461 1:104 1:361	ⒻMd Sp Wt	70	3	3	52¾	88½	810	79½	Antley C W	B 117	14.40	78–12	Contraindication117¾ Devoted Danzig1172½ Fair Kris1173	Gave way	9
26Dec93-2SA	fst	6f	:214 :444 :57 1:093	ⒻMd Sp Wt	53	9	7	73	63¾	1011	1013¾	Antley C W	B 117 b	57.30	75–09	ChoiceClaim1172 MeadowMoon117½ Whatwomn1173	Lugged out badly	12

WORKOUTS: Jly 11 Hol 4f fst :463 H 2/47 Jly 2 Hol 5f fst 1:003 H 8/40 Jun 15 Hol 4f fst :503 H 33/43 Jun 10 Hol 6f fst 1:152 H 9/15 Jun 5 Hol 4f fst :512 H 59/61 May 17 Hol 4f fst :482 H 3/21

134 The first race at Hollywood Park on July 16th is a Maiden Special Weight route with twelve horses entered. The even-money favorite is Starlit Bronze who comes out of the Pharma race we just analyzed. Since her 81 Beyer is above par for the level and Pharma came back to beat winners, Starlit Bronze looks like a legitimate favorite, despite her mediocre breeding.

So who, if anyone, can beat her? The first thing to do once again is throw out all the horses that have already lost on the turf with lower Beyers, as well as the first-time turfers by sires with C or lower turf ratings. This leaves us with three, first-time-on-the-turf contenders in Medinilla, Castle Gardens and Lady Lela. Of these, Lady Lela has the highest sire rating and is the second choice at 4-1. But, she is parked on the extreme outside and has no speed which makes it difficult to back her in this spot. Furthermore, her dam's two previous offspring, according to *Maiden Stats* have both already been tried and lost on the turf.

Medinilla is owned and trained by the powerful Paulson/Hassinger connection, but she is being asked to win her debut at 1 1/16 mile, which few horses can do. Furthermore, the Seattle Slews usually need a race or two and she is dead on the board at 20-1.

This leaves us with Castle Gardens, the public's third choice at 5-1, who has several things going for her. First off, she showed speed and nearly won her debut as a 2-year-old while in Ireland. Secondly, Richard Cross, according to Olmsted's *Trainer Guide*, excels with layoff horses. Having been given an additional eight months to mature and with the stamina indexes to get the distance, this filly, in Chris McCarron's capable hands, could well surprise at an attractive price.

So how to bet? While Starlit Bronze has the highest Beyers and is reunited with Pincay, she still has finished second or third in five of her seven races, which suggests she might not have the winning spirit. At almost even-money she looks like an underlay. Castle Gardens, on the other hand, ran a strong race in her turf debut and is being bet at 5-1. At that price she is worth backing with a Win bet as well as boxing her with Starlit Bronze in the exacta.

Results: On The Creek and It's Ben Freezing dueled for the lead and tired. Crissy Aya tracked them in third and took the lead at the top of the stretch only to get caught by Castle Garden who, under a smart ride by McCarron, shot to the inside and pulled away by a length-and-half. Castle Garden paid $12 to Win and $7.60 to Place, while the $2 exacta (which I didn't have) returned $195. Starlit Bronze made a move for second in the stretch, but faded to third at the wire.

Noteworthy about this race is that sometimes the better bet is the qualifying horse with the second highest pedigree rating because it has other factors going for it. In this case Lady Lela, despite Silver Hawk's A rating, couldn't overcome her outside post, questionable dam and lack of early speed. She fell 15 lengths back and finished midpack.

The fourth race at Hollywood Park on July 20th is a textbook example of the power of good turf ratings prevailing over a false favorite. The first thing to notice about this race is that the even-money choice Step In Toe is quite vulnerable. Even though she has a bullet work, she has mediocre Beyers and has already lost on the grass. The second thing to notice is that only two horses, Lucia's Book and Dixie Duo, have the B or better turf ratings that suggest they might move up on the grass

As a first-time starter in a route, however, Lucia's Book does face a formidable task. But, trainer Charles Stutts wins races with his debut horses and has put a lot of air into his filly, including three consecutive works at a mile, and her SI of 3 says she is bred to get the distance. Of greater concern is the fact that Lucia's Book's dam Nurse Lulu, according to *Maiden Stats*, is 0-2 with her offspring on the turf.

Hollywood Park

4

1 1/16 MILES. (Turf, chute). (1:384) **MAIDEN SPECIAL WEIGHT. Purse $34,000. Fillies and mares, 3-year-olds and upward. Weights: 3-year-olds, 116 lbs. Older, 122 lbs. (Horses that have started for a claiming price of $32,000 or less in their last three starts will have second preference.)**

TURF COURSE 1 1/16 MILES START FINISH

Versailles Vixen

Own: Chatmar Farms & Elm Tree Farm
VALENZUELA F H (299 24 31 35 .08)

Ch. f. 3 (Feb)
Sire: Greinton (Green Dancer)
Dam: Lady Giles (Vaguely Noble)
Br: Rutherford Mike G (Ky)
Tr: Metz Jeff (4 0 0 1 .00)
116

Lifetime Record: 2 M 0 0 $425

1994	2 M 0 0	$425	Turf	0 0 0 0		
1993	0 M 0 0		Wet	0 0 0 0		
Hol (T)	0 0 0 0		Dist (T)	0 0 0 0		

13Jly94-6Hol fst 6½f	:22 :45 1:104 1:173 3↑ (F)Md 32000	43	5 8 815 811 810 511¼	Valenzuela F H	B 117	85.70	70-15	Comeback Kidder1171¼ Gem Quest1175½ Ivonne G1173½	Wide trip	8
27May94-9Hol fst 6f	:222 :453 :573 1:103 3↑ (F)Md 32000	46	9 6 79¾ 913 816 715	Valenzuela F H	B 115	36.20	72-11	Silent Assembly1223½ Rue D'alezan1151¼ Our Cookie1154	No threat	10

WORKOUTS: Jly 6 SA 6f fst 1:163 H *9/13* Jun 30 Fpx 3f fst :36 H *1/3* ●May 24 Fpx 3f fst :342 H *1/7* May 19 Fpx 5f fst 1:02 H *4/7* May 11 Fpx 4f fst :481 Hg*4/13* Apr 29 Hol 6f fst 1:163 H *8/10*

Fair Marryan

Own: Walter Barbara A
ALMEIDA G F (72 3 8 9 .04)

Ro. f. 4
Sire: Runaway Groom (Blushing Groom-Fr)
Dam: Home From the Fair (Northern Dancer)
Br: Walter Barbara A (Cal)
Tr: Stute Melvin F (92 12 17 11 .13)
122

Lifetime Record: 5 M 0 1 $6,800

1994	4 M 0 1	$6,800	Turf	3 0 0 1	$5,950
1993	1 M 0 0		Wet	0 0 0 0	
Hol (T)	1 0 0 0		Dist (T)	1 0 0 0	

3Jly94-9Hol fst 1 1/16	:231 :462 1:111 1:442 3↑ (F)Md Sp Wt	60	4 8 712 78½ 47½ 59½	Sandoval R5	B 117	22.00	70-19	Blue Tess116hd Ladies Cruise1163½ Gold Rule1225½	Saved ground	8
8Jun94-4Hol fm 1 1/16 (T)	:23 :471 1:113 1:421 3↑ (F)Md Sp Wt	57	1 9 99½ 99¾ 99¼ 915¼	Delahoussaye E	B 122	3.10	70-13	Pharma1164½ Starlit Bronze1154½ Blue Tess1151	Saved ground	9
17Apr94-5SA fm 1 (T)	:222 :463 1:112 1:37 3↑ (F)Md Sp Wt	71	8 8 719 79 66¾ 32½	Delahoussaye E	B 123	*2.20	77-14	Crimson Steel123½ On The Cheek1232 Fair Marryan1232½	Late bid	8
3Apr94-5SA fm 1 (T)	:224 :47 1:121 1:37 (F)Md Sp Wt	76	3 7 810 86¼ 64½ 51	Solis A	B 120	25.70	79-21	Sabedoria120½ Crimson Steel120nk Chasse120hd	Far wide into lane	8
28Mar93-6GG gd 6f	:221 :452 :58 1:104 (F)Md Sp Wt	41	2 9 914 814 813 810½	Castaneda M	B 117	22.30	75-13	Vedra117½ Rushin Rocket1171¾ O'let It Snow117nk	Off slowly	9

WORKOUTS: Jly 15 Hol 4f fst :492 H *11/33* Jly 10 Hol 4f fst :474 H *8/56* Jly 1 Hol 4f fst :49 H *10/39* Jun 25 Hol 7f fst 1:30 H *10/10* Jun 20 Hol 7f fst 1:274 H *7/8* Jun 15 Hol 5f fst 1:02 H *21/45*

Step In Toe

Own: Keck Howard B
ATKINSON P (150 11 18 15 .07)

Ch. f. 3 (Apr)
Sire: Ferdinand (Nijinsky II)
Dam: Hidden Light (Majestic Light)
Br: Howard B. Keck (Ky)
Tr: Whittingham Charles (65 3 5 7 .05)
116

Lifetime Record: 5 M 1 0 $7,650

1994	5 M 1 0	$7,650	Turf	1 0 0 0
1993	0 M 0 0		Wet	0 0 0 0
Hol (T)	0 0 0 0		Dist (T)	0 0 0 0

18Jun94-9Hol fst 1 1/16	:223 :452 1:102 1:451 3↑ (F)Md Sp Wt	78	6 4 42¼ 31½ 31½ 21¼	Atkinson P	B 115	28.40	74-18	Et Voila1151¼ Step In Toe115½ Katy's Lady115no	Bumped at wire	9
21May94-3Hol fst 1 1/16	:232 :464 1:114 1:442 3↑ (F)Md Sp Wt	64	8 1 11 1hd 31 57¼	Atkinson P	B 115	28.70	72-18	Neeran115½ Flying Royalty1164½ Fast A Foot110¾		8
Stumbled, veered out start, wide 7/8										
17Apr94-5SA fm 1 (T)	:222 :463 1:112 1:37 3↑ (F)Md Sp Wt	58	2 1 1½ 31½ 45¼ 68¼	Valenzuela P A	B 117	18.40	71-14	Crimson Steel123½ On The Cheek1232 Fair Marryan1232½	Speed, tired	8
5Mar94-6SA fst 6f	:22 :443 :57 1:094 (F)Md Sp Wt	38	3 7 73½ 88¼ 912 1017¼	Gryder A T	B 117	108.00	71-08	Palestrina1171 Magical Avie1173½ Jackie Ramos117½	Bumped, jostled 5/8	12
6Feb94-6SA fst 6½f	:213 :442 1:084 1:15 (F)Md Sp Wt	23	7 4 32½ 57 818 831¼	McCarron C J	B 117	17.80	66-05	Emerald Express1175½ Espadrille1173½ Dancing With Deb1172¾	No factor	8

WORKOUTS: ●Jly 16 Hol 4f fst :464 H *1/46* Jly 11 Hol 5f fst 1:002 H *11/43* Jly 6 Hol 6f fst 1:13 H *8/21* Jly 1 Hol 3f fst :371 H *12/31* Jun 16 Hol 3f fst :361 H *14/32* Jun 8 Hol 5f fst 1:012 H *15/30*

Lucia's Book

Own: Ferraro Maxine A
SORENSON D (84 6 9 8 .07)

B. f. 3 (Feb)
Sire: Mari's Book (Northern Dancer)
Dam: Nurse Lulu (Jacinto)
Br: Ferraro Maxine A & Murty Farm (Ky)
Tr: Stutts Charles R (3 0 1 0 .00)
116

Lifetime Record: 0 M 0 0 $0

1994	0 M 0 0	Turf	0 0 0 0
1993	0 M 0 0	Wet	0 0 0 0
Hol (T)	0 0 0 0	Dist (T)	0 0 0 0

WORKOUTS: Jly 15 SA 1 fst 1:443 H *1/1* Jly 8 SA 1 fst 1:40 H *2/5* Jly 1 SA 1 fst 1:461 H *3/4* Jun 24 SA 7f fst 1:292 H *2/2* Jun 17 SA 6f fst 1:153 Hg*5/10* Jun 13 SA 6f fst 1:17 H *19/19* Jun 8 SA 6f fst 1:162 Hg*7/12* May 28 SA 5f fst 1:033 H *23/25* May 24 SA 4f fst :513 H *29/30*

continued

Dixie Duo

Own: Shields Rebecca N

ANTLEY C W (324 59 54 39 .18)

B. f. 3 (Mar)
Sire: Dixieland Band (Northern Dancer)
Dam: Duo Disco (Spring Double)
Br: Robert N. Verratti (Ky)
Tr: Robbins Jay M (23 3 6 1 .13)

L 116

Lifetime Record :	4 M 1 0	$5,000		
1994	2 M 1 0	$3,600	Turf	0 0 0 0
1993	2 M 0 0	$1,400	Wet	0 0 0 0
Hol Ⓣ	0 0 0 0		Dist Ⓣ	0 0 0 0

18Jun94- 9Hol fst 1 1/16 :223 :452 1:102 1:451 3↑ ⒻMd Sp Wt 63 8 5 56¼ 53¼ 45½ 810 McCarron C J L 115 b 2.70 65-18 Et Voila1151¼ Step In Toe115½ Katy's Lady115no Gave way 9
3Jun94- 6Hol fst 1 1/16 :224 :461 1:112 1:433 3↑ Md c-32000 81 1 8 87½ 52½ 11 2¾ Pedroza M A B 113 b 8.60 82-17 Wamps122¾ Dixie Duo1131¾ Pocomo1131¾ 4 wide into lane 9
Claimed from Takahashi Kensuke, Cross Richard J Trainer
14Nov93- 6SA fst 6f :214 :443 :563 1:084 ⒻMd Sp Wt 61 9 5 910 810 711 512½ Gryder A T B 117 42.90 80-06 C'monLetsDance1173½ CsulMeeting117½ Jcodr'sDevil1124½ Wide into lane 9
31Oct93- 4SA fst 6f :213 :444 :572 1:10 ⒻMd Sp Wt 65 7 9 917 913 98¾ 58¾ Gryder A T B 117 81.50 78-12 Princess Mitterand117hd Whatawoman1174 Gliding Lark1174½ No speed 9

WORKOUTS: Jly 16 Hol 5f fst 1:003 H 11/47 Jly 8 Hol 6f fst 1:144 H 2/12 Jun 29 Hol 5f fst 1:003 H 10/45 Jun 13 Hol 5f fst 1:014 H 15/30 May 29 SA 3f fst :362 H 5/30 May 25 SA 3f gd :363 Hg3/12

Gaby's Flight

Own: Tricar Stable

VALENZUELA P A (241 21 34 29 .09)

Dk. b or br f. 3 (Mar)
Sire: Trempolino (Sharpen Up)
Dam: Our Flight*NZ (Imperial Guard)
Br: Tricar Stables Inc (Cal)
Tr: Silva Jose L (2 1 1 0 .50)

L 116

Lifetime Record :	3 M 0 0	$525			
1994	2 M 0 0	$525	Turf	1 0 0 0	$525
1993	1 M 0 0		Wet	0 0 0 0	
Hol Ⓣ	0 0 0 0		Dist Ⓣ	1 0 0 0	$525

24Jun94- 5GG fm 1 1/16 Ⓣ :232 :473 1:12 1:443 + 3↑ ⒻMd Sp Wt 67 3 4 42 54¼ 54½ 54½ Meza R Q LB 114 14.00 78-14 Valiant Flowers1161 Queen Of The River114hd Hazy Melody1202 No rally 8
26May94- 3GG fst 6f :211 :44 :57 1:102 ⒻMd Sp Wt 58 7 3 57½ 58½ 810 79¼ Meza R Q B 117 14.60 77-14 Stunner1173 Once An Imp1172 Regal Day117nk Showed little 9
18Aug93- 2Dmr fst 5½f :222 :461 :582 1:051 ⒻⓈMd Sp Wt 11 8 5 52½ 64 814 820½ Meza R Q 117 30.30 64-14 L. A. Ballet1172¼ Flying Presence1171¼ Miss Moon Princess117½ 8
Wide, not urged late

WORKOUTS: Jly 16 BM 5f fst 1:012 H 13/43 Jly 6 BM 5f fst 1:034 H 15/21 Jun 10 GG 6f fst 1:133 H 2/10 Jun 3 GG 1 fst 1:461 H 2/2 May 22 GG 4f fst :473 H 3/71 ●May 15 GG 1 fst 1:423 H 1/4

Lady Lou S.

Own: Setzler Elaine

PEDROZA M A (303 30 47 38 .10)

Ch. f. 3 (Feb)
Sire: Red Attack (Alydar)
Dam: Medieval Season (Medieval Man)
Br: Ward William R Jr (WV)
Tr: Armstrong Horace W (7 0 0 0 .00)

L 116

Lifetime Record :	9 M 1 0	$13,040			
1994	9 M 1 0	$13,040	Turf	1 0 0 0	$3,000
1993	0 M 0 0		Wet	0 0 0 0	
Hol Ⓣ	1 0 0 0	$3,000	Dist Ⓣ	1 0 0 0	$3,000

3Jly94- 9Hol fst 1 1/16 :231 :462 1:111 1:442 3↑ ⒻMd Sp Wt 60 3 7 812 89 58½ 49 Pedroza M A LB 116 18.30 70-19 Blue Tess116hd Ladies Cruise1163½ Gold Rule1225½ Wide into lane 8
26Jun94- 5Hol fm 1 1/16 Ⓣ :23 :461 1:101 1:40 + ⒻAlw 40000N2L 71 4 7 717 712 414 415¼ Pedroza M A LB 115 97.80 81-08 Pharma1197½ Magical Avie1175 Just Tops1172¾ Rank, steadied start 7
16Jun94- 1Hol fst 1 1/16 :224 :47 1:112 1:45 ⒻClm 25000N2L 64 7 7 76¾ 66¼ 67¾ 56¾ Linares M G LB 115 70.20 69-24 Buff Duff1175½ Melrose Wine115¾ Sea Of Serenity116nk Wide into lane 7
3Jun94- 2Hol fst 6½f :221 :451 1:102 1:17 3↑ ⒻMd 25000 42 8 7 76 712 711 714 Linares M G LB 116 5.20 70-12 DaytonaBeach1162¾ MrquisDeTerme1164 CurveInTheRod1162¼ Wide trip 8
15May94- 6Hol fst 6½f :214 :45 1:103 1:171 3↑ ⒻMd 32000 54 9 11 1110 89¼ 79 55½ Linares M G LB 115 24.90 78-10 Zippen Miss115hd Lady Kariba117½ That'll Be Fine1171¼ Bumped start 12
17Mar94- 4SA fst 1 1/16 :223 :46 1:104 1:442 ⒻMd 32000 45 3 4 36 48 610 614½ Atherton J E5 LB 112 b 6.50 58-20 Our Passion1172½ Powerful Lad1172¼ Bundy's Sugarbabe1172 Rail trip 7
6Mar94- 6SA fst 1 :223 :463 1:121 1:382 ⒻMd Sp Wt 6 3 7 64½ 79¾ 718 742¼ Tejeira J B 117 b 8.40 34-18 Espadrille117hd Neeran1171 Designatoree1122 Done early 7
23Jan94- 7Aqu fst 6f ⊡ :23 :47 1:00 1:133 ⒻMd Sp Wt 64 7 9 108¾ 87 55½ 21 Brocklebank G V 121 15.90 74-23 Royal Revels1161 Lady Lou S.1212 Amy Be Happy1212¼ Finished well 12
1Jan94- 5Aqu fst 6f ⊡ :231 :473 1:004 1:142 ⒻMd Sp Wt 54 4 9 98½ 94¾ 64¼ 42½ Brocklebank G V 121 40.80 68-27 Forcing Bid116nk Very Careless121nk Sterling Pound1212 Some gain 9

WORKOUTS: Jly 17 SA 5f fst 1:002 H 4/17 Jly 12 SA 4f fst :48 H 3/26 Jun 12 SA 7f fst 1:272 H 5/8 May 31 SA 5f fst 1:012 H 3/14 May 26 SA 6f fst 1:151 H 7/14 May 12 SA 5f fst 1:014 H 18/27

Lady Stonewalk

Own: Lindsey James J

CASTANON A L (74 2 6 8 .03)

B. f. 3 (May)
Sire: Stonewalk (Knightly Manner)
Dam: Green Signal (Roberto)
Br: Heidmann T & Lindsey James (Cal)
Tr: Stepp William T (20 0 0 4 .00)

L 116

Lifetime Record :	6 M 0 0	$0		
1994	5 M 0 0		Turf	0 0 0 0
1993	1 M 0 0		Wet	0 0 0 0
Hol Ⓣ	0 0 0 0		Dist Ⓣ	0 0 0 0

9Jly94- 2Hol fst 1 1/16 :231 :471 1:114 1:451 3↑ ⒻⓈMd 28000 27 5 9 916 919 820 623¾ Trujillo V LB 114 b 112.20 51-25 Astrial116½ Silent Draw11611 Turf Princess1207 By tired ones 9
24Jun94- 4Hol fst 1 1/16 :231 :471 1:124 1:452 3↑ ⒻⓈMd 28000 30 8 10 1015 813 714 625¼ Skelly R V5 L 109 b 78.90 49-19 New Bounty1163½ Silent Draw11513 Turf Princess1202½ Off slowly 10
12Jun94-10Hol fst 6½f :22 :451 1:103 1:171 3↑ ⒻMd 28000 25 9 10 1015 1019 921 923¼ Trujillo V B 114 201.60 59-13 Explosive Illusion1162½ Highly Flirtatious116hd Rue D'alezan116nk Outrun 10
27May94- 9Hol fst 6f :222 :453 :573 1:103 3↑ ⒻMd 28000 34 1 9 815 811 919 920 Castanon E F5 B 109 79.70 67-11 Silent Assembly1223½ Rue D'alezan1151½ Our Cookie1154 Off slowly 10
17Jan94- 1BM fst 1 :23 :473 1:131 1:402 ⒻMd Sp Wt 38 3 7 712 710 69 613½ Vergara O 117 7.10e 58-25 A Trifle To Spare1172¼ Key To V. A.1123 Ali Capote1171½ Off slowly 7
28Dec93- 4SA fst 6f :214 :442 :571 1:094 ⒻMd 28000 35 11 9 1114 1116 813 615¾ Iammarino M P 115 103.60 72-05 Serendipity1172 Saratoga Hope1174½ Quiet Pride1171½ By tired ones 12

WORKOUTS: Jly 1 SA 5f fst 1:014 H 12/29 Jun 20 SA 5f fst 1:002 H 9/20 Jun 10 SA 3f fst :372 H 11/22 Jun 4 SA 4f fst :501 H 16/34 May 25 SA 3f gd :363 H 3/12 May 21 SA 4f fst :491 H 18/33

Youtellemshirley

Own: LaBuda Margaret M

SANDOVAL R (5 0 0 1 .00)

B. f. 4
Sire: Walker's (Jaipur)
Dam: Miss Chatterly (Pass the Glass)
Br: Goemans Peg (Cal)
Tr: Perez Dagoberto L (9 0 0 0 .00)

L 117^5

Lifetime Record :	8 M 1 0	$6,400			
1994	8 M 1 0	$6,400	Turf	0 0 0 0	
1993	0 M 0 0		Wet	1 0 0 0	$2,400
Hol Ⓣ	0 0 0 0		Dist Ⓣ	0 0 0 0	

1Jly94- 9Hol fst 6f :22 :452 :574 1:11 3↑ ⒻⓈMd 28000 44 3 4 31 44½ 58½ 611 Sandoval R5 LB 115 b 24.00 74-12 Accountable1152½ Sand The Coach1163½ Eclipse Del Sol1151¼ Weakened 9
1Jun94- 9Hol fst 6f :214 :451 :571 1:094 3↑ ⒻⓈMd 28000 41 6 5 95½ 89¾ 913 818¼ Castanon A L LB 120 b 60.00 73-11 CboQueen1155 ComebckKidder1154 ChristinMri115¾ Checked 5/8, wide 11
14May94- 9Hol fst 6f :22 :452 :58 1:114 3↑ ⒻⓈMd 28000 46 2 7 41½ 46½ 711 89 Valenzuela P A LB 120 b 8.10 72-09 CopperCoinage1152 Accountable115hd Drona'sStarlet1172¼ Saved ground 11
28Apr94- 6Hol fst 6f :22 :452 :571 1:093 3↑ ⒻMd 32000 37 8 6 75½ 88½ 915 819 Flores D R L 122 5.70 73-08 Time Of Queen1105½ Ruffmeup122½ Chelsea Castle1203 5 wide turn 9
8Apr94- 3SA fst 6f :22 :451 :572 1:10 ⒻⓈMd 35000 71 4 3 41½ 31 32 23½ Flores D R LB 118 6.10 83-12 Royalty Doll1203½ Youtellemshirley118¾ Nancy'sLegend1186 Outfinished 7
25Mar94- 2SA gd 1 1/16 :243 :491 1:141 1:461 ⒻMd 35000 31 3 4 54½ 78 714 719½ Valenzuela F H LB 118 3.80 44-22 Tellalo120½ Dances With Wolves1185½ Sweet Lady Czar1183½ 7
Broke thru gate, pinched start, steadied 7/8
4Feb94- 6SA my 7f :224 :454 1:111 1:25 ⒻMd Sp Wt 34 4 1 31 45 47 413 Martinez F F B 119 13.80 64-16 Cindazanno1193½ Ms. Ukulele1192½ On The Cheek1197 No factor 4
6Jan94- 6SA fst 6f :214 :444 :564 1:094 ⒻMd Sp Wt 50 2 8 55 89½ 814 813¾ Martinez F F B 119 44.40 74-11 Morning Showers119nk Suana1192½ Really Tops1143½ Brief speed 8

WORKOUTS: Jly 16 Hol 5f fst 1:014 H 27/47 Jly 9 Hol 7f fst 1:28 H 5/5 ●Jun 26 Hol 5f fst :591 H 1/44 May 25 Hol 4f fst :48 H 2/51 May 11 Hol 4f fst :47 H 3/37 Apr 24 Hol 5f fst 1:003 H 6/15

While Dixie Duo just lost to Step In Toe, that was a dirt race and she may have bounced off her strong June 3rd comeback. With her move to the grass today, a surface she is bred to like, she has every right to improve. Furthermore, her dam is 1/1 on the turf, so her 9-2 odds look better than fair.

So how to bet? With two route preps under her belt, Dixie Duo should be ready to fire. While Step In Toe comes off a bullet work she has already lost on the turf and at even-money is less attractive than Lucia's Book who has the breeding to surprise. So the best value strategy here is to bet Dixie Duo and Lucia's Book to Win and box them in the exacta.

FOURTH RACE
Hollywood
JULY 20, 1994

1¹⁄₈ MILES. (Turf Chute)(1.384) MAIDEN SPECIAL WEIGHT. Purse $34,000. Fillies and mares, 3-year-olds and upward. Weights: 3-year-olds, 116 lbs. Older, 122 lbs. (Horses that have started for a claiming price of $32,000 or less in their last three starts will have second preference.)

Value of Race: $34,000 Winner $18,700; second $6,800; third $5,100; fourth $2,550; fifth $850. Mutuel Pool $198,960.00 Exacta Pool $156,784.00 Trifecta Pool $151,949.00 Quinella Pool $24,900.00

Last Raced	Horse	M/Eqt. A.Wt	PP	St	¼	½	¾	Str	Fin	Jockey	Odds $1
	Lucia's Book	Lbf 3 116	4	8	7⁴	6⁴	5½	3¹½	1½	Sorenson D	9.90
18Jun94 9Hol⁸	Dixie Duo	LBb 3 116	5	7	5¹	4hd	3¹	2½	2hd	Antley C W	4.60
18Jun94 9Hol²	Step In Toe	B 3 116	3	2	1¹	1²	1¹½	1¹	3⁵	Atkinson P	1.10
24Jun94 5GG⁵	Gaby's Flight	LB 3 116	6	3	2⁴	2²½	2²½	4⁴	4²	Valenzuela P A	4.60
3Jly94 9Hol⁵	Fair Marryan	B 4 122	2	9	9	9	9	6⁵	5²	Almeida G F	8.00
1Jly94 9Hol⁶	Youtellemshirley	Lb 4 117	9	5	3¹	3¹	4³	5½	6¹¹	Sandoval R⁵	56.50
3Jly94 9Hol⁴	Lady Lou S.	LBb 3 116	7	4	4²	5³	6⁴	7⁴	7²½	Pedroza M A	7.20
13Jly94 6Hol⁵	Versailles Vixen	f 3 116	1	1	6¹½	7²½	7³½	8⁶	8¹⁶	Valenzuela F H	58.10
9Jly94 2Hol⁶	Lady Stonewalk	LB 3 116	8	6	8¹½	8¹	8hd	9	9	Castanon A L	176.70

OFF AT 2:27 Start Good. Won driving. Time, :23, :47, 1:11³, 1:37¹, 1:43⁴ Course firm.

$2 Mutuel Prices:

4-LUCIA'S BOOK	21.80	9.60	4.40
5-DIXIE DUO		6.00	3.40
3-STEP IN TOE			2.60

$2 EXACTA 4–5 PAID $135.20 $2 TRIFECTA 4–5–3 PAID $352.40 $2 QUINELLA 4–5 PAID $56.00

B. f, (Feb), by Mari's Book–Nurse Lulu, by Jacinto. Trainer Stutts Charles R. Bred by Ferraro Maxine A & Murty Farm (Ky).

This race is a perfect example of the power of breeding. Lucia's Book and Dixie Duo, the only horses with good turf ratings, made up the $135 exacta. Those astute enough to play them with the favorite in the trifecta would have been rewarded with an additional $352.

The sixth race at Hialeah on May 3rd shows why it is necessary to integrate a horse's turf and stamina ratings. In this 1 1/16 mile Maiden Special Weight race, Al's Memory at even-money and Lottery Lil at 9-5 are the heavy favorites, even though they have lost a combined 34 times. While Pm's Cherokeehoney has already lost twice on the turf at 1 1/8 mile, she has a B rating and is "debuting" at today's shorter distance. Since she is the only speed in the race and held on better despite a pace duel in her second attempt, her 7-1 odds are tempting. But the key handicapping question is: will an easy lead and today's shorter distance help Pm's Cherokeehoney overcome a negative stamina index which suggests a mile is her limit? And if not, what is the best way to profitably play her?

Hialeah Park

6

About 1 1/16 MILES. **(Turf). (1:393) MAIDEN SPECIAL WEIGHT. Purse $14,000 (plus $1,400 FOA). Fillies and mares, 4–year–olds and upward. Weight, 122 lbs.**

TURF COURSE — ABOUT 1 1/16 MILES — START FINISH

Coupled – Alyandover and Andoveraly; Ells Bella Naiola and Diane's Lass

Al's Memory

Own: Green Robert I

DOUGLAS R R (228 44 35 31 .19)

B. f. 4
Sire: Alysheba (Alydar)
Dam: Powder Break (Transworld)
Br: Green Robert I (Ky)
Tr: Olivares Luis (44 9 7 7 .20)

122

Lifetime Record :	22 M 5 4	$29,670			
1994	5 M 2 1	$10,050	Turf	16 0 5 4	$29,320
1993	13 M 3 3	$19,620	Wet	2 0 0 0	$180
Hia Ⓣ	6 0 2 2	$8,865	Dist Ⓣ	7 0 1 2	$10,860

20Apr94-11Hia	fm	*1⅛ Ⓣ	1:501	ⒻMd Sp Wt	72	9	8	67	36½	32½	21¾	Ramos W S	122 b	3.10	79–15	Alinam1221¾ Al's Memory1221¼ Lottery Lil1229	Rallied	9
12Apr94-11Hia	fm	*1 1/16 Ⓣ	1:44	ⒻMd Sp Wt	56	2	9	819	810	58½	38½	Ramos W S	122 b	2.80	73–18	Silly's Philly1221½ Kyle's Pet1227 Al's Memory1222¾	Late rally, outside	10
29Mar94-11Hia	fm	*1⅛ Ⓣ	1:501	ⒻMd Sp Wt	44	7	8	89¾	910	711	714¾	Rodriguez P A	122 b	3.10	66–17	Reigning Lady1221½ Alinam1223¼ Lottery Lil1221¼	No threat	10
14Mar94- 2GP	fm	*1⅛ Ⓣ	:481 1:124 1:373 1:502+	Md Sp Wt	62	10	6	62¾	62¼	53½	47	Santos J A	117 b	2.70	86 —	Kris's Kiss1172½ Fire Festival1223 Lucky Rocky1171½	Belated bid	10
23Feb94-10GP	gd	*1⅜ Ⓣ	:513 1:18 1:43 2:242	Md Sp Wt	71	8	4	42	2hd	1hd	21½	Santos J A	117 b	*2.00	50–29	Mashburn1221½ Al's Memory117½ Sir Newberry122no	Gamely	10
16Dec93- 2Crc	fm	*1⅛ Ⓣ	1:51	3↑ ⒻMd Sp Wt	70	7	4	43	44	22	21	Ramos W S	120 b	3.20	63–30	Speedy Colleen1201 Al's Memory1204 Lottery Lil1203½	Good try	8
7Oct93- 3Bel	fm	1 1/16 Ⓣ	:231 :464 1:121 1:442	3↑ ⒻMd Sp Wt	65	1	10	811	73¾	43	41	Chavez J F	119 b	*2.10	73–19	Bejilla Lass114½ Holly North119nk Helen Ge Ge122nk	Wide turn	10
28Aug93- 1Sar	fm	1 1/16 Ⓣ	:223 :46 1:11 1:432	3↑ ⒻMd Sp Wt	61	11	9	913	75¼	51¾	47	Leon F5	112 b	8.90	70–16	L'heure Bleue122hd Crandall1175 Speedy Colleen1172	Rallied inside	12
12Aug93- 2Sar	fm	1 1/16 Ⓣ	:222 :453 1:112 1:423	3↑ ⒻMd Sp Wt	70	1	10	1119	116¾	64½	44¼	Santos J A	117 b	4.00	77–12	New Account1171¾ Crandall1171½ Holly North1171	Steadied, wide	11
30Jun93- 3Bel	fm	1 1/16 Ⓣ	:232 :47 1:112 1:433	3↑ ⒻMd Sp Wt	57	9	4	43	42½	64½	78¼	Bisono C V5	109 b	3.20	70–15	Sue's Huntress122nk Secretariat's Fire1144¼ Bejilla Lass114nk	Tired	10

WORKOUTS: ● Apr 21 Bel 3f fst :36 H *1/19* Feb 19 Crc 4f sly :482 H (d)*6/44* Feb 12 Crc 7f fst 1:25 H *1/1*

Alyandover

Own: Collier Reginald B

LOPEZ E C (33 3 5 5 .09)

B. f. 4
Sire: Alydar (Raise a Native)
Dam: In the Offing (Hoist the Flag)
Br: Mr. & Mrs. Mark Hardin (Va)
Tr: Williams Theodore (13 2 2 3 .15)

1157

Lifetime Record :	7 M 0 0	$3,380			
1994	1 M 0 0	$300	Turf	2 0 0 0	$320
1993	6 M 0 0	$3,080	Wet	0 0 0 0	
Hia Ⓣ	0 0 0 0		Dist Ⓣ	2 0 0 0	$320

11Jan94- 7Lrl	fst	7f	:232 :47 1:121 1:252	ⒻMd Sp Wt	25	6	7	109½	1014	915	818	Hamilton S D	L 122 b	33.30	64–19	Willie Wood1223¼ Turning Around122¾ What Option1227	Outrun	11
30Dec93- 6Lrl	fst	1 1/16	:25 :493 1:154 1:502	3↑ ⒻMd Sp Wt	35	2	7	712	717	614	514	Skinner K	L 119	3.00	44–41	Lear's Lady119½ Carrantouhill1198 Lady Allen1194	Outrun	7
3Dec93- 3Med	fst	1	:24 :473 1:132 1:401	3↑ ⒻMd Sp Wt	44	2	3	2½	11	21	44½	Ferrer J C	L 121	11.10	70–19	Gotnotickets1212½ Frieda N.1161¾ Eggs Binnedict121nk	Bid, tired	8
18Oct93- 3Pha	fst	170	:22 :472 1:142 1:462	3↑ ⒻMd Sp Wt	28	3	7	78½	56	48½	514½	Molina V H	117 f	6.00	49–29	Wajawonder1172 Pancake House112hd Java Rain11710	Wide, no threat	9
5Oct93- 6Med	yl	1 1/16 Ⓣ	:243 :493 1:142 1:463	3↑ ⒻMd Sp Wt	36	8	7	74½	33	78¾	814	Ferrer J C	117 f	52.30	52–28	Mmlemmle1171¼ RmmbrMidnight1173¾ SpringRunLrk1171	Flattened out	10
22Sep93- 6Med	fst	1 1/16 ⊗	:233 :473 1:13 1:462	3↑ ⒻMd Sp Wt	39	5	2	2½	2½	44	411¾	Ortiz F L	116 f	5.40e	64–17	Feel That Breeze1162¼ Polas Dawn1165¾ Jennifer Sara B.1163¾	Tired	7
8Sep93- 5Med	yl	1 1/16 Ⓣ	:234 :49 1:14 1:461	3↑ ⒻMd Sp Wt	39	4	5	42	97¼	810	814½	Ortiz F L	116 f	20.10	54–27	Dizzy Penny1221½ Polas Dawn116hd Feel That Breeze116no	Steadied, tiring	10

Ells Bella Naiola

Own: Fede Pietro

CHAPMAN K L (140 16 16 9 .11)

Ch. m. 5
Sire: Nijinsky's Secret (Nijinsky II)
Dam: Melody Shower (Diplomat Way)
Br: Smith Mrs J (Fla)
Tr: Fede Pietro (28 3 1 4 .11)

L 1175

Lifetime Record :	30 M 3 6	$13,892			
1994	1 M 0 0	$62	Turf	3 0 1 1	$4,200
1993	9 M 0 2	$2,255	Wet	9 0 0 1	$1,832
Hia Ⓣ	0 0 0 0		Dist Ⓣ	1 0 0 1	$1,540

27Apr94-10Hia	sly	6f	:231 :464 :593 1:13	ⒻMd 10500	16	4	8	914	919	715	715	Paneto W R	L 118 b	36.60	65–14	Pretty Dolly1151¾ Dorado Star1182½ Pierce Bridge1181¾	No threat	9
28Dec93-10Crc	fst	1 1/16	:234 :481 1:143 1:491	3↑ ⒻClm 6250	42	11	9	1016	1015	912	811	Chapman K L7	L 110 b	36.80	64–25	Midnight Fleet1174 Weightless117¾ Naked Daina1121½	Outrun	12
17Dec93- 1Crc	fst	1 1/16	:234 :482 1:152 1:504	3↑ ⒻMd 12500	35	11	9	1014	1012	611	37¼	Chapman K L7	L 115 b	21.90	60–24	LaFemmeD'or120nk CarmenAlin1207 EllsBellNiol1151¼	Slow start, gaining	11
10Dec93- 2Crc	fst	7f	:232 :473 1:133 1:274	3↑ ⒻMd 12500	22	8	5	53	710	714	613	Haldar A	L 122 b	10.10	63–18	Decent Win1213 Miss Rexson1213½ Carmen Alina121nk	Brief speed	11
24Nov93- 2Crc	sly	1	:233 :482 1:164 1:443	3↑ ⒻMd 18000	48	4	3	512	47½	34½	35¼	Haldar A	L 120 b	38.50	63–27	Mangolo1204¾ Mina's Review120½ Ells Bella Naiola1202	Bid inside, hung	7
11Nov93- 5Crc	sly	1⅛ ⊗	:494 1:154 1:433 1:574	3↑ ⒻMd Sp Wt	33	3	3	710	714	718	720½	Haldar A	L 122 b	55.40	49–27	Sweep Over Finish1202¾ Fairy Princess1202 Lottery Lil1204	Faltered	8
22Jly93- 1Crc	fst	1	:243 :484 1:142 1:403	3↑ ⒻMd 18000	16	2	6	611	613	617	624	Reyes A R	120 b	59.70	64–14	Divinenmine1154½ LFmmD'or115¾ CroonMyTun115¾	Away slowly, outrun	6
8Jly93- 2Crc	sly	1	:243 :483 1:134 1:414	3↑ ⒻMd 18000	37	3	7	712	510	510	513½	Portillo D A5	115 b	22.20	68–16	Sunny Wish1133 Divinenmine1209 Moment Of Revenge1221½	No threat	7
25Jun93-10Crc	sly	1 1/16 ⊗	:232 :481 1:143 1:494	3↑ ⒻMd Sp Wt	19	1	9	917	914	914	920½	Reyes A R	121 b	42.00	51–16	Flapper Dancer114½ Tallahatchie1215¼ Little Zip1145½	Outrun	9
30May93- 2Crc	sly	1	:242 :491 1:144 1:441	3↑ ⒻMd 18000	20	6	2	11	54¾	68¾	614	Reyes A R	119 b	19.30	56–17	Fait Accompli108¾ Bird Of Play113nk Sunny Wish112¾	Tired	8

WORKOUTS: Apr 24 Crc 4f fst :533 B *27/27*

Pm's Cherokeehoney

Own: Sierra Cynthia C

NUNEZ E O (84 4 11 10 .05)

B. f. 4
Sire: Great Above (Minnesota Mac)
Dam: Pm's Money Honey (Proud Birdie)
Br: Sierra Cynthia & Michael (Fla)
Tr: O'Connell Kathleen (18 5 0 2 .28)

L 122

Lifetime Record :	7 M 0 0	$3,010			
1994	5 M 0 0	$2,170	Turf	2 0 0 0	$840
1993	2 M 0 0	$840	Wet	1 0 0 0	$950
Hia Ⓣ	2 0 0 0	$840	Dist Ⓣ	0 0 0 0	

20Apr94-11Hia	fm	*1⅛ Ⓣ	1:501	ⒻMd Sp Wt	50	4	1	2hd	22½	22½	412	Nunez E O	L 122 fb	10.30	69–15	Alinam1221¾ Al's Memory1221¼ Lottery Lil1229	Weakened	9
29Mar94-11Hia	fm	*1⅛ Ⓣ	1:501	ⒻMd Sp Wt	60	3	7	78¼	66	56½	57½	Nunez E O	L 122 fb	44.20	73–17	Reigning Lady1221½ Alinam1223¼ Lottery Lil1221¼	Mild bid	10
24Feb94- 4GP	fst	7f	:221 :45 1:103 1:24	ⒻMd Sp Wt	23	10	2	31½	34½	811	821½	Velez J A Jr	L 121 fb	14.30	61–14	Royal Flex1213 Nan's First Girl1214½ Twirly Girly1214	No threat	12
12Jan94- 2GP	sly	6f	:22 :454 :584 1:12	ⒻMd Sp Wt	52	6	3	2½	21½	23½	413	Vasquez J	L 120 fb	9.60	68–19	Just Dance12011 Caro's Beauty120no Beyond The Beyond1201½	Weakened	10
2Jan94- 5Crc	fst	7f	:23 :472 1:143 1:274	ⒻMd Sp Wt	28	8	3	2½	1hd	33	510½	Vasquez J	L 120 fb	2.70	62–20	Shakespeares Dream120no Intriguing1203½ Carmen Alina1206¼	Gave way	8
19Dec93-11Crc	fst	6f	:222 :47 1:00 1:133	3↑ ⒻMd Sp Wt	49	4	8	31½	55	45½	46	Spieth S	L 121 fb	8.00	74–16	RmmbrMidnight121nk TwirlyGirly1212¼ ShksprsDrm1213½	Early foot inside	9
3Mar93- 3GP	fst	6f	:223 :453 :581 1:111	ⒻMd 40000	38	1	11	118	811	713	716	Vasquez J	116	46.30	69–13	Upstate Flyer1163¾ Annie Imp1203¼ Spend The Loot1162½	No threat wide	12

WORKOUTS: Feb 10 Crc 4f fst :532 B *18/18*

Wish Again

Own: Weis Joan A
B. f. 4
Sire: Gonzales (Vaguely Noble)
Dam: Elicabic (Diplomat Way)
Br: Lieb Sandi (Fla)
Tr: Weis Joan A (1 0 0 0 .00)

(handwritten: D 9/4, B-4/3)

BAIN G W (34 0 4 0 .00)

122

Lifetime Record :	20 M 0 0		$3,315
1994	1 M 0 0	$110	Turf	11 0 0 0	$2,000
1993	19 M 0 0	$3,205	Wet	5 0 0 0	$720
Hia ⊤	5 0 0 0	$1,175	Dist ⊤	1 0 0 0	$140

13Apr94-7Hia fm *1¹⁄₁₆ ⊤	1:55²	Clm 35000	54 10 5 5⁶ 5⁷ 9¹² 9¹⁷¼	Bain G W	115	116.00	78-05	Won The Laurel116³ Telegrapher116³ Zee Buck116ⁿᵒ	Faltered 10
16Dec93-3Crc fm *1¹⁄₁₆ ⊤	1:51	3↑ⒻMd Sp Wt	49 5 8 8⁹½ 8¹¹ 8¹¹ 5¹⁰½	Bain G W	120	61.10	54-30	Speedy Colleen120¹ Al's Memory120⁴ Lottery Lil120³½	Without speed 9
11Nov93-5Crc sly 1¹⁄₈ ⊗ :49⁴ 1:15⁴ 1:43³ 1:57⁴	3↑ⒻMd Sp Wt	40 7 8 8¹⁴ 8¹⁶ 6¹⁴ 6¹⁵½	Bain G W	120	87.10	54-27	Sweep Over Finish120²½ Fairy Princess120² Lottery Lil120⁴	No threat 8
24Oct93-5Crc yl 1¹⁄₈ ⊤	1:51²	3↑Md Sp Wt	45 6 10 11¹⁸ 10¹⁷ 8¹³ 7¹⁰½	Bain G W	116	50.10	51-41	Sportin' Charles122½ Lucky Rocky119¾ Sir Newberry119¾	No threat 12
24Sep93-4Crc fst 1¹⁄₁₆	1:50³	3↑Md Sp Wt	50 10 10 9⁸¼ 9⁹½ 5⁹½ 4⁹½	Bain G W	116	40.00	58-29	Mad Mo118¹½ Flying Liane118¹½ Sweep Over Finish118³½	Belated bid 9
1Sep93-4Crc fst 1¹⁄₁₆ :24¹ :48³ 1:13⁴ 1:47³	3↑ⒻClm 12500	17 8 8 8¹⁸ 8²² 8²⁷ 8³³¼	Bain G W	115	83.70	50-15	DHPens Pens Pens112 DHSmiling Tune112¾ I'm For Holme117¹½	Outrun 9
1Aug93-8Crc fst 1¹⁄₈	1:48²	3↑ⒻClm 32000	45 9 3 2²½ 10¹⁰ 10¹⁷ 10¹⁷½	Bain G W	114	122.90	60-17	Heaven's Answer116¼ Mighty Coyote114² Cawarra114ⁿᵏ	Gave way 11
23Jly93-4Crc sly 1¹⁄₁₆ ⊗ :47³ 1:13 1:41¹ 1:56	3↑ⒻAlw 15000N1x	30 2 7 7⁵ 7⁵¼ 7¹¹ 8¹¹	Bain G W	115	63.90	61-15	Flapper Dancer115⅝ Lover's Yarn108³ Tibidabo's Ensign119¾	Outrun 9
18Jly93-7Crc fst 1¹⁄₈ ⊤	1:52⁴	3↑Md Sp Wt	54 2 7 7⁵ 7³¼ 7¹¹ 8¹¹	Bain G W	115	138.20	59-26	Kiko115³ Nick G115ⁿᵒ Alinick117ⁿᵏ	Failed to menace 10
11Jly93-5Crc sly 1¹⁄₈ ⊗ :48¹ 1:13¹ 1:40 1:54²	3↑Md Sp Wt	40 7 7 7²¹ 7²⁸ 7²⁸ 7²³½	Bain G W	115	134.10	63-15	Elite Jeblar115⁴ Ground Level110¹½ Marion Landing115³	Outrun 7

WORKOUTS: Mar 18 PBD 4f gd :49⁴ B 1/1

Fairy Princess

Own: J B J Stable
Ch. f. 4
Sire: Ferdinand (Nijinsky II)
Dam: Parrish Empress (His Majesty)
Br: Parrish Hill Farm (Ky)
Tr: Jennings Lawrence Jr (28 4 0 3 .14)

(handwritten: c 6/2, C 4/1)

VELEZ J A JR (143 21 23 16 .15)

L 122

Lifetime Record :	11 M 2 1		$7,490
1994	3 M 0 1	$2,060	Turf	4 0 1 0	$2,210
1993	8 M 2 0	$5,430	Wet	3 0 1 0	$3,040
Hia ⊤	1 0 0 0	$140	Dist ⊤	2 0 1 0	$1,925

20Apr94-11Hia fm *1¹⁄₈ ⊤	1:50¹	ⒻMd Sp Wt	36 6 3 3²½ 6⁸¼ 7¹¹ 7¹⁸¼	Chapman K L⁵	L 117 fb	9.30	63-15	Alinam122¹¼ Al's Memory122¹¼ Lottery Lil122⁹	Gave way 9
4Mar94-1GP fst 6f	:22² :46¹ :59² 1:12²	ⒻMd 30000	38 5 4 7⁶ 7⁶¾ 6⁶ 3⁸¼	Bravo J	L 119 fb	9.40	70-21	PersonlDncr119⁶¼ Imllttlsowtht119¾ FiryPrincss119¾	Wide top str, rallied 9
2Feb94-10GP sly 1¹⁄₁₆ ⊗ :23² :48¹ 1:15¹ 1:50²	ⒻMd Sp Wt	-0 10 3 5⁸½ 9²⁰ 9³² 8⁴¹	Castillo H Jr	L 120 fb	13.00	— 32	Private Session120³ Gate Princess120⁶ Mrs. Marcos120¹¼	Tired 10
16Dec93-5Crc fst 1¹⁄₁₆	1:51	ⒻMd Sp Wt	48 6 1 11 3²½ 5⁸½ 5¹⁰½	Castillo H Jr	L 120 fb	5.30	53-30	Speedy Colleen120¹ Al's Memory120⁴ Lottery Lil120³½	Tired badly 8
11Nov93-5Crc sly 1¹⁄₈ ⊗ :49⁴ 1:15⁴ 1:43³ 1:57⁴	ⒻMd Sp Wt	60 8 1 1³ 2¹½ 2¹½ 2²½	Ramos W S	L 120 b	9.80	67-27	Sweep Over Finish120²½ Fairy Princess120² Lottery Lil120⁴	Second best 8
3Sep93-10Atl fm 1¹⁄₈ ⊗ :23¹ :47⁴ 1:13² 1:47¹	ⒻMd Sp Wt	56 1 2 2½ 1² 1³ 1⁴½	Lopez C C	L 116 fb	3.10	81-10	Excavating111³ Fairy Princess116⁷ Total Immersion116ⁿᵒ	Best of rest 9
13Aug93-4Mth fm 1¹⁄₈ ⊤ :24 :49² 1:14¹ 1:46³	3↑ⒻMd Sp Wt	49 7 1 11 2¼ 6¾ 6¹⁰¼	Lopez C C	L 113 fb	19.70	60-31	Mon Montreaux114¼½ Lady Gladiator114½ Nine Guns114³½	Gave way 7
23Jly93-4Mth fst 1	:23⁴ 1:14¾ 1:31 1:39⁴	ⒻMd 10000	27 4 2 2² 6⅝ 6¹⁵ 6²³¼	Castillo H Jr	L 114 fb	9.60	55-22	MkinASttement114¼½ DrlinYouSndMe114ⁿᵏ RcklssPlc109²	Broke thru gate 7
11Jly93-11Mth fst 1 70	:23⁴ :47² 1:13⁴ 1:46¹	ⒻMd 10000	26 1 3 3² 4¼½ 3¹⁰ 4¹⁶¼	Castillo H Jr	L 114 fb	4.60	52-25	FriendsRUs114¼ WiseBaroness114¹½ DrlinYouSendMe114¹	Raced 5-wide 8
9Apr93-10Hia sly 6f	:22² :45⁴ :58² 1:11⁴	ⒻMd Sp Wt	37 6 4 4¹½ 6½ 7¹² 8¹⁶	Douglas R R	L 120 b	26.60	70-15	Twenty Second Ave121ʰᵈ Explosive Luck121½ GoldenAutumn114²	Faded 11

WORKOUTS: Apr 30 Hia 4f fst :49⁴ B 11/19 • Apr 16 Hia 6f fst 1:15² H 1/4 • Apr 8 Hia 6f fst 1:14³ H 1/6 Apr 1 Hia 5f fst 1:02 B 3/17 Mar 1 GP 3f fst :37 B 7/18 Feb 21 GP 6f fst 1:15¹ B 11/22

Private Delight

Own: Appleton Arthur I
B. f. 4
Sire: Private Account (Damascus)
Dam: Raja's Delight (Raja Baba)
Br: Foxfield (Ky)
Tr: Jennings Lawrence Jr (28 4 0 3 .14)

(handwritten: C 2/7, C 3/5)

DOUGLAS R R (228 44 35 31 .19)

122

Lifetime Record :	2 M 1 0		$1,020
1994	2 M 1 0	$1,020	Turf	0 0 0 0	
1993	0 M 0 0		Wet	0 0 0 0	
Hia ⊤	0 0 0 0		Dist ⊤	0 0 0 0	

| 8Apr94-4Tam fst 1¹⁄₁₆ :24² :49¹ 1:15 1:49² | ⒻMd Sp Wt | 21 1 4 5⁷½ 4¹⁷ 4¹¹ 3¹⁴ | Warner T | 120 | 4.70 | 61-25 | Imperial Appeal120⁴ DAnkara Mio120¹⁰ Private Delight120¹ | Mild bid 7 |
| Placed second through disqualification. |
| 15Mar94-8Tam fst 7f | :23² :47⁴ 1:15¹ 1:29² | ⒻMd Sp Wt | 21 5 9 9¹⁴ 9¹⁷ 8¹⁵ 6¹¹¾ | Warner T | 120 | 6.70 | 59-22 | BurnAboutBuzz120¹ ImperialAppel120¾ OnScrn120ⁿᵏ | Shuffled back start 9 |

WORKOUTS: Apr 26 Hia 5f fst 1:03 B 4/8 Mar 30 Bdw 5f fst 1:04 B 3/3 Mar 12 Bdw 4f fst :53¹ Bg 1/1 Feb 26 Bdw 4f fst 1:03⁴ Bg 3/3 Feb 19 Bdw 4f fst :52⁴ B 2/2

Lottery Lil

Own: Franks John
B. f. 4
Sire: Lucky North (Northern Dancer)
Dam: Bet Bigger (Court Open)
Br: Franks John (Fla)
Tr: Wolfson Martin D (40 4 6 12 .10)

(handwritten: B 5/4)

NO RIDER (—)

L 122

Lifetime Record :	12 M 0 7		$12,850
1994	4 M 0 2	$3,690	Turf	9 0 0 ⑥	$10,830
1993	8 M 0 5	$9,160	Wet	1 0 0 1	$1,680
Hia ⊤	2 0 0 2	$3,220	Dist ⊤	3 0 0 2	$3,780

20Apr94-11Hia fm *1¹⁄₈ ⊤	1:50¹	ⒻMd Sp Wt	69 8 6 7⁸ 4⁸ 4³½ 3⁹	Castaneda M	L 122 b	5.70	78-15	Alinam122¹¼ Al's Memory122¹¼ Lottery Lil122⁹	Late rally 9
29Mar94-11Hia fm *1¹⁄₈ ⊤	1:50¹	ⒻMd Sp Wt	67 4 9 9⁹¾ 4⁸ 4³½ 3⁴¾	Madrid S O	L 122 b	8.90	76-17	Reigning Lady122¾ Alinam122³¼ Lottery Lil122¹½	Late rally inside 10
15Mar94-2GP fm 1¹⁄₈ ⊗ :47¹ 1:10⁴ 1:35³ 1:47⁴	ⒻMd Sp Wt	62 11 10 10¹² 8¹² 5⁶ 5⁸	Madrid S O	L 121 b	18.70	87-05	Super Chef121¾ Alinam121¾ Silly's Philly121³½	Belated bid 12
7Jan94-1GP fst 1¹⁄₁₆ :23³ :47² 1:12 1:44⁴	ⒻMd Sp Wt	46 6 9 10¹⁰ 10¹² 10¹⁶ 10¹⁴	Velasquez J	L 120 b	25.30	72-10	Wild Wild Robin121⁴ Bland120¹¼ Royalty On Ice120³	Outrun 11
16Dec93-5Crc fst 1¹⁄₁₆	1:51	ⒻMd Sp Wt	61 2 5 5⁴½ 5³ 4⁸ 3⁴½	Madrid S O	L 122 b	3.90	65-27	Speedy Colleen120¹ Al's Memory120⁴ Lottery Lil120³½	Lacked rally 8
11Nov93-5Crc sly 1¹⁄₈ ⊗ :49⁴ 1:15⁴ 1:43³ 1:57⁴	ⒻMd Sp Wt	57 2 7 5⁷⁴ 3⁵ 3⁴ 3⁴½	Madrid S O	L 119 b	4.50	67-27	SweepOverFinish120²½ FiryPrincess120² LottryLil120⁴	Lacked response 8
26Oct93-5Crc yl 1¹⁄₈ ⊤	1:53	ⒻMd Sp Wt	62 9 9 9¹½ 7⁶ 3¹½ 3¹½	Madrid S O	L 116 b	1.50	51-46	Cremeux119²½ Alinam119¼½ Lottery Lil119⁴	Late rally 9
10Sep93-4Crc yl 1¹⁄₈ ⊗	1:54⁴	ⒻMd Sp Wt	68 2 6 7⁷ 4⁴½ 4¼½ 3⁴½	Madrid S O	118 b	2.40	71-14	Recipe For Romance118³½ Mad Mo118ⁿᵏ Lottery Lil118¹½	Late rally 8
8Aug93-5Crc fm 1¹⁄₁₆ ⊗	1:44²	3↑ⒻMd Sp Wt	64 7 10 10⁸¾ 7³¾ 5³½ 1⁸	Madrid S O	117 b	8.90	80-16	Play On The Levee117¹¼ Infinitely Better117½ Lottery Lil117ʰᵈ	11
Steadied leaving bkstr, late rally
24Jly93-12Crc fst *1¹⁄₁₆ ⊤	1:46²	3↑ⒻMd Sp Wt	56 10 9 8⁷½ 6⁴½ 4⁶½ 4¹⁰¾	Rivera J A II	115 b	38.20	61-29	Courageous Belle110⁷ Slew O Tunes115³½ Diane's Lass110ⁿᵏ	Late rally 10

WORKOUTS: Apr 17 Crc 4f fst :50⁴ B 13/24 Apr 9 Crc 4f fst :50⁴ B 32/61 Mar 5 Crc 5f fst 1:03² B 24/39 Feb 26 Crc 4f fst :48⁴ H 4/48 Feb 15 Crc 5f fst 1:03 B 15/22 Feb 12 Crc 3f fst :37² B 11/24

Flashy Current

Own: Moore Joseph B Jr
Ch. f. 4
Sire: Little Current (Sea-Bird)
Dam: Flashy Dance (Dancing Champ)
Br: Harris Nancy (Ky)
Tr: Moore Joseph B Jr (—)

(handwritten: C 6)

ST LEON G (38 3 2 2 .08)

122

Lifetime Record :	12 M 2 0		$2,555
1994	2 M 0 0	$255	Turf	2 0 1 0	$1,655
1993	10 M 2 0	$2,300	Wet	1 0 0 0	$90
Hia ⊤	0 0 0 0		Dist ⊤	1 0 1 0	$1,480

| 8Apr94-4Tam fst 1¹⁄₁₆ :24² :49¹ 1:15 1:49² | ⒻMd Sp Wt | 7 3 7 7¹³ 7²⁶ 5¹¹ 5²²¼ | Henry W T | 120 | 7.50 | 52-25 | Imperial Appeal120⁴ DAnkara Mio120¹⁰ Private Delight120¹ | No threat 7 |
| Placed 4th through disqualification. |
11Mar94-2FG fst 1¹⁄₁₆ :23 :47² 1:13⁴ 1:46¹	ⒻMd 15000	26 4 11 11²¹ 11²¹ 10¹⁸ 8²³¾	Walker B J Jr	L 120	68.00	60-11	Mar Togo120¹ Cautela Gold120⁸ Peace Of Heart120²	No threat 12
10Dec93-2Crc fst 7f	:23² :47³ 1:13³ 1:27¾	3↑ⒻMd 25000	13 5 5 5⁶½ 6⁵¾ 6⁶¼ 6²²¾	Spieth S	121	6.30	54-18	Decent Mark121³ Miss Rexson121½ Carmen Alina121¾	Never close 6
24Oct93-1Kee fst 1¹⁄₁₆ ⊗ :24 :49² 1:15¹ 1:53³	3↑ⒻMd 25000	-0 1 11 11¹⁵ 11¹⁵ 12²⁸ 12²⁰½	Thompson T J	L 113 f	33.10	26-18	Maggie Strong118³½ CountDa'Judge121¼½ HarvestTheGold118⁵	Brief foot 12
2Oct93-1TP gd 1¹⁄₁₆ :24¹ 1:15	3↑ⒻMd Sp Wt	8 1 7 6²¼ 8⁹ 8¹⁹ 8⁴⁷½	Carvalho E J⁵	L 114	13.40	30-25	All Love119⁵ Hanque Bankque119¹² Sasso122ⁿᵏ	Saved ground 8
1Sep93-5Del fm 1¹⁄₁₆ ⊤ :23³ :47 1:12¹ 1:45¹	3↑Md Sp Wt	52 6 5 4³½ 4⁶½ 4⁷¼ 4⁷½	Burnham D M	115	9.40	75-11	Never Ever115⁴ Flashy Current115³ Galaquoit115¹½	Willingly 8
16Aug93-6Tdn fst 1¹⁄₁₆ :24¹ :49 1:14³ 1:47¼	3↑ⒻMd Sp Wt	26 3 5 8⁹½ 6¹³ 5⁸ 6²²½	Luttrell M G⁵	B 105	13.60	57-23	Priceless Freedom115⁸ Queen's Savage122¼ Mayor Of Oak115⁶¼	Outrun 8
26Jly93-2Tdn fst 1¹⁄₁₆ :24¹ :49 1:00⁴ 1:13³	3↑Md Sp Wt	30 2 10 8⁸ 5⁷ 5¹⁰ 5¹¹	Weiler D	B 116	15.30	71-17	Family Lit115²½ Joy South116²¾ Burning Spirit116ⁿᵏ	Belated rally 10
4Jly93-10Mnr fm 7f ⊤ :23² :46¹ 1:10⁴ 1:25²	3↑ⒻAlw 3500NC	26 8 10 10¹⁴ 10¹² 8⁹ 4⁹	Williams D A	B 120	21.90	—	Nellie Fox123¾ Shockaree115²½ Cheap Wheels115⁴	No factor 10
25May93-6CD fst 1¹⁄₁₆	1:37⁴	3↑ⒻMd 30000	26 9 12 11²¹ 11¹¹ 12²⁶¼ 12²⁸¾	Hebert T J	110	45.20	61-19	Go Quietly107¾ Shelbiana Ridge110³ Miss Joe Anne110²	12
Steady sharply, early, outrun

WORKOUTS: Feb 13 FG 3f fst :37 B 4/11

continued

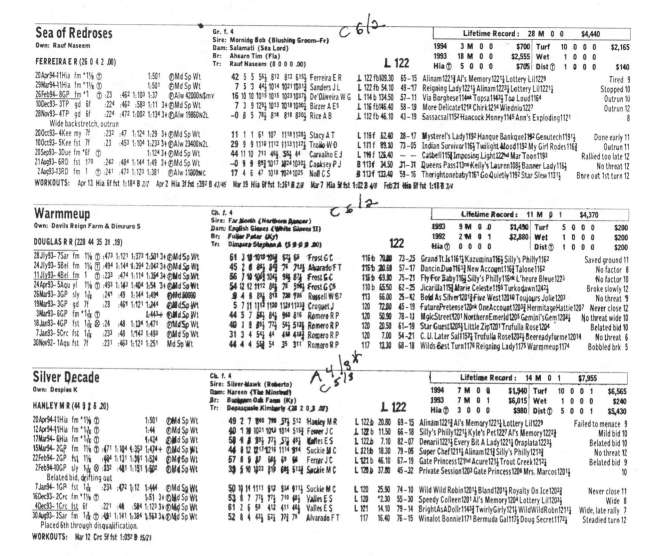

© 1994 DRF, Inc.

Since Pm's Cherokeehoney is a threat to steal the race, she should be bet to Win and played behind both Al's Memory and Lottery Lil in the exacta. Or alternatively, she can be bet to Win and played for second and third in the trifecta with the two favorites. While Silver Decade has an A rating, she is a proven and chronic loser and, since a horse's race record always supercedes its pedigree ratings, she should be thrown out.

Results: Pm's Cherokeehoney opened up a 4 length lead going into the stretch before her negative stamina index and Al's Memory caught up with her at the wire. Lottery Lil finished a distant third. Al's Memory paid $4 on the Win. The $2 exacta returned $26.60, while the $2 trifecta paid $55.20. While Pm's Cherokeehoney didn't win the race, she was worth playing here because lone speed horses are definite threats to outrun the limits of their stamina indexes and, if not win, finish in the money.

Belmont Park

2

1¼ MILES. **(Inner Turf). (1:57³) MAIDEN SPECIAL WEIGHT. Purse $30,000. Fillies and mares, 3-year-olds and upward. Weights: 3-year-olds, 118 lbs. Older, 122 lbs.**

INNER TURF COURSE — 1¼ MILES — START FINISH

Crazy Fling

Own: Skara Glen Stables

BAILEY J D (—)

Ch. f. 3 (May)
Sire: Alysheba (Alydar)
Dam: Some Romance (Fappiano)
Br: Skara Glen Stables (Ky)
Tr: Skiffington Thomas J (—)

118

Lifetime Record: 1 M 1 0 $5,940

1994	1 M 1 0	$5,940	Turf	1 0 1 0	$5,940
1993	0 M 0 0		Wet	0 0 0 0	
Bel Ⓣ	0 0 0 0		Dist Ⓣ	0 0 0 0	

13Aug94- 5Sar fm 1 1/16 [T] :242 :491 1:134 1:45 3↑ Ⓕ Md Sp Wt 63 4 8 84¾ 84¼ 62½ 22½ Bailey J D 117 4.50e 71–23 MsTasso1172½ CrzyFling117hd DnceOfSunshine117nk Bmpd, checked early 10

WORKOUTS: Aug 29 Sar tr.t 5f fst 1:07 B 15/18 Aug 8 Sar tr.t Ⓣ 5f fm 1:01 B 3/22 Aug 1 Sar tr.t Ⓣ 4f fm :504 B 21/23 Jly 25 Sar tr.t Ⓣ 6f fm 1:16 B 9/10 Jly 18 Bel 5f fst 1:04 B 24/25 Jly 13 Bel 4f fst :523 Bg 37/38

Faux

Own: Weinreb Donald

PEZUA J M (—)

Ch. f. 4
Sire: Mining (Mr. Prospector)
Dam: Fabulous Fraud (Le Fabuleux)
Br: Phipps Ogden Mills (Ky)
Tr: Debonis Robert (—)

122

Lifetime Record: 10 M 0 0 $1,530

1994	5 M 0 0		Turf	6 0 0 0	$1,530
1993	5 M 0 0	$1,530	Wet	2 0 0 0	
Bel Ⓣ	5 0 0 0	$1,530	Dist Ⓣ	3 0 0 0	$1,530

20Jly94- 2Bel fm 1 1/16 Ⓣ :232 :462 1:104 1:424 3↑ Ⓕ Md Sp Wt 55 7 6 97½ 75¾ 66½ 68¼ Alvarado F T 122 61.10 76–18 Marigal1161 Nobody Picked Six1162¾ Serene Beauty1162¾ Steadied 1/2 pl 12
19Jun94- 5Bel fm 1¼ [T] :484 1:131 1:382 2:022 3↑ Ⓕ Md Sp Wt 39 11 5 32½ 116 108¾ 917¼ Alvarado F T 122 56.50 60–15 Sudana1143½ Valley Ofthe Jolly114¾ Neon Fairytale114nk Done early 11
3Jun94- 5Bel fm 1¼ [T] :481 1:132 1:38 2:014 3↑ Ⓕ Md Sp Wt 50 1 1 11 53 76½ 813¼ Beckner D V7 115 20.50 68–16 Kyle's Pet1225½ Sudana114½ Viva La Dance114no Used in pace 11
20May94- 5Bel my 1 1/16 ⊗ :23 :46 1:104 1:441 3↑ Ⓕ Md Sp Wt 10 6 9 910 915 921 1038½ Cruguet J 124 44.00 42–25 Muko115hd Sam's Diary1158 La Rima115½ Outrun 10
8May94- 5Bel my 1¼ ⊗ :471 1:123 1:39 2:043 3↑ Ⓕ Md Sp Wt 23 5 5 56½ 75½ 711 726¾ Cruguet J 124 19.70 41–17 Mrs. Marcos1248 Call Today1101 Number Thirty1155¾ No factor 8
11Nov93- 2Aqu yl 1½ Ⓣ :52 1:173 2:101 2:371 3↑ Ⓕ Alw 30000N1X 37 8 8 1215 1216 1122 1023 Alvarado F T 115 72.80 47–34 Talone1157½ Saffronella115hd Amistad1172 Outrun 12
10Oct93- 6Bel fst 1 1/16 :23 :462 1:123 1:48 3↑ Ⓕ Md Sp Wt — 1 9 911 914 — — Cruguet J 119 b 20.40 — 25 HelloHnne1191½ Cro'sBeuty1191½ KeyToThePec1149 Broke slowly, eased 9
12Sep93- 5Bel yl 1 1/16 Ⓣ :232 :462 1:104 1:431 3↑ Ⓕ Md Sp Wt 13 1 6 59 911 917 929¼ Cruguet J 118 11.90 51–23 Arctic Aaria1184 Magical Afleet1181 Genie's Flight118½ No factor 10
27May93- 7Bel fm 1¼ [T] :48 1:133 1:382 2:03 3↑ Ⓕ Md Sp Wt 62 4 2 31½ 52 52¾ 46¼ Cruguet J 115 29.30 68–17 Marie Celeste1191¼ Chic Sheba1152½ Gulch's Ravine1152¾ Saved ground 7
23Apr93- 3Aqu fst 6f :223 :462 :583 1:112 3↑ Ⓕ Md 50000 46 6 5 68½ 76½ 66½ 610¼ Mojica R Jr 115 4.80e 75–15 Miss Bold Appeal1151 Jiving Around1152¾ WavingTheFlag1154 No factor 7

WORKOUTS: Aug 23 Bel tr.t 4f my :53 B 2/3

Dormcat

Own: Bohemia Stable

BECKNER D V (—)

Dk. b or br f. 3 (May)
Sire: Storm Cat (Storm Bird)
Dam: Little Red Robin (Ribot)
Br: Daniel M. Galbreath (Ky)
Tr: Jerkens H Allen (—)

113⁵

Lifetime Record: 6 M 0 0 $3,200

1994	6 M 0 0	$3,200	Turf	0 0 0 0	
1993	0 M 0 0		Wet	2 0 0 0	$1,500
Bel Ⓣ	0 0 0 0		Dist Ⓣ	0 0 0 0	

20Aug94- 2Sar my 6f :223 :463 :591 1:112 3↑ Ⓕ Md Sp Wt 47 4 4 32 3½ 42 510¼ Beckner D V5 112 b 11.40 73–16 Steady As A Cat1174 Demonize1171 Sonnet1173½ Tired 7
8Aug94- 5Sar fst 7f :231 :461 1:11 1:242 3↑ Ⓕ Md Sp Wt 60 6 3 52½ 75½ 63¾ 68¼ Maple E 117 8.90 74–15 Affirmed Peach1171½ Facula1171¼ Radiant Beams1221¾ No threat 10
28Jly94- 2Sar sly 6f :221 :46 :583 1:12 3↑ Ⓕ Md Sp Wt 60 4 2 42½ 24 35 47¾ Maple E 116 f 7.40 73–17 Mismatch1162½ Real Catch1165 Faith In Dreams116½ Tired 7
22Feb94- 6GP fst 1 1/16 :233 :472 1:114 1:444 Ⓕ Md Sp Wt 69 4 1 1hd 1hd 3½ 44½ Maple E 120 13.60 81–19 Red Star120¾ Accent On Gold120nk Ballerina Queen1203½ Weakened 10
15Feb94- 1GP fst 6f :22 :452 :582 1:121 Ⓕ Md Sp Wt 41 4 4 35 35 54¾ 55 Maple E 121 4.80 75–13 Superb Broad121no Our Miz Waki1211½ Cantada's Sword121½ Faded 7
23Jan94- 6GP fst 6f :22 :444 :571 1:094 Ⓕ Md Sp Wt 45 12 1 44 46½ 69¾ 915¼ Maple E 120 7.00 76–08 Classy Pat1205½ Brand Loyalty1203 Code Of Old120¾ Faded 12

WORKOUTS: ●Aug 31 Bel 7f fst 1:28 B 1/8 Aug 17 Sar 5f fst 1:044 B 54/61 Aug 7 Sar 3f fst :36 H 2/21 Aug 4 Sar 5f fst 1:014 H 14/34 Jly 25 Sar 5f fst 1:02 H 26/64 Jly 16 Bel 4f fst :494 B 49/66

Miriams Jewel

Own: Tucker Jeffrey

SANTOS J A (—)

Ch. f. 3 (May)
Sire: Cozzene (Caro)
Dam: Chop It Up (The Axe II)
Br: Penn W E (Ky)
Tr: O'Connell Richard (—)

118

Lifetime Record: 3 M 0 0 $0

1994	3 M 0 0	Turf	3 0 0 0
1993	0 M 0 0	Wet	0 0 0 0
Bel Ⓣ	1 0 0 0	Dist Ⓣ	2 0 0 0

22Aug94- 2Sar sf 1 3/8 Ⓣ :483 1:152 1:421 2:02 3↑ Ⓕ Md Sp Wt 9 12 4 63¾ 106½ 913 830 Davis R G 117 f 50.70 25–41 Highland Tootsie1223½ Polish Treaty1171½ Miami Meadow117hd Tired 12
7Aug94- 10Sar gd 1 3/8 Ⓣ :472 1:124 1:382 1:564 3↑ Ⓕ Md Sp Wt 53 10 6 1013 105¾ 99¾ 819½ Velazquez J R 117 f 9.20 61–18 Daad1175½ Helen Ge Ge1177 Flower Delivery1171½ No factor 11
20Jly94- 2Bel fm 1 1/16 Ⓣ :232 :462 1:104 1:424 3↑ Ⓕ Md Sp Wt 42 3 12 1213 109 1011 1014 Santos J A 116 f 5.40 70–18 Marigal1161 Nobody Picked Six1162¾ Serene Beauty1162¾ Broke slowly 12

WORKOUTS: Aug 31 Sar Ⓣ 4f sf :512 B (d) 2/2 Aug 16 Sar Ⓣ 4f fm :50 B 12/17 Aug 5 Sar tr.t Ⓣ 5f gd 1:034 B (d) 8/11 ●Aug 1 Sar tr.t Ⓣ 3f fm :354 H 1/5 Jly 14 Bel [T] 7f fm 1:282 B (d) 1/1 Jly 7 Bel [T] 7f fm 1:264 H (d) 1/

Superstar J. R.

Own: Joran Stables

CRUGUET J (—)

Ch. f. 3 (Mar)
Sire: Exuberant (What a Pleasure)
Dam: Tryjen (Teddy's Courage)
Br: Winner's Circle Corp. (Ky)
Tr: Ortiz Paulino O (—)

118

Lifetime Record: 6 M 0 1 $1,320

1994	6 M 0 1	$1,320	Turf	1 0 0 0
1993	0 M 0 0		Wet	1 0 0 0
Bel Ⓣ	1 0 0 0		Dist Ⓣ	1 0 0 0

29Jun94- 1Bel fst 7f :223 :461 1:122 1:254 3↑ Ⓕ Md 30000 20 2 4 2½ 3½ 710 720½ Velazquez J R 110 21.60 53–16 Hillraiser114½ Stop Right Here1137 Over Dinner112nk Used up 7
19Jun94- 5Bel fm 1¼ [T] :484 1:131 1:382 2:022 3↑ Ⓕ Md Sp Wt 24 4 1 11½ 2hd 98½ 1024½ Leon F 114 58.60 53–15 Sudana1143½ Valley Ofthe Jolly114¾ Neon Fairytale114nk Used in pace 11
2Jun94- 2Bel fst 1 1/16 :232 :473 1:134 1:473 3↑ Ⓕ Md 35000 –0 5 2 65 713 727 739¾ Mojica R Jr 114 20.60 — 28 Bland1225 Nectaria1112½ Stop Right Here114¾ Done early 7
19May94- 5Bel sly 6f :221 :46 :584 1:121 3↑ Ⓕ Md 35000 35 4 4 55½ 65½ 78 711¼ Chavez J F 115 4.60 69–15 Glory And Grace1151¼ Holiday Holly1152¾ Lady Great111no Outrun 7
10May94- 9Bel fst 7f :224 :462 1:111 1:241 3↑ Ⓕ Md 35000 –0 10 11 96¼ 98½ 1120 1133¼ Smith M E 115 *2.50 48–12 LuckyOldViolet1137½ PremiumSpice1153 CelstilFlm1151 No factor, wide 14
21Apr94- 5Aqu fst 6f :222 :46 :582 1:113 3↑ Ⓕ Md 35000 50 5 5 43½ 34 36 38¾ Chavez J F 115 10.30 75–19 Terra Rubra1157 Passing Magic1151¾ Superstar J. R.1151 Greenly 14

WORKOUTS: Aug 28 Bel tr.t 6f fst 1:153 B 2/6 Aug 21 Bel tr.t 4f fst :502 H 10/16 Jly 14 Bel 4f fst :52 B 31/35 Jun 16 Bel tr.t 4f fst :492 B 8/16

Steadfast Love

Own: Pomerantz Lawrence J

VELAZQUEZ J R (—)

Dk. b or br f. 3 (Apr)
Sire: Chief's Crown (Danzig)
Dam: Martessana (Sir Ivor)
Br: Bertram N. Linder (Ky)
Tr: Dimauro Stephen L (—)

118

Lifetime Record: 2 M 0 0 $0

1994	2 M 0 0	Turf	2 0 0 0
1993	0 M 0 0	Wet	0 0 0 0
Bel Ⓣ	1 0 0 0	Dist Ⓣ	0 0 0 0

13Aug94- 2Sar fm 1 1/16 [T] :231 :47 1:111 1:433 3↑ Ⓕ Md Sp Wt 52 5 8 812 1011 79½ 714 Leon F 117 46.20 67–23 World Predictions1173 Jodi's Land1175½ Powerful Lad117no No factor 10
29Jun94- 2Bel fm 1 Ⓣ :232 :47 1:112 1:364 3↑ Ⓕ Md Sp Wt 47 12 6 94½ 127½ 1210 1213¾ Velazquez J R 114 26.10 64–20 Viva La Dance114½ Miami Meadow1092½ Hoist It Proudly1221½ Done early 12

WORKOUTS: Aug 27 Sar 5f gd 1:031 B 8/10 Aug 8 Sar tr.t Ⓣ 5f fm 1:02 B 17/22 Jly 18 Bel 4f fst :494 H 23/45 Jly 13 Bel tr.t 5f fst 1:031 B 5/7 Jly 7 Bel tr.t 4f fst :504 B 6/6 ●Jun 25 Bel tr.t 4f fst :481 H 1/21

continued

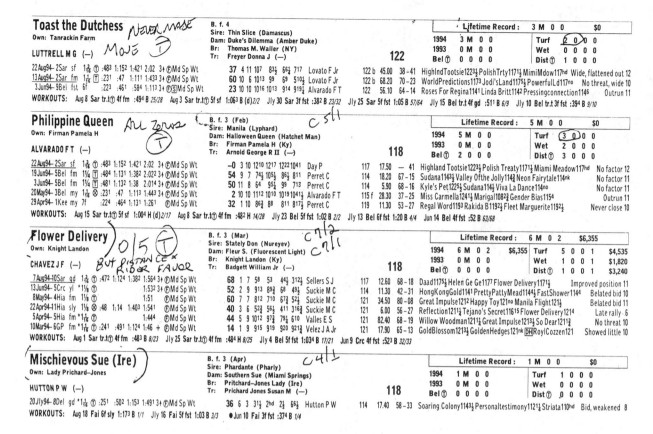

Toast the Dutchess — B. f. 4
Own: Tanrackin Farm
Sire: Thin Slice (Damascus)
Dam: Duke's Dilemma (Amber Duke)
Br: Thomas M. Waller (NY)
Tr: Freyer Donna J (—)
LUTTRELL M G (—) 122

Lifetime Record: 3 M 0 0 $0
1994 3 M 0 0 | Turf 2 0 0 0
1993 0 M 0 0 | Wet 0 0 0 0
Bel (T) 0 0 0 0 | Dist (T) 1 0 0 0

22Aug94- 2Sar sf 1 3/16 (T) :483 1:152 1:421 2:02 3+ (F)Md Sp Wt 37 4 11 107 83½ 66¼ 717 Lovato F Jr 122 b 45.00 38-41 HighIndTootsie1223½ PolishTrty1171½ MimiMdow117hd Wide, flattened out 12
13Aug94- 2Sar fm 1 1/16 [T] :231 :47 1:111 1:433 3+ (F)Md Sp Wt 60 10 6 1013 99 69 510½ Lovato F Jr 122 b 68.20 70-23 WorldPredictions1173 Jodi'sLand1175½ PowerfulLd117no No threat, wide 10
3Jun94- 9Bel fst 6f :223 :461 :584 1:113 3+ (F)(S)Md Sp Wt 23 10 10 1016 1013 914 919¼ Alvarado F T 122 56.10 64-14 Roses For Regina1141 Linda Britt1142 Pressingconnection1146 Outrun 11
WORKOUTS: Aug 8 Sar tr.t(T) 4f fm :494 B 25/28 Aug 3 Sar tr.t(T) 5f sf 1:063 B (d)2/2 Jly 30 Sar 3f fst :382 B 23/32 Jly 25 Sar 5f fst 1:05 B 57/64 Jly 15 Bel tr.t 4f gd :511 B 6/9 Jly 10 Bel tr.t 3f fst :394 B 9/10

Philippine Queen — B. f. 3 (Feb)
Own: Firman Pamela H
Sire: Manila (Lyphard)
Dam: Halloween Queen (Hatchet Man)
Br: Firman Pamela H (Ky)
Tr: Arnold George R II (—)
ALVARADO F T (—) 118

Lifetime Record: 5 M 0 0 $0
1994 5 M 0 0 | Turf 3 0 0 0
1993 0 M 0 0 | Wet 2 0 0 0
Bel (T) 2 0 0 0 | Dist (T) 3 0 0 0

22Aug94- 2Sar sf 1 3/16 (T) :483 1:152 1:421 2:02 3+ (F)Md Sp Wt -0 3 10 1210 1217 1222 1041 Day P 117 17.50 — 41 Highland Tootsie1223½ Polish Treaty1171½ Miami Meadow117hd No factor 12
19Jun94- 5Bel fm 1¼ [T] :484 1:131 1:382 2:022 3+ (F)Md Sp Wt 54 9 7 74½ 105½ 86½ 811 Perret C 114 18.20 67-15 Sudana1143½ Valley Ofthe Jolly114¾ Neon Fairytale114nk No factor 11
3Jun94- 5Bel fm 1¼ [T] :481 1:132 1:38 2:014 3+ (F)Md Sp Wt 50 11 8 64 95¼ 99 713 Perret C 114 5.90 68-16 Kyle's Pet1225½ Sudana114½ Viva La Dance114no No factor 11
20May94- 3Bel my 1 1/16 (X) :231 :47 1:113 1:443 3+ (F)Md Sp Wt 2 10 10 1112 1010 1019 1041½ Alvarado F T 115 f 28.30 37-25 Miss Carmella1241½ Marigal1082¾ Gender Bias1154 Outrun 11
29Apr94- 1Kee my 7f :224 :464 1:131 1:261 (F)Md Sp Wt 32 1 10 86¾ 88 811 817½ Perret C 119 11.30 53-27 Regal Word1193 Rakida B1192½ Fleet Marguerite1192½ Never close 10
WORKOUTS: Aug 15 Sar tr.t(T) 5f sf 1:004 H (d)2/17 Aug 8 Sar tr.t(T) 4f fm :482 H 14/28 Jly 23 Bel 5f fst 1:02 B 2/2 Jly 13 Bel 6f fst 1:20 B 4/4 Jun 14 Bel 4f fst :52 B 63/68

Flower Delivery — B. f. 3 (Mar)
Own: Knight Landon
Sire: Stately Don (Nureyev)
Dam: Fleur S. (Fluorescent Light)
Br: Knight Landon (Ky)
Tr: Badgett William Jr (—)
CHAVEZ J F (—) 118

Lifetime Record: 6 M 0 2 $6,355
1994 6 M 0 2 $6,355 | Turf 5 0 0 1 $4,535
1993 0 M 0 0 | Wet 1 0 0 1 $1,820
Bel (T) 0 0 0 0 | Dist (T) 1 0 0 1 $3,240

7Aug94-10Sar gd 1 3/16 (T) :472 1:124 1:382 1:564 3+ (F)Md Sp Wt 68 1 7 58 53 44½ 312½ Sellers S J 117 12.60 68-18 Daad1175½ Helen Ge Ge1177 Flower Delivery1171½ Improved position 11
13Jun94- 5Crc yl *1⅛ (T) 1:532 3+ (F)Md Sp Wt 52 2 9 913 89¾ 68 49½ Suckie M C 114 11.30 42-31 HongKongGold1141 PrettyPattyMead1144½ FastShower1144 Belated bid 10
8May94- 4Hia fm 1⅛ (T) 1:51 (F)Md Sp Wt 60 7 7 812 710 67¾ 52½ Suckie M C 121 34.50 80-08 Great Impulse1212 Happy Toy121no Manila Flight121½ Belated bid 11
22Apr94-11Hia sly 1⅛ (X) :48 1:14 1:403 1:541 (F)Md Sp Wt 40 3 6 52¾ 56½ 411 316¾ Suckie M C 121 6.00 56-27 Reflection1211¼ Tejano's Secret11615 Flower Delivery1214 Late rally 6
5Apr94- 5Hia fm *1 1/16 (T) 1:444 (F)Md Sp Wt 44 5 9 1012 97¾ 79¼ 610 Valles E S 121 82.40 68-19 Willow Woodman1211¼ Great Impulse1213½ So Dear1211¾ No threat 10
10Mar94- 6GP fm *1 1/16 (T) :241 :491 1:124 1:46 + (F)Md Sp Wt 14 1 9 915 919 920 921¾ Velez J A Jr 121 17.90 65-13 GoldBlossom1213½ GoldenHedges121nk [DH]RoylCozzen121 Showed little 10
WORKOUTS: Aug 1 Sar tr.t(T) 4f fm :483 B 8/23 Jly 25 Sar tr.t(T) 4f fm :484 H 8/25 Jly 4 Bel 5f fst 1:034 B 17/21 Jun 9 Crc 4f fst :523 B 32/33

Mischievous Sue (Ire) — B. f. 3 (Apr)
Own: Lady Prichard-Jones
Sire: Phardante (Pharly)
Dam: Southern Sue (Miami Springs)
Br: Pritchard-Jones Lady (Ire)
Tr: Prichard Jones Susan M (—)
HUTTON P W (—) 118

Lifetime Record: 1 M 0 0 $0
1994 1 M 0 0 | Turf 1 0 0 0
1993 0 M 0 0 | Wet 0 0 0 0
Bel (T) 0 0 0 0 | Dist (T) 0 0 0 0

20Jly94- 8Del gd *1 1/16 (T) :251 :502 1:153 1:491 3+ (F)Md Sp Wt 36 6 3 31½ 2hd 2½ 66½ Hutton P W 114 17.40 58-33 Soaring Colony1143½ Personaltestimony1121¼ Striata110hd Bid, weakened 8
WORKOUTS: Aug 18 Fai 6f sly 1:173 B 1/1 Jly 16 Fai 5f fst 1:03 B 3/3 ●Jun 10 Fai 3f fst :374 B 1/4

The second race at Belmont on September 3rd was sent in by fellow handicapper and colleague Mark Cramer who field tested the ratings in *Exploring Pedigree* at New York, Florida and Maryland tracks. Eliminating the proven losers, Mark narrowed the contenders down to Crazy Fling, Flower Delivery, Mischievous Sue and Dormcat who was making her turf debut.

While Crazy Fling may have deserved to be the mild favorite, she was worth betting against at 4-5 odds because of the fact that she had lower turf speed figures than five-time loser Flower Delivery and was coming off a suspiciously slow 1:07 work on August 29th.

Looking for better value, Mark decided to key on first-time-turfer Dormcat at 4-1 because of her high turf and class ratings as well as the fact that she had the stamina indexes to get the 1 1/4 mile distance. Mark bet Dormcat to Win and also boxed her in the exacta with Crazy Fling, Flower Delivery (who was switching to the redhot Chavez) and the lightly-raced Mischievous Sue who was bred to get the distance and going off at an enticing 14-1.

SECOND RACE 1¼ MILES. (Inner Turf)(1.57^{3}) **MAIDEN SPECIAL WEIGHT. Purse $30,000. Fillies and mares, 3-year-olds and upward. Weights: 3-year-olds, 118 lbs. Older, 122 lbs.**

Belmont

SEPTEMBER 3, 1994

Value of Race: $30,000 Winner $18,000; second $6,600; third $3,600; fourth $1,800. Mutuel Pool $246,915.00 Exacta Pool $491,395.00 Quinella Pool $113,699.00

Last Raced	Horse	M/Eqt.	A.	Wt	PP	¼	½	¾	1	Str	Fin	Jockey	Odds $1
20Aug94 ^{2}Sar^{5}	Dormcat	bf	3	113	3	$1^{1/2}$	$1^{1/2}$	$1^{1/2}$	1^{1}	1^{3}	1^{3}	Beckner D V^{5}	4.30
7Aug94 ^{10}Sar^{3}	Flower Delivery		3	118	9	$8^{1/2}$	9^{10}	$8^{1/2}$	4^{1}	$4^{1\,1/2}$	2^{nk}	Chavez J F	4.10
13Aug94 ^{5}Sar^{2}	Crazy Fling		3	118	1	$7^{1\,1/2}$	$8^{1/2}$	9^{14}	5^{hd}	$5^{3\,1/2}$	$3^{1\,1/4}$	Bailey J D	0.80
22Aug94 ^{2}Sar^{10}	Philippine Queen	b	3	118	8	6^{1}	$5^{1/2}$	4^{hd}	$2^{1/2}$	3^{hd}	4^{nk}	Alvarado F T	28.50
22Aug94 ^{2}Sar^{8}	Miriams Jewel	f	3	118	4	2^{1}	$2^{1/2}$	$2^{1/2}$	$3^{1\,1/2}$	$2^{1/2}$	5^{4}	Santos J A	29.70
22Aug94 ^{2}Sar^{7}	Toast the Dutchess	b	4	122	7	$5^{1/2}$	$6^{1/2}$	7^{1}	$6^{2\,1/2}$	6^{3}	$6^{3\,3/4}$	Luttrell M G	15.50
13Aug94 ^{2}Sar^{7}	Steadfast Love		3	118	6	4^{1}	$3^{1/2}$	$3^{1/2}$	$7^{1/2}$	$7^{1/2}$	7^{1}	Velazquez J R	30.70
20Jly94 ^{2}Bel^{6}	Faux		4	122	2	10	10	10	10	9^{14}	8^{nk}	Pezua J M	27.70
20Jly94 ^{8}Del^{6}	Mischievous Sue-IR		3	118	10	9^{5}	$7^{1/2}$	6^{hd}	8^{4}	8^{1}	9^{17}	Hutton P W	13.90
29Jun94 ^{1}Bel^{7}	Superstar J. R.	b	3	118	5	3^{hd}	4^{1}	5^{1}	$9^{1/2}$	10	10	Cruguet J	30.20

OFF AT 1:30 Start Good. Won driving. Time, :23^{3}, :49^{2}, 1:14^{1}, 1:38^{3}, 2:02^{4} Course firm.

$2 Mutuel Prices:			
3-(C)-DORMCAT	10.60	6.00	2.60
9-(I)-FLOWER DELIVERY		5.40	2.20
1-(A)-CRAZY FLING			2.20

$2 EXACTA 3-9 PAID $44.40 $2 QUINELLA 3-9 PAID $21.60

Dk. b. or br. f, (May), by Storm Cat–Little Red Robin, by Ribot. Trainer Jerkens H Allen. Bred by Daniel M. Galbreath (Ky).

NW1 Allowance and Stakes Races

A number of horses bred for the turf break their maiden on the dirt and are then switched to the grass against winners at the NW1 Allowance level. The third race at Santa Anita on April 7th is such an example with what looks like a legitimate odds-on favorite. The horse to beat here is Wild Impulse who just missed with a high Beyer in her first attempt on the grass. Of the other eight horses, all can be eliminated on pedigree grounds, except for Dynatane with her A rating, and three-time winner Rising Mist. While Baby Diamonds has also won on the turf, Wild Impulse just beat her by more than five lengths.

Though Dynatane looks totally outgunned in the Beyer department, it is worth noting that those figures were earned on the dirt in sprints as a 2-year-old. Furthermore, she won her debut, has had eight months to further mature and McAnally excels with layoff horses. Switched to a surface she is bred to love, she has every right to dramatically improve today for her Wayfare farm owner/breeder connections. On the downside, her dam's offspring are 0-4 on the turf and she is 30-1 on the board.

With the best turf record in the race, Rising Mist is the other possible upsetter. Although she is shortening from a mile to a sprint, this maneuver works well at the Santa Anita downhill course with early speed routers. Nevertheless, she has less speed than Wild Impulse, and has already lost three times at this level since coming from New Zealand.

3

About 6½ Furlongs. (Turf). (1:11³) ALLOWANCE. Purse $37,000. Fillies and mares, 3-year-olds and upward, which are non-winners of $3,000 other than maiden or claiming. Weights: 3-year-olds, 115 lbs. Older, 123 lbs. Non-winners of a race other than claiming, allowed 3 lbs. (Starters for a claiming race of $32,000 or less in their last three starts and maidens that are non-starters for a claiming price have second preference.)

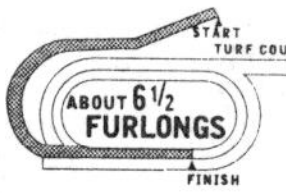

ABOUT 6½ FURLONGS

Dynatane

Own: Kilbride Duane & Roger & Wayfare Fm

GONZALEZ S JR (326 38 30 40 .12)

Dk. b or br f. 3 (Apr)
Sire: Dynaformer (Roberto)
Dam: Matane (Avatar)
Br: Wafare Farm (Ky)
Tr: McAnally Ronald (124 18 16 16 .15)

L 110⁵

handwritten: A 6/1, D 7/2, Dirt only

	Lifetime Record :	2 1 0 0	$10,500		
1993	2 1 0 0	$10,500	Turf	0 0 0 0	
1992	0 M 0 0		Wet	0 0 0 0	
SA	0 0 0 0		Dist	0 0 0 0	

| 22Aug93–8AP | fst 6f | :21⁴ :45 :58 1:12 | ⑤Palatine B C50k | 26 8 7 86½ 8¹² 9²¹ 9¹⁷½ | Sellers S J | L 115 | 27.80 | 67–14 | Traceys Rock115½ Mariah's Storm115²½ Bayou Plans115½ | No rally 11 |
| 28Jly93–1AP | fst 5½f | :22¹ :46¹ :59⁴ 1:07¹ | ⑤Md Sp Wt | 41 1 1 2½ 1hd 12½ 1no | Sellers S J | L 118 | *1.60 | 77–20 | Dynatane118no Alystone118⁶ Rasputin118⁶½ | All out 6 |

WORKOUTS: Apr 2 Hol 5f fst 1:00² H 7/32 • Mar 28 Hol 7f fst 1:29¹ H 12/18 Mar 22 Hol 7f fst 1:28¹ H 3/10 Mar 15 Hol 6f fst 1:16 H 10/14 Mar 9 Hol 6f fst 1:14² H 10/18 Mar 2 Hol 5f fst 1:01¹ H 7/25

Francie's Fancy

Own: Willis Michael T

ANTLEY C W (341 56 32 22 .16)

Ch. f. 4
Sire: Flying Paster (Gummo)
Dam: Pale Leaves (Green Dancer)
Br: Cardiff Stud Farms (Cal)
Tr: Rash Rodney (83 9 10 14 .11)

L 123

handwritten: D4/3, cull

	Lifetime Record :	5 1 1 1	$30,325		
1994	3 0 1 1	$13,950	Turf	0 0 0 0	
1993	2 1 0 0	$16,375	Wet	0 0 0 0	
SA ⑦	0 0 0 0		Dist ⑦	0 0 0 0	

26Feb94–5SA	fst 1	:23 :46¹ 1:10⁴ 1:35²	⑤Alw 40000N1x	75 1 8 78½ 56 57¾	Stevens G L	LB 117	3.70	84–10	Dolce Amore117³ Wende120¾ Explosivmes112³½	No late bid 9
28Jan94–7SA	fst 6f	:21³ :44⁴ :57 1:09³	⑤SAlw 37000N1x	81 4 5 49½ 54 43½ 33	Stevens G L	LB 120	*1.30	86–12	Ms.Something117²½ MemorableTravels120no Francie'sFancy120¹½	Mild bid 6
7Jan94–7SA	fst 6½f	:21² :44¹ 1:08² 1:14⁴	⑤SAlw 37000N1x	81 1 9 77¾ 67 35½ 2⁸	Stevens G L	LB 120	6.70	90–08	Andestine120⁸ Francie's Fancy120¹⅜ No Sacrifice121½	Second best 9
21Mar93–1Hol	fst 1½	:23¹ :46² 1:11 1:44 3+	⑤Alw 39000N1x	69 4 1 1hd 1hd 1hd 53½	Stevens G L	LB 114	*2.90	77–11	Bless Shoe122no Nortena113¾ Alyshena114²½	Weakened a bit 9
13Mar93–9SA	fst 6f	:21⁴ :45¹ :57³ 1:10²	⑤SMd Sp Wt	82 4 7 74 63½ 34 1¹	Stevens G L	B 117	21.90	85–12	Francie's Fancy117¹ January Jeanie117¹½ Malojen117²½	4 Wide stretch 12

WORKOUTS: Mar 28 SA 7f fst 1:28² H 6/7 Mar 15 SA 3f fst :39² H 23/23 Feb 24 SA 3f fst :37¹ H 15/21 Feb 18 SA 7f gd 1:32¹ H 2/4 Feb 11 SA 3f fst :36 H 15/37 Jan 26 SA 3f gd :36⁴ H 3/11

Northern Wool

Own: Pinner John

BLACK C A (320 26 37 36 .08)

Dk. b or br f. 4
Sire: Northern Baby (Northern Dancer)
Dam: Kissapotamus (Illustrious)
Br: LeBus C L (Ky)
Tr: Velasquez Danny (27 2 2 1 .07)

L 120

handwritten: N.W. training on turf; BS 1/1; BS 1/4

	Lifetime Record :	12 2 1 3	$42,013		
1994	5 1 0 0	$11,475	Turf	0 0 0 0	
1993	6 1 1 2	$24,350	Wet	0 0 0 0	
SA	0 0 0 0		Dist	0 0 0 0	

16Mar94–5SA	fst 1	:23⁴ :47³ 1:11³ 1:37⁴	⑤Clm c–12500	68 8 4 3³ 33½ 3⁶ 53½	Valenzuela F H	LB 118 b	3.80	77–20	Crossoverthebridge116½ Fatih'sFntsy118⅜ ElicecAloh116½	Mild late bid 10
	Claimed from Pegram Michael, Baffert Bob Trainer									
9Mar94–2SA	fst 1½	:23² :47² 1:12 1:43⁴	⑤Clm 20000	53 7 4 3³ 4² 7¹⁰ 8¹⁷½	Valenzuela F H	LB 116 b	11.30	58–24	Gordy's Dancer115½ Moms Baby116²½ Charisma116¼	Gave way 8
23Feb94–7SA	fst 1½	:23⁴ :48 1:12³ 1:44	⑤Clm 16000	88 1 1 1¹½ 1¹½ 2hd 1²	Valenzuela F H	LB 115 b	7.70	75–22	Northern Wool115² Moms Baby115no Madame Bovary112no	8
	Brushed 1/16, game on rail									
11Feb94–1SA	fst 1	:23¹ :47² 1:11⁴ 1:37⁴	⑤Clm 20000	65 6 6 72¾ 85½ 7¹⁰ 58	Antley C W	LB 115	11.10	72–19	Bazillions117no Her Mink Coat117½ Oh Sweet Thing117²	Raced wide 8
15Jan94–1SA	fst 1	:23 :47 1:11² 1:36⁴	⑤Clm 16000	73 4 2 2½ 2hd 2½ 4²½	Flores D R	LB 115	*3.50	82–14	LeadoffPosition113² LottaRelity115½ BlushAndBshfull115no	Outfinished 9
29Dec93–2SA	fst 6f	:22 :45 :57 1:09³	⑤Clm 32000	64 6 2 62½ 64¾ 84½ 8⁶	Flores D R	LB 115	17.90	83–11	Angie'sTreasure115hd SilkySndSmmy115hd TenCrtDimond117no	Gave way 8
11Jun93–3Hol	fst 1½	:24 :48² 1:13¹ 1:44²	⑤Clm 40000	66 6 2 3² 3¹½ 2⁵ 37½	Nakatani C S	LB 115	3.70	72–17	Pretty Neat Gal116⅔ Fowler Empire119⅞ Northern Wool116²½	Weakened 7
21May93–1Hol	fst 7f	:22³ :46¹ 1:11⁴ 1:24³ 3+	⑤Md 40000	74 3 1 2½ 2½ 1¹ 1¹	Nakatani C S	LB 115	2.60	86–10	NorthernWool115¹ BbyTkThGold116² FuturGust115¹⁹	Wide backstretch 5
23Apr93–9Hol	fst 1½	:23¹ :46³ 1:11⁴ 1:46 3+	⑤Md 40000	60 9 3 4¹ 3¹ 2³½ 3²	Lopez A D	L 115	*2.60	69–20	Fleur Frau121½ Princess Waquoit115½ Northern Wool115²½	4 Wide 1st turn 10
1Apr93–6SA	fst 6½f	:22 :45² 1:10¹ 1:17	⑤Md 50000	70 9 2 5³ 4² 2³½ 2¹²½	Lopez A D	LB 117	8.40	83–13	Yazma117²¾ Northern Wool117¹½ Lovely Explosion117²½	Wide backstretch 9

WORKOUTS: Apr 2 SA 5f fst 1:01³ H 23/43 Mar 27 SA 5f fst :59¹ H 3/68 Mar 3 SA 5f fst 1:00 H 6/41 Feb 6 SA 5f fst 1:01 H 18/108 Jan 29 SA 6f fst 1:17 H 30/31 Jan 23 SA 4f fst :49³ H 28/37

Rising Mist (NZ)

Own: Karim & Perez & Tillett

STEVENS G L (381 58 71 66 .15)

B. m. 6
Sire: Kreisler (Icecapade)
Dam: Rivermist (Riverton)
Br: Walker Mrs N A (NZ)
Tr: Abrams Barry (52 7 4 3 .13)

L 123

handwritten: 3-time winner

	Lifetime Record :	21 3 2 2	$19,360		
1994	3 0 0 0	$3,000	Turf	21 3 2 2	$19,360
1993	1 0 0 1	$6,000	Wet	0 0 0 0	
SA ⑦	4 0 0 1	$9,000	Dist ⑦	0 0 0 0	

9Mar94–8SA	fm 1 ⑦	:23¹ :48 1:12⁴ 1:37²	⑤Alw 40000N1x	78 10 1 1¹ 1¹ 1hd 41½	Nakatani C S	LB 118	4.80	76–22	Smart Patch118no Ka Lae118¹ Freezelin118½	Outfinished 10
13Feb94–1SA	fm 1⅛ ⑦	:47³ 1:12¹ 1:37² 1:49⁴	⑤Alw 40000N1x	– 5 1 1hd 53½ 7¹⁵	Nakatani C S	LB 118	*1.50	– 30	LrkInThMedow118½ SmrtPtch117nk Explosivms112²¾	Dueled, tired badly 7
7Jan94–5SA	fm 1 ⑦	:23¹ :46⁴ 1:10⁴ 1:35²	⑤Clm 80000	78 9 4 52½ 52½ 65 76½	Desormeaux K J	LB 116	6.40	81–10	Sun Finger117¹½ Fantastic Kim116¹ Alieria113¾	No rally 10
31Dec93–5SA	fm 1 ⑦	:22³ :47⁴ 1:12 1:36¹ 3+	⑤Alw 40000N1x	85 2 2 3² 32½ 3² 3²	Nakatani C S	LB 118	14.70	82–16	Wendy's Daughter115½ Waitryst118½ Rising Mist118²	Rail trip 10
21Nov92♦ Pukekohe(NZ)	fm *6f ⑦RH 1:09² 3+ Franklin Long Roofing 1200 Alw 3000			34½	Vance R D	122	–		Enhancer121³½ Al Comber121⅜ Rising Mist122½	10
	Settled in 9th, strong late outside rally, 3rd 50y out									
7Nov92♦ Ellerslie(NZ)	fm *7f ⑦RH 1:23² 3+ Mabey & Wallace 1400 Alw 3400			2⅔	Vance R D	121	–		Naypokyani124½ Rising Mist121hd Morgan Le Fey119nk	8
	Saved ground in 4th, 5th 4f out, finished strongly									
20Oct92♦ Ellerslie(NZ)	gd *6f ⑦RH 1:10³ 3+ Kreisler Sprint Alw 3700			118¾	Cossill C J	121	–		Pure Lust118hd Morgan Le Fey118³ Primary School118nk	12
2May92♦ Ellerslie(NZ)	gd *1 ⑦RH 1:38² Endeavour Mile Alw 3700			14¹⁵½	Tims P G	121	–		Ocean Liner121²⅜ Aree Lass121½ Zaruka121½	15
	Settled towards rear, last 2f out, never a factor									
20Apr92♦ Ellerslie(NZ)	fm *1 ⑦RH 1:35⁴ 3+ Firestone Firehawk Mile Alw 10500			6¹½	Tims P G	124	–		Rocketry119¾ Normandy River121nk Mystic Cache119no	17
	Tracked in 2nd on rail to 3f out, stopped									
4Apr92♦ Te Aroha(NZ)	fm *7f ⑦RH 1:24¹ Waikato Stud Plate Alw 3300			1⅜	Johnson P D	121	–		Rising Mist121⅜ Carrhill121¹¼ King Kardene121½	14
	Rated in 6th, improved turn halfway, led 1f out, driving									

WORKOUTS: Apr 3 SA 4f fst :48⁴ H 13/39 Mar 22 SA 5f fst 1:03⁴ H 84/93 Mar 4 SA 5f fst :59³ H 2/49 Feb 27 SA 4f fst :49² H 38/51 Feb 3 SA 6f fst 1:16³ H 25/35 Jan 28 SA 5f fst 1:02¹ H 68/80

So how to bet? Since Wild Impulse not only comes off a good turf performance, but also broke her maiden in a sprint, she deserves to be the heavy favorite. Still, like Rising Mist, she is shortening from a mile and may not get the lead from Little Luxuries or

Dynatane, who showed :22.1 speed in her debut. At 30-1 Dynatane is worth a moderate Win bet and should also be boxed with Wild Impulse in the exacta, along with a saver exacta, Rising Mist to Dynatane.

Results: Dynatane and Little Luxuries dueled for the lead with Wild Impulse tracking in fourth. Dynatane shook loose in the stretch and returned $63.60 to Win with Wild Impulse closing for second. The $2 exacta returned $143.80.

The second race at Santa Anita on April 10th is a slightly more complicated NW1 sprint. The co-favorites are Crowning Decision and Daggett Peak who are both going off at less than 3-2. Daggett Peak is dropping back into the NW1 level after being competitive with NW 2 turf horses. Crowning Decision is the only horse to have won on the turf.

Santa Anita Park

2 ***About 6½ Furlongs.*** **(Turf). (1:11 3) ALLOWANCE. Purse $37,000. 3–year–olds which are non–winners of $3,000 other than maiden or claiming. Weight, 120 lbs. Non–winners of a race other than claiming, allowed 3 lbs. (Starters for a claiming races of $32,000 or less in their last three starts and maidens that are non–starters for a claiming price have second preference.)**

START TURF COURSE — ABOUT 6½ FURLONGS — FINISH

Daggett Peak

Own: Evergreen Farm
ANTLEY C W (353 58 33 23 .16)

Ch. c. 3 (Mar) E4/4. A4/1
Sire: It's Freezing (T. V. Commercial)
Dam: Etheldreda (Diesis)
Br: Mill Ridge Farm Ltd (Ky)
Tr: Sahadi Jenine (45 7 4 6 .16)

120

Lifetime Record:	10 1 4 1	$48,100			
1994	3 0 0 1	$7,150	Turf	4 0 2 1	$18,950
1993	7 1 4 0	$40,950	Wet	0 0 0 0	
SA (T)	3 0 1 1	$12,550	Dist (T)	2 0 1 1	$12,550

5Mar94- 1SA	fm 1 (T)	:222 :46 1:103 1:351	Pirate Cove60k	71 6 2 22 31½ 32 78	Black C A	116 b	11.00	81–07	Eagle Eyed113¾ Majestic Style1141¼ Soul Of The Matter1152		8
Carried out lst turn, steadied 5/16											
13Feb94- 7SA	fm *6½f (T)	:213 :441 1:074 1:141	Alw 41000N2X	83 5 2 31 32 5¾ 31	Black C A	114 b	5.10	86–12	SharpTry117½ HeavenlyCrusade114½ DaggettPeak114no	Bumped 1/4 pole	8
2Jan94- 4SA	fst 1 1/16	:233 :472 1:104 1:41	Alw 40000N1X	76 3 5 43½ 43½ 410 513¾	Antley C W	B 115	10.60	76–10	Wekiva Springs1183 Ferrara1153½ CrowningDecision1186	Stumbled start	6
10Dec93- 2Hol	fm 1 1/16 (T)	:234 :481 1:133 1:444	Alw 32000N1X	79 7 4 42½ 21½ 1hd 21¼	Antley C W	B 119 b	1.90	71–27	Duca1191¼ Daggett Peak1194 Sundays Cream1171	4 Wide 7/8	7
5Nov93- 5SA	fm *6½f (T)	:214 :442 1:081 1:142	Alw 32000N1X	80 4 4 2hd 1hd 1½ 21	Desormeaux K J	B 118 b	7.70	85–14	Devon Port1151 Daggett Peak1181 Bullamatic1152¾	Inside duel	11
23Sep93-12Fpx	fst 6½f	:213 :453 1:111 1:173	(R)Barrts Juv110k	56 5 5 3½ 41½ 73½ 1010½	Black C A	B 115 b	*1.70	77–11	Stepoutoftheboat1133 Bullamatic1141 Ramanujan116no	Gave way	10
11Sep93- 6Dmr	fst 6f	:214 :45 :573 1:102	Md Sp Wt	74 7 2 31½ 31½ 1½ 13	Black C A	B 118 b	5.00	89–12	DaggettPek1183 DnnyTheDemon118hd RiverFlyer1182¼	Lugged in, driving	11
21Aug93- 4Dmr	fst 6f	:214 :45 :573 1:104	Md Sp Wt	66 5 2 31 3½ 22½ 23	Black C A	B 117	3.80	82–09	Winning Pact1173 Daggett Peak1173½ Saltgrass117hd	Wide backstretch	8
10Jly93- 2Hol	fst 5½f	:22 :453 :581 1:044	Md Sp Wt	34 5 6 41¼ 53 59½ 613	Black C A	117	*.70	75–12	Prenup117no Gracious Ghost1178½ Prevasive Force1172¾	Wide backstretch	7
19Jun93- 6Hol	fst 5f	:22 :453 :581	Md Sp Wt	63 3 3 3nk 2hd 1hd 22	Black C A	117	5.70	91–10	Swift Walker1172 Daggett Peak117¾ Devil's Mirage1172	Good effort	10

WORKOUTS: Apr 4 SA 5f fst 1:023 H *49/53* Mar 29 SA 3f fst :37 H *10/19* Mar 23 SA 5f fst 1:002 H *20/112* Mar 16 SA 3f fst :35 H *2/22* Mar 1 SA 5f fst 1:00 H *8/48* Feb 22 SA 3f fst :371 H *13/18*

Win The Case

Own: Brooks & King & Landsburg
VALENZUELA F H (383 42 24 38 .11)

Dk. b or br g. 3 (Apr) L7/9 D8/5
Sire: Kalim (Hotfoot)
Dam: Legal Protection (Maheras)
Br: King & Landsburg (Cal)
Tr: Cerin Vladimir (53 10 7 9 .19)

L 117

Lifetime Record:	8 1 5 1	$29,670		
1994	3 1 2 0	$17,100	Turf	0 0 0 0
1993	5 M 3 1	$12,570	Wet	0 0 0 0
SA (T)	0 0 0 0		Dist (T)	0 0 0 0

16Mar94- 9SA	fst 6½f	:222 :453 1:104 1:172	(S)Md 28000	68 4 4 2hd 11½ 13½ 1no	Valenzuela F H	LB 116 b	3.00	85–12	Win The Case116no Al Monsoon1184 Golden Century1182¾	Just held	9
19Feb94- 2SA	fst 6½f	:22 :444 1:091 1:16	(S)Md 32000	65 7 5 21½ 22½ 25 29	Valenzuela F H	LB 118 b	*1.80	83–08	CusingRumors1189 WinTheCs1181¾ LightningN'ic1181¼	Lugged in stretch	7
2Feb94- 9SA	fst 7f	:223 :454 1:112 1:242	(S)Md 32000	61 6 6 74¼ 5¾ 1½ 21¼	Valenzuela F H	LB 118 b	*1.70	79–13	Natural Charmer1181¼ Win The Case1182¾ Victory Sun118¾		10
Ducked in sharply 1/16											
15Dec93- 9Hol	gd 6½f	:221 :453 1:122 1:192	(S)Md 32000	62 4 8 94½ 42½ 31 2¾	Valenzuela F H	LB 119 b	*2.40	71–19	Real Smart119¾ Win The Case1195 Coniston1175½	Finished well	11
3Dec93- 6Hol	fst 6f	:22 :452 :581 1:112	(S)Md 28000	64 2 7 42 76¾ 46 31¼	Valenzuela F H	LB 117 b	3.90	82–15	Go To Bat119no Tam Rise1191¼ Win The Case1171¾		9
Bumped, steadied 3/8, in tight late											
15Nov93- 6SA	fst 7f	:232 :47 1:114 1:243	(S)Md 28000	59 3 3 4¾ 51¾ 32½ 21¾	Valenzuela F H	LB 115 b	*2.10	78–15	He's Fabulous1171¾ Win The Case115hd Al's Sunshine1173½	Finished well	10
29Oct93- 4SA	fst 7f	:221 :451 1:112 1:25	Md 28000	53 1 10 63 45½ 52 21¼	Valenzuela F H	LB 116 b	54.00	77–19	Ketchum Z's1181¼ Win The Case116hd Al Khedive1132¾	Finished well	12
2Oct93- 4Fpx	fst 6f	:223 :461 :584 1:121	(S)Md 32000	31 3 7 86½ 67½ 914 711¼	Valenzuela F H	B 118 b	3.90	75–09	Endless Dust1182 Al Khedive113no Umpscious1132	Bid, tired	10

WORKOUTS: Apr 7 SA 3f fst :351 H *3/24* Apr 1 SA 4f fst :473 H *9/37* Mar 27 SA 3f fst :36 H *9/33* Mar 11 SA 4f fst :484 H *16/27* Mar 5 SA 5f fst 1:002 H *11/57* Feb 13 SA 3f fst :371 B *17/24*

Crowning Decision

Own: Golden Eagle Farm
BLACK C A (333 27 37 37 .08)

Only turf winner

B. c. 3 (Mar) C5/3
Sire: Chief's Crown (Danzig)
Dam: Lady Pastor (Flying Paster) D4/3
Br: Mabee Mr & Mrs John C (Ky)
Tr: Rash Rodney (86 9 10 15 .10)

Router low odds

L 120

Lifetime Record:	5 1 0 3	$34,250			
1994	2 0 0 2	$12,000	Turf	1 1 0 0	$15,950
1993	3 1 0 1	$22,250	Wet	0 0 0 0	
SA (T)	0 0 0 0		Dist (T)	0 0 0 0	

30Jan94- 3SA	fst 1 1/16	:234 :474 1:114 1:414	Alw 40000N1X	83 2 3 31½ 1hd 31½ 36½	Black C A	LB 118	4.70	79–12	Pollock's Luck115no Numerous1166½ Crowning Decision1183½	Inside bid	5
2Jan94- 4SA	fst 1 1/16	:233 :472 1:104 1:41	Alw 40000N1X	89 2 3 32 33 34 36½	Stevens G L	LB 118	9.40	83–10	Wekiva Springs1183 Ferrara1153½ Crowning Decision1186	Best of rest	6
8Dec93- 6Hol	fm 1 (T)	:242 :491 1:14 1:383	Md Sp Wt	82 1 2 2½ 1½ 14 16½	Stevens G L	LB 118	*1.20	72–26	CrowningDecision1186½ Dancethetides1183½ CrolinBlew118hd	Clearly best	10
13Nov93- 6SA	fst 6½f	:222 :452 1:093 1:154	Md Sp Wt	78 5 3 3½ 1hd 11 32¼	Stevens G L	LB 118	2.10	90–11	Numerous118no Amarisingstar1182¼ Crowning Decision1182¾	Outfinished	7
24Oct93- 6SA	fst 6f	:213 :444 :573 1:101	Md Sp Wt	69 5 2 52¼ 51¾ 61½ 43	Atkinson P	B 117	1.40e	83–12	Egaynt1172 GrndCherokee117½ WindwoodLd117½	Lacked room 1/4-3/16	8

WORKOUTS: Mar 31 SA 7f fst 1:25 H *1/3* Mar 23 SA 5f fst 1:031 H *99/112* Mar 17 SA 4f fst :464 H *3/32* Feb 13 SA 4f fst :484 H *34/54* Jan 26 SA 4f gd :494 H *28/44* Jan 16 SA 4f fst :49 H *38/58*

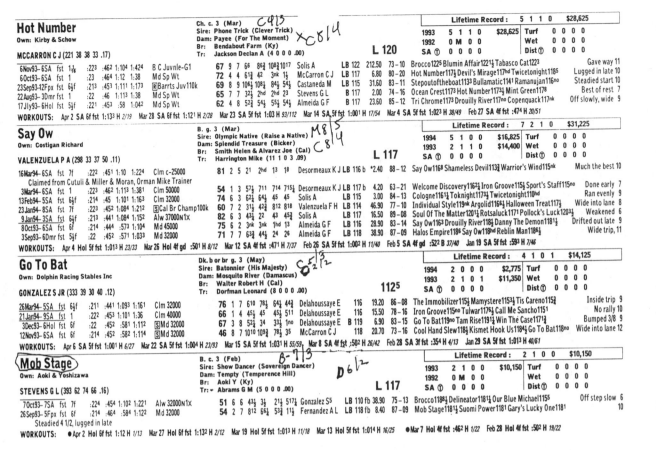

The only other horse with the turf breeding to be competitive here is Mob Stage with his B- rating. While the minus sign suggests horses by Show Dancer normally need a race on the turf, this is a sprint and Mob Stage comes off a bullet 1:12 work. Furthermore, he won his sprint debut as a 2-year-old and Abrams, according to Olmsted's *Trainer Guide*, wins races with layoff horses. With six months off to mature and the switch to Gary Stevens and the grass, Mob Stage has every right to improve. On the downside, he broke his maiden in claiming company and is dead on the board at 14-1.

So how to bet? Betting a Maiden Claiming winner in its first start on the turf against a horse that has already won on the surface against better is a risky proposition. But, it is important to note that neither Crowning Decision nor Daggett Peak have as yet beat winners. The former broke his maiden at a mile on a slow lead, which he is not likely to get today. The latter has already lost three times at this level. Still, Mob Stage is a bit of a stab and should only be moderately played to Win and behind the two favorites in the exacta. An argument could also be made for a small Mob Stage-Daggett Peak exacta, since Daggett Peak finished close to the winner in his two turf sprints on this course.

SECOND RACE
Santa Anita
APRIL 10, 1994

ABOUT 6½ FURLONGS. (Turf)(1.11³) ALLOWANCE. Purse $37,000. 3-year-olds which are non-winners of $3,000 other than maiden or claiming. Weight, 120 lbs. Non-winners of a race other than claiming, allowed 3 lbs. (Starters for a claiming races of $32,000 or less in their last three starts and maidens that are non-starters for a claiming price have second preference.)

Value of Race: $37,000 Winner $20,350; second $7,400; third $5,550; fourth $2,775; fifth $925. Mutuel Pool $375,135.00 Exacta Pool $372,691.00 Quinella Pool $70,781.00

Last Raced	Horse	M/Eqt. A.Wt	PP St	¼	½	Str	Fin	Jockey	Odds $1
5Mar94 1SA7	Daggett Peak	b 3 120	1 4	3hd	5²½	5¹½	1¾	Antley C W	1.40
7Oct93 7SA5	Mob Stage	LBb 3 117	7 5	5²½	4½	4¹½	2½	Stevens G L	13.80
30Jan94 3SA3	Crowning Decision	LB 3 120	3 3	4hd	3hd	3hd	3¹½	Black C A	1.30
16Mar94 6SA1	Say Ow	LBb 3 117	5 1	2²½	2²	1hd	4¹¾	Valenzuela P A	7.20
16Mar94 9SA1	Win The Case	LBb 3 117	2 7	6¹½	6²½	6¹½	5¹½	Valenzuela F H	31.20
6Nov93 6SA10	Hot Number	LB 3 120	4 2	1hd	1hd	2¹½	6²	McCarron C J	10.90
26Mar94 5SA4	Go To Bat	B 3 112	6 6	7	7	7	7	Gonzalez S Jr5	26.80

OFF AT 1:01 Start Good. Won driving. Time, :22, :44², 1:08¹, 1:14¹ Course firm.

$2 Mutuel Prices:				
1-DAGGETT PEAK		4.80	3.20	2.40
7-MOB STAGE			8.60	3.80
3-CROWNING DECISION				2.60

$2 EXACTA 1-7 PAID $50.80 $2 QUINELLA 1-7 PAID $32.20

Ch. c, (Mar), by It's Freezing-Etheldreda, by Diesis. Trainer Sahadi Jenine. Bred by Mill Ridge Farm Ltd (Ky).

© 1994 DRF, Inc.

The fourth race at Hollywood Park on May 22nd is a NW1 turf route. The heavy favorite here at 6-5 is Timbalier, who is dropping with high Beyers from a competitive effort in a Grade 3 turf race. Also dropping from a cheaper stakes, with an even higher Beyer, is Saltgrass at 3-1. Of the rest of the field, Crowning Decision just got creamed by Timbalier and none of the first-time grass horses, except Vaudeville, are bred for the turf.

Vaudeville, however, is very interesting. With Theatrical on top and Riverman on the bottom of his pedigree, he not only has a spectacular turf pedigree, but also the stamina to run all day. Furthermore, his :44.3/1:09.2 fractions give him plenty of tactical speed for Gary Stevens to use and Vaudeville's sharp Crockett/Fierce connections, according to the Form, win races when they ship to southern California.

Hollywood Park

4

1 MILE. (Turf). (1:32⁴) ALLOWANCE. Purse $40,000. 3-year-olds, which have never won two races. Weight: 119 lbs. Non-winners of a race other than claiming at one mile or over, 2 lbs. Such a race at any distance, 4 lbs. (Winners that have started for a claiming price of $32,000 or less in their last three starts and maidens that are not starters for claiming price have second preference.)

TURF COURSE
1 MILE
FINISH START

Count Con						Lifetime Record:	6 1 1 1	$22,650

Own: Hawn W R

Dk. b or br g. 3 (May)
Sire: Fast Account (Private Account)
Dam: Con's Sister (Summer Time Guy)
Br: Hawn W R (Cal)
Tr: MacDonald Mark (3 0 1 0 .00)

C9|3
D7|4

1994	4 1 1 1	$22,650	Turf 0 0 0 0
1993	2 M 0 0		Wet 0 0 0 0
Hol①	0 0 0 0		Dist① 0 0 0 0

L 115

ATKINSON P (41 3 7 4 .07)

27Apr94–3Hol	fst 1¹⁄₁₆	:24 :47⁴ 1:12¹ 1:44³	Clm 50000	78	5 4	4²	2hd	1hd	2¾	Atkinson P	LB 116	9.70	77–18	Veracity116¾ Count Con116⁴ Baja Bill116¹½	4 wide 2nd turn 5
27Feb94–3SA	fst 1	:22² :46³ 1:11² 1:37³	Clm 32000	63	1 7	6⁴½	5²½	5⁴	3⁶½	Atkinson P	LB 115	6.60	74–18	Set On Cruise115¹½ Gary's Lucky One114⁵ Count Con115nk	7
	Lacked room 5/16, altered path														
10Feb94–4SA	fst 1¹⁄₁₆	:23¹ :47³ 1:12³ 1:45¹	Md 32000	72	8 9	9⁷¾	7⁵½	3²½	11¾	Atkinson P	LB 117	74.00	69–27	Count Con117¹¾ Apurado115¹ Pacific Edition117²	9
	Broke very slow, bumped hard, circled field, lugged in 1/16														
28Jan94–2SA	fst 6f	:21³ :44⁴ :57¹ 1:10¹	Md 32000	57	4 11	11¹³	11¹²	9¹³ 7⁹	Esposito M	LB 119	95.40	77–12	Heartless Wager118⁴½ Double Sec118¹½ Bobs Bikini118hd	11
	Bore in, steadied start														
4Dec93–4Hol	fst 1¹⁄₁₆	:23³ :47² 1:12¹ 1:45	Md 32000	33	8 3	3¹	4²½	7¹⁰ 7²⁰	Gryder A T	LB 119 b	17.60	56–15	Sierra Real114¹ My Ideal119² Stage Door Tyler119¹	Wide both turns 8
15Nov93–6SA	fst 7f	:23² :47 1:11⁴ 1:24³	⑤Md 32000	23	2 10	10¹⁰	10¹⁰	9¹⁷ 8¹⁸½	Antley C W	B 117	15.60	61–15	He's Fabulous117¹¾ Win The Case115hd Al's Sunshine117³½	Off slow, green 10

WORKOUTS: May 19 SA 4f gd :48³ H 4/25 May 13 SA 6f fst 1:14¹ H 9/17 May 7 SA 5f gd 1:03 B 15/20 Apr 25 SA 3f fst :37⁴ H 8/12 Apr 21 SA 4f fst :48⁴ H 20/43 Apr 15 SA 7f fst 1:29¹ H 5/8

Wildly Joyous
Own: Horowitz Elliot J
DESORMEAUX K J (57 14 10 7 .25)

Dk. b or br c. 3 (Apr)
Sire: Wild Again (Icecapade)
Dam: Joyous Pirouette (Nureyev)
Br: Axmar Stable (Ky)
Tr: Mandella Richard (14 2 2 2 .14)

117

Lifetime Record :	2 1 0 1	$23,150			
1994	2 1 0 1	$23,150	Turf	0 0 0 0	
1993	0 M 0 0		Wet	0 0 0 0	
Hol Ⓣ	0 0 0 0		Dist Ⓣ	0 0 0 0	

29Apr94- 8Hol fst 6f :221 :451 :571 1:093 3↑ Alw 37000N1X 93 8 3 5½ 6¾ 5¾ 3¾ Desormeaux K J 115 6.10 89-13 Honest Happiness117½ I'm Huge119² Wildly Joyous115¾ 8
Forced 6 wide into lane
8Apr94- 6SA fst 5½f :214 :444 :57 1:031 3↑ Md Sp Wt 88 3 1 1½ 1½ 1½ 1nk Desormeaux K J 116 3.40 94-12 Wildly Joyous116nk Emilion1144 Dion The Tyrant123¾ Game inside 7
WORKOUTS: May 15 Hol 5f fst 1:043 H 59/60 Apr 24 SA 5f fst 1:011 H 11/27 Apr 18 SA 3f fst :374 B 20/22 Apr 6 SA 4f fst :491 H 25/31 Mar 28 SA 7f fst 1:274 H 4/7 Mar 20 Hol 7f fst 1:29 H 3/3

Hippiater
Own: Artful Racing Group & Bauer & Emery
PEDROZA M A (72 8 10 11 .11)

B. g. 3 (Mar)
Sire: L'Emigrant (The Minstrel)
Dam: Gallic Fable (Le Fabuleux)
Br: Tulsia S R L (Ky)
Tr: Marquez Alfredo (3 1 0 0 .33)

L 115

Lifetime Record :	9 1 2 0	$15,825			
1994	9 1 2 0	$15,825	Turf	0 0 0 0	
1993	0 M 0 0		Wet	0 0 0 0	
Hol Ⓣ	0 0 0 0		Dist Ⓣ	0 0 0 0	

5May94- 2Hol fst 5½f :22 :45 :57 1:031 3↑ Md 25000 83 2 1 1½ 11 11½ 1½ Antley C W LB 115 2.30 95-08 Hippiater115½ Sharp Warrior115²½ Knight's Intrusion1154 Gamely 8
22Apr94- 4SA fst 6f :22 :451 :572 1:10 Md 32000 80 4 2 2½ 1hd 1½ 2¾ Antley C W LB 118 10.60 86-11 Hot Papa Tom118¾ Hippiater118½ Scottish Slew118hd Drifted out 1/16 11
14Apr94- 9SA fst 5½f :213 :45 :571 1:034 Md 28000 53 6 6 74½ 76½ 66 68¾ Gonzalez S5 LB 111 12.40 82-09 Flaming George118½ Two Way Fort118²½ It's Pete Again118nk 12
Steadied hard 3 1/2, wide into lane
24Mar94- 4SA fst 6f :214 :444 :57 1:10 Md 32000 67 3 4 31½ 33 33½ 53½ Castanon A L LB 118 9.20 83-09 Dash Of Vanilla116½ One SharpChip118½ FlamingGeorge118¾ No late bid 7
11Mar94- 4SA fst 6½f :213 :45 1:102 1:17 Md 32000 31 8 3 21½ 3½ 710 719 Valenzuela P A LB 118 b *2.60 68-14 Kismet Hook Us116hd Top Gear118nk Flaming George1184 Took up 5/16 11
24Feb94- 9SA fst 1$\frac{1}{16}$:23 :471 1:121 1:433 Md 32000 58 8 1 11 11½ 44½ 614½ Valenzuela P A LB 117 b 2.90 62-20 Mint Green117nk Pacific Edition1176½ Emir Al Nasr1175½ Broke in, bumped 11
10Feb94- 4SA fst 1$\frac{1}{16}$:231 :473 1:123 1:451 Md 32000 64 9 1 11 1½ 21½ 54½ Gonzalez S5 LB 112 b *2.00 64-27 Count Con117½ Apurado1151 Pacific Edition1172 Blew last turn 9
28Jan94- 6SA fst 6f :213 :443 :57 1:101 Md 32000 78 5 3 22 21½ 2½ 2¾ Gonzalez S5 LB 113 b 7.70 85-12 Light Sky118¾ Hippiater113½ Tomslongwait1186 Drifted out lane 12
5Jan94- 4SA fst 6½f :212 :442 1:093 1:162 Md 32000 35 3 7 32 56 713 816½ Gonzalez S5 B 113 b 45.60 74-09 Grand Cherokee118³½ Hunters Way118² Bobs Bikini118hd Saved ground 12
WORKOUTS: Apr 9 SA 4f fst :464 H 4/45

Crowning Decision
Own: Golden Eagle Farm
NAKATANI C S (2 1 0 1 .50)

B. c. 3 (Mar)
Sire: Chief's Crown (Danzig)
Dam: Lady Pastor (Flying Paster)
Br: Mabee Mr & Mrs John C (Ky)
Tr: Rash Rodney (14 3 1 1 .21)

L 119

Lifetime Record :	7 1 0 4	$39,800			
1994	4 0 0 3	$17,550	Turf	3 1 0 1	$21,500
1993	3 1 0 1	$22,250	Wet	0 0 0 0	
Hol Ⓣ	2 1 0 0	$15,950	Dist Ⓣ	2 1 0 0	$15,950

7May94- 8Hol fm 1 Ⓣ :23 :464 1:104 1:35 Sptlgt BC H-G3 68 8 2 21 21 33½ 712¼ Valenzuela F H LB 114 6.10 77-11 Fumo Di Londra119³ Unfinished Symph116½ Timbalier113no Bumped 1/8 8
10Apr94- 2SA fm *6½f Ⓣ :22 :442 1:081 1:141 Alw 37000N1X 83 3 3 42¾ 32 31½ 31¼ Black C A LB 120 *1.30 86-13 Daggett Peak120¾ Mob Stage117½ Crowning Decision120½ Game try 7
30Jan94- 3SA fst 1$\frac{1}{16}$:234 :472 1:114 1:414 Alw 40000N1X 83 2 3 31½ 1hd 31½ 36½ Black C A LB 118 4.70 79-12 Pollock's Luck115no Numerous1166½ Crowning Decision118³½ Inside bid 5
2Jan94- 4SA fst 1$\frac{1}{16}$:233 :472 1:104 1:41 Alw 40000N1X 89 2 3 32 33 34 36½ Stevens G L LB 118 9.40 83-10 Wekiva Springs118³ Ferrara115³½ Crowning Decision1186 Best of rest 6
8Dec93- 6Hol fm 1 Ⓣ :242 :491 1:14 1:383 Md Sp Wt 82 1 2 2½ 1½ 14 16½ Stevens G L LB 118 *1.20 72-26 CrowningDecision1186½ Dancethetides118³½ CrolinBlew118hd Clearly best 10
13Nov93- 6SA fst 6½f :222 :452 1:093 1:154 Md Sp Wt 78 5 3 3½ 1hd 11 32½ Stevens G L LB 118 2.10 90-11 Numerous118no Amarisingstar118²½ Crowning Decision118²¾ Outfinished 7
24Oct93- 6SA fst 6f :213 :444 :573 1:101 Md Sp Wt 69 5 2 52½ 51¾ 61½ 43 Atkinson P B 117 1.40e 83-12 Egaynt117² GrndCherokee117½ WindwoodLd117½ Lacked room 1/4-3/16 8
WORKOUTS: May 17 Hol 4f fst :501 H 16/21 Apr 29 Hol 7f fst 1:271 H 3/9 Apr 21 SA 4f fst :484 H 20/43 Mar 31 SA 7f fst 1:25 H 1/3 Mar 23 SA 5f fst 1:031 H 99/112 Mar 17 SA 4f fst :464 H 3/32

Hello Chicago
Own: Lanni & Sloan
BLACK C A (92 10 17 14 .11)

B. c. 3 (May)
Sire: Broad Brush (Ack Ack)
Dam: Party Worker (Speak John)
Br: Manning Family Trust & Honeagle Farm (Ky)
Tr: Hess R B Jr (20 4 2 1 .20)

L 119

Lifetime Record :	2 1 0 1	$23,500			
1994	2 1 0 1	$23,500	Turf	0 0 0 0	
1993	0 M 0 0		Wet	1 1 0 0	$18,700
Hol Ⓣ	0 0 0 0		Dist Ⓣ	0 0 0 0	

25Apr94- 9SA my 1 :234 :48 1:141 1:40 Md Sp Wt 88 4 5 41 3nk 1hd 1nk Black C A LB 117 10.20 69-31 Hello Chicago117nk Too Many Notes1177 Prince Dahar1176½ 5 wide 7/8 8
6Mar94- 2SA fst 6f :22 :451 :572 1:10 Md Sp Wt 65 2 7 97½ 86¾ 59 310½ Flores D R LB 118 24.80 76-09 De Vito1184 Minero1186½ Hello Chicago118²½ Broke out, bumped 11
WORKOUTS: May 18 Hol 4f fst :474 H 10/57 May 12 Hol 4f fst :493 H 22/27 Apr 19 Hol 5f fst 1:04 H 19/20 Apr 13 Hol 4f fst :472 H 2/15 Apr 7 Hol 4f fst :474 H 3/17 Mar 29 Hol 5f fst 1:024 H 17/21

Vaudeville
Own: Crockett Ron
STEVENS G L (84 16 11 18 .19)

B. c. 3 (Mar)
Sire: Theatrical*Ire (Nureyev)
Dam: S'Nice (Riverman)
Br: Foxfield (Ky)
Tr: Fierce Fordell (—)

117

Lifetime Record :	1 1 0 0	$11,000			
1994	1 1 0 0	$11,000	Turf	0 0 0 0	
1993	0 M 0 0		Wet	0 0 0 0	
Hol Ⓣ	0 0 0 0		Dist Ⓣ	0 0 0 0	

16Apr94- 1GG fst 6f :22 :443 :564 1:092 Md Sp Wt 91 5 2 2½ 1hd 2hd 1hd Lopez A D B 118 *2.00 92-09 Vaudeville118hd Knight Of Blue118²½ Sir Cutter Slew11811 Held gamely 7
WORKOUTS: May 16 GG 5f fst 1:011 H 17/25 May 10 GG 5f fst 1:00 H 2/46 May 3 GG 5f fst 1:02 H 7/15 Apr 29 GG 5f gd 1:034 H 49/72 ●Apr 13 GG 4f fst :463 H 1/38 Apr 6 GG 5f fst 1:003 Hg 2/38

Saltgrass
Own: Hibbert R E
ANTLEY C W (70 16 12 7 .23)

B. c. 3 (May)
Sire: Woodman (Mr. Prospector)
Dam: Papochino (Apalachee)
Br: Hibbert R E (Ky)
Tr: Rash Rodney (14 3 1 1 .21)

L 119

Lifetime Record :	10 1 2 2	$54,050			
1994	5 1 1 1	$43,700	Turf	2 0 2 0	$21,000
1993	5 M 1 1	$10,350	Wet	0 0 0 0	
Hol Ⓣ	1 0 1 0	$6,000	Dist Ⓣ	1 0 1 0	$6,000

20Apr94- 8SA fm 1⅛ Ⓣ :461 1:103 1:36 1:483 Ⓡ La Puente79k 94 4 5 56½ 55 3½ 2no Antley C W LB 114 5.40 76-24 MajesticStyle114no Saltgrass114²½ Mkinnhonestbuck1157 Awkward start 5
3Apr94- 3SA fst 1 :231 :464 1:12 1:373 Alw 40000N1X 77 4 5 52¾ 4½ 41½ 47 Antley C W LB 119 3.50 74-19 LittleLiteBud116½ CollegeTown116¾ BeutifulCrown1164½ 4 wide 2nd turn 5
5Mar94- 3SA fst 1 :224 :461 1:102 1:35 Alw 40000N1X 84 4 7 65 63½ 55 58¼ Antley C W B 120 17.90 86-07 Strodes Creek117½ Irgun117³½ Sharp Phase1174 7
Checked early, 5 wide into lane
12Feb94- 7SA fst 1 :23 :462 1:103 1:35 Alw 40000N1X 95 4 6 54½ 42 35 36½ Antley C W B 118 24.70 87-14 Argolid116² Strodes Creek1154½ Saltgrass118½ Came up empty 6
9Jan94- 7SA fst 1$\frac{1}{16}$:231 :47 1:111 1:432 Md Sp Wt 86 5 9 96¾ 85¼ 3½ 11½ Antley C W B 117 11.20 78-17 Saltgrass117½ Popular Leader117nk Al Gaucho117hd 10
Rank early, 5 wide into lane
26Dec93- 9SA fst 1 :221 :45 1:101 1:363 Md Sp Wt 62 6 5 55½ 42½ 56½ 713¾ Gryder A T B 117 3.80 73-10 Almaraz117½ Dramatic Gold117½ Advantage Miles1176 Took up 3/16 10
17Nov93- 4Hol fm 1 Ⓣ :463 1:111 1:361 Md Sp Wt 70 1 3 31½ 21½ 21½ 23½ Gryder A T B 118 4.60 80-14 Valiant Nature118³½ Saltgrass118³½ Unrivaled1181 Second best 9
Quarter time unavailable
31Oct93- 6SA fst 1$\frac{1}{16}$:223 :461 1:111 1:432 Md Sp Wt 34 9 4 1hd 21 710 932 McCarron C J B 117 b 4.90 55-15 WekivSprings11711 CollectbleWine117²¾ Let'sBeCurious1172 Rank, wide 9
11Sep93- 6Dmr fst 6f :214 :45 :573 1:102 Md Sp Wt 55 1 3 1hd 2hd 42 67¼ Desormeaux K J B 118 b 5.40 81-12 Daggett Peak118³ Danny The Demon118hd River Flyer118²½ Gave way 11
21Aug93- 4Dmr fst 6f :214 :45 :573 1:104 Md Sp Wt 57 7 6 41 2hd 33 36½ Desormeaux K J B 117 b 4.00 78-09 Winning Pact117³ Daggett Peak117³½ Saltgrass117hd Wide backstretch 8
WORKOUTS: May 18 Hol 4f fst :52 H 55/57 May 12 Hol 7f fst 1:28 H 3/4 May 6 Hol 5f fst :591 H 5/35 Apr 29 SA tr.t 3f gd :394 H 12/13 Apr 17 SA 4f fst :482 H 14/55 Apr 10 SA 3f fst :371 B 22/25

continued

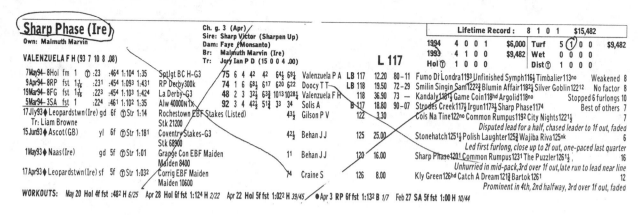

Sharp Phase (Ire)
Own: Malmuth Marvin

Ch. g. 3 (Apr)
Sire: Sharp Victor (Sharpen Up)
Dam: Faye (Monsanto)
Br: Malmuth Marvin (Ire)
Tr: Jory Ian P D (15 0 0 4 .00)

VALENZUELA F H (93 7 10 8 .08) **L 117**

Lifetime Record :	8 1 0 1	$15,482			
1994	4 0 0 1	$6,000	Turf	5 1 0 0	$9,482
1993	4 1 0 0	$9,482	Wet	0 0 0 0	
Hol (T)	1 0 0 0		Dist (T)	1 0 0 0	

7May94- 8Hol fm 1 (T) :23 :464 1:104 1:35 Sptlgt BC H-G3 75 6 4 42 42 64½ 69¼ Valenzuela P A LB 117 12.20 80-11 Fumo Di Londra1193 Unfinished Symph116½ Timbalier113no Weakened 8
9Apr94- 8RP fst 1 1/16 :231 :454 1:093 1:431 RP Derby300k 74 1 6 68½ 617 620 622 Doocy T T LB 118 19.50 72-29 Smilin Singin Sam1222¾ Blumin Affair1182¼ Silver Goblin12212 No factor 8
19Mar94- 8FG fst 1 1/16 :223 :454 1:103 1:424 La Derby-G3 48 2 3 32½ 63¾ 1013 1028¼ Valenzuela F H 118 36.90 73 — Kandaly1181½ Game Coin118hd Argolid118no Stopped 6 furlongs 10
5Mar94- 3SA fst 1 :224 :461 1:102 1:35 Alw 40000N1X 92 3 4 42½ 51¾ 33 34 Solis A B 117 18.80 90-07 Strodes Creek117½ Irgun1173½ Sharp Phase1174 Best of others 7
17Jly93♦ Leopardstwn(Ire) gd 6f (T)Str 1:14 Rochestown EBF Stakes (Listed) 43½ Gilson P V 122 3.30 Cois Na Tine122no Common Rumpus1192 City Nights1221½ 7
Tr: Liam Browne Stk 21200 *Disputed lead for a half, chased leader to 1f out, faded*
15Jun93♦ Ascot(GB) yl 6f (T)Str 1:181 Coventry Stakes-G3 42½ Behan J J 125 25.00 Stonehatch1251½ Polish Laughter125¾ Wajiba Riva125nk 6
Stk 68900 *Led first furlong, close up to 2f out, one-paced last quarter*
1May93♦ Naas(Ire) gd 5f (T)Str 1:01 Grange Con EBF Maiden 11 Behan J J 120 16.00 Sharp Phase1201 Common Rumpus1231 The Puzzler1261½ 16
Maiden 8400 *Unhurried in mid-pack,3rd over 1f out,late run to lead near line*
17Apr93♦ Leopardstwn(Ire) sf 5f (T)Str 1:032 Corrig EBF Maiden 74 Craine S 126 8.00 Kly Green126hd Catch A Dream121¾ Bartok1261 12
Maiden 10600 *Prominent in 4th, 2nd halfway, 3rd over 1f out, faded*

WORKOUTS: May 20 Hol 4f fst :482 H *6/25* Apr 28 Hol 6f fst 1:124 H *2/22* Apr 22 Hol 5f fst 1:022 H *39/45* ●Apr 3 RP 6f fst 1:132 B *1/7* Feb 27 SA 5f fst 1:00 H *10/44*

Timbalier
Own: Lord & Lady White

B. c. 3 (Mar)
Sire: Sovereign Dancer (Northern Dancer)
Dam: Chaleur (Rouge Sang)
Br: Dr. & Mrs. Emler A. Neuman (Ky)
Tr: Speckert Christopher (8 1 2 1 .13)

MCCARRON C J (54 14 8 8 .26) **119**

Lifetime Record :	3 1 0 2	$38,500			
1994	3 1 0 2	$38,500	Turf	1 0 0 1	$15,000
1993	0 M 0 0		Wet	0 0 0 0	
Hol (T)	1 0 0 1	$15,000	Dist (T)	1 0 0 1	$15,000

Entered 21May94- 8 HOL
7May94- 8Hol fm 1 (T) :23 :464 1:104 1:35 Sptlgt BC H-G3 87 7 7 73¾ 74¼ 75½ 33½ Black C A 113 2.70 85-11 Fumo Di Londra1193 Unfinished Symph116½ Timbalier113no Late bid 8
2Apr94- 4SA fst 1 :23 :464 1:111 1:36 Md Sp Wt 97 1 2 21 2hd 1hd 11½ Delahoussaye E B 117 *1.70 89-15 Timbalier1171½ Minero1125½ Rindo Indy1171 Inside trip 8
12Mar94- 6SA fst 6f :212 :444 :57 1:091 Md Sp Wt 82 2 9 813 66½ 54¼ 33 Goldberg S F 118 62.30 88-05 Wolf Bait118½ Tobin Ruler1182½ Timbalier118no Off very slow, green 9

WORKOUTS: May 17 SA 4f fst :492 H *18/26* May 3 SA 5f fst 1:003 H *8/25* Apr 29 SA tr.t 6f gd 1:162 H *3/5* ●Apr 22 SA 7f fst 1:271 H *1/11* ●Apr 15 SA 5f fst :58 H *1/42* Apr 11 SA 3f fst :383 H *17/20*

© 1994 DRF, Inc.

So how to bet? Since this looks like a three horse race and the favorite is a short price, the best thing to do is bet Vaudelville to Win and key him in exacta boxes with Timbalier and Saltgrass. Additionally, just in case Vaudeville proves third best, he can be singled in the trifecta in the third spot behind a box of the two favorites. Or, more boldly, boxed in the trifecta with Timbalier and Saltgrass.

FOURTH RACE
Hollywood
MAY 22, 1994

1 MILE. (Turf)(1.324) ALLOWANCE. Purse $40,000. 3–year–olds, which have never won two races. Weight: 119 lbs. Non–winners of a race other than claiming at one mile or over, 2 lbs. Such a race at any distance, 4 lbs. (Winners that have started for a claiming price of $32,000 or less in their last three starts and maidens that are not starters for claiming price have second preference.)

Value of Race: $40,000 Winner $22,000; second $8,000; third $6,000; fourth $3,000; fifth $1,000. Mutuel Pool $414,743.00 Exacta Pool $366,159.00 Trifecta Pool $202,625.00 Quinella Pool $39,448.00

Last Raced	Horse	M/Eqt.	A.Wt	PP	St	¼	½	¾	Str	Fin	Jockey	Odds $1
16Apr94 1GG1	Vaudeville	B	3 117	6	4	41	41	31	1hd	1½	Stevens G L	6.70
20Apr94 8SA2	Saltgrass	LB	3 119	7	7	71½	76	5½	2½	22	Antley C W	3.10
7May94 8Hol3	Timbalier	B	3 119	8	5	53½	2hd	2½	32½	33½	McCarron C J	1.20
25Apr94 9SA1	Hello Chicago	LB	3 119	5	6	61½	6½	65	5½	44	Black C A	16.70
29Apr94 8Hol3	Wildly Joyous		3 117	2	2	21½	31½	1hd	41½	52¼	Desormeaux K J	5.60
7May94 8Hol7	Crowning Decision	LB	3 119	4	3	3hd	51	4hd	610	611	Nakatani C S	7.20
27Apr94 3Hol2	Count Con	LB	3 115	1	8	8	8	8	73½	710	Atkinson P	36.60
5May94 2Hol1	Hippiater	LB	3 115	3	1	1hd	1hd	72½	8	8	Pedroza M A	50.30

OFF AT 2:47 Start Good. Won driving. Time, :23, :462, 1:104, 1:35 Course firm.

$2 Mutuel Prices:				
	6–VAUDEVILLE	15.40	7.20	3.60
	7–SALTGRASS		4.40	2.60
	9–TIMBALIER			2.20

$2 EXACTA 6–7 PAID $57.20 $2 TRIFECTA 6–7–9 PAID $151.80 $2 QUINELLA 6–7 PAID $33.00

B. c, (Mar), by Theatrical*Ire–S'Nice, by Riverman. Trainer Fierce Fordell. Bred by Foxfield (Ky).

© 1994 DRF, Inc.

The eighth race at Golden Gate Fields on April 24th is another powerful example where a knowledge of the good turf sires paid off with a double digit mutuel. Going over the field, the 3-2 favorite is Fittobetried who has already won a stakes on the grass and has the highest turf Beyers. He also has some good works. On the downside, Fittobetried hasn't raced in two months, seems to need slow fractions to close on, and Jimenez is a cold trainer.

But who can beat him? Log Buster and Margie's Boy are by lousy turf sires and can be thrown out. Dancethetides and Suomi Power have both already lost on the turf, with lower Beyers than Fittobetried has, and look no better than second best.

Golden Gate Fields

8

$1\frac{1}{16}$ MILES. (Turf, chute). (1:39^{2}) **GREEN ALLIGATOR HANDICAP. $30,000 Added. 3-year-olds. Closed with a fee of $30, with $30,000 added, of which $6,000 to second, $4,500 to third, $2,250 to fourth and $750 to fifth. Weights: Thursday, April 21, 1994. Horses must re-enter by closing time of entries. High weights preferred. Closed Thursday, April 21 with 7 nominations.**

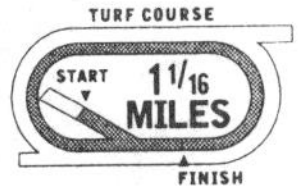

Log Buster

Own: Franks John R

CHAPMAN T M (231 36 28 26 .16)

Ro. c. 3 (Apr)
Sire: Hatchet Man (The Axe II)
Dam: Aunt Stel (Advocator)
Br: Franks John R (Ky)
Tr: Hollendorfer Jerry (183 52 33 25 .28)

117

Lifetime Record:	6 2 0 0	$23,075			
1994	3 2 0 0	$21,725	Turf	0 0 0 0	
1993	3 M 0 0	$1,350	Wet	1 1 0 0	$12,650
GG [T]	0 0 0 0		Dist [T]	0 0 0 0	

19Feb94-9GG	sly	$1\frac{1}{16}$	:23^{2}	:47^{2}	1:12	1:46^{1}	Alw 23000N1X	74	2	5	1^{1}	2^{hd}	1^{hd}	1^{2}	Chapman T M	L 120	4.70	66-29	Log Buster120^{2} Suomi Power117$^{2\frac{1}{2}}$ Errante117^{16}	Drifted out stretch	5
2Feb94-8GG	fst	1	:22^{3}	:45^{4}	1:10^{1}	1:35^{2}	Alw 23000N1X	57	7	6	5$^{4\frac{1}{2}}$	7$^{8\frac{1}{4}}$	7^{13}	7$^{18\frac{1}{2}}$	Gonzalez R M	L 120	*1.30e	75-17	Double Jab117^{10} Notellumnothing117$^{1\frac{1}{2}}$ Errante117^{1}	Dull try	7
14Jan94-3BM	fst	1	:22^{4}	:47^{1}	1:12^{2}	1:38	Md Sp Wt	75	2	2	2^{5}	2$^{\frac{1}{2}}$	1$^{1\frac{1}{2}}$	1$^{1\frac{1}{2}}$	Baze R A	L 118	2.00	83-22	Log Buster118$^{1\frac{1}{2}}$ Bensred118$^{\frac{3}{4}}$ Warning Label118^{6}	Closed gamely	5
18Dec93-7BM	fst	6f	:22^{1}	:45^{1}	:57^{4}	1:10^{4}	Md Sp Wt	62	3	8	7$^{7\frac{1}{2}}$	7^{10}	7$^{8\frac{1}{4}}$	7^{6}	Judice J C	L 118	18.50	76-18	Reality'sConquest118^{3} LuthierFever118$^{\frac{1}{2}}$ LypheorCstle118^{2}	Raced wide	10
20Nov93-1BM	fst	6f	:22^{3}	:45^{3}	:58	1:10^{4}	Md Sp Wt	60	4	6	7$^{5\frac{1}{2}}$	6^{7}	5$^{8\frac{1}{2}}$	4$^{3\frac{1}{4}}$	Judice J C	L 118	22.80	79-14	WriterToPass118$^{\frac{3}{4}}$ ParlayAegean118^{2} GardenShdow118$^{\frac{1}{2}}$	Very wide trip	7
4Nov93-4BM	fst	6f	:22^{3}	:45^{3}	:58	1:10^{3}	Md 20000	28	8	8	8$^{7\frac{1}{4}}$	8^{10}	7^{14}	5^{17}	Chapman T M	L 118	16.30	66-16	LFbuluxFort118^{6} SuprfnWn118^{hd} Notnthslftm113^{5}	Broke out start, wide	8

WORKOUTS: Apr 17 GG 1 fst 1:39 H *1/3* Apr 10 GG 1 fst 1:40^{3} H *2/4* Apr 5 GG 7f fst 1:25^{3} H *1/2* Mar 30 GG [T] 6f fm 1:15^{3} H (d) *1/3* ●Mar 23 GG 5f fst 1:00^{3} H *1/30* Mar 18 GG 4f fst :50 H *17/24*

Dancethetides

Own: Farm Haras Libertad

BOULANGER G (342 43 62 50 .13)

Ch. g. 3 (Apr)
Sire: Vanlandingham (Cox's Ridge)
Dam: Lisa's Song (Cutlass)
Br: Cavey D Michael (Md)
Tr: Jory Ian P D (10 1 2 0 .10)

113

Lifetime Record:	8 1 1 0	$24,500			
1994	3 1 0 0	$14,050	Turf	1 0 1 0	$5,800
1993	5 M 1 0	$10,450	Wet	1 0 0 0	$850
GG [T]	0 0 0 0		Dist [T]	0 0 0 0	

25Mar94-6SA	fst	1	:22^{4}	:46^{1}	1:11	1:36^{3}	Md 50000	93	7	3	3$^{\frac{1}{2}}$	2^{hd}	1^{hd}	1$^{\frac{1}{2}}$	Valenzuela P A	LB 118 b	5.60	86-22	Dncethetides118$^{\frac{1}{2}}$ HillsboroRod118$^{7\frac{1}{2}}$ PcificEdition118$^{4\frac{1}{2}}$	Bumped, gamely	7
19Mar94-6SA	sly	$1\frac{1}{16}$	:23^{1}	:47^{3}	1:13^{2}	1:48^{1}	Md Sp Wt	76	2	4	4$^{2\frac{1}{2}}$	3^{3}	4^{5}	5$^{5\frac{1}{4}}$	Atkinson P	LB 117 b	28.80	49-41	Wild Invader117$^{2\frac{3}{4}}$ Carolina Blew117$^{\frac{1}{2}}$ Island Sport117^{2}	Weakened	7
5Mar94-9SA	fst	1	:22^{1}	:45^{3}	1:10	1:35^{2}	Md Sp Wt	60	8	6	9^{10}	10^{12}	9^{17}	6^{19}	Nakatani C S	LB 117	20.80	73-07	Marconi117^{4} Scenic Route117$^{1\frac{3}{4}}$ Collectable Wine117^{1}	Very wide trip	10
26Dec93-9SA	fst	1	:22^{1}	:45	1:10^{1}	1:36^{3}	Md Sp Wt	70	9	8	9^{13}	9$^{6\frac{1}{2}}$	6$^{6\frac{1}{2}}$	4$^{9\frac{1}{2}}$	Pincay L Jr	LB 117	13.40	77-10	Almaraz117$^{1\frac{3}{4}}$ DramaticGold117$^{1\frac{3}{4}}$ AdvantgeMiles117^{6}	Wide, steadied 3/16	10
8Dec93-6Hol	fm	1 [T]	:24^{2}	:49^{1}	1:14	1:38^{3}	Md Sp Wt	68	3	4	4^{2}	3^{2}	3^{4}	2$^{6\frac{1}{2}}$	Valenzuela P A	B 118	30.60	65-26	Crowning Decision118$^{6\frac{1}{2}}$ Dancethetides118$^{3\frac{1}{2}}$ Carolina Blew118^{hd}		10
Lugged in, hit rail upper stretch																					
31Oct93-6SA	fst	$1\frac{1}{16}$	:22^{3}	:46^{1}	1:11^{1}	1:43^{2}	Md Sp Wt	50	6	6	8$^{6\frac{1}{2}}$	7$^{5\frac{1}{4}}$	6$^{8\frac{1}{2}}$	6^{23}	Solis A	B 117 b	12.30	64-15	WekivaSprings117^{11} CollectableWine117$^{2\frac{3}{4}}$ Let'sBeCurious117^{2}	No threat	9
10Oct93-9SA	fst	6f	:21^{4}	:44^{3}	:57	1:09^{4}	Md Sp Wt	68	1	11	8$^{5\frac{1}{4}}$	7$^{3\frac{3}{4}}$	6^{4}	4$^{6\frac{1}{2}}$	Gonzalez S5	112 b	51.50	82-16	Sky Kid117^{hd} Gracious Ghost117^{5} Al's River Cat117$^{1\frac{1}{4}}$	Off step slow	11
11Sep93-6Dmr	fst	6f	:21^{4}	:45	:57^{3}	1:10^{2}	Md Sp Wt	16	4	6	6$^{4\frac{1}{2}}$	7$^{8\frac{1}{4}}$	11^{16}	11^{23}	Pincay L Jr	B 118	65.70	66-12	Daggett Peak118^{3} Danny The Demon118^{hd} River Flyer118$^{2\frac{1}{4}}$	Faltered	11

WORKOUTS: Apr 17 Hol 5f fst 1:00^{4} H *11/26* Apr 11 Hol 4f fst :51 H *19/19* Apr 6 Hol 4f fst :49^{1} H *14/25* Mar 14 Hol 5f fst 1:03^{1} H *22/22* Feb 28 Hol 5f fst 1:00^{1} H *5/24* Feb 23 Hol 6f fst 1:14 H *3/8*

Fittobetried

Own: Capehart T R & Masanovich M

MERCADO P (202 14 18 26 .07)

Dk. b or br g. 3 (Mar)
Sire: Unpredictable (Tri Jet)
Dam: Chemise De Nuit*Fr (Viceregal)
Br: Capehart Tom & Masanovich Mike (Cal)
Tr: Jimenez Luis A (22 1 2 4 .05)

119

Lifetime Record:	10 2 0 2	$91,600			
1994	3 0 0 1	$38,750	Turf	1 1 0 0	$27,500
1993	7 2 0 1	$52,850	Wet	1 0 0 1	$11,250
GG [T]	0 0 0 0		Dist [T]	0 0 0 0	

21Feb94-8GG	sly	1	:22^{3}	:46^{3}	1:11^{3}	1:37^{2}	Golden Bear82k	69	5	7	7^{9}	7^{6}	4^{8}	3^{11}	Mercado P	LB 118 b	48.90	73-21	Double Jab116^{2} Halloween Treat114^{9} Fittobetried118^{hd}	Wide stretch	7
22Jan94-3BM	fst	$1\frac{1}{16}$	:23^{1}	:47	1:11^{3}	1:42^{3}	El Cm Rl Dy-G3	69	6	2	2^{1}	2^{1}	4^{6}	6$^{11\frac{3}{4}}$	Mercado P	LB 113 b	69.70	78-13	Tabasco Cat113^{1} Flying Sensation115^{6} Robannier115^{nk}	Gave way	7
1Jan94-8BM	fst	1	:23	:46^{3}	1:11^{2}	1:36^{2}	Cal Juvnl-G3	72	2	3	4^{3}	5^{3}	5$^{5\frac{1}{4}}$	4^{8}	Mercado P	LB 113 b	22.60	83-18	Robannier115$^{4\frac{1}{2}}$ Gracious Ghost113$^{2\frac{1}{2}}$ Delineator118^{1}	Steadied 2nd turn	5
26Nov93-8BM	fm	1 [T]	:23^{1}	:47	1:11^{4}	1:38^{1}	Foster City50k	81	5	4	5^{7}	4$^{5\frac{1}{2}}$	3$^{2\frac{1}{2}}$	1$^{2\frac{1}{2}}$	Mercado P	LB 113 b	14.50	82-15	Fittobetried113$^{2\frac{1}{2}}$ Al Renee115^{no} Saratoga Bandit114$^{1\frac{1}{2}}$	Strong finish	6
6Nov93-6BM	fst	1	:22^{3}	:46^{3}	1:11^{1}	1:37	L Stanford50k	69	9	8	7$^{5\frac{3}{4}}$	5$^{6\frac{1}{4}}$	4^{6}	4^{9}	Sorenson D	LB 116 b	104.90	76-17	Delineator115^{4} Al Renee115^{1} Arezzo116^{4}	Raced wide	9
26Sep93-8BM	fst	6f	:22^{3}	:45^{4}	:57^{4}	1:10^{2}	[S]S Mateo Juv38k	60	6	5	6^{6}	6$^{5\frac{1}{2}}$	6$^{6\frac{1}{2}}$	6$^{6\frac{1}{2}}$	Belvoir V T	LB 115 b	26.10	77-15	Ricks Ebony Star112^{no} Halos Empire114^{1} Shadowood118^{1}	No rally	7
17Sep93-7Fpx	fst	$6\frac{1}{2}$f	:22^{3}	:47^{2}	1:12^{4}	1:19^{2}	[S]Md Sp Wt	55	4	3	3^{1}	5$^{2\frac{1}{2}}$	2$^{1\frac{1}{2}}$	1^{2}	Gomez E A	LB 118 b	*.70	79-13	Fittobetried118^{2} ATributeToMax118$^{1\frac{1}{2}}$ StaffOfLife118$^{1\frac{1}{2}}$	Lacked room 1/4	7
29Aug93-8BM	fst	6f	:22	:44^{1}	:56^{3}	1:09^{4}	Atherton40k	69	7	5	6^{5}	6^{9}	6^{9}	4^{8}	Hernandez R A	LB 115 b	123.60	79-17	Al Renee116^{5} One Happy Fella114^{1} Subtle Trouble112^{2}	Even try	10
12Aug93-7Bmf	fst	$5\frac{1}{2}$f	:21^{3}	:45^{2}	:58	1:04^{3}	Md Sp Wt	58	4	2	2$^{\frac{1}{2}}$	2^{2}	3$^{1\frac{1}{2}}$	3$^{2\frac{3}{4}}$	Gomez E A	LB 118 b	29.40	89-10	R. B. Joe118$^{1\frac{1}{4}}$ Prevent Defense118$^{1\frac{1}{2}}$ Fittobetried118^{2}	Tiring late	8
12Jun93-3GG	fst	5f	:21^{2}	:45		:58	Md Sp Wt	32	7	12	12^{14}	11^{14}	11^{16}	9$^{9\frac{1}{4}}$	Gomez E A	B 118 b	91.50	83-10	Gold Bet118^{1} Dinadidit118^{no} Florida Wings118$^{1\frac{3}{4}}$	Shuffled back start	12

WORKOUTS: Apr 20 GG [T] 3f fm :37^{3} H (d) *1/2* ●Apr 16 GG 1 fst 1:41^{2} H *1/5* Apr 11 GG 7f fst 1:28^{4} H *1/1* Apr 6 GG [T] 5f fm 1:04^{2} H (d) *4/6* Mar 19 GG 6f fst 1:15^{2} H *9/17* ●Mar 14 GG 3f fst :36^{4} H *1/5*

continued

Gold Bet A3/2 C1+14*

Own: Olsen Cindy & Retzloff Jack

JUDICE J C (62 5 11 9 .08)

Dk. b or br g. 3 (Mar)
Sire: Lord At War (Arg) (General)
Dam: Madame Gold (Mr. Prospector)
Br: Olsen & Retzloff (Cal)
Tr: Bonde Jeff (67 9 11 10 .13)

114

Lifetime Record:	6 2 0 1	$25,850			
1994	1 0 0 0	$750	Turf	0 0 0 0	
1993	5 2 0 1	$25,100	Wet	0 0 0 0	
GG Ⓣ	0 0 0 0		Dist Ⓣ	0 0 0 0	

23Mar94-8GG	fst	6f	:214 :443 :57 1:102	Kfar Tov H30k	71	4 3 31 34 58 57¼	Baze R A	LB 116 b	11.90	80–18	Al Renee121½ Kingsburg Guy1161¼ Kluggie115½	Gave way	5
26Sep93-8BM	fst	6f	:223 :454 :574 1:102	[S]S Mateo Juv38k	64	7 6 54½ 52½ 53½ 55	Hansen R D	LB 115 b	4.30	79–15	RicksEbonyStar112no HalosEmpire1141 Shadowood1181	Raced well wide	7
12Sep93-7BM	fst	6f	:22 :443 :572 1:102	Alw 20000N1X	68	6 3 44½ 33 3nk 1no	Hansen R D	LB 116 b	7.10	84–14	Gold Bet116no Crazy Tale1132½ Frank C1151	Steadied sharply early	10
9Jly93-11Pln	fst	5½f	:22 :46 :574 1:043	[S]Alamda C Fty26k	21	3 3 33½ 32½ 58 512¾	Gonzalez R M	LB 117 b	*.70	77–10	Breaking Up1155 Jeep Shot1171 Piazza115¾	Bobbled start, wide	6
12Jun93-3GG	fst	5f	:212 :45 :58	Md Sp Wt	61	9 3 21 2hd 1hd 11	Gonzalez R M	LB 118 b	2.90	92–10	Gold Bet1181 Dinadidit118no Florida Wings1181¾	Steady drive	12
28May93-2GG	gd	5f	:22 :463 :592	Md Sp Wt	57	8 5 53½ 43 33 32½	Gonzalez R M	LB 118 b	4.20	82–20	Ramanujan118no Smoke In Motion1182½ Gold Bet1183	Wide trip	8

WORKOUTS: Apr 20 GG Ⓣ 4f fm :503 H (d) *2/5* Apr 14 GG 6f fst 1:133 H *3/13* Apr 8 GG 4f fst :493 H *18/40* Apr 2 GG 1 fst 1:364 H *1/1* Mar 20 GG 3f fst :363 Hg *13/30* Mar 14 GG 7f fst 1:252 H *1/1*

Margie's Boy U814

Own: Bailey Donald A & Marilyn

LOPEZ A D (200 32 29 33 .16)

Ch. g. 3 (Apr)
Sire: The Carpenter (Gummo)
Dam: Courageous Thing (Gin Tour)
Br: Bailey Mr & Mrs Donald (Cal)
Tr: McCutcheon James R (1 0 0 0 .00)

113

Lifetime Record:	11 2 0 2	$16,926			
1994	4 2 0 1	$15,350	Turf	0 0 0 0	
1993	7 M 0 1	$1,576	Wet	2 1 0 1	$9,850
GG Ⓣ	0 0 0 0		Dist Ⓣ	0 0 0 0	

19Mar94-3SA	sly	1 1/16	:24 :48 1:131 1:483	Alw 39000N1X	74	2 2 23 24 34½ 34¾	Atkinson P	LB 117 b	6.60	47–41	HeartlessWager117¾ Almaraz1204 Mrgie'sBoy1177½	Drifted out 2nd turn	4
5Mar94-3GG	fst	1 1/16	:232 :474 1:12 1:45	Clm 10000	63	7 1 11 11 11 14	Noguez A M	LB 117 b	8.00	72–19	Margie's Boy1174 Isidro Y Martine1172 Sir Slewy117nk	Steady drive	7
18Feb94-6GG	my	1 1/16	:23 :471 1:121 1:452	Md 12500	67	4 1 1½ 12 11½ 16	Noguez A M	LB 118 b	40.60	70–22	Margie's Boy1186 Polo Magic1181½ Jewel OfNostalgia118nk	Steady urging	10
28Jan94-4BM	gd	6f	:223 :461 :584 1:114	Md 12500	34	10 8 87¼ 99½ 914 716½	Mercado P	LB 118 b	7.00	65–17	Golden Possibility1186 Shine Brite1181½ Polo Magic1185	Wide trip	11
15Oct93-3Fno	fst	5½f	:22 :462 :584 1:051	Md 12500	40	5 6 42½ 42 43 42	Dalton S L	LB 118	*1.20	79–11	I'm Gunna Do115no Icy Rapids1181½ Mr. Nasty John118½	Wide, mild rally	6
9Oct93-6Fno	fst	6f	:224 :46 :584 1:111	[S]Md 16000	47	3 8 84 64 34½ 33½	Dalton S L	LB 118	9.10	80–27	Piazza1182½ Prince Spiked Tea1181 Margie's Boy118½	Stumbled break	8
3Oct93-8Fpx	fst	6f	:222 :461 :591 1:121	Md 32000	42	9 7 86¾ 85 98½ 76¾	Fernandez A L	LB 118	27.10	79–09	Jury Trial1181¼ Bobs Bikini118nk Smooth Shot1183½	Wide trip	10
22Sep93-8Fpx	fst	6f	:22 :454 :582 1:111	[S]Md 32000	47	10 7 99¼ 811 712 711¼	Fernandez A L	B 118	93.40	80–12	Smoke In Motion1181½ Bon Oeuf1186 Golden Explo1181½	No threat	10
27Jun93-6Stk	fst	4½f	:223 :453 :513	Md 12500	11	3 4 79½ 611 613½	Cruz J B	B 118 b	9.40	— —	CrowningsCricket1186 ClydeTheGlide118½ PrinceSpikdT118½	No threat	8
20May93-4Hol	fst	5f	:213 :451 :58	Md 35000	34	2 8 810 88½ 812 613¾	Castaneda M	116 b	77.00	80–11	American Original1183 Troyalty1184 Valiant Pirate1184½	Steadied early, wide	10

WORKOUTS: Apr 20 GG Ⓣ 4f fm :512 H (d) *4/5* Apr 14 SA 5f fst 1:042 H *30/30* Mar 16 SA 5f fst 1:05 H *42/44* Mar 1 BM 4f fst :511 H *9/10* Feb 24 BM 3f fst :371 H *4/14* Feb 10 BM 7f fst 1:292 H *1/1*

Suomi Power B4/3

Own: Biszantz Gary & Enberg Dick

WARREN R J JR (244 32 29 33 .13)

Ch. g. 3 (Apr)
Sire: Explodent (Nearctic)
Dam: Yrmika Fr (Crystal Palace)
Br: Mandysland Farm & Enemy Stable (Ky)
Tr: Greenman Walter (54 10 10 7 .19)

114

Lifetime Record:	14 1 5 2	$33,940			
1994	4 0 2 1	$13,325	Turf	2 0 1 1	$8,725
1993	10 1 3 1	$20,615	Wet	1 0 1 0	$4,600
GG Ⓣ	1 0 1 0	$4,600	Dist Ⓣ	1 0 1 0	$4,600

Entered 23Apr94- 6 GG

3Apr94-9GG	fm	1 1/16 Ⓣ	:221 :462 1:113 1:444+	Alw 23000N1X	70	7 8 76¾ 75 612 210	Warren R J Jr	LB 117 b	11.70	72–18	Still Swingin'11710 Suomi Power1171 Dinner Affair117½	Lacked room into stretch	8
12Mar94-9GG	fst	1 1/16	:224 :462 1:102 1:413	Alw 23000N1X	68	2 6 68½ 68 612 615	Boulanger G	LB 117 b	16.10	74–17	Bai Brun1174 Warning Label1173½ Juan's Boy1171	Outrun	7
19Feb94-9GG	sly	1 1/16	:232 :472 1:12 1:461	Alw 23000N1X	71	5 4 42½ 1hd 2hd 22	Grabowski J A	LB 117 b	5.80	64–29	Log Buster1202 Suomi Power1172½ Errante11716	Bumped stretch	5
22Jan94-1BM	fm	7f Ⓣ	:23 :472 1:134 1:264	Crdf Stk F H27k	70	4 4 48½ 45 33½ 35½	Grabowski J A	LB 114 b	9.20	— —	Makinanhonestbuck1152 Flying Kajo1163½ Suomi Power11411	Even try	4
17Dec93-1BM	gd	1	:23 :471 1:123 1:392	Md 20000	67	2 2 21 1hd 15 13½	Warren R J Jr	LB 118 b	*1.10	73–25	Suomi Power1183½ Healbequick1161 Cordial Hello1182	Steady drive	7
21Nov93-3BM	fst	1	:223 :464 1:124 1:382	Md 20000	62	5 4 41½ 3½ 1½ 22	Boulanger G	LB 118 b	3.60	76–19	Barts Demon1182 Suomi Power1183 Notinthislifetime1162½	Held well	8
13Nov93-1BM	fst	1	:23 :473 1:123 1:374	Md Sp Wt	53	2 3 32 32 26 310¼	Warren R J Jr	LB 118 b	12.50	71–23	River Flyer11810 Catch The Native118nk Suomi Power1183	Even late	6
22Oct93-6SA	fst	1	:224 :464 1:123 1:384	Md 32000	53	4 2 1hd 2hd 3½ 46½	Desormeaux K J	LB 117 b	4.00	70–21	Veracity117hd Prevasive Force1172¾ Sierra Real1123½	Inside duel	9
11Oct93-2SA	fst	6f	:212 :443 :574 1:112	Md 32000	53	6 1 41 51¾ 41½ 42½	Solis A	LB 118 b	1.80	77–11	A Tribute To Max118hd Beau Finesse1181 Toknight1181½	4 Wide into lane	9
26Sep93-5Fpx	fst	6f	:214 :464 :584 1:122	Md 32000	50	4 2 32½ 41¼ 41½ 21½	Nakatani C S	LB 118 b	*1.30	85–09	Mob Stage1181½ Suomi Power1181 Gary's Lucky One1181	Split rivals	10

WORKOUTS: Apr 17 GG 5f fst 1:023 H *48/76* Apr 12 GG 3f fst :371 H *3/5* Mar 27 GG 5f fst 1:011 H *20/74* Mar 21 GG 3f fst :354 H *2/4* Mar 6 GG 5f fst 1:024 H *20/34* Feb 28 GG 3f fst :381 H *5/13*

This leaves Gold Bet as the most likely upsetter. Besides his excellent A3 turf rating, his stamina index of 2 suggests he won't have any trouble getting the 1 1/16 mile distance. He also comes into today's race with four works including a mile stamina builder on April 2nd and a turf workout on April 20th. Furthermore, Gold Bet has never been run for a tag and he needed his last race where he chased :21.4/:44.3 fractions and tired. But, today he looks like he can get a fairly easy lead and is a definite wire to wire threat. On the downside, with all this going for him why has leading rider Russell Baze gotten off of him and why is Gold Bet almost 10-1? Probably because the public is simply not in tune with pedigree handicapping.

So how to bet? Sure wish the obvious horse would be 10-1 more often! Gold Bet has the best turf breeding in the race, has the SI's to get the distance, and with the sprint to route maneuver is almost certain to control the pace. With speed and breeding in his corner, maybe Baze is off the horse because Gold Bet's owner/breeder connections didn't want him to kill the price. So why not relish the 10-1 odds? Time for a prime Win wager on Gold Bet as well an exacta box with the favorite.

EIGHTH RACE

Golden Gate
APRIL 24, 1994

1$\frac{1}{16}$ MILES. (Turf Chute)(1.41[1]) GREEN ALLIGATOR HANDICAP. $30,000 Added. 3-year-olds. Closed with a fee of $30, with $30,000 added, of which $6,000 to second, $4,500 to third, $2,250 to fourth and $750 to fifth. Weights: Thursday, April 21, 1994. Horses must re-enter by closing time of entries. High weights preferred. Closed Thursday, April 21 with 7 nominations.

Value of Race: $30,210 Winner $16,710; second $6,000; third $4,500; fourth $2,250; fifth $750. Mutuel Pool $288,721.00 Exacta Pool $340,639.00 Quinella Pool $38,988.00

Last Raced	Horse	M/Eqt.	A.Wt	PP	St	$\frac{1}{4}$	$\frac{1}{2}$	$\frac{3}{4}$	Str	Fin	Jockey	Odds $1
23Mar94 8GG5	Gold Bet	LBb	3 116	4	2	1^1	1^{hd}	1^1	1^5	1$^{1\frac{1}{2}}$	Judice J C	9.60
25Mar94 6SA1	Dancethetides	LBb	3 113	2	4	5^{hd}	5$^{1\frac{1}{2}}$	3^{hd}	2$\frac{1}{2}$	2^5	Boulanger G	3.50
21Feb94 8GG3	Fittobetried	LBb	3 119	3	1	4$^{1\frac{1}{2}}$	6	6	5^4	3^1	Mercado P	1.50
19Feb94 9GG1	Log Buster	L	3 117	1	5	6	4^{hd}	4$\frac{1}{2}$	4$\frac{1}{2}$	4^1	Chapman T M	2.80
3Apr94 9GG2	Suomi Power	LBb	3 115	6	3	2$\frac{1}{2}$	3^2	2^2	3$^{1\frac{1}{2}}$	5^4	Warren R J Jr	6.00
19Mar94 3SA3	Margie's Boy	LBbf	3 116	5	6	3^1	2$\frac{1}{2}$	5^2	6	6	Lopez A D	11.70

OFF AT 5:09 Start Good. Won driving. Time, :24^2, :49^1, 1:13^1, 1:38^3, 1:45^2 Course good.

$2 Mutuel Prices:

4–GOLD BET	....	21.20	10.40	5.00
2–DANCETHETIDES	....		5.20	3.40
3–FITTOBETRIED	....			2.80

$5 EXACTA 4–2 PAID $269.50 $4 QUINELLA 2–4 PAID $95.20

Dk. b. or br. g, (Mar), by Lord At War (Arg)–Madame Gold, by Mr. Prospector. Trainer Bonde Jeff. Bred by Olsen & Retzloff (Cal).

Turf Claiming Races

Horses with good turf breeding that have broken their maiden but aren't good enough to win on the dirt at the allowance or stakes level sometimes surprise when they are dropped into grass claiming races. The eighth race at Hollywood Park on June 3rd is a spectacular example of the mutuels that are sometimes possible.

The first thing to notice about this race is that none of the horses has as yet won on the turf and that Heartless Wager at 3-1 is a lukewarm favorite. Surveying the field, Win The Case, One Rich Runner, Hot Papa Tom, Jabbawat and Baja Bill have all already lost turf routes at the claiming level and are unlikely winners. While Hot Papa Tom has yet to run on the turf and might improve, his broodmare sire Fort Calgary has a D turf rating and mostly gets sprinters.

Of the class droppers, even though Sky Kid has speed, he is not bred for the turf and has already lost twice on that surface. Heartless Wager is dropping out of a stakes and shortening to a mile. But, he is not bred for the turf, has never passed horses and is unlikely to get an easy lead from Sky Kid. Come crunch time, he is likely to once again fold in the stretch. While Dancethetides also isn't bred for the turf, he has finished second twice on the grass, is dropping, and has the right closing style to get up in this race.

This leaves us with Jade Master and Beau Finesse, two more class droppers, as the only horses that have B or better turf ratings in the race. While Beau Finesse is coming off a two month rest, he has two seven-furlong workouts and Brad MacDonald, according to Olmsted's *Trainer Guide*, wins with layoff horses. Furthermore, he has thus far only sprinted on the turf and his stamina indexes indicate he should love the added distance. In his only attempt at a mile, he broke slowly and still was only beaten by less than four lengths. Since Beau Finesse is "debuting" at this distance on the turf, he should be given the benefit of the doubt and assigned Val de l'Orne's B rating. At 54-1, he is a huge overlay.

Jade Master, the last class dropper, is equally enticing. He has the best turf breeding in the race and finished right behind Beau Finesse in the Baldwin. While he faded in both his routes, they were on the dirt and, in his only win, he closed from off the pace. Since his stamina indexes say he can route and this is Jade Master's turf "debut" at the mile distance, he has to be given a chance, despite his 10-1 odds, to run to his breeding. Solis seems to be the only right jockey for the horse, too, since Jade Master won at 19-1 when switching to Solis, implying that his sole turf race with Atkinson is not to be treated seriously.

So how to bet? Despite their long prices, Beau Finesse and Jade Master are the only two horses dropping in the race that are truly bred for the turf and that have never lost at the distance. Furthermore, their turf tries were on Santa Anita's downhill turf course and some horses dramatically improve when switching to Hollywood Park. Given that there

Hollywood Park

8

1 MILE. (Turf). (1:32 3) CLAIMING. Purse $45,000. 3-year-olds. Weight, 122 lbs. Non-winners of two races at a mile or over since April 5, allowed 3 lbs. Such a race since then, 6 lbs. Claiming price $62,500; for each $2,500 to $55,000, allowed 1 lb. (Races when entered for $50,000 or less not considered.)

TURF COURSE — 1 MILE — FINISH START

Heartless Wager

Own: Kirkwood Al & Sandee
PINCAY L JR (112 16 15 14 .14) $62,500

Ch. c. 3 (May)
Sire: Mt. Livermore (Blushing Groom)
Dam: Catherine's Bet (Grey Dawn II)
Br: Thunderhill Farm (Ky)
Tr: Puhich Michael (8 1 1 0 .13)

L 116

Lifetime Record:	9 2 0 1	$44,752			
1994	7 2 0 0	$40,777	Turf	1 0 0 0	$1,627
1993	2 M 0 1	$3,975	Wet	1 1 0 0	$22,000
Hol Ⓣ	0 0 0 0		Dist Ⓣ	0 0 0 0	

4May94-3Hol fst 1 1/16 :224 :454 1:102 1:43 [R]Sunday Slnce56k 85 1 1 18 11½ 2hd 46¾ Black C A LB 118 b 19.10 79-18 Little Lite Bud120nk College Town1162 Western Trader1204½ Weakened 7
1Apr94-7Kee fst 7f :222 :453 1:113 1:243 Alw 29500N2X 82 6 7 2hd 2hd 41¼ 46¼ Steiner J J L 113 b 9.40 72-22 Upping The Ante118¾ Clev Er Irish1181½ Sierra Cat1134 Dueled, tired 9
9Apr94-8Kee fm 1 1/16 Ⓣ :224 :47 1:112 1:424 Transylvania54k 87 1 1 1hd 1hd 3½ 54¼ Steiner J J L 118 b 20.60 88-12 Star Of Manila1212 Prix De Crouton1182 Carpet118no Pace, tired 6
9Mar94-3SA sly 1 1/16 :24 :48 1:131 1:483 Alw 39000N1X 82 1 1 13 14 13 1¾ Black C A LB 117 b 3.70 52-41 Heartless Wager117¾ Almaraz1204 Margie's Boy1177½ Speed, held 4
27Feb94-6SA fst 6f :212 :442 :564 1:094 Clm c-50000 57 7 2 41½ 51¾ 74¾ 710½ Valenzuela F H LB 115 b 5.60 77-10 Smoke In Motion1152 Cold CoolN'bold1161¾ MarkNinety109¾ 5 wide trip 7
Claimed from Three Sisters Stable, Chambers Mike Trainer
8Jan94-2SA fst 6f :213 :444 :571 1:101 Md 32000 80 11 2 42 2hd 12½ 14½ Valenzuela F H LB 118 b *2.10 86-12 Heartless Wager1184½ Double Sec1181¾ Bobs Bikini118hd Cleary best 11
4Jan94-6SA fst 6f :214 :443 :562 1:081 Md 50000 68 6 5 52½ 44½ 47½ 412½ Antley C W LB 118 4.90 83-09 Unfinished Symph1182 Mr Peter P.11810 Greattobeking113½ No raily 8
1Dec93-6SA fst 6f :213 :442 :57 1:102 Md 50000 68 9 1 2½ 2hd 2½ 32½ Valenzuela F H LB 118 3.20 82-10 Freezit1181¼ Mr Peter P.1181¼ Heartless Wager118hd Outfinished 9
20Dec93-1Hol fst 6f :221 :444 :564 1:09 Md Sp Wt 67 6 7 78¾ 77¾ 67½ 58¾ Delahoussaye E LB 118 *2.10e 86-09 Pollock's Luck1181¾ Superpak1184 Native Blast118½ Wide into lane 8

WORKOUTS: May 30 SA 4f fst :474 H 6/26 May 12 Hol 3f fst :374 H 20/23 May 6 Hol 3f fst :372 H 16/16 Apr 5 Kee Ⓣ 5f fm 1:021 B (d) 1/1 Mar 27 SA 3f fst :364 B 17/33 Mar 10 SA 5f fst 1:004 H 14/46

Beau Finesse

Own: Equine Thoroughbred Associates
FLORES D R (95 7 5 10 .07) $62,500

B. c. 3 (Mar)
Sire: Val de l'Orne (Val de Loir)
Dam: Anna Geranios (Raja Baba)
Br: Solstice (Md)
Tr: MacDonald Brad (2 0 0 0 .00)

L 116

Lifetime Record:	7 1 1 0	$15,300		
1994	2 0 0 0		Turf	2 0 0 0
1993	5 1 1 0	$15,300	Wet	0 0 0 0
Hol Ⓣ	0 0 0 0		Dist Ⓣ	0 0 0 0

30Mar94-8SA fm *6½f Ⓣ :211 :432 1:071 1:133 Baldwin82k 67 4 7 813 812 812 610½ Black C A LB 116 b 76.60 79-08 Silver Music114½ Eagle Eyed1171¼ Makinanhonestbuck117nk No rally 8
13Feb94-7SA fm *6½f Ⓣ :213 :441 1:074 1:141 Alw 41000N2X 75 2 8 811 812 76¾ 74¾ Delahoussaye E LB 116 27.30 82-12 Sharp Try117½ Heavenly Crusade114½ Daggett Peak114no No factor 8
30Dec93-6SA fst 6f :211 :443 :572 1:103 Md 32000 66 4 8 1012 77¼ 43 12 Delahoussaye E LB 118 *1.80 84-15 Beau Finesse1182 Chuppatti118hd Kismet Hook Us118hd Lugged in 1/8 12
26Nov93-5Hol fst 6½f :214 :444 1:102 1:164 Md 32000 65 10 2 65¼ 42½ 31½ 43½ McCarron C J LB 119 *1.30 81-09 Sundays Cream119hd Warrior's Wind1123 Don Icecapade119½ 10
Veered out start, 5 wide turn
28Oct93-2BM fst 1 :232 :473 1:123 1:381 Md Sp Wt 60 6 5 54 42½ 54 53½ Kaenel J L LB 118 5.10 75-20 Blowin De Turn1181 Catch The Native1181 Barts Demon118½ 6
Ducked out sharply start
11Oct93-2SA fst 6f :212 :443 :574 1:112 Md 32000 60 2 8 76¼ 74 62½ 2hd McCarron C J LB 118 15.40 80-11 A Tribute To Max118hd Beau Finesse1181 Toknight1181½ Lacked room 1/8 9
3Sep93-6Dmr fst 5½f :22 :452 :571 1:033 Md 32000 50 2 11 1010 813 511 513½ McCarron C J LB 118 41.90 80-09 Halos Empire1186 Say Ow118hd Réblin Man1186½ Broke slowly 11

WORKOUTS: May 29 SA 7f fst 1:284 H 4/6 May 23 SA 7f fst 1:29 H 3/5 May 22 SA 3f fst :39 H 19/21 May 17 SA 6f fst 1:121 H 2/19 May 12 SA 3f fst :38 B 16/19 Apr 18 SA 5f fst 1:032 H 30/31

Sky Kid

Own: A R Stable Inc & Loughran
DESORMEAUX K J (95 20 14 16 .21) $62,500

Ch. c. 3 (Apr)
Sire: Naked Sky (Al Hattab)
Dam: Passionate Baby (Full Pocket)
Br: Don Dee Farm (Fla)
Tr: Hess R B Jr (29 4 5 3 .14)

L 116

Lifetime Record:	6 1 0 0	$20,825			
1994	2 0 0 0	$1,725	Turf	2 0 0 0	$800
1993	4 1 0 0	$19,100	Wet	0 0 0 0	
Hol Ⓣ	0 0 0 0		Dist Ⓣ	0 0 0 0	

17Apr94-9GG fst 6f :213 :442 :564 1:093 Alw 23000N1X 81 6 1 42¼ 42½ 34 46½ Lopez A D LB 117 3.50 84-18 Juan's Boy1204½ Iron Groove1171 Wise And Bold1171 Even late 8
16Mar94-2SA fm *6½f Ⓣ :214 :44 1:073 1:14 Alw 37000N1X 79 8 1 78 710 79 64½ Delahoussaye E LB 120 7.00 83-12 Solar Wagon1201 Beautiful Crown1201¼ Superfluously120no Mild bid 9
5Nov93-5SA fm *6½f Ⓣ :214 :442 1:081 1:142 Alw 32000N1X 69 1 8 3½ 31 32 55¼ Lopez A D LB 118 8.50 81-14 Devon Port1151 Daggett Peak1181 Bullamatic1152¾ Inside trip 11
10Oct93-9SA fst 6f :214 :443 :57 1:094 Md Sp Wt 84 4 3 1hd 1½ 11½ 1hd Lopez A D LB 117 7.50 88-16 Sky Kid117hd Gracious Ghost1175 Al's River Cat1171¼ Inside duel 11
28Aug93-2Dmr fst 6½f :221 :453 1:102 1:164 Md Sp Wt 73 5 1 41½ 61¾ 53 45¾ Lopez A D LB 117 19.90 81-12 Brocco1173½ Sir Harry Bright1171¾ Dawson Ridge117½ Troubled trip 8
28Jly93-6Dmr fst 5f :22 :454 :583 Md Sp Wt 57 5 2 4¾ 31½ 44 56 Desormeaux K J LB 117 8.70 98-07 Showdown1172½ Shepherd's Field117hd Drouilly River1172 Weakened 8

WORKOUTS: May 30 Hol 4f fst :501 H 52/63 May 25 Hol 4f fst :502 H 42/51 May 18 GG 5f fst 1:031 H 35/36 May 12 GG Ⓣ 5f fm 1:02 H (d) 2/2 May 4 GG 4f fst :493 H 24/34 Apr 12 Hol 3f fst :381 H 13/14

Win The Case

Own: Brooks & King & Landsburg
VALENZUELA F H (144 12 16 14 .08) $57,500

Dk. b or br g. 3 (Apr)
Sire: Kalim (Hotfoot)
Dam: Legal Protection (Maheras)
Br: King & Landsburg (Cal)
Tr: Cerin Vladimir (22 5 3 3 .23)

L 114

Lifetime Record:	11 1 6 1	$39,770			
1994	6 1 3 0	$27,200	Turf	2 0 0 0	$4,300
1993	5 M 3 1	$12,570	Wet	0 0 0 0	
Hol Ⓣ	1 0 0 0	$3,375	Dist Ⓣ	1 0 0 0	$3,375

18May94-4Hol fm 1 Ⓣ :234 :472 1:12 1:371 Clm 57500 74 7 6 52½ 64¼ 53 43½ Valenzuela F H LB 114 b 13.70 74-19 Set On Cruise1162½ Baja Bill1171 Veracity116nk Finished well 7
1May94-4Hol fst 1 1/16 :231 :462 1:11 1:433 Clm 30000 71 6 6 45½ 43 32 26 Valenzuela F H LB 114 b 6.40 77-14 Tulwar1166 Win The Case114½ Tis Careno117½ 6
6 wide 7 1/2, rail after
10Apr94-2SA fm *6½f Ⓣ :22 :442 1:081 1:141 Alw 37000N1X 75 2 7 65½ 65¼ 64¾ 54½ Valenzuela F H LB 117 b 31.20 82-13 Daggett Peak120¾ MobStage117½ CrowningDecision1201½ Wide into lane 7
16Mar94-9SA fst 6½f :222 :453 1:104 1:172 [S]Md 28000 68 4 4 2hd 11½ 13½ 1no Valenzuela F H LB 116 b 3.00 85-12 Win The Case116no Al Monsoon1184 Golden Century1182¾ Just held 9
19Feb94-2SA fst 6½f :22 :444 1:091 1:16 [S]Md 32000 65 7 5 21½ 22½ 25 29 Valenzuela F H LB 118 b *1.80 83-08 CusingRumors1189 WinTheCs1181¾ LightningN'ic1181¼ Lugged in stretch 7
2Feb94-9SA fst 7f :223 :454 1:112 1:242 [S]Md 32000 61 6 6 74¼ 5¾ 1½ 21¼ Valenzuela F H LB 118 b *1.70 79-13 Natural Charmer1181¼ Win The Case1182¾ Victory Sun118¾ 10
Ducked in sharply 1/16
15Dec93-9Hol gd 6½f :221 :453 1:122 1:192 [S]Md 32000 62 4 8 94½ 42½ 31 2¾ Valenzuela F H LB 119 b *2.40 71-19 Real Smart119¾ Win The Case1195 Coniston1175½ Finished well 11
3Dec93-6Hol fst 6f :22 :452 :581 1:112 [S]Md 28000 64 2 7 42 76¾ 46 31¼ Valenzuela F H LB 117 b 3.90 82-15 Go To Bat119no Tam Rise1191¼ Win The Case1171¾ 9
Bumped, steadied 3/8, in tight late
15Nov93-6SA fst 7f :232 :47 1:114 1:243 [S]Md 28000 59 3 3 4¾ 51¾ 32½ 21¾ Valenzuela F H LB 115 b *2.10 78-15 He's Fabulous1171¾ Win The Case115hd Al's Sunshine1173½ Finished well 10
29Oct93-4SA fst 7f :221 :451 1:112 1:25 Md 28000 53 1 10 63 45½ 52 21¼ Valenzuela F H LB 116 b 54.00 77-19 Ketchum Z's1181¼ Win The Case116hd Al Khedive1132¾ Finished well 12

WORKOUTS: May 30 SA 4f fst :482 H 9/26 May 11 SA 4f fst :48 H 6/45 Apr 23 SA 6f fst 1:151 Hg 12/17 Apr 18 SA 3f fst :361 H 10/22 Apr 7 SA 3f fst :351 H 3/24 Apr 1 SA 4f fst :473 H 10/38

Hot Papa Tom

Own: Dante Charles T
ANTLEY C W (120 27 23 10 .23) $62,500

Gr. c. 3 (Apr)
Sire: Incinderator (Northern Dancer)
Dam: Thisonesforyoudad (Fort Calgary)
Br: Dante Charles T (Cal)
Tr: Threewitt Noble (4 0 0 0 .00)

116

Lifetime Record:	4 1 1 0	$15,625		
1994	4 1 1 0	$15,625	Turf	0 0 0 0
1993	0 M 0 0		Wet	0 0 0 0
Hol Ⓣ	0 0 0 0		Dist Ⓣ	0 0 0 0

22May94-9Hol fst 7f :221 :45 1:10 1:231 Clm 40000 73 4 9 1116 1112 65¼ 51¾ Desormeaux K J 116 9.00 86-08 Chief Brody116¾ HighDesertStar116nk OneHappyFella116no Pinched start 11
22Apr94-4SA fst 6f :22 :451 :572 1:10 Md 32000 82 2 4 31 41½ 4¾ 1¾ Desormeaux K J 118 *1.10 87-11 Hot Papa Tom118¾ Hippiater118½ Scottish Slew118hd Rail trip 11
8Apr94-9SA fst 6½f :214 :45 1:102 1:17 Md 32000 72 9 6 43¼ 53½ 31 2hd Delahoussaye E 118 6.60 87-12 OneSharpChip118hd HotPapaTom118½ FortunteAnswer1181 Led, caught 12
11Mar94-4SA fst 6½f :213 :45 1:102 1:17 Md 32000 65 1 9 711 67½ 44 44½ Delahoussaye E 118 4.20 83-14 Kismet Hook Us116hd Top Gear118nk Flaming George1184 Wide into lane 11

WORKOUTS: May 30 SA 4f fst :464 H 2/26 May 19 SA 4f gd :483 H 4/25 May 13 SA 6f fst 1:143 H 10/17 May 7 SA 5f gd 1:02 H 9/20 May 2 SA 4f fst :482 H 14/64 Apr 13 SA 3f fst :36 H 4/28

continued

One Rich Runner

Own: Stacey Richard & Sally & Scott

ATHERTON J E (78 7 5 5 .09) — $55,000

Dk. b or br g. 3 (Feb)
Sire: A Run (Empery)
Dam: Rich Indian (Raise a Native)
Br: Golden State Stables Inc (Cal)
Tr: Bunn Thomas M Jr (7 0 1 1 .00)

L 1085

Lifetime Record: 10 1 1 1 $20,765

1994 6 1 1 1 $18,250 | Turf 2 0 1 1 $7,800
1993 4 M 0 0 $2,515 | Wet 0 0 0 0
Hol Ⓣ 0 0 0 0 | Dist Ⓣ 0 0 0 0

1May94-8GG fm 1 1/16 Ⓣ :233 :482 1:123 1:45 + Clm 25000 75 5 4 42½ 32 1hd 21½ Boulanger G LB 117 b *2.30 79-19 Sir Dunoon1171½ One Rich Runner1171 Sheer Ambition1151 Game try 7
10Apr94-9GG fm 1 1/16 Ⓣ :23 :472 1:123 1:451 + Clm 30000 79 4 4 517 511 34½ 32½ Lopez A D LB 116 b 12.00 78-20 Dinner Affair1172 Champagne Shane117nk One Rich Runner1166 5
Rank early, mild late bid
16Mar94-6SA fst 7f :222 :451 1:10 1:224 Clm 25000 56 7 10 106½ 107½ 810 711½ Castanon A L LB 115 b 35.20 76-12 Say Ow1168 Shameless Devil113¾ Warrior's Wind115nk No rally 10
23Feb94-2SA fst 7f :222 :451 1:103 1:233 ⓈClm 25000 63 4 7 76½ 64½ 52½ 56½ Castanon A L LB 116 b 10.40 77-13 Tis Careno1151¾ Cajero1163½ Sparkling Brooks115½ No rally 7
2Feb94-4SA fst 7f :223 :452 1:102 1:233 ⓈMd 32000 66 5 9 64 63½ 32 11¾ Castanon A L LB 118 b 33.90 84-13 OneRichRunner1181¾ KismetHookUs116¾ Uronurown1182¾ Circled horses 9
20Jan94-2SA fst 1 :231 :472 1:121 1:374 ⓈMd 35000 55 7 6 65½ 64½ 66 69 Stevens G L LB 115 b 18.60 71-20 TwoMinuteDrill1151½ DshOfVnill1153 HevnlyStr117hd Steadied 3/4 & 5 1/2 8
30Dec93-2SA fst 1 :221 :46 1:111 1:374 ⓈMd 32000 56 5 5 52½ 53 43 55½ Delahoussaye E LB 118 b 5.00 75-14 Canyon Passage1182½ Heavenly Star1181½ Petrinsic118hd No late bid 9
15Dec93-9Hol gd 6½f :221 :453 1:122 1:192 ⓈMd 32000 36 6 6 83¾ 53½ 43 512½ Castanon A L LB 119 b 18.30 60-19 Real Smart119¾ Win The Case1195 Coniston1175½ Wide into lane 11
22Oct93-6SA fst 1 :224 :464 1:123 1:384 Md 32000 50 2 3 2hd 3nk 2½ 58 Black C A LB 117 b 13.10 68-21 Veracity117hd Prevasive Force1172¾ Sierra Real1123½ Steadied 1/8 9
2Oct93-4Fpx fst 6f :223 :461 :584 1:121 ⓈMd 32000 49 10 9 911 710 58½ 44 Black C A LB 118 16.40 84-09 Endless Dust1182 Al Khedive113no Umpscious1132 Steady gain 10

WORKOUTS: May 30 Hol 5f fst 1:023 H 47/56 May 24 Hol 5f fst 1:031 H 13/19 May 18 Hol 3f fst :364 H 14/26 May 13 Hol 4f fst :522 H 31/32 Mar 11 Hol 4f fst :511 H 10/11 Mar 5 Hol 4f fst :533 B 17/17

Jabbawat

Own: Colvin & Gould

BLACK C A (138 13 21 20 .09) — $62,500

Dk. b or br g. 3 (May)
Sire: Slewpy (Seattle Slew)
Dam: Aerturas (Manado)
Br: Gould William (Ky)
Tr: Hofmans David (15 2 3 4 .13)

L 116

Lifetime Record: 6 1 1 0 $33,150

1994 4 1 1 0 $30,425 | Turf 1 0 0 0
1993 2 M 0 0 $2,725 | Wet 0 0 0 0
Hol Ⓣ 0 0 0 0 | Dist Ⓣ 0 0 0 0

12May94-7Hol fst 6f :213 :442 :563 1:093 Clm 50000 79 2 6 43 31½ 22 21½ Black C A LB 116 b 3.60 90-09 Card Connection1161½ Jabbawat1164½ Royally Right1171 Inside bid 7
16Mar94-2SA fm *6½f Ⓣ :214 :44 1:073 1:14 Alw 37000N1X 76 3 7 55 56 67½ 75½ Black C A LB 120 15.70 82-12 Solar Wagon1201 Beautiful Crown1201½ Superfluously120no Weakened 9
2Mar94-8SA fst 6f :212 :442 :563 1:091 ⓇBolsa Chica79k 74 2 4 44½ 43½ 42½ 47 Stevens G L LB 114 12.50 84-13 King's Blade1165 Troyalty122no Subtle Trouble1192 No rally 5
30Jan94-4SA fst 6f :213 :443 :564 1:094 Md Sp Wt 80 7 1 3nk 2hd 1hd 1nk Black C A LB 118 9.20 88-07 Jabbawat118nk Wolf Bait1181¾ Superpak1181½ Came in start 7
20Dec93-1Hol fst 6f :221 :444 :564 1:09 Md Sp Wt 74 2 2 1hd 2hd 32 46½ Nakatani C S LB 118 4.10 89-09 Pollock's Luck1181¾ Superpak1184 Native Blast118½ Weakened 8
28Nov93-4Hol fst 6f :214 :444 :57 1:091 Md Sp Wt 64 2 7 31 22 24 511½ McCarron C J LB 118 8.80 83-09 Al's River Cat1187½ Defrocker118hd Fraley1181¾ 9
Off step slow, checked 1/2

WORKOUTS: May 30 Hol 5f fst 1:014 H 38/56 May 25 Hol 5f fst 1:031 H 29/38 May 7 Hol 6f fst 1:142 H 6/10 May 1 Hol 6f fst 1:14 H 5/17 Apr 24 Hol 5f fst 1:021 Hg 10/15 Apr 17 Hol 5f fst 1:014 H 21/26

Baja Bill

Own: Burke Gary W & Timothy R

NAKATANI C S (19 2 5 4 .11) — $62,500

Ch. g. 3 (Jan)
Sire: Soviet Lad (Nureyev)
Dam: Nickle Lady (Plugged Nickle)
Br: Racetime Inc (Ky)
Tr: Mitchell Mike (47 7 9 5 .15)

L 116

Lifetime Record: 7 1 1 3 $35,175

1994 3 0 1 2 $19,950 | Turf 2 0 1 1 $14,400
1993 4 1 0 1 $15,225 | Wet 0 0 0 0
Hol Ⓣ 1 0 1 0 $9,000 | Dist Ⓣ 2 0 1 1 $14,400

18May94-4Hol fm 1 Ⓣ :234 :472 1:12 1:371 Clm 57500 77 1 5 1½ 11 1½ 22½ Desormeaux K J LB 117 *2.80 76-19 Set On Cruise1162½ Baja Bill1171 Veracity116nk Outfinished 7
27Apr94-3Hol fst 1 1/16 :24 :474 1:121 1:443 Clm 50000 71 1 1 11 1hd 2hd 34½ Stevens G L LB 116 b *1.10 73-18 Veracity116½ Count Con1164 Baja Bill1161½ Weakened 5
15Apr94-7SA fst 1 :223 :462 1:112 1:371 Clm 40000 81 7 7 77 52½ 31½ 31 Stevens G L LB 115 b 5.10 82-18 Diamond Relaunch116½ Apurado116½ Baja Bill1158½ Best of rest 7
9Sep93-7Dmr fm 1 Ⓣ :231 :473 1:124 1:374 Alw 36000N2L 71 2 6 63¾ 51¾ 31½ 32¾ Solis A LB 117 b 12.40 79-17 Arezzo1171½ Change Tables1171½ Baja Bill1171¾ Boxed in 3 1/2-5/16 9
6Aug93-6Dmr fst 6f :221 :452 :58 1:112 Md 32000 70 8 5 54 34½ 22 1½ Nakatani C S LB 117 b *2.30 82-15 Baja Bill117½ Mark Ninety117½ Cologne1173½ Got up 10
15Jly93-2Hol fst 5½f :221 :453 :582 1:05 Md 50000 44 6 2 31½ 33½ 55 69½ Nakatani C S LB 118 b 3.20 78-12 Mi Profe116½ Passiano1182½ High Desert Star1182 Faltered 9
5Jly93-6Hol fst 5½f :22 :453 :582 1:05 Md c-40000 46 6 5 78½ 79½ 67 57½ Nakatani C S B 118 7.00 80-11 LegendryIde1182¾ Stpoutofthbot1183 Showdown1181½ Wide backstretch 9
Claimed from Baffert Bob, Baffert Bob Trainer

WORKOUTS: May 30 Hol 4f fst :504 H 56/63 May 14 Hol 4f fst :482 H 12/48 Apr 11 Hol 6f fst 1:16 H 12/13 Apr 4 Hol 3f fst :371 H 8/18 Mar 29 SLR tr.t 6f fst 1:13 H 3/7 Mar 24 SLR tr.t 6f fst 1:144 H 3/6

Jade Master

Own: Lee China

SOLIS A (159 29 15 24 .18) — $62,500

Ch. c. 3 (May)
Sire: Jade Hunter (Mr. Prospector)
Dam: Mama Coca (Secretariat)
Br: Belford David (Ky)
Tr: Sterling Patty (6 1 0 0 .17)

L 116

Lifetime Record: 7 1 0 0 $24,175

1994 7 1 0 0 $24,175 | Turf 1 0 0 0
1993 0 M 0 0 | Wet 1 1 0 0 $17,600
Hol Ⓣ 0 0 0 0 | Dist Ⓣ 0 0 0 0

12May94-8Hol fst 6½f :214 :441 1:09 1:152 Alw 37000N2L 76 3 7 810 64½ 43 45½ Solis A LB 121 b 7.80 87-09 Teasing Sea1183 Bird County121½ Tri Chrome1211¾ 4 wide into lane 8
16Apr94-7SA fst 1 :223 :461 1:11 1:363 Alw 40000N1X 79 1 6 65¾ 44 46½ 49½ Delahoussaye E LB 116 b 7.00 77-17 R Friar Tuck1178½ Eagle's Deed116¾ Paster's Caper116no Rail trip 7
30Mar94-8SA fm *6½f Ⓣ :211 :432 1:071 1:133 Baldwin82k 67 6 6 68½ 68½ 69½ 710¾ Atkinson P LB 115 b 29.70 79-08 Silver Music114½ Eagle Eyed1171½ Makinanhonestbuck117nk No threat 8
20Mar94-6SA my 6½f :22 :454 1:12 1:184 Md Sp Wt 83 9 6 79¾ 58½ 3½ 17 Solis A LB 118 b 19.00 78-19 Jade Master1187 Nucay1183 Le Dome118¾ Clear, driving 9
5Mar94-9SA fst 1 :221 :453 1:10 1:352 Md Sp Wt 53 5 4 41½ 43½ 611 1022¾ Bailey J D LB 117 b 13.50 69-07 Marconi1174 Scenic Route1171¾ Collectable Wine1171 Bid, gave way 10
12Feb94-9SA fst 6f :213 :442 :563 1:092 Md Sp Wt 76 10 6 74½ 65 69 57½ Atkinson P LB 118 b 21.90 82-11 Meadow Mischief1184½ Gold Miner's Slew118¾ Wolf Bait1181 Raced wide 11
28Jan94-3SA fst 6½f :22 :45 1:093 1:16 Md c-50000 48 3 4 2hd 2½ 67½ 617 Gryder A T B 118 b 7.00 75-12 Sacrifice1183½ Island Sport1182½ Capo1181¾ Dueled, tired 6
Claimed from Suffolk Racing Partnership Ltd, Gregson Edwin Trainer

WORKOUTS: ●Jun 1 Hol 3f fst :35 Hg 1/33 May 26 Hol 4f fst :48 H 10/36 May 5 Hol 5f fst 1:022 H 36/40 Apr 28 Hol 5f fst 1:01 H 27/63 Apr 9 Hol 7f fst 1:292 H 9/10 Mar 28 Hol 4f fst :481 H 2/28

Dancethetides

Own: Farm Haras Libertad

VALENZUELA P A (107 6 16 12 .06) — $62,500

Ch. g. 3 (Apr)
Sire: Vanlandingham (Cox's Ridge)
Dam: Lisa's Song (Cutlass)
Br: Cavey D Michael (Md)
Tr: Jory Ian P D (21 0 0 4 .00)

L 116

Lifetime Record: 11 1 2 0 $31,875

1994 6 1 1 0 $21,425 | Turf 3 0 2 0 $11,800
1993 5 M 1 0 $10,450 | Wet 1 0 0 0 $850
Hol Ⓣ 2 0 1 0 $5,800 | Dist Ⓣ 1 0 1 0 $5,800

14May94-3Hol fst 1 1/16 :224 :454 1:102 1:43 ⓇSunday Slnce56k 84 3 4 28 21½ 51¾ 57½ Valenzuela P A LB 118 b 15.50 78-18 LittleLiteBud120nk CollegeTown1162 WesternTrader1204½ Bid, weakened 7
8May94-5Hol fm 1 1/16 Ⓣ :23 :46 1:103 1:421 Alw 40000N1X 63 7 4 42 32½ 65 911¾ Valenzuela P A LB 117 b 5.00 73-12 Hot Number119½ Muirfield Village1171½ Mint Green1151 Saved ground 11
24Apr94-8GG gd 1 1/16 Ⓣ :242 :491 1:131 1:452 + GrnAlligtorH30k 87 2 5 52¾ 33 25 21½ Boulanger G LB 113 b 3.50 77-21 Gold Bet1161½ Dancethetides1135 Fittobetried1191 Closed well 6
25Mar94-6SA fst 1 :224 :461 1:11 1:363 Md 50000 93 7 3 3½ 2hd 1hd 1½ Valenzuela P A LB 118 b 5.60 86-22 Dncethetides118½ HillsboroRod1187½ PcificEdition1184½ Bumped, gamely 7
19Mar94-6SA sly 1 1/16 :231 :473 1:132 1:481 Md Sp Wt 76 2 4 42½ 33 45 55½ Atkinson P LB 117 b 28.80 49-41 Wild Invader1172¾ Carolina Blew117½ Island Sport1172 Weakened 7
5Mar94-9SA fst 1 :221 :453 1:10 1:352 Md Sp Wt 60 8 6 910 1012 917 619 Nakatani C S LB 117 20.80 73-07 Marconi1174 Scenic Route1171¾ Collectable Wine1171 Very wide trip 10
26Dec93-9SA fst 1 :221 :45 1:101 1:363 Md Sp Wt 70 9 8 913 96½ 66½ 49½ Pincay L Jr LB 117 13.40 77-10 Almaraz1171¾ DramaticGold1171¾ AdvantgeMiles1176 Wide, steadied 3/16 10
8Dec93-6Hol fm 1 Ⓣ :242 :491 1:14 1:383 Md Sp Wt 68 3 4 42 32 34 26½ Valenzuela P A B 118 30.60 65-26 Crowning Decision1186½ Dancethetides1183½ Carolina Blew118hd 10
Lugged in, hit rail upper stretch
31Oct93-6SA fst 1 1/16 :223 :461 1:111 1:432 Md Sp Wt 50 6 6 86½ 75½ 68½ 623 Solis A B 117 b 12.30 64-15 WekivaSprings11711 CollectableWine1172¾ Let'sBeCurious1172 No threat 9
10Oct93-9SA fst 6f :214 :443 :57 1:094 Md Sp Wt 68 1 11 85½ 73¾ 64 46½ Gonzalez S5 112 b 51.50 82-16 Sky Kid117hd Gracious Ghost1175 Al's River Cat1171½ Off step slow 11

WORKOUTS: May 30 Hol 4f fst :493 H 46/63 Apr 17 Hol 5f fst 1:004 H 11/26 Apr 11 Hol 4f fst :51 H 19/19 Apr 6 Hol 4f fst :491 H 14/25 Mar 14 Hol 5f fst 1:031 H 22/22

is no clear favorite in the race and that Heartless Wager is likely to succumb to a pace duel, anything can happen. The presence of a false favorite is always an encouragement to bet on whatever insight we have. The best strategy here is to bet Jade Master and Beau Finesse to Win and Place and, if you're feeling bold, box them in the exacta.

EIGHTH RACE

Hollywood

JUNE 3, 1994

1 MILE. (Turf)(1.32³) CLAIMING. Purse $45,000. 3-year-olds. Weight, 122 lbs. Non-winners of two races at a mile or over since April 5, allowed 3 lbs. Such a race since then, 6 lbs. Claiming price $62,500; for each $2,500 to $55,000, allowed 1 lb. (Races when entered for $50,000 or less not considered.)

Value of Race: $45,000 Winner $24,750; second $9,000; third $6,750; fourth $3,375; fifth $1,125. Mutuel Pool $309,986.00 Exacta Pool $224,949.00 Trifecta Pool $187,526.00 Quinella Pool $32,295.00

Last Raced	Horse	M/Eqt. A.Wt	PP	St	¼	½	¾	Str	Fin	Jockey	Cl'g Pr	Odds $1
12May94 8Hol⁴	Jade Master	LBb 3 116	9	9	10	10	10	8²	1½	Solis A	62500	10.50
30Mar94 8SA⁶	Beau Finesse	LBb 3 116	2	10	8⁵	8³	7ʰᵈ	6¹	2ⁿᵏ	Castanon A L	62500	53.90
1May94 8GG²	One Rich Runner	LBb 3 109	5	8	9²½	9²	9¹½	7½	3¹¾	Atherton J E⁵	55000	27.10
12May94 7Hol²	Jabbawat	LBb 3 116	7	3	3½	4½	5¹	3¹	4ʰᵈ	Black C A	62500	7.50
22May94 9Hol⁵	Hot Papa Tom	3 116	6	6	7²½	6½	6²	4ʰᵈ	5¹½	Antley C W	62500	5.00
14May94 3Hol⁵	Dancethetides	LBb 3 118	10	4	4¹½	3¹½	2¹	11	6¹½	Valenzuela P A	62500	13.30
18May94 4Hol⁴	Win The Case	LBb 3 114	4	7	6½	7²	8³	9⁴	7³	Valenzuela F H	57500	13.50
14May94 3Hol⁴	Heartless Wager	LBb 3 116	1	2	1¹	1ʰᵈ	1½	2¹	8ⁿᵒ	Gonzalez S Jr	62500	3.10
18May94 4Hol²	Baja Bill	LB 3 116	8	5	5½	5¹½	3½	5ʰᵈ	9¹⁰	Nakatani C S	62500	3.90
17Apr94 9GG⁴	Sky Kid	LB 3 116	3	1	2¹	2ʰᵈ	4ʰᵈ	10	10	Desormeaux K J	62500	4.60

OFF AT 10:33 Start Good. Won driving. Time, :22⁴, :46¹, 1:10³, 1:36 Course firm.

$2 Mutuel Prices:

9-JADE MASTER		23.00	14.20	9.20
2-BEAU FINESSE			39.60	16.80
5-ONE RICH RUNNER				10.00

$2 EXACTA 9-2 PAID $994.60 $2 TRIFECTA 9-2-5 PAID $10,691.60 $2 QUINELLA 2-9 PAID $520.60

Ch. c, (May), by Jade Hunter–Mama Coca, by Secretariat. Trainer Sterling Patty. Bred by Belford David (Ky).

The fourth race at Golden Gate Fields on May 20th is another example of a horse excelling at the claiming level once given the opportunity to run on the surface it was bred for. Looking over the field, Belle Centaine and Miscreat Miss are the only horses with wins on the turf and the two favorites. Neither of them, however, looks like a lock. Belle Centaine hasn't won a race in more than two years. Miscreant Miss's only win in four tries was run near the lead on a slow pace, which she is not likely to get from Intersquaw today. Furthermore, she won at a mile and her SI's of 4 and 5 suggest she might not be able to get the extra 1/16 mile.

Sew Original and Money Crunch are proven losers on the surface and at the distance and can be eliminated. This leaves us with Intersquaw and Sweet But Bold as the other possible winners in the race. While Intersquaw has an A- rating, the minus sign after it suggests she might need a run over the surface before being ready to fire. Furthermore, her SI's of 4 indicate she might have trouble getting the 1 1/16 mile distance. Still, Intersquaw could get loose on an easy lead in this spot and, despite her 34-1 odds, cannot be thrown out.

While Sweet But Bold is bred top and bottom for the turf, it is worrisome that she has yet to be tried on that surface. After 32 races, the Form asks its readers why hasn't Sweet But Bold been tried on the grass, which probably accounts for her nearly 12-1 odds. Nevertheless, Sweet But Bold is bred for the turf, likes to win, and has proven she can get the distance.

Golden Gate Fields

4

1 1/16 MILES. **(Turf, chute). (1:411) CLAIMING. Purse $13,000. Fillies and mares, 3-year-olds and upward. Weights: 3-year-olds, 114 lbs. Older, 121 lbs. Non-winners of two races at a mile or over since February 1, allowed 3 lbs. One such race since then, 5 lbs. Claiming price $12,500, if for $10,500, allowed 2 lbs. (Maiden, claiming, starter and classified handicap races for $10,000 or less not considered.)**

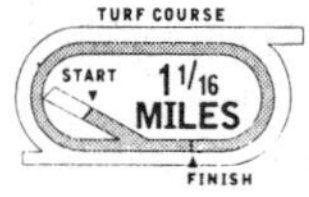

Sew Original

Own: Green Valley Ranch

BAZE R A (487 133 93 71 .27) **$12,500**

Dk. b or br f. 4
Sire: Eleven Stitches (Windy Sands)
Dam: Old Classic (Olden Times)
Br: Old English Rancho (Cal)
Tr: Arterburn Lonnie (50 13 11 7 .26)

116

Lifetime Record :	16 4 3 2	$46,575			
1994	5 1 1 0	$12,750	Turf	5 0 2 1	$12,375
1993	11 3 2 2	$33,825	Wet	1 0 0 0	
GG Ⓣ	2 0 1 1	$5,650	Dist Ⓣ	2 0 1 1	$5,650

13May94- 7GG fst 1 :223 :462 1:12 1:39 ⒻClm 10000 72 8 5 54 31 2hd 1½ Chapman T M LB 116 7.70 76-26 Sew Original116½ Mia Magdalena1161½ K.A.'sWindyOne117hd Rallied wide 8
13Apr94- 5GG fst 1 1/16 :231 :463 1:111 1:443 ⒻClm c-8000 46 1 2 1½ 42½ 711 714¾ Chapman T M LB 116 *2.30 59-25 Mangia Mangia1161½ Felonius Flight1181½ Mia Magdalena1161 Stopped 8
Claimed from Silverado Springs Stable, Pederson Dean Trainer
16Mar94- 8GG fst 1 :224 :46 1:094 1:361 ⒻAlw 23000N1X 61 2 2 1hd 35 47½ 413 Belvoir V T L 118 4.90 77-19 Finest City1182½ Soap Opera Queen1184½ Queens Casino1186 Weakened 6
10Feb94- 6GG gd 1 1/16 :223 :464 1:111 1:44 ⒻAlw 23000N1X 64 3 2 2hd 31½ 46 58½ Chapman T M LB 118 *2.30 69-21 ⒹFinest City121hd Lady Glorious1184 Queens Casino118¾ Weakened 7
23Jan94- 5BM yl 1 Ⓣ :24 :484 1:142 1:413 ⒻAlw 22000N1X 74 3 2 1½ 1hd 2hd 23½ Chapman T M LB 115 *3.40 61-32 Miscreant Miss1163½ Sew Original1152 Gold Conde116nk Second best 8
31Dec93- 6BM fst 1 :222 :461 1:11 1:36 ⒻClm c-16000 82 6 2 22 2½ 2hd 1hd Chapman T M LB 116 *1.40 90-11 Sew Original116hd Lottaleese1116 But Perfect116½ Closed gamely 6
Claimed from Hoffman Al H F T, Jenda Charles J Trainer
7Nov93- 6BM fm 1 Ⓣ :233 :474 1:134 1:392 ⒻClm 20000 59 5 3 44½ 3nk 33 45 Chapman T M LB 116 6.80 77-17 Melusive1162 Doulabella1161 Temperavie1162 Wide 2nd turn 7
11Oct93- 3BM fst 1 1/16 :231 :471 1:121 1:453 ⒻClm 16000 78 6 3 2½ 2hd 1hd 11¼ Chapman T M LB 116 6.70 75-29 Sew Original1161¼ Red Chimes1161 Ackful1187 Wide trip 6
11Sep93- 4BM fm 1 Ⓣ :224 :461 1:113 1:383 ⒻClm 25000 61 4 2 24 44 45½ 46½ Chapman T M LB 116 2.70 79-10 Beautiful Eagle1161½ Doulabella1161 Our Fast Ball1154 Weakened 7
19Aug93- 9Bmf fst 1 :223 :462 1:12 1:373 3+ ⒻMd 16000 69 4 1 1½ 11 11½ 11¼ Chapman T M LB 117 *1.70 86-14 Sew Original1171¼ Nifty Magic1217 Fatih's Fantasy1211 Held well 9

WORKOUTS: May 6 **GG 7f fst 1:282 H** *2/5* Apr 29 **GG 6f gd 1:152 H** *4/16* Apr 22 **GG 5f fst 1:013 H** *10/33* Apr 6 **BM 6f fst 1:144 H** *1/1* Mar 30 **BM 5f fst 1:04 H** *8/8* Mar 8 **BM 6f fst 1:153 H** *3/5*

Intersquaw

Own: Grayson Bobby W

TOSCANO P R (22 3 2 1 .14) **$10,500**

Dk. b or br f. 4
Sire: Interco (Intrepid Hero)
Dam: Indian Bend (Triple Bend)
Br: Warwick George M (Cal)
Tr: Grayson Bobby W (8 0 2 3 .00)

114

Lifetime Record :	10 1 2 1	$6,935		
1994	2 0 0 0	$488	Turf	0 0 0 0
1993	6 1 2 1	$6,447	Wet	0 0 0 0
GG Ⓣ	0 0 0 0		Dist Ⓣ	0 0 0 0

4May94- 1GG fst 5½f :212 :442 :562 1:03 ⒻClm 6250N2X 52 2 6 65 67½ 610 410 Hummel C R B 118 b 13.50 85-09 Regal Frenchie1188 Under Full Sail1182 Sinfirly Good118hd No rally 7
14Apr94- 6GG fst 6f :221 :452 :58 1:11 ⒻClm 12500N2L 49 4 4 52½ 53¼ 66½ 68¾ Hummel C R 118 b 22.10 75-16 Gospel Music118½ Endangered1164½ Dryer's Illusion118hd Outrun 6
15Oct93-11Fno fst 5f :22 :452 :571 3+ ⒻClm 10000N2X 56 4 5 2hd 2½ 1½ 22½ Gomez E A B 114 b 3.10 89-11 Adam's Pick1182½ Intersquaw114nk Mezquita113no Outfinished 6
7Oct93- 8Fno fst 5½f :222 :463 :583 1:044 3+ ⒻMd 8000 56 3 1 13 18 18 110 Gomez E A B 116 b *.40 83-15 Intersquaw11610 SweepyLaMoon1215 PickAPocketful1165 Crushed foes 7
6Sep93- 3Sac fst 6f :223 :452 :58 1:12 3+ ⒻⓈMd 12500 49 10 3 42 44 45½ 45 Jauregui L H B 116 b 3.80 74-16 Jintana1211 Guided Dancer1163 Precocious Day1161 No late bid 12
27Aug93- 4BM fst 5½f :213 :452 :582 1:051 ⒻMd 12500 50 2 4 44½ 55 65 55 Jauregui L H 117 b 12.30 80-18 Fighting Mary117½ Wapiti Baby1172 Our Distant Star117½ No rally 7
12Aug93-13Bmf fst 6f :221 :453 :582 1:114 ⒻⓈMd 12500 49 3 8 42 3½ 23 33 Jauregui L H B 117 b 5.90 83-10 Sticky Candy1173 Truce In Balance117hd Intersquaw117½ Wide trip 12
18Jly93- 4Sol fst 5½f :222 :46 :591 1:054 ⒻⓈMd 12500 54 6 1 1hd 2hd 2hd 2½ Jauregui L H B 117 b 11.40 84-10 Tell's Vision112½ Intersquaw1171 Fit Model117hd Held gamely 9
6Sep92- 6Sac fst 6f :221 :462 :592 1:122 ⒻⓈMd 12500 16 5 2 31 2½ 43 610 White T C B 118 17.70 67-14 Janskite1181 Bodanelli1185 Deadly Darling118 Drifted out lane 10
17May92- 5TuP fst 5f :222 :454 :582 ⒻMd Sp Wt -0 8 9 98½ 913 925 935¾ Arnold A D 119 b 27.60 58-08 Merits Misty1191¼ Cabanal119½ Purity Pinker119 Outrun 9

WORKOUTS: May 13 **Pln 4f fst :522 H** *8/8* Apr 11 **Fpx 5f fst 1:01 H** *2/4* Apr 9 **Fpx 4f fst :473 H** *2/12* Mar 9 **Fpx 5f fst 1:04 H** *3/6* Feb 25 **Fpx 5f fst 1:022 H** *7/11*

Money Crunch

Own: Snow Daryl

HUMMEL C R (184 16 23 17 .09) **$10,500**

Ch. m. 5
Sire: Explodent (Nearctic)
Dam: Marginal Money (Plugged Nickle)
Br: Evergreen Thoroughbreds Inc (Ky)
Tr: Snow Daryl (26 4 2 3 .15)

114

Lifetime Record :	38 4 5 7	$32,945			
1994	9 2 1 1	$10,925	Turf	1 0 0 0	
1993	17 0 2 4	$6,037	Wet	8 2 0 2	$11,056
GG Ⓣ	1 0 0 0		Dist Ⓣ	1 0 0 0	

6May94- 7GG sly 1 1/16 :231 :464 1:113 1:452 ⒻAlw 6250s 63 6 4 48 46 53½ 44½ Belvoir V T L 121 b 6.60 65-25 S Maxine1212 Iona Prospector1161 Duchess Of Dignity1181½ Wide trip 6
24Apr94- 4GG gd 1 1/16 :23 :464 1:114 1:46 ⒻClm 10000 71 5 8 67 52½ 42½ 43½ Belvoir V T LB 116 b 8.50 63-33 Our Girlfriend1162 Felonius Flight1161½ Innerquest116no Even late 9
13Apr94- 5GG fst 1 1/16 :231 :463 1:111 1:443 ⒻClm 8000 64 8 7 712 65½ 44 44 Davenport C L LB 116 b 9.80 70-25 Mangia Mangia1161½ Felonius Flight1181½ Mia Magdalena1161 Wide trip 8
1Apr94- 8GG fm 1 1/16 Ⓣ :233 :473 1:111 1:43 ⒻAlw 23000N1X 50 5 10 109¾ 912 916 915 Olguin G L LB 118 b 39.10 76-09 Mangaki's Girl1182½ Hillwalker118hd Wild Again Miss1182 No threat 12
27Feb94- 6GG gd 1 1/16 :241 :481 1:13 1:453 ⒻClm 6250 67 2 4 43½ 32½ 11½ 1nk Olguin G L LB 118 b 4.60 69-28 Money Crunch118nk Miss Juliet1184 Iona Prospector1182½ Rallied inside 8
18Feb94- 5GG my 1 1/16 :23 :47 1:12 1:46 ⒻClm 6250N1Y 62 9 4 49 45 32 1¾ Olguin G L LB 118 b 6.00 67-33 Money Crunch118¾ MiaMagdalena1182 NoraOfClare113hd Closed gamely 11
2Feb94- 1GG fst 1 :223 :463 1:121 1:384 ⒻClm 6250N1Y 55 1 4 43 32 31½ 2¾ Olguin G L LB 118 b 6.50 76-17 Outbound'nDown113¾ MonyCrunch118nk SlwStorm118nk Finished gamely 6
21Jan94- 3BM fst 6f :223 :454 :583 1:112 ⒻClm 4000N1Y 54 7 7 78½ 77½ 66½ 34¾ Chapman T M LB 120 b 11.10 79-14 Uclassy Won120nk Turn To Heaven1204½ Money Crunch120no Mild late bid 8
8Jan94- 5BM fst 5½f :213 :451 :58 1:044 ⒻClm 4000 56 4 8 89½ 87¾ 88¼ 93¾ Olguin G L LB 116 b 17.30 83-17 Rachael's Prospect1181 Louise Fay116hd Sparkin Miss116hd No threat 10
12Dec93- 9BM my 6f :231 :464 :592 1:121 3+ ⒻClm 4000 47 5 7 77 66 66¼ 65¼ Olguin G L B 116 10.30 69-20 Golden Grace116½ Vagabond Lace116nk Far Too Thrilling116½ No rally 9

WORKOUTS: Apr 8 **BM 5f fst 1:013 H** *10/16* Mar 27 **BM 4f fst :501 H** *5/8* Mar 21 **BM 5f fst 1:031 H** *8/13* Mar 15 **BM 4f fst :491 H** *6/10*

Belle Centaine (NZ)

Own: Chaiken & Chaiken & Heller & Heller

B. m. 6
Sire: Centaine (Century)
Dam: Belle Promenade (Sir Godfrey)
Br: Chittick G J (NZ)
Tr: Sherman Art (72 9 8 9 .13)

CHAPMAN T M (311 46 33 35 .15) $12,500 116

Lifetime Record :	20 2 4 5	$41,144			
1994	1 0 0 0	$412	Turf	18 2 4 5	$40,732
1993	2 0 0 0		Wet	0 0 0 0	
GG Ⓣ	1 0 0 0		Dist Ⓣ	3 1 0 2	$27,750

13May94- 4GG fst 6f :214 :45 :573 1:101 ⒻClm 16000 70 3 8 811 811 78½ 57¾ Chapman T M LB 116 f 28.00 80-18 Labiblica1164½ Imprimis118½ Gypsy Jones116nk No rally 8
23Apr93- 9GG fm 1⅛ Ⓣ :483 1:13 1:374 1:501 ⒻHcp 12500s 63 3 6 54½ 53½ 63½ 76¼ Baze R A LB 115 f 3.80 84-08 Saavy Caavy1142 Truly Needy1171 Regal Eve112no Dull try 10
14Apr93- 6GG fst 1 :23 :463 1:111 1:374 ⒻClm 12500 63 5 3 32 65¼ 77 67 Warren R J Jr LB 116 f 4.10 74-24 Gum Writer116no Nicole's Red Lady1161¼ Moms Baby1183½ Gave way 8
15Nov92- 1BM fm 1 1/16 Ⓣ :234 :473 1:123 1:444 3↑ ⒻClm c-32000 76 5 3 31½ 31½ 42½ 32¼ Castaneda M LB 116 *1.70 84-15 Sweet Leader1162 Bally's Starlet117nk Belle Centaine-NZ116 Wide trip 7
Claimed from Bloomer Robert L, Cross Richard J Trainer
18Oct92- 2SA fm 1⅛ Ⓣ :49 1:13 1:372 1:493 3↑ ⒻClm 62500 84 3 4 43½ 42½ 53½ 32¾ Desormeaux K J LB 118 7.00 68-22 TropiclStephni115hd SummrGlory-NZ1162¾ BllCntin-NZ118 Rough start 7
2Sep92- 7Dmr fm 1⅜ Ⓣ :481 1:132 1:382 2:15 3↑ ⒻHoney Fx H 56k 74 2 3 33 41¾ 67½ 713 Stevens G L LB 114 26.10 76-09 Silvered117nk Fantastic Ways1151¼ Campagnarde-Ar119 No mishap 7
23Aug92- 5Dmr fm 1⅛ Ⓣ :484 1:124 1:371 1:493 3↑ ⒻAlw 40000 79 2 3 42½ 51¾ 45½ 45¼ Stevens G L LB 118 6.50 83-12 SouthernTruce116nk QueenSize-NZ116hd ShrwdVixn116 Not enough late 9
6Aug92- 7Dmr fm 1 1/16 Ⓣ :243 :484 1:131 1:432 3↑ ⒻAlw 36000 85 6 2 21 2½ 11 1nk Stevens G L LB 119 6.80 89-11 BelleCntin-NZ119nk DH LyinToThMoon116 DH RivrPtrol-GB122 Gamely 7
17Jly92- 5Hol fm 1 1/16 Ⓣ :23 :463 1:101 1:402 ⒻAlw 34000 80 8 5 46 35 36½ 36 Stevens G L LB 116 14.80 87-12 Visible Gold1193½ RiverPatrol-GB1192½ BelleCentaine-NZ116 No mishap 10
7Dec91 ♦ Trentham(NZ) fm *7f ⓉLH 1:242 3↑ Pukerua Bay Handicap Hcp 3500 2½ Robson G J 111 - Dempsey115¾ Belle Centaine111¾ - 13
Rallied final furlong, not good enough. 3rd unavailable

WORKOUTS: ●May 6 GG 7f fst 1:27 H *1/5* Apr 29 GG 7f fst 1:30 H *2/3* Apr 21 GG 6f fst 1:142 H *4/10* Apr 14 GG 5f fst 1:01 H *5/33* ●Apr 4 Pln 5f fst 1:001 H *1/9*

Miscreant Miss

Own: Maycock Robert C

B. m. 5
Sire: Fatih (Icecapade)
Dam: Walkin West (Walker's)
Br: Maycock Bob (Cal)
Tr: Silva Jose L (135 32 24 14 .24)

MEZA R Q (303 41 42 37 .14) $12,500 116

Lifetime Record :	29 5 4 6	$58,510			
1994	7 2 0 2	$22,500	Turf	4 1 0 2	$18,225
1993	6 1 1 1	$9,000	Wet	1 0 1 0	$2,200
GG Ⓣ	1 0 0 0	$575	Dist Ⓣ	2 0 0 1	$3,875

11May94- 7GG fm 1 1/16 Ⓣ :223 :462 1:103 1:432+ 3↑ ⒻClm 22500 72 5 4 42½ 43 55¼ 55½ Meza R Q L 114 b 14.50 83-11 Lady Rhythm1211 Lottaleese118½ Eminere1162 Even late 8
15Apr94- 1GG fst 1 1/16 :231 :471 1:121 1:454 ⒻAlw 6250s 59 5 3 43 42½ 56½ 57½ Meza R Q L 121 b 3.00 60-23 S Maxine121hd Miss Juliet1163½ Marv's Prospect1183½ No rally 6
1Apr94- 6GG fst 1 1/16 :232 :47 1:11 1:433 ⒻAlw 6250s 77 1 2 31½ 31½ 2hd 1¾ Meza R Q LB 118 b 2.60 79-25 Miscreant Miss118¾ Skywalker Wilkes1162½ S Maxine1215 Rallied inside 7
23Mar94- 6GG fst 6f :214 :444 :573 1:111 ⒻAlw 6250s 71 4 7 87½ 710 68 33 Meza R Q L 118 b 5.50 80-18 Timely Bidder1163 Short Stories121hd Miscreant Miss1181½ Rallied inside 9
10Feb94- 8GG gd 1 :222 :452 1:102 1:37 ⒻAlw 26000N2X 55 2 5 68 812 916 913 Meza R Q L 121 b 7.20 73-21 Spectacular Fort1182 Cargosita118no Waslaw's Daughter1182 Dull try 9
23Jan94- 5BM yl 1 Ⓣ :24 :484 1:142 1:413 ⒻAlw 22000N1X 82 4 3 31½ 2hd 1hd 13½ Meza R Q L 116 b 4.10 65-32 Miscreant Miss1163½ Sew Original1152 Gold Conde116nk Drew clear 8
~~9Jan94- 6BM fm 1 1/16 Ⓣ :232 :471 1:122 1:462 ⒻAlw 22000N1X 74 3 4 45½ 23 22 3¾~~ Meza R Q L 116 b *2.60 75-23 Lottaleese115nk Tooty Brush116½ Miscreant Miss1163½ Mild late bid 7
19Dec93- 6BM fm 1 Ⓣ :232 :482 1:133 1:401 3↑ ⒻAlw 20000N1X 75 10 5 41½ 43 45½ 33↓ Mercado P L 116 b 40.10 75-22 SunshinTrt1163 Hrcomsththrll114no DH GtoGoFlow114 Wide stretch run 10
18Nov93- 6BM fst 1 :233 :481 1:124 1:374 3↑ ⒻClm 9000 76 4 2 2hd 1hd 12 14½ Meza R Q L 114 b 21.40 81-22 Miscreant Miss1144½ Road ToRomance116½ QuiteRegal1161 Steady drive 7
30Oct93- 1BM fst 1 :23 :471 1:131 1:401 3↑ ⒻClm 6250 71 3 3 33 2½ 2½ 2hd Espinoza V5 L 111 b 19.70 69-25 MemorblLss118hd MiscrntMiss1115 SkywlkrWilks1165 Drifted wide drive 7

WORKOUTS: May 4 GG Ⓣ 5f fm 1:014 H (d) *2/2* Apr 28 GG 5f gd 1:012 H *2/47* Mar 20 GG 3f fst :354 Hg*4/30* Mar 13 GG 6f fst 1:123 H *2/13* Mar 5 GG 6f fst 1:134 H *3/7* Feb 25 GG 6f gd 1:154 H *12/19*

Sweet But Bold

Own: Ho & Moger Jr & Wong

Dk. b or br f. 4
Sire: Hail Bold King (Bold Bidder)
Dam: Miss Moya (Windy Sands)
Br: Old English Rancho (Cal)
Tr: Moger Ed Jr (76 13 6 12 .17)

WARREN R J JR (338 49 42 41 .14) $10,500 114

Lifetime Record :	32 6 4 5	$40,053			
1994	7 2 0 1	$10,188	Turf	0 0 0 0	
1993	16 3 2 3	$19,190	Wet	7 0 0 2	$4,825
GG Ⓣ	0 0 0 0		Dist Ⓣ	0 0 0 0	

6May94- 7GG sly 1 1/16 :231 :464 1:113 1:452 ⒻAlw 6250s 63 2 3 36 24 32½ 54¾ Warren R J Jr LB 121 15.80 65-25 S Maxine1212 Iona Prospector1161 Duchess Of Dignity1181½ Gave way 6
1Apr94- 6GG fst 1 1/16 :232 :47 1:11 1:433 ⒻAlw 6250s 57 7 5 76½ 46½ 56½ 511¼ Baze R A LB 121 f 3.40 68-25 Miscreant Miss118¾ Skywalker Wilkes1162½ S Maxine1215 No rally 7
24Mar94- 7GG fst 1 1/16 :241 :484 1:132 1:454 ⒻClm 8000 67 1 3 42 41¾ 2hd 13½ Baze R A LB 116 f 3.20 68-32 Sweet But Bold1163½ Mangia Mangia1161 Saddles1162 Closed well 9
12Mar94- 7GG fst 1 :222 :462 1:113 1:373 ⒻClm 6250 70 1 6 68½ 54 31½ 12 Baze R A LB 116 3.90 83-17 Sweet But Bold1162 Truffles N Cream1124 In A Song1161½ Rallied wide 9
23Feb94- 5GG gd 1 :234 :474 1:13 1:394 ⒻClm 6250 63 1 5 56 53¼ 44 45¼ Baze R A LB 116 2.90 67-28 Innerquest1183 D' Amelia116nk Outbound 'n Down1182 Even try 6
9Feb94- 9GG fst 1 1/16 :233 :472 1:12 1:44 ⒻClm 6250 69 6 4 45½ 42½ 42 35 Baze R A LB 116 3.60 72-25 Innerquest1113½ Foreign Word1181½ Sweet But Bold116½ Even late 8
16Jan94- 1BM fst 1 :231 :473 1:124 1:401 ⒻClm 10500 54 5 5 65 64¾ 56 55 Beckner D5 LB 109 14.40 67-16 Forbes Sparkler1162½ But Perfect116hd Victoria's Derby1161½ No rally 6
30Dec93- 1BM fst 6f :23 :462 :583 1:111 3↑ ⒻClm c-8000N2X 48 7 6 77¾ 76½ 64¾ 66¼ Baze R A LB 114 2.90 74-19 ForbesSparkler1192 Cataluna114no FastSqueeze1162½ Steadied briefly 1/8 8
Claimed from Brown Don & Guy, Brown Guy Trainer
8Dec93- 4BM my 6f :222 :452 :58 1:104 3↑ ⒻClm 8000N2X 68 3 8 86½ 65¼ 42 3nk Baze R A LB 116 10.20 82-17 Toni Regina116nk Barraba Express117no Sweet But Bold1162 Mild late bid 9
21Nov93- 1BM fst 1 :23 :474 1:124 1:382 ⒻClm 10000 51 6 6 62½ 51¾ 54¾ 67½ Belvoir V T LB 116 14.80 70-19 ButPerfect1144 Victoria'sDerby1161 Knight'sSecretary116hd No excuses 7

WORKOUTS: May 16 GG 4f fst :491 H *14/27* May 3 GG 4f fst :53 H *22/25* Apr 28 GG 4f gd :553 H *52/52* Apr 22 GG 4f fst :511 H *24/27* Apr 13 GG 4f fst :502 H *28/38* Mar 21 GG 4f fst :523 H *15/15*

160   So how to bet? Betting horses that have never won on the turf against those that have is normally a losing proposition. But, in this case, both favorites look vulnerable and moderate win bets on Intersquaw and Sweet But Bold are worth making.

FOURTH RACE
Golden Gate
MAY 20, 1994

1 1/16 MILES. (Turf Chute)(1.41[1]) CLAIMING. Purse $13,000. Fillies and mares, 3–year–olds and upward. Weights: 3–year–olds, 114 lbs. Older, 121 lbs. Non–winners of two races at a mile or over since February 1, allowed 3 lbs. One such race since then, 5 lbs. Claiming price $12,500, if for $10,500, allowed 2 lbs. (Maiden, claiming, starter and classified handicap races for $10,000 or less not considered.)

Value of Race: $13,000 Winner $7,150; second $2,600; third $1,950; fourth $975; fifth $325. Mutuel Pool $84,518.00 Exacta Pool $82,852.00 Quinella Pool $16,644.00

Last Raced	Horse	M/Eqt.	A.Wt	PP	St	1/4	1/2	3/4	Str	Fin	Jockey	Cl'g Pr	Odds $1
6May94 7GG5	Sweet But Bold	LB	4 115	6	5	6	5hd	41	31½	1½	Warren R J Jr	10500	11.70
13May94 4GG5	Belle Centaine–NZ	LBf	6 116	4	3	41	41½	31½	11½	22	Chapman T M	12500	2.10
13May94 7GG1	Sew Original	L	4 116	1	4	31½	3hd	51	42	3nk	Baze R A	12500	2.50
11May94 7GG5	Miscreant Miss	LBb	5 116	5	1	21	21½	11½	2hd	44½	Meza R Q	12500	1.50
6May94 7GG4	Money Crunch	LBb	5 116	3	6	5½	6	6	56	511	Hummel C R	10500	12.20
4May94 1GG4	Intersquaw	Bb	4 114	2	2	12	1½	2hd	6	6	Toscano P R	10500	34.70

OFF AT 2:34 Start Good. Won driving. Time, :224, :472, 1:121, 1:374, 1:441 Course firm.

$2 Mutuel Prices:

6–SWEET BUT BOLD	25.40	10.00	4.20
4–BELLE CENTAINE–NZ		4.60	3.20
1–SEW ORIGINAL			3.20

$5 EXACTA 6–4 PAID $287.00 $4 QUINELLA 4–6 PAID $59.60

Dk. b. or br. f, by Hail Bold King–Miss Moya, by Windy Sands. Trainer Moger Ed Jr. Bred by Old English Rancho (Cal).

© 1994 DRF, Inc.

The foregoing examples all illustrate how powerful turf ratings can be when used in conjunction with pace analysis and a horse's stamina indexes. A major appeal of playing turf races is that horses bred for it regularly go off as significant overlays in their first grass starts because the public overemphasizes past dirt form and is generally afraid to bet that a horse bred for the turf will move up because of its breeding, or is ignorant of breeding factors.

While most of our turf examples have come from California, it is important to note that Eastern, Midwestern and Florida racing provides handicappers with even more turf betting opportunities, since they offer many more maiden turf events. The same principles I have applied to California have proven validity at these other racing circuits as well.

9

Handicapping for Off-track Performance

While the pedigree ratings for first-time starters and horses making their debut on the turf can be powerful handicapping tools, the situation is far more ambiguous with regard to off-track performance. This is so because, while a horse's genetic proclivity toward running in the mud or slop remains constant, the actual conditions under which it must run vary considerably. The mud, for example, at Saratoga is not the same as the mud at Belmont. But even leaving differences between tracks aside, the texture of the racing surface at the *same* track can vary from day to day and sometimes even within the same card. Depending upon such factors as the composition of the soil, the temperature, the amount of moisture, sunlight and wind, post position, and the pitch of the track, horses will either "move up" or decline in their off-track performance.

And, as if it that were not enough, to further confound any rating system, a horse's off-track performance can also be dramatically manipulated by its trainer, depending upon the addition or subtraction of a variety of stickers and mud caulks.

Even further, from a handicapping perspective, horses that are bred to hate the slop are often scratched when it rains, leaving shorter fields with several "qualifiers" which make it difficult to isolate off-track breeding as a primary factor in one's selection.

Given the absence of consistent racing conditions, one is tempted to throw out the notion of creating a useful off-track (OT) rating system all together. But, that would be to throw the baby out with the bathwater. Even though OT ratings are not generally reliable enough to be used as a *primary* handicapping factor, they can be helpful in further narrowing the list of contenders that one might like for other handicapping reasons.

Furthermore, the OT ratings in *Exploring Pedigree* may work better than other attempts because the criteria for identifying the superior *wet* off-track sires is more demanding: with sires whose progeny win 22 percent or more of the time receiving an A+ off-track rating; those winning between 20-21 percent getting an A rating; 19 percent receiving a B rating; and those winning between 17-18 percent earning a C+ off-track rating. Surveying the top-rated wet off-track sires, it quickly becomes apparent that most of them descend from Mr. Prospector and In Reality, two sires noted for transmitting the early speed (and perhaps little feet) that plays so well on fast sloppy surfaces.

Another innovation in our OT ratings is the identification of those sires whose progeny seem to improve on *tiring* surfaces. While much more research is needed in this area, the preliminary evidence suggests that descendants of Damascus, and particularly of Private Account, often move up on slow, tiring surfaces. No doubt, there are other sires whose progeny improve on tiring surfaces and these will be included in future editions of *Exploring Pedigree*. Meanwhile, handicappers are urged to develop their own additional list of T-rated sires whose progeny locally win races on tiring surfaces. With this as background let's turn to a few races where off-track breeding was a factor.

Golden Gate Fields

7

6 Furlongs. **(1:07^3) CLAIMING. Purse $12,000. Fillies 3-year-olds. Weight, 122 lbs. Non-winners of two races since December 1, allowed 3 lbs. A race since then, 5 lbs. Claiming price $12,500; if for $10,500, 2 lbs. (Maiden races, claiming, starter and classified handicap races for $10,000 or less not considered.)**

START / 6 FURLONGS / FINISH

Sarah's Boop

Own: Franco Richard A

WARREN R J JR (7 1 0 2 .14) — $12,500

Ch. f. 3 (Feb)
Sire: Whitesburg (Crimson Satan)
Dam: Floating Home (Our Native)
Br: Rancho Jonata (Cal)
Tr: Sherman Art (3 1 0 1 .33)

116

Lifetime Record:	6 1 1 1	$10,228			
1994	1 1 0 0	$3,850	Turf	0 0 0 0	
1993	5 M 1 1	$6,378	Wet	0 0 0 0	
GG	0 0 0 0		Dist	5 1 1 1	$10,228

12Jan94-2BM	fst	6f	:23	:46^1	:58^4	1:11^4	ⒻⓈMd 12500	62	12	1	41¾	41¼	11½	13	Warren R J Jr	LB 117	8.70	82-16	Sarah'sBoop117^3 ToCroWithLove118^4 Thrillofwin117¾	Raced well wide	12
26Dec93-2BM	fst	6f	:23	:46^3	:59^2	1:12^2	ⒻⓈMd 12500	37	10	4	5^2	52¾	44½	46	Warren R J Jr	LB 117	2.70	73-17	Silent Daughter117^1 Cup And A Half117^{no} Talinear117^5	Wide trip	12
8Dec93-1BM	my	1	:23^1	:47^1	1:11^4	1:38	ⒻMd 20000	37	3	4	53½	57½	613	616	Warren R J Jr	LB 117	4.30	64-19	Annual Dance117½ Berg Rosa1173½ Valnesian1123½	Gave way	8
18Nov93-8BM	fst	6f	:22^1	:45^3	:58^2	1:12	ⒻPresidio25k	48	6	4	45½	56	46½	35½	Warren R J Jr	LB 115	10.60	70-21	Nellie's Bargain115½ Royal Boutique115^5 Sarah's Boop1151½	Wide trip	7
10Nov93-5BM	fst	6f	:22^3	:46^1	:59^1	1:12^4	ⒻⓈMd 12500	56	10	6	63¼	42½	32	21	Warren R J Jr	LB 117	11.10	71-18	Easy To Follow117^1 Sarah's Boop117^2 Bev117^2	Forced wide early	12
28Oct93-5BM	fst	6f	:22^3	:46	:58^4	1:12	ⒻMd 12500	43	10	2	3½	2½	31½	44	Warren R J Jr	LB 117	8.00	72-17	Skeptic LadyLee117^{nk} DesertMystique1173½ K.L.'sImagene117^{nk}	Wide trip	12

WORKOUTS: Jan 30 **GG 5f fst 1:00^2 H** *4/29* Jan 21 **BM 4f fst :49^1 H** *22/62* Jan 5 **BM 4f fst :49^2 H** *17/24* Dec 18 **BM 4f gd :48^4 H** *4/23* Dec 4 **BM 3f gd :36^3 H** *2/18* Nov 28 **BM 4f fst :51 H** *40/44*

Annie Pearl

Own: Sens Du Cheval Farm

CHAPMAN T M (—) — $12,500

B. f. 3 (May)
Sire: Parramon (Lyphard)
Dam: Northern Lass (Canadian Gil)
Br: Sens du Cheval Farm (Cal)
Tr: Mason Lloyd C (1 0 0 0 .00)

116

Lifetime Record:	7 1 0 3	$7,337			
1994	1 0 0 1	$1,650	Turf	0 0 0 0	
1993	6 1 0 2	$5,687	Wet	1 1 0 0	$3,575
GG	0 0 0 0		Dist	5 1 0 3	$7,175

7Jan94-6BM	fst	6f	:22^3	:46^1	:59	1:12	ⒻClm 12500	58	7	8	8^9	8^5	5^3	33½	Eide B M	LB 116	10.60	77-16	JanuaryHussy116^3 EnumclawRockette118½ AnniePerl116^5	Mild late bid	9
8Dec93-5BM	my	6f	:22^3	:45^4	:58^2	1:11^3	ⒻⓈMd 12500	59	4	4	41½	2½	11½	13½	Eide B M	LB 117	14.70	78-17	Annie Pearl1173½ Construct A Tune117^2 Funtime HadByAll117^{no}	Drew off	12
18Nov93-2BM	fst	1	:22^3	:47^1	1:13^2	1:39^3	ⒻMd 12500	20	8	1	1^1	2^{hd}	47½	720¾	Baze R A	LB 117	13.90	51-22	Crystals N'ice117^8 Tie And Dye117^1 Bonito Grande1173½	Gave way	10
4Nov93-1BM	fst	1	:22^4	:47^4	1:14	1:42^1	ⒻMd 12500	37	7	7	77¼	64½	54	57¼	Boulanger G	B 117	13.20	52-31	Quick Babe112¾ Dedicated Half117^2 Elegant Design1172½	Wide trip	7
21Oct93-1BM	fst	6f	:22^4	:46^4	:59^4	1:13^1	ⒻⓈMd 12500	33	4	4	45½	46	36	37½	Boulanger G	B 117	9.70e	62-23	Delta Summer117^5 Tie And Dye1172½ Annie Pearl117½	Even try	7
3Oct93-3BM	fst	6f	:23^1	:47^2	1:00^1	1:13^3	ⒻMd 12500	38	2	4	1½	1^1	2^1	3^5	Eide B M	B 117	44.30	63-23	Choice Cookie117^3 Daring Dame117^2 Annie Pearl117¾	Weakened	11
19Sep93-5BM	fst	6f	:23	:46^2	:58^4	1:12	ⒻⓈMd 12500	28	2	8	108½	89¼	8^{12}	710½	Belvoir V T	B 117	20.90	65-14	Alohi Treasure117^1 Boni Betty1172½ Dedicated Half117^1	Showed little	11

WORKOUTS: Jan 30 **BM 5f fst 1:01^3 H** *22/53* Jan 22 **BM 4f fst :48^4 H** *26/49* Jan 16 **BM 4f fst :48^2 H** *13/43* Jan 1 **BM 5f fst 1:00^1 H** *3/15* Dec 26 **BM 4f fst :50 H** *9/24* Dec 19 **BM 3f fst :36^4 H** *4/26*

Olive H.

Own: Burke Gary W & Timothy R

ESPINOZA V (13 1 2 3 .08) — $12,500

B. f. 3 (Jan)
Sire: Danebo (Bold Forbes)
Dam: Olive Mill (Gummo)
Br: Harris Farms Inc (Cal)
Tr: Miyadi Steven (2 0 0 0 .00)

111^5

Lifetime Record:	7 2 1 2	$17,850			
1994	1 0 0 0		Turf	0 0 0 0	
1993	6 2 1 2	$17,850	Wet	0 0 0 0	
GG	0 0 0 0		Dist	5 2 0 2	$15,950

16Jan94-9SA	fst	6½f	:21^2	:44^1	1:09^2	1:16^1	ⒻClm 25000	51	8	3	44½	58¾	6^{12}	613½	Stevens G L	LB 115	10.40	77-09	MissMoonPrincess1151¾ DesertButtrfly115½ VidoMnu1171¼	Saved ground	11
1Dec93-1Hol	fst	6f	:22^1	:46^1	:59	1:12	ⒻClm 20000	54	7	2	2½	2^{hd}	1^2	11¾	Stevens G L	LB 118	2.10	80-13	OliveH.1181¾ TrackCheerleder1152½ HoistTheWinner115^{hd}	Veered out start	7
17Nov93-6BM	fst	6f	:22^2	:46	:58^4	1:12	ⒻClm c-16000	59	4	5	1^{hd}	1^{hd}	11½	3^1	Meza R Q	LB 113	*1.40	75-17	Skeptic Lady Lee113^{no} January Hussy115^1 Olive H.1132½	Weakened late	7
Claimed from Harris Farms Inc, Gaines Carla Trainer																					
28Oct93-4BM	fst	6f	:22^2	:45^2	:57^4	1:10^4	ⒻAlw 20000N1X	55	2	4	31½	4^3	78½	66	Meza R Q	LB 113	9.40	76-17	Dancing Fir115¾ Alconnie116^{hd} Clever Covers1152½	Drifted in early	7
7Oct93-2BM	fst	6f	:23	:46^1	:58^3	1:11^2	ⒻMd 20000	60	7	1	2^{hd}	1^{hd}	1^5	1^7	Meza R Q	LB 117	*.90	79-15	Olive H.117^7 Skeptic Lady Lee117^{hd} Call Me The Star118½	Handy win	7
23Sep93-6BM	fst	6f	:22^2	:46	:59^1	1:12^1	ⒻMd 20000	60	7	3	2½	3^2	1^{hd}	3½	Meza R Q	LB 117	*1.70	74-17	My Blue Genes117½ Skeptic Lady Lee117^{no} Olive H.117^3	Wide trip	8
21Aug93-3Bmf	fst	5½f	:22^1	:46	:58^2	1:05	ⒻMd 20000	53	3	3	3^1	31½	2^3	22½	Kaenel J L	LB 117	2.10	87-09	Knight Blues1172½ Olive H.1172¼ Vail Link117½	Second best	7

WORKOUTS: Jan 29 **Hol 6f fst 1:14^4 Hg***8/8* Jan 23 **Hol 4f fst :48^4 H** *8/15* Jan 12 **Hol 3f fst :37^4 H** *7/11* Jan 7 **Hol 5f fst 1:01^2 H** *9/27* Dec 29 **Hol 4f fst :48 H** *3/20* Dec 16 **Hol 4f fst :50 H** *33/54*

Soft Hits

Own: Wiles Dennis A

GONZALEZ R M (11 0 0 1 .00) $12,500

Dk. b or br f. 3 (Mar)
Sire: Knights Choice (Drum Fire)
Dam: Magical Market (Rising Market)
Br: Richard Landeis (Fla)
Tr: Roberts Craig (—)

116

Lifetime Record:	8 1 0 1	$13,525			
1994	1 0 0 0		Turf	0 0 0 0	
1993	7 1 0 1	$13,525	Wet	0 0 0 0	
GG	0 0 0 0		Dist	4 1 0 0	$10,975

16Jan94-9SA	fst 6½f	:212 :441 1:092 1:161	ⒻClm 22500	34	10 2 55½ 712 1018 1120½	Gonzalez S5	LB 109 b	54.90	70-09	Miss MoonPrincess1151¾ DesertButterfly115½ VideoMenu1171¼	Wide trip	11
31Dec93-2SA	fst 6f	:211 :441 :564 1:10	ⒻClm 28000	31	10 6 88½ 1115 1213 1115¾	Pedroza M A	LB 113 b	102.50	71-10	Video Menu1171¼ Desert Butterfly115no Nice Account115nk	Gave way	12
1Dec93-1Hol	fst 6f	:221 :461 :59 1:12	ⒻClm 20000	38	6 3 41½ 51¾ 43 56½	Valenzuela P A	LB 118 b	8.10	74-13	Olive H.1181¾ Track Cheerleader1152½ HoistTheWinner115hd	Steadied 3 1/2	7
23Oct93-5SA	fst 7f	:223 :453 1:111 1:243	ⒻClm 32000	24	5 4 4¾ 5¾ 88½ 919¾	Desormeaux K J	LB 117 b	11.90	60-14	Rushing Attack117½ Native Lace1123½ Windy Perch1161	Dueled, tired	10
10Oct93-2SA	fst 6f	:214 :451 :581 1:114	ⒻMd 32000	56	1 5 2hd 1hd 2hd 11½	Desormeaux K J	LB 117 b	*1.60	78-16	Soft Hits1171½ Fowl Whether1172 Chromogenic1171¼	Inside duel	12
13Sep93-2Dmr	fst 5½f	:22 :452 :573 1:041	ⒻMd 32000	57	5 4 1hd 2½ 33 35½	Valenzuela P A	LB 118 b	33.60	84-10	Bella Jessica1185 Just Leavin Ladies118½ Soft Hits1181	Weakened	12
30Aug93-2Dmr	fst 6f	:22 :452 :58 1:112	ⒻMd 32000	55	7 1 42 34 45½ 46	Black C A	L 117 b	21.50	76-14	Rabiadella117¼ Storm Celine1173 Crystals N'ice1101¾	Wide trip	10
9Jly93-4Hol	fst 5½f	:222 :461 :591 1:06	ⒻMd c-32000	36	11 8 84½ 86½ 810 710	Lopez A D	B 118 b	7.80	72-15	Reconquista1186 Superbe Slew113½ Mahim Tova118½	Wide trip	11
Claimed from Landeis Gaylin Or Richard, Ward Wesley A Trainer												

WORKOUTS: Jan 30 BM 4f fst :50 H 30/50 Jan 10 SA 1 fst 1:44 H 4/4 Dec 26 SA 6f fst 1:133 H 3/24 Dec 19 SA 5f fst 1:01 H 7/16 Dec 11 SA 4f fst :474 H 15/44 Nov 26 Hol 4f fst :474 H 7/35

Fowl Whether

Own: Three Sisters Stable

MEZA R Q (8 3 1 1 .38) $12,500

Dk. b or br f. 3 (Mar)
Sire: Proud Birdie (Proud Clarion)
Dam: Nodouble's Queen (Nodouble)
Br: Carl Bowling (Fla)
Tr: Chambers Mike (—)

116

Lifetime Record:	7 1 1 1	$15,238			
1994	1 0 0 0		Turf	0 0 0 0	
1993	6 1 1 1	$15,238	Wet	0 0 0 0	
GG	0 0 0 0		Dist	4 1 1 1	$14,113

16Jan94-9SA	fst 6½f	:212 :441 1:092 1:161	ⒻClm 25000	50	3 6 67½ 611 712 814	Valenzuela F H	LB 115 b	26.00	77-09	MissMoonPrincess1151¾ DesrtButtrfly115½ VidoMnu1171¼	Wide into lane	11
31Dec93-2SA	fst 6f	:211 :441 :564 1:10	ⒻClm 32000	46	8 9 1111 1014 99½ 710	Stevens G L	LB 115 b	12.40	77-10	VideoMenu1171¼ DesertButtrfly115no NicAccount115nk	Improved position	12
11Dec93-1Hol	fst 5½f	:213 :45 :581 1:05	ⒻClm 25000	59	1 4 34 36 34 41¾	Valenzuela F H	LB 118 b	4.90	85-13	We Say Do118no Storm Celine118½ Spiritual Path1161¼	Outfinished	6
28Nov93-10Hol	fst 6f	:22 :451 :574 1:104	ⒻMd 32000	71	6 1 2hd 1hd 2hd 1hd	Valenzuela F H	LB 119 b	2.60	86-09	Fowl Whether119hd Exclusive Elegance1173½ Serendipity1195½	Gamely	7
24Oct93-2SA	fst 6f	:214 :451 :58 1:112	ⒻMd 32000	40	5 2 3nk 3nk 32 38¼	Stevens G L	LB 117 b	*1.20	72-12	Ashtabula1172½ Desert Butterfly1175½ DH Right Blend117	Weakened	8
10Oct93-2SA	fst 6f	:214 :451 :581 1:114	ⒻMd 32000	52	9 1 1hd 2hd 1hd 21½	Stevens G L	LB 117 b	3.30	76-16	Soft Hits1171½ Fowl Whether1172 Chromogenic1171¼	Outfinished	12
13Sep93-2Dmr	fst 5½f	:22 :452 :573 1:041	ⒻMd 32000	44	4 7 72½ 85¾ 119½ 810	Stevens G L	B 118	15.10	80-10	Bella Jessica1185 Just Leavin Ladies118½ Soft Hits1181		12
Took up 5/16, dropped whip final yards												

WORKOUTS: Feb 1 GG 5f fst 1:034 H 31/32 Dec 27 SA 5f fst 1:003 H 10/41 Dec 20 SA 4f fst :481 H 7/21 Nov 21 SA 5f fst 1:041 H 39/39 Nov 15 SA 5f fst 1:003 H 16/29 Nov 9 SA 5f fst 1:013 B 21/42

Skeptic Lady Lee

Own: Silverado Springs Stable

BAZE R A (—) $12,500

Dk. b or br f. 3 (Jan)
Sire: Our Michael (Bolero)
Dam: Skeptic Lady (Olden Times)
Br: Rancho Jonata (Cal)
Tr: Pederson Dean (—)

116

Lifetime Record:	8 2 4 0	$18,563			
1994	2 0 1 0	$2,938	Turf	1 0 0 0	
1993	6 2 3 0	$15,625	Wet	1 0 1 0	$1,700
GG	0 0 0 0		Dist	6 2 3 0	$17,263

23Jan94-1BM	sly 6f	:224 :46 :584 1:122	ⒻClm c-10000	61	5 5 65¾ 65½ 54 21	Baze R A	LB 116 b	*1.70	78-22	Knight Blues1161 SkepticLadyLee1161 HalfCopper1121½	Rallied far wide	6
Claimed from John H Deeter Trust, Benedict Jim Trainer												
5Jan94-6BM	fst 6f	:223 :461 :59 1:12	ⒻClm 20000	55	2 6 79½ 76 53½ 43	Warren R J Jr	LB 116 b	6.10	78-16	UniversalAffir1141½ FlyToVictory1161 HoistTheWinner116½	Mild late bid	8
4Dec93-6BM	fm 1 Ⓣ	:233 :48 1:132 1:394	ⒻAlw 20000N1X	54	4 5 65½ 710 711 710½	Schacht R	LB 114 b	22.10	70-17	PrincssMittrnd1141½ PrincssInChrg1142½ RoylBoutiqu1131	Showed little	8
17Nov93-6BM	fst 6f	:222 :46 :584 1:12	ⒻClm 16000	62	1 6 79½ 76½ 43 1no	Boulanger G	LB 113 b	11.40	76-17	Skeptic Lady Lee113no January Hussy1151 Olive H.1132½	Rallied far wide	7
28Oct93-5BM	fst 6f	:223 :46 :584 1:12	ⒻMd 12500	53	1 6 52½ 32½ 21 1nk	Baze R A	LB 117 b	*1.30	76-17	SkepticLdyLee117nk DesertMystique1173½ K.L.'sImgene117nk	Rallied wide	12
7Oct93-2BM	fst 6f	:23 :461 :583 1:112	ⒻMd 20000	42	5 7 76½ 65½ 36 27	Tohill K S	LB 117 b	3.10	72-15	OliveH.1177 SkepticLadyLee117hd CllMeTheStr118½	Shuffled back start	7
23Sep93-6BM	fst 6f	:222 :46 :591 1:121	ⒻMd 20000	60	5 6 56½ 46 3½ 2½	Tohill K S	LB 117 b	8.70	74-17	My Blue Genes117½ Skeptic Lady Lee117no Olive H.1173	Wide late bid	8
29Aug93-5BM	fst 5f	:22 :46 :59	ⒻⓈMd 12500	56	4 8 66 55½ 44 2¾	Tohill K S	LB 117 b	24.40	88-17	Hazy Star117¾ Skeptic Lady Lee1172 La Campesina117nk	Wide late run	9

WORKOUTS: Jan 17 BM 5f fst 1:03 H 28/32 Dec 29 BM 5f fst 1:034 H 39/43 Dec 19 BM 4f fst :492 H 22/45 Nov 28 BM 3f fst :39 H 11/12 Nov 13 BM 3f fst :36 H 3/27

Knight Blues

Own: Klokstad Billie

BOULANGER G (18 0 5 5 .00) $12,500

Dk. b or br f. 3 (Apr)
Sire: Knights Choice (Drum Fire)
Dam: Monday Blues (Blue Ensign)
Br: Klokstad Billie A (WA)
Tr: Klokstad Bud (2 0 1 1 .00)

116

Lifetime Record:	9 2 2 0	$14,807			
1994	2 1 0 0	$5,087	Turf	0 0 0 0	
1993	7 1 2 0	$9,720	Wet	1 1 0 0	$4,675
GG	1 0 0 0		Dist	3 1 1 0	$6,687

23Jan94-1BM	sly 6f	:224 :46 :584 1:122	ⒻClm 10000	63	4 1 11 11½ 11 11	Boulanger G	LB 116	2.00	79-22	Knight Blues1161 Skeptic Lady Lee1161 Half Copper1121½	Steady drive	6
5Jan94-6BM	fst 6f	:223 :461 :59 1:12	ⒻClm 18000	45	6 3 11 2hd 31½ 57	Boulanger G	LB 114	7.20	74-16	Universal Affair1141½ FlyToVictory1161 HoistTheWinner116½	Weakened	8
18Dec93-1BM	gd 6f	:223 :46 :583 1:114	ⒻClm 10000	61	4 1 1½ 1hd 11½ 21	Boulanger G	LB 116	2.70	76-18	Hoist The Winner1161 Knight Blues1162½ Half Copper1111	Second best	7
23Sep93-3BM	fst 5½f	:214 :453 :583 1:051	ⒻClm 20000	55	6 5 3½ 3nk 33 44	Boulanger G	LB 113	*1.30e	81-17	Clever Covers1151½ Jetting Trip116½ Eightskateandonate1172	Weakened	6
4Sep93-6BM	fst 5½f	:211 :452 :58 1:043	ⒻClm 25000	39	3 4 42½ 44½ 67½ 712	Boulanger G	LB 116	2.70e	76-14	Road Buddy1142½ Clever Covers116½ Jetting Trip1161	Fractious gate	7
21Aug93-3BM	fst 5½f	:221 :46 :582 1:05	ⒻMd 20000	60	4 1 11 1½ 13 12½	Boulanger G	LB 117	*1.40	90-09	Knight Blues1172½ Olive H.1172½ Vail Link117½	Ridden out	7
6Jly93-7Pln	fst 5½f	:22 :46 :58 1:044	ⒻMd 16000	56	10 1 11½ 1hd 2hd 22	Boulanger G	LB 117	5.50	87-17	Delightful Year1172 Knight Blues1177 Cold Justice1172½	Second best	10
29Jun93-9Pln	fst 5½f	:213 :46 :573 1:042	ⒻMd Sp Wt	36	4 2 22 57 812 712½	Kaenel J L	LB 117	7.30e	78-09	Saucy Trick1173 Gentle Rainfall1172½ Jilly's Halo1171½	Gave way	10
16May93-1GG	fst 5f	:21 :45 :581	ⒻMd Sp Wt	8	2 3 54 913 919 820	Warren R J Jr	LB 117	7.10	71-13	Dancing Fir117¾ Saucy Trick1178 Yearly Tour1172	Lugged out 3/8	10

WORKOUTS: Dec 31 BM 5f fst 1:004 H 10/56 Dec 12 BM 5f my 1:044 H (d) 18/46 Dec 2 BM 4f gd :494 H 19/41 Nov 24 BM 4f fst :493 H 23/36

January Hussy

Own: Weigel M J

BELVOIR V T (15 6 2 2 .40) $12,500

B. f. 3 (Jan)
Sire: Bold T. Jay (Island Agent)
Dam: Dance Hall Hussy (Noble Dancer)
Br: Marshall Frank Mr & Mrs (Cal)
Tr: Arterburn Tim (—)

118

Lifetime Record:	7 2 2 0	$15,737			
1994	1 1 0 0	$6,050	Turf	0 0 0 0	
1993	6 1 2 0	$9,687	Wet	0 0 0 0	
GG	0 0 0 0		Dist	6 2 2 0	$15,575

7Jan94-6BM	fst 6f	:223 :461 :59 1:12	ⒻClm 12500	67	8 7 75 51½ 1½ 13	Belvoir V T	LB 116	3.00	81-16	January Hussy1163 EnumclawRockette118½ AnniePearl1165	Closed well	9
18Dec93-1BM	gd 6f	:223 :46 :583 1:114	ⒻClm 10000	53	6 4 54 32 42½ 44½	Baze R A	LB 116	*2.10	72-18	Hoist The Winner1161 Knight Blues1162½ Half Copper1111	Wide trip	7
2Dec93-3BM	gd 6f	:224 :461 :581 1:113	ⒻClm 12500	46	5 5 53 56 49½ 48	Baze R A	L 116	*.50	70-24	EnumclwRockett1125 LorrinsLov1181 ChoicCooki1132	Raced well wide	6
17Nov93-6BM	fst 6f	:222 :46 :584 1:12	ⒻClm 16000	62	2 7 65½ 63½ 21½ 2no	Baze R A	LB 115	3.60	76-17	Skeptic Lady Lee113no January Hussy1151 Olive H.1132½	Wide late bid	7
25Oct93-7BM	fst 6f	:222 :454 :582 1:122	ⒻClm c-12500	59	9 2 53½ 42½ 45 2nk	Baze R A	LB 115	4.60	74-20	HazyStar115nk JanuryHussy1151 Eightsktendonte115no	Rallied well wide	9
Claimed from Marshall Mr & Mrs Frank, Benedict Jim Trainer												
11Oct93-4BM	fst 6f	:23 :47 1:00 1:124	ⒻMd 12500	59	3 3 42 41¾ 2hd 16	Baze R A	LB 117	4.10	72-19	January Hussy1176 Dolly Duck1175 Champagnefortrish117hd	Ridden out	8
5Sep93-5BM	fst 5½f	:213 :454 :584 1:051	ⒻMd 12500	31	3 9 810 88½ 78 58½	Warren R J Jr	LB 117	29.50	77-15	Bubble Dance1171¼ Bev1171½ Hub City Dancer1173	No rally	9

WORKOUTS: Jan 29 BM 3f gd :371 H 6/20 Jan 14 BM 3f fst :362 H 11/37 Jan 2 BM 4f fst :472 H 5/44 ●Dec 27 BM 3f fst :34 H 1/15 Dec 12 BM 3f my :392 H (d) 17/18 Nov 10 BM 4f fst :474 H 5/29

164   The seventh race on February 6th at Golden Gate Fields is a typical example of the complexity of applying off-track ratings to a fast sloppy surface. Going over the entries, Annie Pearl, Skeptic Lady Lee and Knight Blues have all run well in the slop and must be considered. Interestingly, even though Skeptic Lady Lee just lost to Knight Blues in the slop, today she is a slight favorite over that rival at 5-2. Meanwhile the public is less enamored of Annie Pearl and is sending her off at nearly 8-1. Apparently they feel that her mud win won't be replicated in the slop. Sarah's Boop failed miserably in her only off-track effort and is being dismissed at better than 13-1.

Of the "first-time-in-the-slop" horses, Olive H. with her A rating on both the top and bottom of her pedigree is a standout. She also has several other angles going for her; she's already beaten winners in southern California, needed her last race, is dropping in class and is getting blinkers which should give her more speed today. Given all these advantages, she is a bargain at 5-1.

Of the other first-time-sloppers, Soft Hits and Fowl Weather (despite her name) are not particularly bred for the slop, have yet to beat winners, and are not being agressively bet. January Hussy, on the other hand, just won at this level and with her BT rating might come right back.

So how to bet? Olive H. is an attractive Win and Place bet. Alternatively, she can be boxed with Knight Blues, Skeptic Lady Lee, January Hussy and Annie Pearl, all of whom have a shot to complete the exacta.

SEVENTH RACE

Golden Gate

FEBRUARY 6, 1994

6 FURLONGS. (1.07³) CLAIMING. Purse $12,000. Fillies 3–year–olds. Weight, 122 lbs. Non–winners of two races since December 1, allowed 3 lbs. A race since then, 5 lbs. Claiming price $12,500; if for $10,500, 2 lbs. (Maiden races, claiming, starter and classified handicap races for $10,000 or less not considered.)

Value of Race: $12,000 Winner $6,600; second $2,400; third $1,800; fourth $900; fifth $300. Mutuel Pool $110,862.00 Exacta Pool $113,580.00

Last Raced	Horse	M/Eqt.	A.Wt	PP	St	¼	½	Str	Fin	Jockey	Cl'g Pr	Odds $1
16Jan94 9SA6	Olive H.	LBb	3 111	3	2	11½	11½	14	12	Espinoza V5	12500	5.40
7Jan94 6BM1	January Hussy	LB	3 118	8	3	51	41	3½	2hd	Belvoir V T	12500	4.70
23Jan94 1BM2	Skeptic Lady Lee	LBb	3 116	6	4	7½	62	53	32	Baze R A.	12500	2.50
23Jan94 1BM1	Knight Blues	LB	3 116	7	1	23	2hd	2hd	4nk	Boulanger G	12500	3.10
12Jan94 2BM1	Sarah's Boop	LB	3 116	1	7	8	8	8	5nk	Warren R J Jr	12500	13.50
16Jan94 9SA8	Fowl Whether	Lb	3 116	5	5	3hd	3½	4½	6hd	Meza R Q	12500	8.10
7Jan94 6BM3	Annie Pearl	LB	3 116	2	8	6½	73	71	73	Chapman T M	12500	7.50
16Jan94 9SA11	Soft Hits	LBb	3 116	4	6	4½	51	61	8	Gonzalez R M	12500	27.00

OFF AT 4:45 Start Good. Won driving. Time, :22, :45², :58, 1:11³ Track sloppy.

$2 Mutuel Prices:

3–OLIVE H.	12.80	7.00	4.60
8–JANUARY HUSSY		5.60	3.80
6–SKEPTIC LADY LEE			3.40

$5 EXACTA 3–8 PAID $167.50

B. f, (Jan), by Danebo–Olive Mill, by Gummo. Trainer Miyadi Steven. Bred by Harris Farms Inc (Cal).

While Olive H. ran to her excellent slop breeding, it is important to note that her pedigree might not have been the determining factor in her win. She was also taking a significant drop in class from the tougher southern California circuit, had already beaten winners, and was getting an important equipment change. Olive H. was an attractive play in this spot not only because of her breeding, but because of these other factors. Since all starters wore mud caulks in this race, that was a non-factor.

Saratoga

1

1⅛ MILES. (Inner Turf). (1:47) MAIDEN SPECIAL WEIGHT. Purse $27,000. 3-year-olds and upward. Weights: 3-year-olds, 117 lbs. Older, 122 lbs.

INNER TURF COURSE 1⅛ MILES START FINISH

According To Cole

D11

B. c. 3 (Feb)
Sire: D'Accord (Secretariat)
Dam: Sweet Country Miss (Dr. Giddings)
Br: Iacovelli Lawrence E (NY)
Tr: Jayko John C (5 0 1 0 .00)

Own: Iacovelli Lawrence
HERNANDEZ R (27 1 3 3 .04)

117

Lifetime Record :	2 M 1 0	$5,500			
1994	2 M 1 0	$5,500	Turf	0 0 0 0	
1993	0 M 0 0		Wet	1 0 0 0	
Sar Ⓣ	0 0 0 0		Dist Ⓣ	0 0 0 0	

21Aug94- 9Sar sly 7f :224 :461 1:123 1:263 3↑ [S]Md Sp Wt	-0 7 6 52 77½ 912 1028½	Hernandez R	117 *2.30 43-20	Roscoe Turner1121 But Anyway1173 Doubtful Debt1171	Bumped break	11
30Jly94-10Sar gd 6f :223 :463 :591 1:121 3↑ [S]Md Sp Wt	58 7 7 64½ 42½ 24 29	Hernandez R	116 22.70 71-13	Royal Judge1119 According To Cole1163 Stinky Dinky1161¼	Mild rally	9

WORKOUTS: Aug 16 Sar tr.t 4f fst :504 B *13/40* Aug 11 Sar tr.t 5f fst 1:043 B *11/15* Jly 29 Sar 3f sly :363 H *7/14* Jly 25 Sar 5f fst 1:01 Hg*9/64* Jly 12 Sar tr.t 5f fst 1:05 B *2/3* Jly 7 Sar tr.t 5f fst 1:05 B *1/1*

Vigan

1/2

B. c. 3 (Apr)
Sire: Blushing John (Blushing Groom)
Dam: Northern Aspen (Northern Dancer)
Br: Paulson Allen E (Ky)
Tr: Zito Nicholas P (52 7 7 9 .13)

Own: Paulson Allen E
SELLERS S J (71 11 6 13 .15)

117

Lifetime Record :	3 M 0 0	$0		
1994	3 M 0 0		Turf	1 0 0 0
1992	0 M 0 0		Wet	0 0 0 0
Sar Ⓣ	0 0 0 0		Dist Ⓣ	0 0 0 0

19Jly94- 5Bel fst 7f :224 :454 1:112 1:243 3↑ Md Sp Wt	26 8 3 53 86 915 925½	Beckner D V5	111 50.70 54-19	Crafty Mist1161¼ Level Land1165½ Fleet Stalker1163½	Done early	9
4Jly94-10Bel fm 1 1/16 [T] :234 :471 1:104 1:411 3↑ Md Sp Wt	30 8 2 31 75½ 815 924¾	Cruz E F	116 47.10 65-10	Dutchess First1164¼ Sophie's Friend1161¼ Thirty Good Ones1161¾	Tired	10
5Jun94- 3Bel fst 7f :222 :453 1:101 1:222 3↑ Md Sp Wt	50 2 10 1012 108 1011 1019	Luzzi M J	114 53.40 72-11	PartyMnners1145 Convince1142 RoyceJoseph114hd	Broke slowly, outrun	10

WORKOUTS: Aug 23 Sar tr.t 4f fst :54 B *27/29* Aug 14 Sar tr.t 5f gd 1:042 B *2/2* Aug 7 Bel 5f fst 1:022 B *14/22* Jly 31 Bel 4f fst :483 B *8/32* Jly 12 Bel 5f fst 1:031 B *30/40* Jly 2 Bel 4f fst :504 B *40/48*

Silver Safari

f/4

Ro. g. 3 (Mar)
Sire: Northern Jove (Northern Dancer)
Dam: Sans Jacques (Dom Alaric)
Br: El Batey Bloodstock (NY)
Tr: O'Connell Richard (17 4 2 1 .24)

Own: Very Un Stable
DAVIS R G (115 8 22 14 .07)

Faces open company

117

Lifetime Record :	10 M 1 4	$24,520			
1994	3 M 0 1	$4,860	Turf	4 0 0 1	$9,300
1993	7 M 1 3	$19,660	Wet	0 0 0 0	
Sar Ⓣ	0 0 0 0		Dist Ⓣ	0 0 0 0	

20Jly94- 1Bel fm 1 1/16 [T] :243 :481 1:123 1:442 3↑ [S]Md Sp Wt	65 8 9 94¾ 52¼ 73½ 43¾	Davis R G	116 b 2.50e 70-18	North Forty Four117½ Lord Basil1163 Advance Warning116nk	Four wide	10
26Jun94- 5Bel fm 1 Ⓣ :23 :46 1:094 1:34 3↑ Md Sp Wt	36 3 3 42½ 74½ 1012 1026	Luzzi M J	114 b 9.20 66-16	Holy Mountain1146½ Dole Raider1141¾ Haagbah1141¼	Gave way	12
4Jun94- 9Bel fm 1 1/16 Ⓣ :223 :452 1:10 1:422 3↑ [S]Md Sp Wt	68 9 1 11 11½ 2½ 34	Luzzi M J	114 b *2.10 82-10	Mr. Baba1143 Watrals Sea Trip1141 Silver Safari1143	Weakened	10
22Nov93- 3Aqu fst 6f :223 :471 1:001 1:132 [S]Md Sp Wt	56 11 1 21 21 31 34½	Davis R G	118 b 3.50 70-22	To Ta Roo1184½ Tamara R.118hd Silver Safari1181¼	Willingly	11
13Nov93- 8Aqu yl 1 Ⓣ :241 :483 1:143 1:414 [S]Damon Runyon74k	74 12 3 31½ 22 21½ 4½	Davis R G	113 b 44.50 67-33	SminolSpirt117no Popol'sGold117no SprtnVictory122nk	Drifted, weakened	12
2Nov93- 2Aqu fst 6f :23 :472 :592 1:12 [S]Md Sp Wt	29 6 2 42 53 58 819½	Migliore R	118 b 8.40 62-18	Concordes Prospect1186¼ Bit Of Puddin118nk Night Trap1187	Tired	10
6Aug93- 3Sar fst 5f :22 :453 :582 [S]Md Sp Wt	51 9 3 31½ 32 34 35¼	Antley C W	118 b 6.60 88-07	Gulliviegold1184 Frozen Ammo1181¼ Silver Safari118nk	Lacked rally	10
21Jly93- 5Bel fst 5½f :23 :462 :582 1:044 [S]Md Sp Wt	50 9 6 51½ 2½ 35 311½	Antley C W	118 b 2.00e 79-14	Direct Satellite1185¼ Gulliviegold1186¼ Silver Safari1185	Bid, tired	9
25Jun93- 3Bel fst 6f :224 :471 1:002 1:143 [S]Md Sp Wt	45 7 3 11½ 1½ 2hd 2nk	Antley C W	118 b 9.40 67-22	Private Deal118nk Silver Safari118¾ Legasus1181½	Gamely	10
14Jun93- 3Bel fst 5½f :224 :463 :591 1:06 [S]Md Sp Wt	50 1 1 21½ 33 45½ 46½	Antley C W	118 11.70 78-14	Crafty Harold1186 Seminole Spirt118hd Background Artist118½	Tired	9

WORKOUTS: Aug 24 Sar tr.tⓉ 4f gd :503 B *10/14* ●Aug 17 Sar tr.tⓉ 5f fm :591 H *1/20* Aug 9 Sar Ⓣ 5f fm 1:004 H *2/19* Aug 2 Sar Ⓣ 5f fm 1:031 B *15/16* Jly 29 Sar tr.tⓉ 4f sf :502 B *7/14* Jly 14 Bel [T] 5f fm 1:05 B (d) *19/*

Raise The Devil

c+11

Ch. c. 3 (Apr)
Sire: Devil's Bag (Halo)
Dam: Raise the Market (Raise a Native)
Br: Tuttle Mrs Wylie F L (Md)
Tr: Reynolds Patrick L (5 0 1 0 .00)

Own: Collier Reginald B
LEON F (48 1 4 7 .02)

117

Lifetime Record :	4 M 0 0	$510		
1994	4 M 0 0	$510	Turf	0 0 0 0
1993	0 M 0 0		Wet	0 0 0 0
Sar Ⓣ	0 0 0 0		Dist Ⓣ	0 0 0 0

24Jly94- 1Lrl fst 6f :231 :472 1:00 1:123 Md Sp Wt	7 2 6 75 812 819 823	Douglas F G	120 68.60 54-11	Thrisk1203 Folignos Flight1201¼ South West Hostage120nk	Outrun	8
9Jly94- 3Lrl fst 7f :23 :461 1:121 1:253 Md Sp Wt	24 11 2 56 612 1017 1120¼	Carle J D	120 b 77.30 61-18	Roys Call120nk Thrisk120no Brains120¾	Fell back	12
15Mar94-11GP fst 1 1/16 :231 :472 1:122 1:443 Md Sp Wt	-0 8 2 51¾ 1018 930 1060¼	Turner T G	122 fb 68.10 27-15	Andover Scholar12210 In A Schocking Way122¾ ChasinGold1226	Stopped	10
16Feb94- 3GP fst 6f :22 :45 :571 1:094 Md Sp Wt	17 6 3 53¾ 711 718 729	Ferrer J C	122 fb 13.50 63-12	A Firm Mister1221¾ Ten Star Fleet1226¼ Chasin Gold122no	No factor	7

WORKOUTS: Aug 22 Sar tr.tⓉ 4f sf :531 B (d) *10/13* Aug 9 Sar 4f fst :50 B *13/21* ●Aug 4 Sar tr.tⓉ 3f fm :354 H *1/5* Jly 21 Pim 5f fst 1:02 B *4/9* Jly 6 Pim 5f fst 1:051 B *3/3*

Crafty Truce

A14

Ro. g. 3 (Mar)
Sire: Crafty Prospector (Mr. Prospector)
Dam: Reason for Truce (Turn to Reason)
Br: Iselin James H & Little Jr Marvin (Ky)
Tr: Serpe Philip M (27 5 4 2 .19)

Own: Kaufman Robert
KRONE J A (125 21 28 14 .17)

117

Lifetime Record :	3 M 0 0	$820		
1994	2 M 0 0	$660	Turf	0 0 0 0
1993	1 M 0 0	$160	Wet	0 0 0 0
Sar Ⓣ	0 0 0 0		Dist Ⓣ	0 0 0 0

17Jly94- 9Bel fst 6f :222 :453 :581 1:113 3↑ Md 35000	47 5 2 21½ 33 37 412¾	Smith M E	116 3.80 70-16	Lightnin' Cat112nk Stantorian1137 Super Twenty Five1165½	Tired	8
17Jun94- 9Bel fst 6f :221 :452 :58 1:11 3↑ Md Sp Wt	65 7 9 63½ 63½ 96¾ 87¼	Smith M E	114 10.00 79-15	Thru 'n Thru1141¼ [D]Ball's Bluff1141¼ Changing Rahy114nk	Four wide	10
9Oct93- 5Med fst 6f :222 :462 :592 1:131 Md Sp Wt	38 5 2 32 54½ 77½ 78¾	Bravo J	118 4.80 66-25	Powerful Patch1181¼ Settle Quick118hd Drop Em118no	Tired	9

WORKOUTS: Aug 21 Sar 5f fst 1:014 H *9/24* Aug 12 Sar tr.tⓉ 5f fm 1:003 H *3/9* Jly 12 Bel 5f fst 1:023 B *22/40* Jly 7 Bel [T] 5f fm 1:03 B (d)*4/7* Jly 1 Bel 4f fst :481 B *4/38* Jun 10 Bel 7f fst 1:281 B *1/1*

Sup With The Devil

c+11 c/4*

B. g. 3 (Apr)
Sire: Devil's Bag (Halo)
Dam: Cloud of Music (Northern Dancer)
Br: Paul Mellon (Va)
Tr: Miller MacKenzie (17 3 3 2 .18)

Own: Rokeby Stables
BECKNER D V (83 11 11 7 .13)

112[5]

Lifetime Record :	2 M 0 0	$0		
1994	2 M 0 0		Turf	0 0 0 0
1992	0 M 0 0		Wet	1 0 0 0
Sar Ⓣ	0 0 0 0		Dist Ⓣ	0 0 0 0

29Jly94- 3Sar my 6f :222 :462 :584 1:122 3↑ Md Sp Wt	70 6 6 68 63¼ 54¼ 55	Beckner D V5	111 8.40 74-19	Its A Star1161¾ Ball's Bluff1162 Crown Pearl116nk	Broke slowly	6
9Jly94- 3Bel fst 6f :233 :471 :584 1:104 3↑ Md Sp Wt	45 2 5 52¾ 52¾ 58 515¼	Beckner D V5	111 6.70 72-24	Convince1161½ Darien Cowboy1162½ Count On Broadway1169½	In traffic	6

WORKOUTS: Aug 16 Sar tr.t 5f fst 1:053 B *12/15* Aug 10 Sar tr.tⓉ 4f fm :494 B *14/20* Aug 5 Sar tr.t 4f fst :51 B *14/30* Jly 27 Sar tr.t 3f sly :38 B *4/5* ●Jly 22 Sar 4f sly :48 H *1/7* Jly 16 Bel 4f fst :48 H *11/66*

166 The first race at Saratoga on August 26th was a 1 1/8 mile Maiden Special Weight route that was taken off the turf and run on a surface labeled sloppy. Going over the entries, the clear pedigree standout is Crafty Truce with his A off-track rating. But Crafty Truce has to overcome two strong negatives in that he is being bumped back up in class after losing to maiden claimers and his stamina index of 4 raises doubts about his ability to get the 1 1/8 mile distance.

The actual betting favorite at 3-2 is Sup With The Devil who has a decent C+ rating, the highest off-track Beyers, has never run for a tag, and is bred to get the distance. The surprise second choice is Silver Safari at 5-2, who is moving from the normally easier state-bred level into an open Maiden Special Weight event and has an F off-track rating. More positively, Silver Safari does have some speed, has run competitively in three of his four turf routes and may be the only horse legged up to get the distance.

According To Cole isn't bred for the slop and ran like it on August 21st when he lost to New York-breds. Vigan with his L rating might move up on an off-track, but he has yet to pass a horse in his career and can't be backed with any confidence.

Raise the Devil looks little better. He is shipping in from Laurel, where he competed against a lower class of straight maiden company and lost by 23 lengths. He wasn't bet then, isn't being bet today, and can be tossed out.

So how to bet? Like most off-track events, this is an ambiguous race. Crafty Truce is the only legitimate off-track pedigree play, but he really isn't bred to get the distance. Besides Crafty Truce's short stamina index of 4, his dam's progeny, according to *Maiden Stats 1993*, have an average winning distance of only 7.2 furlongs. Still, Crafty Truce looks like the speed of the race and may get an easy enough lead here to wire this field. He should be played moderately to Win and put behind Sup With The Devil in the exacta.

FIRST RACE
Saratoga
AUGUST 26, 1994

1⅛ MILES. (1.47) MAIDEN SPECIAL WEIGHT. Purse $27,000. 3-year-olds and upward. Weights: 3-year-olds, 117 lbs. Older, 122 lbs. (Day 31 of a 34 Day Meet. Clear. 83.)(ORIGINALLY SCHEDULED FOR TURF.)

Value of Race: $27,000 Winner $16,200; second $5,940; third $3,240; fourth $1,620. Mutuel Pool $159,670.00 Exacta Pool $286,930.00

Last Raced	Horse	M/Eqt. A.Wt	PP	St	¼	½	¾	Str	Fin	Jockey	Odds $1
20Jly94 1Bel4	Silver Safari	b 3 117	3	1	2⁶	2²	2⁶	2²	1¹½	Davis R G	2.30
17Jly94 9Bel4	Crafty Truce	3 117	5	3	1½	1¹½	1¹	1¹	2²¾	Krone J A	3.00
29Jly94 3Sar5	Sup With The Devil	b 3 112	6	4	6	6	5⁸	3⁷	3¹⁴	Beckner D V5	1.40
21Aug94 9Sar10	According To Cole	3 117	1	5	3½	4¹	3¹	4¹	4²	Hernandez R	8.10
19Jly94 5Bel9	Vigan	3 117	2	6	5⁴	5⁴	4hd	5²⁰	5	Sellers S J	11.80
24Jly94 1Lrl8	Raise The Devil	bf 3 117	4	2	4hd	3½	6	6	—	Leon F	16.40

Raise The Devil: Eased

OFF AT 1:00 Start Good. Won driving. Time, :22⁴, :47, 1:12³, 1:39², 1:53¹ Track sloppy.

$2 Mutuel Prices:

5-(E)–SILVER SAFARI	6.60	3.20	2.20
7-(G)–CRAFTY TRUCE		3.80	2.20
8-(H)–SUP WITH THE DEVIL			2.20

$2 EXACTA 5-7 PAID $24.00

Ro. g, (Mar), by Northern Jove–Sans Jacques, by Dom Alaric. Trainer O'Connell Richard. Bred by El Batey Bloodstock (NY).

© 1994 DRF, Inc.

Oh, well. Crafty Truce couldn't get the distance and we got beat by an F-rated horse. On to the next race.

The sixth race at Santa Anita on March 20th is an example of a race run on a tiring track that further reflects the complexity of handicapping for an off-track surface. Looking over the contenders, the favorite at 2-1 is Major Discovery, who comes off a highly-rated effort in the slop, which the public assumes will be replicated on today's more tiring surface in which horses have been running six or more ticks slower than usual. Also being bet at less than 3-1 are Devon Dancer and Le Dome, who is making his first start as a 3-year-old for Gary Jones after being shipped in from England.

Santa Anita Park

6

6½ Furlongs. (1:14) MAIDEN SPECIAL WEIGHT. Purse $32,000. 3-year-olds. Weight, 118 lbs. (Non-starters for a claiming price of $32,000 or less in their last three starts preferred.)

Major Discovery
Own: McCaffery Trudy & Toffan John

GONZALEZ S JR (281 33 24 35 .12)

Dk. b or br c. 3 (Mar)
Sire: Seeking the Gold (Mr. Prospector)
Dam: Anything for Love (Seattle Slew)
Br: Brant Peter M (Ky)
Tr: Gonzalez J Paco (47 9 9 3 .19)

1135

Lifetime Record :		4 M 2 0		$12,850	
1994	2 M 1 0	$6,400	Turf	0 0 0 0	
1993	2 M 1 0	$6,450	Wet	1 0 1 0	$6,400
SA	3 0 1 0	$7,250	Dist	0 0 0 0	

20Feb94- 6SA (wf) 6f .214 .444 .571 1.092 Md Sp Wt (90) 4 1 2½ 2hd 1hd 2½ Gonzalez S⁵ B 113 b 4.50 89-09 Beautiful Crown118½ Major Discovery113⁶ Island Sport118½ Game try 6
23Jan94- 5SA fst 1 .224 .462 1:101 1:363 Md Sp Wt 78 4 5 53 47 510 613 Valenzuela P A B 117 8.90 80-13 Dramatic Gold117nk MuirfieldVillge117½ Devil'sMirge117½ Forced in start 10
26Dec93- 9SA fst 1 .221 .45 1:101 1:363 Md Sp Wt 68 5 7 711 65½ 77 510½ Valenzuela P A B 117 *3.20 76-10 Almaraz117½ Dramatic Gold117½ Advantage Miles117⁶ 10
    Off slow, took up 3/16
28Nov93- 6Hol fst 6f .22 .451 .573 1:103 Md Sp Wt 69 8 4 4½ 42½ 22 22 Valenzuela P A B 118 1.90 85-09 Windwood Lad118² Major Discovery118² Rindo Indy118½ 4 Wide early 8
WORKOUTS: Mar 15 SA 5f fst 1:00³ H 21/59 Mar 9 SA 5f fst 1:03 H 51/59 ●Mar 2 SA 5f fst :59 H 1/38 Feb 16 SA 3f fst :36³ H 17/30 Feb 11 SA 5f fst :59³ Hg 7/62 Feb 3 SA 5f fst 1:00⁴ H 15/62

Star Doctor
Own: Oldknow & Phipps

FLORES D R (103 5 13 17 .05)

Ch. c. 3 (Apr)
Sire: Dr. Carter (Caro)
Dam: Retsina Star (Pia Star)
Br: William H. Oldknow & Robert W. Phipps (Ky)
Tr: Sadler John W (71 15 11 8 .21)

FM
B

118

Lifetime Record :		0 M 0 0		$0	
1994	0 M 0 0		Turf	0 0 0 0	
1993	0 M 0 0		Wet	0 0 0 0	
SA	0 0 0 0		Dist	0 0 0 0	

WORKOUTS: ●Mar 15 Hol 5f fst 1:01 H 1/23 Mar 8 Hol 6f fst 1:13³ H 3/11 Mar 2 Hol 6f fst 1:14⁴ H 6/11 Feb 24 Hol 6f fst 1:14¹ H 6/15 ●Feb 18 Hol 5f fst 1:00 H 1/20 Feb 12 Hol 5f fst 1:03¹ H 39/56
    Feb 6 SA 4f fst :50³ H 65/83 Jan 28 SA 6f fst 1:14³ H 32/48 Jan 21 SA 5f fst 1:00⁴ H 9/48 Feb 15 SA 5f fst 1:01² H 28/49 Jan 9 SA 4f fst :50⁴ H 51/55 Jan 3 SA 4f fst :47⁴ H 11/38

Granby
Own: Rancho San Miguel & Vienna & Vienna

MCCARRON C J (178 31 30 27 .17)

Ch. g. 3 (Jan)
Sire: Groovy (Norcliffe)
Dam: For Safekeeping (Steward)
Br: Prestonwood Farm Inc (Ky)
Tr: Vienna Darrell (67 8 13 7 .12)

DM

L 118

Lifetime Record :		0 M 0 0		$0	
1988	0 M 0 0		Turf	0 0 0 0	
1987	0 M 0 0		Wet	0 0 0 0	
SA	0 0 0 0		Dist	0 0 0 0	

WORKOUTS: Mar 14 SA 6f fst 1:13⁴ H 13/23 Mar 9 SA 6f fst 1:13 H 3/42 Mar 2 SA 6f fst 1:13³ H 6/16 Feb 24 SA 6f fst 1:15¹ H 20/22 Feb 19 SA 5f fst 1:01 H 37/75 Feb 14 SA 4f fst :49² H 14/21

Devon Dancer
Own: St George's Farm

STEVENS G L (331 47 65 57 .14)

B. c. 3 (Feb)
Sire: Herat (Northern Dancer)
Dam: Devon Lass (Secretariat)
Br: St. George's Farm (Ky)
Tr: Shoemaker Bill (45 6 4 6 .13)

FM
BT

L 118

Lifetime Record :		2 M 0 2		$9,600	
1994	2 M 0 2	$9,600	Turf	0 0 0 0	
1993	0 M 0 0		Wet	0 0 0 0	
SA	2 0 0 2	$9,600	Dist	2 0 0 2	$9,600

5Feb94- 4SA fst 6½f .221 .45 1:092 1:154 Md Sp Wt 79 4 1 2hd 2hd 2hd 34½ Nakatani C S LB 118 *2.00 89-11 Irgun118⁴ Scenic Route118nk Devon Dancer118¹³ Caught on wire 6
15Jan94- 2SA fst 6½f .213 .441 1:092 1:154 Md Sp Wt 76 2 5 1hd 1hd 1hd 36 Pincay L Jr B 118 12.20 87-08 Strodes Creek118⁵ Native Blast118¹ Devon Dancer118⁵½ Steadied 3/4 7
WORKOUTS: Mar 17 SA 4f fst :48 H 7/32 ●Mar 11 SA 6f fst 1:12 H 1/13 Mar 5 SA 4f fst :48⁴ H 25/62 Feb 22 SA 3f fst :36⁴ H 11/18 Feb 3 SA 3f fst :36² H 7/28 Jan 28 SA 5f fst 1:00 H 13/80

Nucay
Own: Cubanacan Stables

PINCAY L JR (279 43 39 32 .15)

Ch. c. 3 (Jan)
Sire: Lord At War (General)
Dam: Must Ask (Naskra)
Br: Wimborne Farm Inc (Ky)
Tr: Marshall Robert W (25 3 3 4 .12)

BT
B

118

Lifetime Record :		0 M 0 0		$0	
1994	0 M 0 0		Turf	0 0 0 0	
1993	0 M 0 0		Wet	0 0 0 0	
SA	0 0 0 0		Dist	0 0 0 0	

WORKOUTS: Mar 16 Hol 5f fst 1:01 Hg 4/17 Mar 11 Hol 7f fst 1:28² H 1/3 Mar 5 Hol 6f fst 1:14² H 10/14 Feb 28 Hol 6f fst 1:13⁴ H 2/8 Feb 22 Hol 6f fst 1:13² H 2/18 Feb 16 Hol 5f fst 1:02¹ H 21/37
    Feb 12 Hol 5f fst 1:00 H 5/56 Feb 6 Hol 4f fst :49¹ H 10/34 Jan 26 Hol 4f fst :47⁴ H 4/19 Jan 21 Hol 3f fst :37¹ H 4/9 Jan 16 Hol 3f fst :38² H 9/10 Oct 16 Hol 4f fst :48 H 5/24

continued

Mint Chocolate　　C
Own: Stenger Mr & Mrs Richard

Dk. b or br c. 3 (Jan)
Sire: Key to the Mint (Graustark)
Dam: So Tenderlee (Forcten)
Br: Stenger Cally & Richard (Ky)
Tr: Fanning Jerry (62 8 7 10 .13)

VALENZUELA F H (308 32 19 30 .10)　　L 118

	Lifetime Record :		1 M 0 0		$800				
1994	1 M 0 0		$800	Turf	0 0 0 0
1993	0 M 0 0			Wet	0 0 0 0
SA	1 0 0 0		$800	Dist	0 0 0 0

6Mar94–2SA fst 6f　:22　:451　:572 1:10　Md Sp Wt　55 5 5 2½ 3² 3⁷ 5¹4½ Valenzuela F H LB 118 b 3.70 72–09 De Vito118⁴ Minero118⁶½ Hello Chicago118²½　Weakened 11

WORKOUTS: Mar 15 SA 6f fst 1:12 H 2/19 Mar 1 SA 5f fst 1:00¹ H 10/48 Feb 23 SA 6f fst 1:12⁴ H 4/26 Feb 13 SA 5f fst :59³ H 3/60 Feb 3 SA 5f fst 1:00² H 7/62 Jan 28 SA 5f fst :59 H 3/80

Le Dome (Ire)　LASIX　— A . for GRANDSIRE
Own: Lord White

Ch. c. 3 (May)
Sire: Salt Dome (Blushing Groom)
Dam: La Courant (Little Current)
Br: G N Clark & Airlie Stud (Ire)
Tr: Jones Gary (60 14 6 11 .23)

DESORMEAUX K J (269 59 47 31 .22)　　118

	Lifetime Record :		1 M 0 0		$0				
1993	1 M 0 0			Turf	1 0 0 0
1992	0 M 0 0			Wet	0 0 0 0
SA	0 0 0 0			Dist	0 0 0 0

15Oct93♦ Newmarket(GB) yl 6f ①Str 1:15　EBF Snailwell Maiden Stk(Div2)　12¹⁶ Cochrane R　126 14.00 Luana121³½ Ramani126⁶ Calypso Monarch126¹　17
Tr: Michael Stoute　Maiden 12900　　Slowly into stride, never a factor

WORKOUTS: •Mar 10 SA 6f fst 1:12¹ Hg 1/30 Mar 4 SA 6f fst 1:12² H 2/36 Feb 27 SA 6f fst 1:13¹ Hg 17/46 Feb 22 SA 5f fst 1:03 H 71/77 Feb 9 SA 6f gd 1:16³ Hg 3/8

Goldie Glenn　　BM　(57-?)
Own: Paulson Allen E

B. c. 3 (Feb)
Sire: Jade Hunter (Mr. Prospector)
Dam: Tangaroa (Tan Pronto) → chert?
Br: Paulson Allen E (Ky)
Tr: Hassinger Alex L Jr (14 2 3 1 .14)

GRYDER A T (187 7 14 25 .04)　　118

	Lifetime Record :		0 M 0 0		$0				
1994	0 M 0 0			Turf	0 0 0 0
1993	0 M 0 0			Wet	0 0 0 0
SA	0 0 0 0			Dist	0 0 0 0

WORKOUTS: Mar 14 SA 6f fst 1:13³ H 11/23 Mar 9 SA 6f fst 1:16 H 30/42 Mar 4 SA 6f fst 1:15² H 32/36 Feb 27 SA 5f fst 1:01³ H 35/44 Feb 22 SA 5f fst 1:00⁴ H 29/77 Feb 15 SA 5f fst 1:01³ H 23/33
Feb 10 SA 4f fst :50⁴ H 50/67 Jan 29 SA 3f fst :38¹ H 23/25 Jan 24 SA 4f fst :49³ H 27/39 Sep 19 SLR 4f fst :49⁴ H 2/2

Jade Master　　BM BT
Own: Lee China

Ch. c. 3 (May)
Sire: Jade Hunter (Mr. Prospector)
Dam: Mama Coca (Secretariat)
Br: Belford David (Ky)
Tr: Van Berg Jack C (166 24 36 19 .14)

SOLIS A (311 50 50 48 .16)　　L 118

	Lifetime Record :		3 M 0 0		$800				
1994	3 M 0 0		$800	Turf	0 0 0 0
1993	0 M 0 0			Wet	0 0 0 0
SA	3 0 0 0		$800	Dist	1 0 0 0

5Mar94–9SA fst 1　:22¹　:45³ 1:10　1:35²　Md Sp Wt　53 5 4 4¹½ 4³½ 6¹¹ 10²²½ Bailey J D LB 117 b 13.50 69–07 Marconi117⁴ Scenic Route117¹½ Collectable Wine117¹　Bid, gave way 10
12Feb94–9SA fst 6f　:21³　:442　:56³ 1:09²　Md Sp Wt　76 10 6 7⁴½ 65 69 57½ Atkinson P LB 118 b 21.90 82–11 Meadow Mischief118⁴ Gold Miner's Slew118² Wolf Bait118¹　Raced wide 11
28Jan94–3SA fst 6½f　:22　:45 1:09³ 1:16　Md c–50000　48 3 4 2ʰᵈ 2½ 67½ 6¹⁷ Gryder A T B 118 b 7.00 75–12 Sacrifice118³½ Island Sport118²½ Capo118¹½　Dueled, tired 6
Claimed from Suffolk Racing Partnership Ltd, Gregson Edwin Trainer

WORKOUTS: Mar 14 Hol 5f fst 1:00¹ H 4/22 Feb 28 Hol 7f fst 1:27² H 3/3 Feb 21 Hol 7f fst 1:26³ H 1/2 Feb 16 Hol 4f fst :47³ H 4/39 Feb 6 Hol 5f fst 1:01⁴ H 17/38 Jan 22 SA 4f fst :49¹ H 40/50

While Major Discovery's 90 Beyer is impressive, that was earned on a wet-fast track, not a tiring track, and Devon Dancer will most likely be pressing him early in this spot. Still, even though Major Discovery might be a pace casualty, he can't be eliminated. Devon Dancer, on the other hand, will have to fight for the lead on a surface he is bred to hate and is worth throwing out at a short price. Le Dome, on the other hand, is more intriguing. He is being bet, is a grandson of slop-loving Blushing Groom, out of a mud-loving Little Current mare, and might well move up on a tiring surface.

Of the rest of the experienced horses, Mint Chocolate — based on his first race and mediocre off-track breeding — is a toss out. Jade Master, however, is more interesting. A recent claim by cagey trainer Jack Van Berg, he is bred top and bottom to love an off-track and has Secretariat with his BT rating as his broodmare sire, which should move him up on a tiring track. Furthermore, Jade Master is shortening from a mile conditioning effort and gets Alex Solis in the irons. On the downside, the colt is dead on the board at 19-1.

Of the first-time starters, Star Doctor and Granby have F and D off-track ratings, aren't getting bet, and can be tossed out. Nucay, however, is worth further consideration. He is by Lord At War who has a BT rating and is out of a B-rated Naskra mare. Furthermore, trainer Robert Marshall is competent with first-time starters, Nucay has a nice seven-furlong work, and Laffit Pincay has agreed to ride. On the downside, Lord At War is only a mediocre debut sire and Nucay is dead on the board at 18-1. While Goldie Glenn is by the same sire as Jade Master, he's out of a mare by an unknown broodmare sire, is 59-1 on the board, and gets a cold rider for his debut.

So how to bet? Based on our analysis, there are four legitimate contenders; Major Discovery, Le Dome, Nucay and Jade Master. While Major Discovery might put Devon's Dancer away early in the mud, he isn't worth a Win bet at 2-1. Similarly, Le Dome, who has never raced on the dirt, let alone on a tiring off-track, is poor value at 5-2. This leaves us with Nucay and Jade Master as longshot possibilities. While Nucay's BT rating makes him an automatic Win bet, he can't be a prime bet because Lord At War is only a mediocre debut sire. Similarly, Jade Master might move up on this tiring surface because of Secretariat's BT influence. But, since he has already lost three races, he too is only worth a modest Win bet. Given their high odds, it is also worthwhile playing both Nucay and Jade Master behind Major Discovery and Le Dome in the exacta as a Place bet.

	SIXTH RACE		6½ FURLONGS. (1.14) MAIDEN SPECIAL WEIGHT. Purse $32,000. 3-year-olds. Weight, 118 lbs. (Non-starters for a claiming price of $32,000 or less in their last three starts preferred.)									

Santa Anita
MARCH 20, 1994

Value of Race: $32,000 Winner $17,600; second $6,400; third $4,800; fourth $2,400; fifth $800. Mutuel Pool $562,886.00 Exacta Pool $557,982.00 Quinella Pool $86,185.00

Last Raced	Horse	M/Eqt. A.Wt	PP	St	¼	½	Str	Fin	Jockey	Odds $1
5Mar94 9SA10	Jade Master	LBb 3 118	9	6	7⁴	5¹½	3½	1⁷	Solis A	19.00
	Nucay	B 3 118	5	3	3hd	4⁴	4²½	2³	Pincay L Jr	18.00
15Oct93 New12	Le Dome-Ir	LBb 3 118	7	5	6hd	8⁵	6hd	3¾	Desormeaux K J	2.40
5Feb94 4SA3	Devon Dancer	LB 3 118	4	4	1hd	2hd	1hd	4⁴	Stevens G L	2.70
20Feb94 6SA2	Major Discovery	Bb 3 113	1	1	2½	1hd	2½	5¹½	Gonzalez S Jr⁵	2.10
	Star Doctor	B 3 118	2	8	8²	7hd	7⁴	6²	Flores D R	23.80
6Mar94 2SA5	Mint Chocolate	LBb 3 118	6	2	4⁸	3⁴	5⁴	7²½	Valenzuela F H	8.50
	Granby	B 3 118	3	7	5¹	6¹	8⁴	8hd	McCarron C J	20.10
	Goldie Glenn	B 3 118	8	9	9	9	9	9	Gryder A T	59.00

OFF AT 3:07 Start Good. Won driving. Time, :22, :45⁴, 1:12, 1:18⁴ Track muddy.

$2 Mutuel Prices:

9-JADE MASTER	40.00	14.80	8.60
5-NUCAY		16.00	9.00
7-LE DOME-IR			4.60

$2 EXACTA 9-5 PAID $556.00 $2 QUINELLA 5-9 PAID $266.00

Ch. c, (May), by Jade Hunter–Mama Coca, by Secretariat. Trainer Van Berg Jack C. Bred by Belford David (Ky).

© 1994 DRF, Inc.

Major Discovery and Devon Dancer did indeed get hooked in a speed duel and tired in a race that was won in a slow 1:18.4. Both Jade Master and Nucay came from off the pace to complete the exacta (which I didn't have) for over five hundred dollars. Still, this was a difficult race to more aggressively bet because of all the conflicting information about the contenders. Furthermore, because both Nucay and Jade Master raced with mud caulks, while none of the bet-down horses had them, the addition of caulks might have been an even more important factor in the outcome of the race than the two horses' pedigrees. It appears as if good mud breeding *plus* mud caulks is a potent combination.

Santa Anita Park

4

1 1/16 MILES. **(1:39) 57th Running of THE SAN FELIPE STAKES. Purse $200,000 Added (Grade II). 3-year-olds (Foals of 1991). By subscription of $100 each to accompany the nomination, $1,500 additional to start, with $200,000 added, of which $40,000 to second, $30,000 to third, $15,000 to fourth and $5,000 to fifth. Weight, 122 lbs. Non-winners of $60,000 twice at any time or $70,000 once since December 25 at one mile or over, allowed 3 lbs; Of $50,000 any distance, 6 lbs. Starters to be named through the entry box by the closing time of entries. A trophy will be presented to the owner of the winner. Closed Wednesday, March 9, with 10 nominations.**

1 1/16 MILES
START ▲ ▲ FINISH

Pollock's Luck

Own: Lucky Me Stable

MCCARRON C J (178 31 30 27 .17)

Dk. b or br c. 3 (May)
Sire: Polish Navy (Danzig)
Dam: Lucky Us (Nijinsky II)
Br: Virginia Knott Bender (Ky)
Tr: Cross Richard J (13 3 1 1 .23)

116

Lifetime Record:	3 2 0 1	$42,400			
1994	2 1 0 1	$27,550	Turf	0 0 0 0	
1993	1 1 0 0	$14,850	Wet	0 0 0 0	
SA	2 1 0 1	$27,550	Dist	1 1 0 0	$22,000

30Jan94-3SA	fst	1 1/16	:234 :474 1:114 1:414	Alw 40000N1X	95 4 1 1½ 2hd 1hd 1no	Stevens G L	B 115	1.60	86-12	Pollock'sLuck115no Numrous1166½ CrowningDcison1183½	Brushed, gamely	5
9Jan94-3SA	fst	6½f	:213 :441 1:084 1:152	Alw 37000N1X	91 4 2 23 32 33 32¼	McCarron C J	B 120	2.70	93-08	Soul Of The Matter1201¼ Rotsaluck1171 Pollock's Luck1203½	Best of rest	6
20Dec93-1Hol	fst	6f	:221 :444 :564 1:09	Md Sp Wt	90 4 4 31 1hd 11 11¾	McCarron C J	B 118	2.50	95-09	Pollock's Luck1181¾ Superpak1184 Native Blast118½	Driving	8

WORKOUTS: ●Mar 17 SA 3f fst :334 H *1/20* Mar 12 SA 1 fst 1:373 H *1/2* Mar 7 SA 1 fst 1:411 H *2/6* Mar 1 SA 6f fst 1:141 H *11/16* Feb 24 SA 5f fst 1:02 H *20/31* Feb 16 SA 4f fst :482 B *25/71*

Numerous

Own: Keck Howard B

VALENZUELA P A (231 24 24 41 .10)

B. c. 3 (Mar)
Sire: Mr. Prospector (Raise a Native)
Dam: Number (Nijinsky II)
Br: Claiborne Farm & The Gamely Corp (Ky)
Tr: Whittingham Charles (82 7 11 10 .09)

116

Lifetime Record:	4 1 2 0	$40,575			
1994	2 0 2 0	$23,000	Turf	0 0 0 0	
1993	2 1 0 0	$17,575	Wet	0 0 0 0	
SA	3 1 2 0	$38,400	Dist	1 0 1 0	$8,000

Entered 19Mar94- 3 SA

16Feb94-8SA	fst	1⅛	:461 1:102 1:352 1:482	®Bradbury75k	97 4 5 54 33 25 22	McCarron C J	B 115	*.70	89-17	Dramatic Gold1152 Numerous1154½ Almaraz1168	Too late	6
30Jan94-3SA	fst	1 1/16	:234 :474 1:114 1:414	Alw 40000N1X	95 5 2 2½ 31 2hd 2no	Delahoussaye E	B 116	*1.00	86-12	Pollock's Luck115no Numerous1166½ Crowning Decision1183½		5
Lugged in late, brushed												
13Nov93-6SA	fst	6½f	:222 :452 1:093 1:154	Md Sp Wt	83 6 4 51¼ 42¼ 31 1no	McCarron C J	B 118	*1.20	92-11	Numerous118no Amrisingstr1182¼ CrowningDecision1182¾	4 Wide into lane	7
11Sep93-6Dmr	fst	6f	:214 :45 :573 1:102	Md Sp Wt	60 8 11 111¼ 98¾ 65 45½	McCarron C J	B 118	8.20	82-12	Daggett Peak1183 Danny The Demon118hd River Flyer1182¼		11
Off slowly, wide, rush far turn, greenly lane												

WORKOUTS: ●Mar 15 SA 5f fst :581 H *1/59* ●Mar 9 SA 7f fst 1:25 H *1/8* Mar 3 SA 5f fst 1:003 H *8/41* Feb 25 SA 3f fst :36 H *5/19* Feb 11 SA 5f fst 1:001 H *16/62* Jan 28 SA 3f fst :35 B *3/28*

Brocco

Own: Broccoli Mr & Mrs Albert

STEVENS G L (331 47 65 57 .14)

Ch. c. 3 (Mar)
Sire: Kris S. (Roberto)
Dam: Anytime Ms. (Aurelius II)
Br: Meadowbrook Farms Inc (Fla)
Tr: Winick Randy (2 0 0 0 .00)

L 119

Lifetime Record:	4 3 1 0	$653,550			
1993	4 3 1 0	$653,550	Turf	0 0 0 0	
1992	0 M 0 0		Wet	0 0 0 0	
SA	2 2 0 0	$537,600	Dist	2 1 1 0	$620,000

19Dec93-8Hol	fst	1 1/16	:231 :461 1:093 1:403	Hol Futy-G1	105 5 3 43½ 42 22 2¾	Stevens G L	LB 121	*.40	97-12	Valiant Nature121¾ Brocco1212½ Flying Sensation121½	Lacked room 1/2	6
6Nov93-6SA	fst	1 1/16	:223 :462 1:104 1:424	B C Juvnle-G1	97 5 5 56 53¾ 1hd 15	Stevens G L	LB 122	3.00	90-10	Brocco1225 Blumin Affair1221½ Tabasco Cat1223	4-Wide 2nd turn	11
7Oct93-7SA	fst	7f	:224 :454 1:102 1:221	Alw 32000N1X	89 4 1 22 2hd 11½ 18½	Stevens G L	LB 118	*.60	92-13	Brocco1188½ Delineator1181½ Our Blue Michael1155	Crushed foes	6
28Aug93-2Dmr	fst	6½f	:221 :453 1:102 1:164	Md Sp Wt	86 6 2 3½ 4¾ 1hd 13½	Stevens G L	LB 117	*1.00	87-12	Brocco1173½ Sir Harry Bright1171¾ Dawson Ridge117½	Wide, easily	8

WORKOUTS: Mar 18 SA 3f fst :364 B *13/23* ●Mar 13 SA 7f fst 1:251 H *1/10* ●Mar 7 SA 1 fst 1:391 H *1/6* ●Feb 28 SA 7f fst 1:244 H *1/7* Feb 22 SA 6f fst 1:143 H *33/44* Feb 16 SA 5f fst 1:01 B *36/76*

Gracious Ghost

Own: Whispering Woods Farm

SOLIS A (311 50 50 48 .16)

Gr. c. 3 (Mar)
Sire: Relaunch (In Reality)
Dam: Ambo (Olden Times)
Br: Hunter Barbara (Ky)
Tr: Van Berg Jack C (166 24 36 19 .14)

L 116

Lifetime Record:	11 1 5 0	$107,625			
1994	4 0 3 0	$56,875	Turf	1 0 0 0	$6,250
1993	7 1 2 0	$50,750	Wet	0 0 0 0	
SA	5 0 3 0	$44,575	Dist	2 0 1 0	$27,500

Entered 19Mar94- 8 GG

2Mar94-8SA	fst	6f	:212 :442 :563 1:091	®Blsa Chca75k	72 1 5 33 31½ 32½ 57½	Desormeaux K J	LB 116	*1.20	83-13	King's Blade1165 Troyalty122no Subtle Trouble1192	Lacked room 3/16	5
13Feb94-6SA	fst	7f	:22 :444 1:091 1:221	S Vicente BCG3	91 1 3 54½ 31 21½ 22	Desormeaux K J	LB 114	5.50	89-13	Fly'n J. Bryan1142 Gracious Ghost1142¼ Cois Na Tine1161	2nd best	6
22Jan94-3SA	fst	1 1/16	:234 :473 1:11 1:414	®S Catalina75k	90 4 2 21 41½ 32½ 24	Delahoussaye E	LB 116	7.80	82-12	Wekiva Springs1214 Gracious Ghost116¾ Dream Trapp1171¼	No match	5
1Jan94-8BM	fst	1	:23 :463 1:112 1:362	Cal Juvnl-G3	79 1 4 32 41½ 32 24½	Meza R Q	LB 113	3.40	86-18	Robannier1154½ Gracious Ghost1132½ Delineator1181	Steadied 2nd turn	5
19Dec93-8Hol	fst	1 1/16	:231 :461 1:093 1:403	Hol Futy-G1	84 3 2 23 31½ 56 512¼	Desormeaux K J	LB 121	8.20	86-12	Valiant Nature121¾ Brocco1212½ Flying Sensation121½	Gave way	6
27Nov93-8Hol	fm	1 Ⓣ	:23 :464 1:101 1:343	Generous-G3	80 6 8 76½ 76¾ 55 53	Delahoussaye E	LB 114	6.50	89-10	Delineator1181¾ Devon Port116hd Ferrara114no	Late gain	8
20Nov93-2Hol	fst	7f	:211 :433 1:083 1:213	Md Sp Wt	85 5 2 47 47 2½ 1hd	Desormeaux K J	LB 118	2.50	96-04	Gracious Ghost118hd Devil's Mirage1187 Diamond Ball1183½	Driving	7
30Oct93-6SA	fst	6½f	:212 :444 1:094 1:161	Md Sp Wt	63 1 3 43½ 42 55½ 410¼	Almeida G F	B 117	2.10	79-12	Soul Of The Matter1174 Al's River Cat1173½ Elaine'sLove1173	Weakened	7
10Oct93-9SA	fst	6f	:214 :443 :57 1:094	Md Sp Wt	84 3 4 41½ 41½ 21½ 2hd	Almeida G F	B 117	*2.40	88-16	Sky Kid117hd Gracious Ghost1175 Al's River Cat1171¼	Too late	11
26Jly93-8Hol	fst	6f	:213 :443 :57 1:10	Hol Juv ChmpG2	68 5 8 63¼ 52¼ 56 57	Almeida G F	B 115	16.00	83-11	Ramblin Guy1172 Swift Walker1171¾ Individual Style1172¾		8
Awkward start, steadied 1/4												

WORKOUTS: Feb 11 Hol 4f fst :473 H *5/32* ●Jan 19 Hol 5f fst 1:011 H *1/10* Jan 13 Hol 5f fst 1:011 H *10/32*

Soul Of The Matter
Own: Bacharach Burt
DESORMEAUX K J (269 59 47 31 .22)

Dk. b or br c. 3 (Apr)
Sire: Private Terms (Private Account)
Dam: Soul Light (T. V. Commercial)
Br: Blue Seas Music, Inc. (WV)
Tr: Mandella Richard (66 11 13 12 .17)

116

Lifetime Record:	4 2 0 2	$59,750			
1994	2 1 0 1	$29,350	Turf	1 0 0 1	$9,000
1993	2 1 0 1	$30,400	Wet	0 0 0 0	
SA	2 2 0 0	$35,750	Dist	0 0 0 0	

5Mar94- 1SA fm 1 Ⓣ :22^{2} :46 1:10^{3} 1:35^{1} Pirate Cove60k 85 4 6 76½ 65 54 32 Desormeaux K J B 115 2.70 87-07 Eagle Eyed113¾ Majestic Style1141¼ Soul Of The Matter115^{2} 8
Lacked room 2nd turn & 1/8
9Jan94- 3SA fst 6½f :21^{3} :44^{1} 1:08^{4} 1:15^{2} Alw 37000N1X 96 3 4 54¾ 44 22½ 11¼ Desormeaux K J B 120 *.50 95-08 Soul OfTheMatter1201¼ Rotsaluck117^{1} Pollock'sLuck1203½ Drifted in 1/16 6
26Nov93-11Hol fst 7f :22 :44^{3} 1:08^{1} 1:21 H Prevue BC-G3 96 2 5 41½ 41½ 21½ 31½ McCarron C J B 115 2.00 97-09 Individual Style1211¼ Egayant117^{nk} Soul Of The Matter1151¼ 6
Drifted in, checked 1/8
30Oct93- 6SA fst 6½f :21^{2} :44^{4} 1:09^{4} 1:16^{1} Md Sp Wt 88 6 5 54 31 2^{hd} 14 Desormeaux K J B 117 3.60 90-12 SoulOfTheMatter117^{4} Al'sRiverCat1173½ Elaine'sLove117^{3} Clear, driving 7
WORKOUTS: Mar 17 SA Ⓣ 4f fm :49^{2} H (d) *1/2* Mar 4 SA Ⓣ 3f fm :37^{1} H (d) *2/2* Feb 27 SA Ⓣ 6f fm 1:17 H (d) *1/1* Feb 23 SA 5f gd 1:03^{3} B (d) *1/1* Feb 6 SA 1 fst 1:41^{3} H *5/6* Jan 31 SA 1 fst 1:40 H *3/3*

Valiant Nature
Own: V H W Stables
PINCAY L JR (279 43 39 32 .15)

B. c. 3 (Mar)
Sire: His Majesty (Ribot)
Dam: Premium Win (Lyphard)
Br: Winchell Verne H (Ky)
Tr: McAnally Ronald (106 17 13 13 .16)

119

Lifetime Record:	3 2 0 0	$291,500			
1993	3 2 0 0	$291,500	Turf	1 1 0 0	$16,500
1992	0 M 0 0		Wet	0 0 0 0	
SA	1 0 0 0		Dist	1 1 0 0	$275,000

19Dec93- 8Hol fst 1$\frac{1}{16}$:23^{1} :46^{1} 1:09^{3} 1:40^{3} Hol Futy-G1 106 2 1 13 11½ 12 1¾ Pincay L Jr B 121 16.00 98-12 Valiant Nature121¾ Brocco1212½ Flying Sensation121½ Speed, held 6
17Nov93- 4Hol fm 1 Ⓣ :46^{3} 1:11^{1} 1:36^{1} Md Sp Wt 78 4 1 1^{hd} 11½ 11½ 13½ Pincay L Jr B 118 *2.50 84-14 Valiant Nature1183½ Saltgrass1183½ Unrivaled118^{1} Ducked in 1/16 9
Quarter time unavailable
10Oct93- 9SA fst 6f :21^{4} :44^{3} :57 1:09^{4} Md Sp Wt 47 6 7 64½ 87¼ 89 1014¼ Pincay L Jr B 117 13.20 74-16 Sky Kid117^{hd} Gracious Ghost117^{5} Al's River Cat1171¼ Lugged out late 11
WORKOUTS: ●Mar 18 SA 4f fst :46^{1} B *1/41* ●Mar 13 SA 1 fst 1:37^{3} H *1/5* Mar 7 SA Ⓣ 1 fm 1:43^{3} H (d) *1/1* Feb 28 SA Ⓣ 7f fm 1:27^{4} H *1/1* Feb 22 SA 7f fst 1:26^{4} B *13/16* Feb 16 SA Ⓣ 7f fm 1:31^{4} H (d) *4/4*

The fourth race at Santa Anita on March 20th run on a Tiring surface is a more formful example of a horse running to his OT breeding. Looking at the field, Pollock's Luck is coming off a two month layoff, is bred to hate an off-track and can be thrown out. While Valiant Nature beat Broco in December, he has been off for three months and is not bred, especially on his bottom side, to like an off-track. Though Brocco also hasn't raced since losing to Valiant Nature, he won off works as a first-time starter and has decent C+ off-track breeding. Gracious Ghost has recency and superior sloppy off-track breeding in his favor, but he has already been trounced by Valiant Nature and Brocco and is unlikely to move up enough on a tiring track to defeat either of those two rivals, in spite of the mud caulks he is getting.

This leaves us with Soul Of The Matter who, at 6-1, has several things in his favor: first off, he is by Private Terms whose progeny, like other sons of Private Account, do well on tiring tracks; secondly, he is the only horse in the race with a recent route prep; and thirdly, his 2-year-old Beyers compare quite favorably with those of Valiant Nature and Brocco.

So how to bet? Given the likely speed duel between Polluck's Luck and Valiant Nature, both of whom are also unlikely to relish the off-going, first Brocco and then Soul Of The Matter are likely to pick up the pieces. A prime Win bet on Soul Of The Matter is in order, as well as keying him in an exacta box with Brocco, the only other horse that might like the tiring surface.

172

FOURTH RACE

Santa Anita
MARCH 20, 1994

1¹⁄₁₆ MILES. (1.39) 57th Running of THE SAN FELIPE STAKES. Purse $200,000 Added (Grade II). 3–year–olds (Foals of 1991). By subscription of $100 each to accompany the nomination, $1,500 additional to start, with $200,000 added, of which $40,000 to second, $30,000 to third, $15,000 to fourth and $5,000 to fifth. Weight, 122 lbs. Non–winners of $60,000 twice at any time or $70,000 once since December 25 at one mile or over, allowed 3 lbs; Of $50,000 any distance, 6 lbs. Starters to be named through the entry box by the closing time of entries. A trophy will be presented to the owner of the winner. Closed Wednesday, March 9, with 10 nominations.

Value of Race: $208,500 Winner $118,500; second $40,000; third $30,000; fourth $15,000; fifth $5,000. Mutuel Pool $621,885.00 Exacta Pool $573,865.00 Quinella Pool $71,448.00 Minus Show Pool $108.17

Last Raced	Horse	M/Eqt.	A.Wt	PP	St	¼	½	¾	Str	Fin	Jockey	Odds $1
5Mar94 ¹SA³	Soul Of The Matter	B	3 116	4	5	5	5	5	37	12¼	Desormeaux K J	6.30
19Dec93 ⁸Hol²	Brocco	LB	3 119	2	2	3²	2hd	25	1hd	2hd	Stevens G L	0.90
19Dec93 ⁸Hol¹	Valiant Nature	B	3 119	5	4	1¹½	12	11	22	316	Pincay L Jr	1.30
2Mar94 ⁸SA⁵	Gracious Ghost	LB	3 116	3	3	4³	44½	41½	45	48	Solis A	26.40
30Jan94 ³SA¹	Pollock's Luck	B	3 116	1	1	2¹	33½	3½	5	5	McCarron C J	12.40

OFF AT 2:02 Start Good. Won driving. Time, :23⁴, :47³, 1:12¹, 1:37⁴, 1:44³ Track muddy.

$2 Mutuel Prices:

5–SOUL OF THE MATTER	14.60	3.60	2.20
3–BROCCO		2.60	2.10
6–VALIANT NATURE			2.10

$2 EXACTA 5–3 PAID $32.20 $2 QUINELLA 3–5 PAID $11.20

Dk. b. or br. c, (Apr), by Private Terms–Soul Light, by T. V. Commercial. Trainer Mandella Richard. Bred by Blue Seas Music, Inc. (WV).

While Soul Of The Matter ran to his T rating, it is important to once again emphasize that he also had good current form, a beneficial pace duel, and competitive Beyers going for him. Without these other handicapping factors, his off-track breeding might not have been enough to get him to the winners circle.

Handicappers should primarily use the OT ratings in *Exploring Pedigree* to supplement other insights they have about races run on sloppy or tiring surfaces. In many instances, these ratings will be more useful in toning down a handicapper's enthusiasm for a horse bred to hate an off-track, than in identifying the likely winner.

Epilogue

By now the pari-mutuel rewards of creatively applying a knowledge of pedigree to the handicapping process should be obvious. Horseplayers tired of trying to grind out a modest profit with more conventional approaches will also find themselves intellectually refreshed by the challenge of becoming more complete horseplayers and going beyond dependence on past performance lines.

This is not to say, of course, that pedigree handicapping is some kind of panacea. It isn't. Without respect for speed and pace analysis, a keen sense of value and the ability to pass unplayable races, the profits that come from intelligent pedigree play will be quickly depleted.

Even more, during the writing of this book, I was once again reminded of the tendency for things set in print tend to take on a life and authority of their own. While the great majority of the sire ratings that I started out with at the beginning of 1994 remain unchanged and are reliable projectors of how a young horse is likely to perform when trying something new, some of the ratings have been adjusted in the light of a keener insight into the dynamics of how specific sires are being bred. This is especially true of the precocity ratings for some of the younger stallions, as well as the stamina indexes for a few of the more established sires, both of which have been further tested and refined. Hopefully, readers will not be unduly confused by the fact that some of the sire ratings used in the analysis of various races in the text have subsequently been adjusted in the back of the book.

Rather than foster the illusion of infallibility with regard to the sire ratings, I want to stress again that breeding is an ongoing dynamic process, variously influenced by both nature's whimsy and human intervention. Instead of accepting even the adjusted sire ratings in some rote-like fashion, handicappers are encouraged to ask themselves what a breeder might have had in mind when he or she decided on the particular mating which produced the horse in front of them that they are now considering betting. The more a handicapper learns to think like a breeder, the more effective the utility of the ratings will become.

174   Even further, as handicappers use the ratings, they are likely to discover local track biases that move some sire's ratings up and others down. The more a handicapper sees him or herself as a participant in the creation of these ratings, the more effective his handicapping and decision-making is likely to become.

Beyond the prospect of a big score, what is particularly exciting about the broader dissemination of pedigree knowledge is that as horseplayers with successful businesses become horse owners and breeders they will be in a position to influence the future of the breed.

For too long have horseplayers been condescended to by commercial breeders and racetrack operators. With a deeper understanding of the foundations of the breed, it will be interesting to see if horseplayers will demand a greater variety in the kinds of horses that are bred, as well as the types of races that are carded.

A horseplayer knowledgeable enough about stamina to cash a ticket on a stretchout, or aware of who the good turf sires are, isn't likely, when he or she becomes an owner, to be satisfied with just breeding for speed and precocity. My guess is that a number of them will also want to breed for greater soundness and stamina and to elevate the general tenor of the game. In an age when the degenerate gambler, the fixed race and the drugged horse are the most accepted popular images of our sport, the opportunity exists for informed handicappers to become major players in restoring the Thoroughbred's noble legacy.

Stallion	Sire/Broodmare Sire	FTS	CL	Trf	OT	SI	AB
Abri Fiscal*	The Minstrel/King Emperor	M	8	M	U	5	2-3
Academy Award^	Secretariat/Mr. Prospector	M	7	M	M	3*	3-4
Acallade*	Mr. Prospector/Sir Ivor	F	9	U	U	4	<u>3</u>-4
Acaroid	Big Spruce/Intentionally	D	7	B-	C	2	3-4
Accused	Alleged/Admiral's Voyage	C	9	F	C	4	<u>3</u>-4
Ack Ack	Battle Joined/Turn-to	D	3	C	C	3*	3-4
Adbass*	Northern Dancer/What A Pleasure	M	8	M	M	4	3-4
Admiral's Flag	Raise a Native/Good Counsel	M	8	U	U	3	3-4
Admiral's Shield	Crozier/Crafty Admiral	D	8	U	F	2	<u>3</u>-4
Advocator	Round Table/Bull Lea	U	6	C	M	<u>4</u>	3-4
Advocatum	Advocator/Bold Bidder	F	9	F	D	4	3-4
Aegean's Bolger	Bolger/Aegean Isle	M	9	U	M	4	3-4
Affirmed	Exclusive Native/Crafty Admiral	C	3*	C	C	2	3-4
Afleet*	Mr. Prospector/Venetian Jester	D	4	C	C	4	2-3
African Man	African Sky/Mandamus	D	8	U	F	4	3-4
African Sky	Sing Sing/Nimbus	C	8	B	D	4	3-4
Again Tomorrow*	Honest Pleasure/Northern Dancer	F	9	U	F	5	3-4
Age Quod Agis	Al Hattab/Pavot	D	9	F	F	4	3-4
Aggravatin'	Silent Screen/Damascus	C	8	F	F	3	<u>3</u>-4
Agitate	Advocator/Swaps	C	8	C	F	5	<u>3</u>-4
Ahonoora	Lorenzaccio/Martial	C	2	B	C	3*	3-4
Air Forbes Won <u>c</u>	Bold Forbes/Tobin Bronze	C	5	C	C+	4	3-4
Akarad	Labus/Abdos	C	5	C		1	<u>3</u>-4
Al Hattab	The Axe II/Abernant	C	3	C	T	1	<u>3</u>-4
Alleged	Hoist The Flag/Prince John	C+	4	C	C	<u>1</u>	<u>3</u>-4
Allen's Prospect	Mr. Prospector/Swaps	B	6	F	C	4	2-3
All For Fun	One For All/Galoot	C	8	B-	F	<u>4</u>	3-4
All Kings	King's Bishop/Dead Ahead	C	8	C	C	5	3-4
Ally Runner*	Alydar/Round Table	U	7	U	L	4	3-4
Al Mamoon*	Believe It/Secretariat	B	7	D	C+	4	3-4
Al Nasr	Lyphard/Caro	D	6	C	D	3*	3-4
Aloha Prospector^	Native Prospector/Hawaii	U	8	M	U	5	3-4
Aloma's Ruler	Iron Ruler/Native Charger	C	7	D	C	4	3-4
Alphabatim	Verbatim/Grey Dawn II	C	8	D	F	2	3-4
Alwasmi*	Northern Dancer/Bustino	D	7	M	T	3	<u>2</u>-3
Always Gallant	Gallant Romeo/Sadair	C	8	F	C	4	3-4
Always Run Lucky	What Luck/Delta Judge	D	8	F	D	4	3-4
Alwuhush^	Nureyev/King Emperor	M	5	L	U	1	3-4

Stallion	Sire/Broodmare Sire	FTS	CL	Trf	OT	SI	AB
Alybrave*	Alydar/Bold and Brave	U	9	U	U	4	3-4
Alydar	Raise a Native/On and On	D	1	C	B	2	3-4
Aly Dark	Alydar/Jig Time	C+	9	B-	A	4	<u>3-4</u>
Alydar's Prophecy*	Alydar/Cyane	U	6	U	L	4	3-4
Aly Rat*	Alydar/Best Turn	U	9	U	M	4	3-4
Alysheba*	Alydar/ Lt. Stevens	F	5	C	F	1	3-4
Alzao	Lyphard/Sir Ivor	C	5	C	D	1*	3-4
Am All Charged Up^	Native Charger/	M	7	M	M	4	2-3
Amazing Prospect	Fappiano/Lucky Debonair	F	8	F	D	5	3-4
Amber Pass	Pass Catcher/Somerset	C	8	F	F	4	<u>3-4</u>
American Standard	In Reality/Bald Eagle	C	6	D	C		2-3
Amorelu^	Conquistador Cielo/Nijinsky II	M	8	M	M	4	3-4
A Native Danzig	Danzig/Raise a Native	C	8	L	F	4	<u>3-4</u>
Ankara	Northern Dancer/Forli	C	7	C	D	4	3-4
Another Reef*	Plum Bold/Jig Time	C	7	L	C	4	2-3
Apalachee <u>c</u>	Round Table/Nantallah	C	6	C	C	4	3-4
April Axe	The Axe II/Bold Lad	D	10	<u>B</u>	F	2	<u>3-4</u>
Arabian Sheik^	Nijinsky II/Damascus	M	9	M	M	4	3-4
Aras an Uachtarain*	Habitat/Nijinsky II	B	10	F	B	5	<u>3-4</u>
Arctic Tern <u>c</u>	Sea Bird/Hasty Road	D	6	C	F	1*	3-4
Ariva	Riva Ridge/Bold Ruler	B	8	U	F	4	<u>2-3</u>
Artichoke	Jacinto/Hill Prince	F	7	C	F	3	3-4
Arts and Letters	Ribot/Battlefield	F	6	D	D	4	<u>3-4</u>
A Run*	Empery/Jacinto	M	9	U	U	4	3-4
Ascot Knight	Danzig/Better Bee	D	6	F	F	4*	2-3
Ask Me	Ack Ack/Tom Rolfe	C+	8	D	F	4	<u>3-4</u>
Assault Landing	Buckfinder/Solo Landing	F	8	F	D	4	2-3
Assert <u>f</u>	Be My Guest/Sea Bird	D	6	C	D	<u>1</u>	<u>3-4</u>
A Title*	Sir Ivor/Swaps	D	9	U	F	4	<u>3-4</u>
Atmosphere*	Nijinsky II/Round Table	F	9	M	F	5	<u>3-4</u>
A Toast to Junius	Raise Your Glass/Prince John	C	8	U	F	5	2-3
At the Threshold <u>c</u>	Norcliffe/Vertex	C	6	C	C	2	2-3
Au Point	Lyphard/Princequillo	C	8	C	C	<u>4</u>	<u>3-4</u>
Aurium*	Mr. Prospector/Creme dela Creme	U	9	U	M	4	3-4
Avatar <u>c</u>	Graustark/Mount Marcy	F	7	D	C	<u>4</u>	3-4
Avenger M.	Staunch Avenger/Gallant Native	C+	9	F	F	5	3-4
Avenue of Flags^	Seattle Slew/Pass the Glass	M	6	M	L	3	3-4
Avie's Copy*	Lord Avie/Iron Ruler	U	8	U	M	2	3-4
Avodire	Nijinsky II/Princequillo	D	8	C	C	1	3-4
Aye's Turn	Best Turn/Mito	C	8	F	C	4	2-3

Stallion	Sire/Broodmare Sire	FTS	CL	Trf	OT	SI	AB
Azzardo*	Raise a Native/Tom Rolfe	C	8	U	C+	5	3-4
Baby Slew*	Seattle Slew/Exclusive Native	M	9	L	L	4	2-3
Badger Land	Codex/Racing Room	D	5	D	C+	2	2-3
Baederwood	Tentam/Northern Dancer	C	6	C	C+	4	2-3
Bailjumper	Damascus/Royal Vale	D	8	C	F	4	3-4
Baillamont	Blushing Groom/Shoemaker	C	3	B	L	I	3-4
Bairn	Northern Baby/Sir Ivor	C	7	D	U	I	3-4
Baldski c	Nijinsky II/Bald Eagle	C	4*	C	F	4*	2-3
Ballydoyle c	Northern Dancer/New Providence	F	8	F	F	4	3-4
Balzac	Buckpasser/Double Jay	D	7	C	C	3	3-4
Banker's Special^	Hawkin's Special/Space Commander	U	9	U	M	5	3-4
Banner Bob	Herculean/Bolinas Boy	F	7	F	C	4	3-4
Banquet Table	Round Table/Barbizon	C	8	C	C	3	3-4
Barbaric Spirit	Barbizon/Fleet Nasrullah	D	8	F	F	4	3-4
Barberstown	Gummo/Bolinas Boy	C	8	B	F	4	3-4
Bargain Day	Prove it/Toulouse Lautrec	C	7	B	T	4	3-4
Baron O'Dublin	Irish Ruler/Gyro	C	8	B	C	4	3-4
Barrera	Raise a Native/Chieftain	C	7	C	C	4	3-4
Basic Rate^	Valdez/Foolish Pleasure	M	9	M	M	5	2-3
Basket Weave*	Best Turn/Buckpasser	C	8	U	M	4	2-3
Bates Motel	Sir Ivor/TV Lark	D	5	C	C	2	3-4
Batonnier	His Majesty/Dumpty Humpty	C	6	B-	C	3	3-4
Battle Call	Native Charger/The Axe II	D	8	B-	F	4	3-4
Battle Launch*	Relaunch/Battle Joined	M	7	U	L.	4	2-3
Bayou Hebert	Hoist the Flag/Damascus	C+	7	D	C+	4	2-3
Be a Native	Exclusive Native/Cavan	D	8	C	F	4	3-4
Be a Prospect c	New Prospect/Cyane	B	8	D	A	4	2-3
Bear Branch^	Alydar/Forli	U	7	M	M	3	3-4
Be a Rullah	Raise a Native/Hail to Reason	D	8	D	F	4	3-4
Beat Inflation	Crozier/Quibu	C	6	F	C	5	2-3
Beaudelaire	Nijinsky II/Habitat	F	7	D	F	3	3-4
Beau Genius^	Bold Ruckus/Viceregal	L	6	M	M	2	2-3
Beau Groton	Groton/Beau Purple	B	7	D	F	4	3-4
Beau's Eagle	Golden Eagle/Maribeau	B	6	B	F	5	3-4
Bel Bolide	Bold Bidder/Graustark	C	7	A-	C	4	3-4
Believe It c	In Reality/Buckpasser	D	6	C	C	4	3-4
Believe the Queen	Believe It/Raise a Native	D	7	D	D	4	3-4
Belted Earl	Damascus/Nantallah	B+	8	F	C	4	3-4
Be My Guest	Northern Dancer/Tudor Minstrel	D	6	C	F	3*	3-4
Benalidar	Alydar/Maribeau	U	10	U	M	3	3-4

Stallion	Sire/Broodmare Sire	FTS	CL	Trf	OT	SI	AB
Bering	Arctic Tern/Lyphard	C+	6	B	U	1*	3-4
Best Native	Exclusive Native/Get Around	A	8	F	C	5	2-3
Best Turn	Turn-to/Swaps	C	3	D	B	2	3-4
Bet Big	Distinctive/Majestic Prince	D	8	D	D	4	2-3
Better Arbitor	Better Bee/Nearctic	C	7	F	D	4	3-4
Bet Twice	Sportin' Life/Dusty Canyon	C	7	C	C	3	2-3
Beyond the Mint*	Key to the Mint/Nijinsky II	U	7	M	M	1	3-4
Bicker	Round Table/Court Martial	D	8	C	D	4	3-4
Big Bold Sefa	Bold Reasoning/Dark Star	A	7	F	C	5	3-4
Big Burn	Never Bend/Sailor	D	6	C	C	5	2-3
Big Chill*	It's Freezing/Cornish Prince	M	8	U	M	4	3-4
Big John Taylor	Speak John/Hasty Road	C	7	F	F	4	3-4
Big Kohinoor	Irish Ruler/Crozier	C+	8	B-	C	5	3-4
Big Leaguer*	Bold Bidder/Greek Ship	D	10	L	F	5	3-4
Big Mukora^	Mr Prospector/Rambunctious	M	8	M	U	5	3-4
Big Pistol	Romeo/Whitesburg	C	7	U	B	4	3-4
Big Presentation	In Reality/Tatan	C	7	D	C	5	3-4
Big Spruce	Herbager/Prince John	D	6	C	C	1	3-4
Big Stanley^	Distinctive Pro/Grey Dawn	L	6	M	L	2	2-3
Bimini Captain	Hail to Reason/My Babu	C	9	C	F	4	2-3
Bionic Light	Majestic Light/Best turn	D	8	D	F	4	3-4
Black Mackee	Captain Courageous/Six Fifteen	B	8	A-	C	4	2-3
Black Prospector^	Mr. Prospector/Distinctive	M	9	U	L	4	2-3
Black Tie Affair^	Miswaki/Al Hattab	M	4	U	L	4*	2-3
Blade	Bold Ruler/Princequillo	C+	6	D	C	4	2-3
Blazing Bart^	The Bart/Prove Out	U	9	M	M	2	3-4
Blazing Ryder*	Red Ryder/Tampa Trouble	B	8	A-	F	5	3-4
Blind Spot*	Majestic Light/Buckpasser	U	9	M	M	4	3-4
Blini*	Nureyev/Hail to Reason	M	5	L	U	3*	3-4
Blood Royal	Ribot/Dedicate	D	7	D	F	2	3-4
Bluebird	Storm Bird/Sir Ivor	L	4	L	M	3	2-3
Blue Buckaroo^	Buckaroo/Sir Ivor	M	6	U	U	3	3-4
Blue Ensign	Hoist the Flag/Bold Ruler	D	6	C	C+	4	2-3
Blue Eyed Davy	Viking Spirit/Blue Prince	C	8	F	C	5	3-4
Blushing Groom	Red God/Wild Risk	C	3	B-	A	3*	3-4
Blushing John*	Blushing Groom/Prince John	C	6	L	M	1	2-3
Bob's Dusty	Bold Commander/Count Fleet	D	7	D	C	3	3-4
Boitron	Faraway Sun/Prudent	U	6	B	M	4	3-4
Bold Agent	Bold Bidder/Rasper 11	M	6	M	M	4	3-4
Bold Arrangement	Persian Bold/Floribunda	D	6	C	B	1	3-4
Bold Badgett*	Damascus/Never Bend	C	7	U	B	4	3-4

Stallion	Sire/Broodmare Sire	FTS	CL	Trf	OT	SI	AB
Bold Bidder	Bold Ruler/To Market	C	4	B	C	4	3-4
Bold Ego c	Bold Tactics/Bullin	C+	7	F	C	5	2-3
Bold Forbes c	Irish Castle/Commodore	C	4	D	B	4	3-4
Bold Hour	Bold Ruler/Mr. Music	C	5	D	B	4	3-4
Bold Laddie	Boldnesian/Vertex	D	7	D	C+	5	3-4
Boldnesian	Bold Ruler/Polynesian	C	3	C	B	3	3-4
Bold Play	Chieftain/Needles	D	9	F	F	4	2-3
Bold Reason	Hail To Reason/Djeddah	C	5	B-	C	4	3-4
Bold Relic	Bold Monarch/Jet Traffic	C	7	D	C+	5	3-4
Bold Revenue	Bold Ruckus/I'm for More	D	5	U	C	4	2-3
Bold Ruckus	Boldnesian/Raise a Native	C	3	C	C+	4	2-3
Bold Ruler	Nasrullah/Discovery	B	1	C	C	4*	2-3
Bold T Jay*	Island Agent/Bold Combatant	C	8	U	T	5	2-3
Bold Tropic	Plum Bold/Herculaneum	C	8	C	C	4	3-4
Bolger	Damascus/Round Table	C	6	C	C	4	3-4
Boss Koss^	Dr. Blum/Stella Aurata	L	10	U	M	5	2-3
Bounding Basque	Grey Dawn/Jean Pierre	D	5	C	C	2	2-3
Boys Nite Out	Cutlass/Bolinas Boy	D	8	F	F	4	3-4
Brave Lad*	Damascus/Protanto	B	7	U	C+	5	2-3
Brave Shot	Bold Bidder/Sir Gaylord	C+	5	C	C	4	3-4
Brazen Brother	Boldnesian/Spy Song	D	7	C	F	5	2-3
Breezing On	Stevward/Raise a Native	C	4	F	C	5	2-3
Breezing On In*	Breezing On/Hoist the Flag	M	9	U	M	5	3-4
Brent's Danzig*	Danzig/Crozier	C	9	M	U	5	2-3
Briarctic	Nearctic/Round Table	C	5	C	F	4	2-3
Brilliant Leader*	Irish River/Northern Dancer	L	8	M	M	5	2-3
Brilliant Protege	Secretariat/Ribot	U	7	M	U	2	3-4
Brilliant Sandy	Crimson Satan/Rasper II	D	8	C	D	5	3-4
Broad Brush	Ack Ack/Hoist the Flag	B+	3	D	C+	2	2-3
Brogan	Nijinsky II/Round Table	B	8	B-	F	2	3-4
Brother Liam*	Hail to Reason/Coco La Terreur	M	9	U	U	4	3-4
Buckaroo	Buckpasser/No Robbery	C	5	D	C	2	2-3
Buckfinder	Buckpasser/Native Dancer	D	4	D	C	3	3-4
Buck Island	Buckpasser/Olden Times	F	10	C	D	4	3-4
Buckley Boy	Alydar/Quack	D	6	D	C	2	3-4
Buckpasser	Tom Fool/War Admiral	C	1	C	F	1	3-4
Bucksplasher	Buckpasser/Northern Dancer	C	5	B-	F	2	3-4
Buen Jefe*	Delaware Chief/No Prevue	C	9	F	M	5	2-3
Burnt Hills^	Conquistador Cielo/Bold Hour	L	7	U	L	4	2-3
Burts Star*	Star de Naskra/Bold Monarch	L	8	U	M	4	3-4

Stallion	Sire/Broodmare Sire	FTS	CL	Trf	OT	SI	AB
Busted	Crepello/Vimy	U	3	L	M	1	3-4
Cabrini Green c	King's Bishop/Tudor Minstrel	B	6	D	F	4	3-4
Cactus Road	Kennedy Road/Bold Bidder	A	8	F	F	5	3-4
Cad	Timeless Moment/Three Bagger	D	9	D	C	5	3-4
Caerleon	Nijinsky II/Round Table	C	1	A	A	1	3-4
Cahill Road^	Fappiano/Le Fabuleux	M	5	U	L	2	2-3
Cajun Prince	Ack Ack/Candy Spots	D	8	C	F	4	3-4
Candi's Gold*	Yukon/Captain Nash	L	7	A-	C	2	2-3
Cannonade	Bold Bidder/Ribot	D	7	C	C	2	3-4
Cannon Dancer*	Explodent/Rattle Dancer	D	8	M	F	4	3-4
Cannon Shell	Torsion/Promise	F	8	B-	C	5	3-4
Capital Idea	Iron Ruler/Arrogate	F	7	C	F	4	3-4
Capote	Seattle Slew/Bald Eagle	D	5	D	B	4	2-3
Captain Courageous	Sailor/Bold Ruler	C+	7	D	C+	4	2-3
Captain James	Captain's Gig/Alcide	F	10	C	F	4	3-4
Captain Nick	Sharpen Up/Double Jump	F	8	B	F	5	3-4
Captain Valid^	Valid Appeal/Northerly	M	8	M	M	3	2-3
Capt. Don	Don B./My Host	C	8	M	C	3	3-4
Capulet's Song	In Reality/Hasty Road	C+	8	F	F	5	2-3
Carborundum^	Best Turn/Papa Fourway	M	8	U	M	4	2-3
Carload*	Relaunch/Pretense	M	8	M	L	4	3-4
Carnivalay	Northern Dancer/Cyane	C	6	B-	C+	3*	3-4
Caro	Fortino/Chamossaire	F	3	B	C	1	3-4
Carolina Ridge*	Cox's Ridge/Groton	U	8	U	U	3	2-3
Caros Love*	Caro/Northern Jove	M	8	M	M	4	2-3
Carr de Naskra	Star de Naskra/Cornish Prince	A	5	C	C+	4	2-3
Carson City^	Mr. Prospector/Blushing Groom	L	6	M	L	4	2-3
Casa Dante*	Exclusive Native/Delta Judge	D	8	M	U	4	3-4
Case the Joint*	The Pruner/Nashville	U	8	M	U	4	3-4
Cassaleria	Pretense/Verbatim	D	7	D	T	4	3-4
Castle Guard	Slady Castle/Chompion	D	6	F	C	4	3-4
Catane*	Hatchet Man/Round Table	U	9	U	M	3	3-4
Cathedral Bells*	Honest Pleasure/Hail to Reason	U	10	U	M	4	3-4
Caucasus	Nijinsky II/Princequillo	D	8	B	C	3	3-4
Cause For Pause*	Baldski/Aureole	D	7	L	A	4	3-4
Caveat	Cannonade/The Axe II	D	6	B-	F	1	3-4
Cefis^	Caveat/Dancing Champ	U	8	L	M	2	3-4
Cee's Tizzy^	Relaunch/Lyphard	M	7	L	L	3	3-4
Century Prince	Rollicking/Bold Effort	D	8	D	C+	5	3-4
Champagneforashley^	Track Barron/Alleged	M	7	U	M	3	2-3
Charging Falls	Taylor's Falls/Space Commander	M	7	M	A	4	3-4

Stallion	Sire/Broodmare Sire	FTS	CL	Trf	OT	SI	AB	181
Charlie Barley^	Affirmed/Dancing Champ	U	7	L	M	2	2-3	
Charming Turn	Best Turn/Poker	C	6	D	C	4	2-3	
Chas Conerly	Big Burn/Fair Ruler	C	8	B-	F	4	3-4	
Cheque Froid*	Icecapade/Damascus	B	9	U	LT	4	2-3	
Cherokee Colony*	Pleasant Colony/Nijinsky II	D	6	C	C	2	2-3	
Cherokee Fellow	Funny Fellow/Francis S.	C	5	C	C	4	2-3	
Chicanery Slew*	Seattle Slew/Pretense	M	9	U	LT	3	3-4	
Chief's Crown c	Danzig/Secretariat	C	5	C	C	2	3-4	
Chief Steward	Chieftain/Deck Hand	C	6	U	M	5	2-3	
Chieftain	Bold Ruler/Roman	C	4	C	C+	4	2-3	
Chinati^	Blushing Groom/Reliance	M	8	L	L	4	3-4	
Chiromancy*	Alydar/Forli	U	8	M	M	4	2-3	
Chisos	Alydar/High Perch	B	7	M	C+	4	3-4	
Chivalry	Nijinsky II/Sir Ivor	F	8	C	F	2	3-4	
Christopher R	Loom/Cavan	C	7	D	C+	4	3-4	
Chumming	Alleged/Sea-Bird	D	8	D	D	4	3-4	
Chromite*	Mr. Prospector/Buckpasser	D	8	U	F	4	3-4	
Circle	Round Table/Nasrullah	B	7	A	D	3	2-3	
Circle Home	Bold Bidder/Ribot	C	8	B	C	3	3-4	
Citidancer^	Dixieland Band/Tentam	L	8	L	L	4	2-3	
Claim*	Mr. Prospector/Vertex	F	9	M	M	4	3-4	
Claramount^	Policeman/Cornish Prince	M	8	M	U	3	2-3	
Classic	Hoist the Flag/Speak John	D	7	F	F	4	3-4	
Classic Account*	Private Account/Arts and Letters	U	9	U	T	2	3-4	
Classic Go Go	Pago Pago/Never Bend	B	7	D	F	4	2-3	
Classic Trial	In Reality/Moslem Chief	M	7	U	M	4	3-4	
Clever Allemont	Clever Trick/Carlemont	D	8	U	F	5	2-3	
Clever Champ	Clever Trick/Venetian court	C	7	F	C+	5	2-3	
Clever Secret*	Secretariat/Pia Star	C	8	M	C	4	3-4	
Clever Trick c	Icecapade/Better Bee	C	5	B-	C	3	2-3	
Coach's Call*	Alydar/Velvet Cap	U	10	U	M	4	3-4	
Coax Me Chad	L'Enjoleur/First Landing	D	8	C	D	3	3-4	
Codex	Arts and Letters/Minnesota Mac	C	4	C	C	4	3-4	
Cognizant	Explodent/Dr. Fager	D	8	M	B	4	2-3	
Cohoes	Mahmoud/Blue Larkspur	D	4	F	F	3	3-4	
Cojak	Cohoes/Dark Star	D	6	F	C	4	3-4	
Col. Denning	Rock Talk/Grey Monarch	D	7	F	F	4	3-4	
Cold Reception	Secretariat/Nearctic	C	7	D	C	1	3-4	
Collier	Mr. Prospector/Sir Ivor	B	7	D	C+	4	3-4	
Colonel Power	Diplomat Way/Vertex	C	8	C	F	5	2-3	
Colonel Stevens	Lt. Stevens/Bupers	F	7	U	M	3	3-4	

Stallion	Sire/Broodmare Sire	FTS	CL	Trf	OT	SI	AB
Come On Mel*	Lucky Mel/Sisters Prince	M	8	U	M	5	3-4
Come Summer*	Junius/Sir Gaylord	M	10	U	U	4	3-4
Comet Kat	Foreign Comet/Crafty Admiral	C	7	D	C	3	2-3
Commadore C.	Mongo/Call Over	F	8	C	C	4	2-3
Commemorate	Exclusive Native/Never Bend	D	6	D	C	3	2-3
Commissioner	Crozier/Jovial Love	C+	7	C	B	5	3-4
Common Grounds	Kris/Lyphard	C	4	B	F	4*	2-3
Compliance	Northern Dancer/Buckpasser	D	6	C	D	2	3-4
Compliment	Nijinsky II/Graustark	U	8	M	M	4	3-4
Comrade In Arms	Brigadier Gerard/Birdbrook	U	4	L	U	1	3-4
Concorde Bound	Super Concorde/Iron Ruler	L	3	M	M	4	2-3
Conduction	Graustark/Prince John	U	9	U	M	4	3-4
Conquistador Cielo c	Mr. Prospector/Bold Cmmander	C+	4	C	B	4*	3-4
Contare	Naskra/One Count	C+	7	D	C	4	3-4
Contortionist	Torsion/Tatan	U	8	M	U	4	3-4
Cool*	Bold Bidder/Northern Dancer	C	10	U	F	4	3-4
Cool Frenchy	French Policy/Persia	D	9	U	C	5	3-4
Cool Halo	Halo/Northern Dancer	B+	8	M	B	4	2-3
Cool Victor*	Tentam/Nearctic	C	3	F	A+	4	2-3
Copelan	Tri Jet/Quadrangle	B	5	D	C+	4	2-3
Cormorant	His Majesty/Tudor Minstrel	C	3*	C	C	1*	3-4
Cornish Prince	Bold Ruler/Eight Thirty	C+	5	D	B	4	2-3
Corporate Report^	Private Account/Key to the Mint	U	6	U	T	2	3-4
Corridor Key	Danzig/Prince John	C	7	D	C+	4	2-3
Cortan	Illustrious/Sailor's Guide	D	8	D	C	4	3-4
Corwyn Bay^	Caerleon/Crowned Prince	M	6	L	L	4	3-4
Cosmic Voyager*	Gummo/Acroterian	F	10	U	L	5	3-4
Cost Conscious*	Believe It/Herbager	U	8	M	M	3	3-4
Counsellors Image	Dancer's Image/Princequillo	D	9	C	F	2	3-4
Count Eric*	Riverman/TV Lark	M	8	M	M	4	3-4
Country Light	Majestic Light/Hoist the Flag	C	7	D	F	3	3-4
Country Manor*	Irish Castle/Hitting Away	F	10	U	F	5	3-4
Country Pine	His Majesty/Vaguely Noble	C	6	C	C	2	2-3
Count the Dots^	Dancing Count/John Williams	U	10	L	L	4	3-4
Coup de Kas*	Kakskaskia/Villamor	B	8	U	C	5	2-3
Court Ruling	Traffic Judge/The Doge	D	8	D	C+	4	3-4
Court Trial c	In Reality/Moslem Chief	D	6	B	C	2	2-3
Covert Operation*	True Colors/Prince John	U	10	M	U	1	3-4
Cox's Ridge c	Best Turn/ Ballydonnell	C	4	C	C	2	3-4
Cozzene	Caro/Prince John	F	4	B-	F	1	3-4
Crafty Native	Native Born/Crafty Admiral	D	8	F	C	4	2-3

Stallion	Sire/Broodmare Sire	FTS	CL	Trf	OT	SI	AB
Crafty Prospector	Mr. Prospector/In Reality	B+	4	D	A	4	2-3
Crawford Special*	Bold Street/Crewman	B	8	U	A	4	3-4
Creamette City	In Reality/Native Dancer	C	8	U	C	5	3-4
Creole Dancer*	Dancing Dervish/Wig Out	L	9	U	U	5	2-3
Criminal Type^	Alydar/No Robbery	U	4	U	L	2	3-4
Crimson Satan	Spy Song/Requiebro	B	3	C	B	5	2-3
Critique	Roberto/Sicambre	F	8	C	F	2	3-4
Crowning*	Raise a Native/Round Table	U	9	U	M	4	3-4
Crusader Sword*	Damascus/Nijinsky II	C+	5	M	U	3	2-3
Crusoe*	The Minstrel/Sir Gaylord	C+	8	F	C+	3	3-4
Cryptoclearance	Fappiano/Hoist The Flag	D	6	U	A+	2	2-3
Crystal Glitters	Blushing Groom/Donut King	C	3	B-	C	3*	3-4
Crystal Run^	Table Run/Anyoldtime	M	8	M	M	4	2-3
Crystal Tas^	Crystal Water/Run of Luck	U	9	U	U	4	3-4
Crystal Water	Windy Sands/T V Lark	D	7	D	F	4	3-4
Cure The Blues	Stop the Music/Dr. Fager	C+	5	D	A	4	2-3
Current Classic*	Little Current/Delta Judge	U	10	M	T	3	3-4
Current Hope	Little Current/Warfare	D	8	C	F	2	2-3
Cutlass	Damascus/Dunce	B	4	C	T	4	2-3
Cutlass Reality*	Cutlass/In Reality	C	6	F	T	3	3-4
Cut Throat	Sharpen Up/Zimone	B	7	C	C	4	3-4
Czaravich	Nijinsky II/Linacre	D	7	D	F	2	3-4
D'Accord	Secretariat/Northern Dancer	C	5	C	D	2	3-4
Dactylographer	Secretariat/Ribot	D	7	C	F	1	2-3
Daily Review*	Damascus/Reviewer	U	9	U	T	3	3-4
Dahar	Lyphard/Vaguely Noble	D	6	C	C	1	3-4
Dallasite	Vaguely Noble/Forli	U	9	M	M	4	3-4
Damascus	Sword Dancer/My Babu	C	2	C	BT	2	3-4
Dambay*	Damascus/Bold Ruler	M	8	U	M	4	3-4
Damister*	Mr. Prospector/Roman Line	M	5	L	L	4*	3-4
Dance Bid	Northern Dancer/Bold Bidder	C	6	B	D	4	3-4
Dance Centre	Mr. Prospector/Olympia	B	9	U	M	5	2-3
Dance Furlough^	Nijinsky II/Riva Ridge	U	9	M	M	1	3-4
Dance Hall*	Assert/Mr. Prospector	U	7	L	M	3	3-4
Dance in Time	Northern Dancer/Chop Chop	D	8	C	B	4	3-4
Dancing Again	Nijinsky II/Round Table	C	8	F	C	2	3-4
Dancing Champ	Nijinsky II/Tom Fool	C	3	C	D	4	2-3
Dancing Count	Northern Dancer/King's Bench	C	4*	B	C	2	3-4
Dancing Czar*	Nijinsky II/Hail to Reason	B	6	U	C	4	2-3
Dancing Wizard	Northern Dancer/Bold Ruler	C	9	D	C	3	3-4

Stallion	Sire/Broodmare Sire	FTS	CL	Trf	OT	SI	AB
Dandy Binge	Draconic/Dandy K.	C	8	F	C	5	2-3
Danebo	Bold Forbes/Sea Hawk	D	9	C	A+	4	3-4
Danski*	Danzig/Pago Pago	L	8	M	L	5	2-3
Danzatore*	Northern Dancer/Raise a Native	D	9	C	F	4	3-4
Danzig c	Northern Dancer/Admiral's Voyage	A	I	B	B	4*	2-3
Danzig Connection	Danzig/Sir Ivor	C	6	C	A+	2	3-4
Danzig Dancer*	Danzig/Tom Rolfe	U	9	L	L	3	3-4
Darby Creek Road	Roberto/Olympia	C	5	B	C	2	2-3
Daring Groom*	Blushing Groom/Prince Dare	B	6	U	C	2	2-3
Darn That Alarm	Jig Time/Blazing Count	C+	5	F	C	3	2-3
Darshan	Shirley Heights/Abdos	D	3	B	M	I	3-4
Dauphin Fabuleux	Le Fabuleux/Vice Regal	D	8	C	F	4	3-4
Dave's Reality*	In Reality/Knightly Manner	M	10	U	L	5	2-3
Dawn of Creation	Raise a Native/Prince John	C	8	U	F	3	2-3
Dawn Quixote^	Grey Dawn II/Seattle Slew	M	7	M	T	4	2-3
Dayjur^	Danzig/Mr. Prospector	L	3	L	L	4	2-3
Debonair Roger	Raja Baba/Lucky Debonair	D	7	D	C	4	3-4
De Braak*	Mr. Prospector/Dance Spell	M	10	M	M	4	2-3
Dee Lance*	Blade/Swoon's Son	L	7	U	M	5	3-4
Defense Verdict	In Reality/ Moslem Chief	B	8	F	F	4	2-3
Defiance	What a Pleasure/The Axe II	F	9	C	F	4	3-4
De Jeau*	Private Thoughts/Rattle Dancer	U	7	U	M	4	2-3
Delaware Chief	Chieftain/Double Jay	C	8	F	C	4	3-4
Delinsky^	Nijinsky II/Graustark	U	10	M	U	I	3-4
Demon's Begone*	Elocutionist/Halo	F	6	D	F	2	3-4
Deposit Ticket^	Northern Baby/Mr. Prospector	M	7	M	M	5	2-3
Deputed Testamony	Traffic Cop/Prove It	D	6	D	C+	2	3-4
Deputy Minister c	Vice Regent/Bunty's Flight	B	3	C	C+	2	3-4
Deputy Regent*	Vice Regent/Canadian Champ	D	9	M	F	4	2-3
Derby Wish*	Lyphard's Wish/Crozier	L	8	M	M	4	2-3
Desert Wine	Damascus/Never Bend	B	7	D	C	3	2-3
Determinant*	Norcliffe/Green Ticket	M	6	U	L	5	2-3
Deuces Are Loose*	Zen/Pampered King II	M	9	M	U	4	3-4
Devil's Bag	Halo/Herbager	C	3	C	C+	2	3-4
Dewan	Bold Ruler/Sun Again	C	3	B	B	3	3-4
Diamondhawke*	Meritable/Drum Fire	L	9	M	L	5	2-3
Diamond Prospect	Mr. Prospector/Social Climber	C	7	C	D	4	2-3
Diamond Shoal	Mill Reef/Graustark	D	6	B	C	2	3-4
Diamond Sword*	Danzig/Herbager	M	7	M	M	4	2-3
Diesis	Sharpen Up/Reliance	B	4	A	F	I	3-4
Digamist^	Blushing Groom/Northern Dancer	L	5	L	L	4*	2-3

Stallion	Sire/Broodmare Sire	FTS	CL	Trf	OT	SI	AB	185
Dimaggio	Bold Hitter/Indian Hemp	D	8	C	C	4	3-4	
Din's Dancer^	Sovereign Dancer/Olden Times	U	7	M	M	2	3-4	
Diplomat Way	Nashua/Princequillo	D	4	B-	D	3	3-4	
Discretion	Bold Lad/Worden	D	9	F	F	4	3-4	
Distant Day	Fleet Nasrullah/Summer Tan	C	7	C	C+	4	3-4	
Distant Land	Graustark/Bold Ruler	D	7	B-	C+	4	3-4	
Distant Ryder	Red Ryder/Tex Courage	D	9	B-	F	4	3-4	
Distinctive	Never Bend/Requested	B	4	D	B	4	2-3	
Distinctive Pro	Mr. Prospector/Distinctive	B	4	D	C	5	2-3	
Dixieland Band	Northern Dancer/Delta Judge	B+	3	B-	B	4*	2-3	
Dixieland Brass*	Dixieland Band/Gallant Romeo	C	6	B-	F	3	2-3	
D.J. Trump*	Raise a Native/Norcliffe	U	6	L	L	4	3-4	
Doc's Leader*	Mr. Leader/Nodouble	U	7	U	T	2	2-3	
Dogwood Passport	Silent Screen/Aureole	D	8	C	D	4	3-4	
Do It Again Dan*	Mr. Leader/Our Michael	C+	7	D	B	4	3-4	
Domasca Dan*	Same Direction/Gold and Myrrh	M	7	M	M	3	2-3	
Dominated*	Exclusive Native/Crafty Admiral	L	7	M	F	4	2-3	
Dominator	Secretariat/Windy Sea	F	10	L	U	4	3-4	
Domremy*	What a Pleasure/Sir Ivor	B	8	U	A	5	2-3	
Don B.	Fleet Nasrullah/Prince John	M	7	U	L	4	3-4	
Don's Choice	Private Account/Stevward	D	7	D	F	2	3-4	
Don't Fool With Me^	Tanthem/Forward Pass	M	9	U	L	4	3-4	
Don't Forget Me	Ahonoora/African Sky	U	4	L	U	4*	3-4	
Don't Hesitate*	Native Uproar/Time Tested	M	7	M	M	5	3-4	
Doonesbury	Matsadoon/Vaguely Noble	C	7	C	C	4	3-4	
Double Edge Sword	Sword Dancer/Discovery	F	8	F	T	3	3-4	
Double Jay	Balladier/Whiskbroom II	B	2	C	D	4*	2-3	
Double Negative^	Mr. Prospector/Distinctive	L	8	U	L	5	3-4	
Double Quick*	To the Quick/Creme dela Creme	U	6	U	M	3	2-3	
Double Sonic	Nodouble/Palestinian	B	6	F	C	2	2-3	
Double Zeus	Spring Double/Ridan	C	8	D	C	4	3-4	
Dover Ridge*	Riva Ridge/Bold Ruler	B	8	D	D	5	2-3	
Dr. Blum	Dr. Fager/Sir Gaylord	C+	4*	D	C	4	2-3	
Dr. Carter	Caro/Chieftain	D	6	C	F	3	3-4	
Dr. Fager	Rough'n Tumble/Better Self	B	1	A	A	4	2-3	
Dr. Koch*	In Reality/Bold Ruler	L	7	U	L	5	2-3	
Dr. McGuire	Dr. Fager/Sir Ivor	F	10	F	D	5	2-3	
Dr. Nauset*	Irish Castle/Malicious	M	10	U	M	3	3-4	
Drone	Sir Gaylord/Tom Fool	B	3	F	B	4	2-3	
Drouilly	Mill Reef/Gun Shot	C	8	D	C+	2	3-4	
Dr. Schwartzman	Flourescent Light/Graustark	D	9	D	C	2	3-4	

Stallion	Sire/Broodmare Sire	FTS	CL	Trf	OT	SI	AB
Drum Fire	Never Bend/Ambiorix	C+	5	C	C+	4	2-3
Drum Score	Drum Fire/Envoy	U	9	U	T	4	3-4
Dr. Valeri	Gunflint/Andy's Glory	F	9	B-	F	3	3-4
Dual Honor*	Seattle Slew/Le Fabuleux	D	9	C	F	2	3-4
Duck Dance	Water Prince/Swoon's Son	C	6	F	D	4	2-3
Due Diligence	Stevward/Third Brother	F	9	F	C	5	3-4
Duluth*	Codex/Key to the Mint	U	8	M	M	4	3-4
Dunant*	Sallust/African Sky	U	8	U	M	5	3-4
Dunham's Gift	Winged T./Dead Ahead	C+	7	B	C	4	3-4
Duns Scotus	Buckpasser/Northern Dancer	F	6	B-	F	3	3-4
Dust Commander	Bold Commander/Windy City II	F	7	D	D	4	3-4
Dynaformer*	Roberto/His Majesty	B	6	A	C	1	2-3
Dynamo Mac	Dynastic/Raise a Native	B	9	F	F	5	3-4
Eager Eagle	T.V. Lark/War Admiral	D	7	B	F	4	3-4
Eager Native	Restless Native/Fleet Nasrullah	B	6	C	C	2	3-4
Eastern Echo^	Damascus/Northern Dancer	M	5	U	M	2	3-4
Easy Goer^	Alydar/Buckpasser	U	3	U	M	1	3-4
Ecliptical*	Exclusive Native/Chieftain	B	8	U	L	4	2-3
Ecole Militaire	Quadrangle/Mongo	U	9	U	C	4	3-4
El Baba	Raja Baba/Hail to All	C+	6	D	C+	4	3-4
Eldorado Bob*	Thatch/Red God	B	8	B	C	5	3-4
El Dorado Kid	Crazy Kid/Flash O' Night	F	8	U	C	5	3-4
Electric Blue*	Cyane/Tibaldo	C+	8	B-	C+	3	2-3
Elegant Life	Distinctive/Bold Hour	C+	9	U	A	5	3-4
Eleven Stitches	Windy Sands/My Host	C	7	C	C	4	3-4
El Gran Senor	Northern Dancer/Buckpasser	C+	4	B	C	3*	3-4
Elmaamul^	Diesis/Roberto	M	6	L	U	1	2-3
Elocutionist	Gallant Romeo/Fleet Nasrullah	C	4	C	C	4	3-4
El Raggaas	NorthernDancer/MajesticPrince	D	7	D	F	2	2-3
Embrace the Wind	Regal Embrace/Speedy Zephyr	C+	8	U	C+	5	2-3
Eminency	Vaguely Noble/Chietain	D	8	C	F	4	3-4
Empire Glory	Nijinsky II/Fleet Nasrullah	C	8	C	C	4	3-4
Encino	Nijinsky II/Crimson Satan	C+	7	C	C	4	3-4
Endow^	Flying Paster/Exclusive Native	M	7	M	M	2	2-3
Ends Well	Lyphard/Stage Door Johnny	C	7	B-	F	1	3-4
Enough Reality	In Reality/Vent Du Nord	M	7	U	L	4	2-3
Entropy	What a Pleasure/Intentionally	C	6	D	F	4	2-3
Erin's Isle	Busted/Shantung	F	8	B-	T	2	3-4
Eskimo	Northern Dancer/Dr. Fager	C	6	C	F	4	2-3
Estate*	Singh/Damascus	M	9	U	M	4	3-4
Eternal Prince	MajesticPrince/Fleet Nasrullah	C	7	C	C	2	2-3

Stallion	Sire/Broodmare Sire	FTS	CL	Trf	OT	SI	AB
Evening Kris^	Kris S./Double Hitch	U	5	M	M	3	2-3
Exact Duplicate	Chateaugay/Summa Cum	F	10	U	F	4	3-4
Exactly Sharp^	Sharpen Up/Caro	U	7	L	M	2	3-4
Exceller	Vaguely Noble/Bald Eagle	D	6	C	D	3	3-4
Exclusive Encore*	Exclusive Native/Court Martial	B	8	U	M	4	2-3
Exclusive Enough*	Exclusive Native/Three Martinis	C	8	U	M	5	2-3
Exclusive Era	Exclusive Native/Nashua	C	6	C	C	2	2-3
Exclusive Gem*	Exclusive Native/Prince John	F	9	U	F	4	3-4
Exclusive Ribot	Ribot/Shut Out	D	8	C	D	4	3-4
Exclusive Native	Raise a Native/Shut Out	C	4	B	C	3*	2-3
Executive Intent*	Secretariat/Buckpasser	D	9	U	U	4	3-4
Executive Order	Secretariat/Sir Ivor	C	7	B-	C	4	2-3
Expect Greatness	Raise a Native/Dance Spell	L	9	U	L	5	2-3
Expensive Decision*	Explosive Bid/Hydrologist	M	7	L	M	3	2-3
Explodent c	Nearctic/Mel Hash	C	4*	B	C	2	2-3
Explosive Bid	Explodent/Diplomat Way	C	7	B	D	2	2-3
Explosive Wagon	Explodent/Conestoga	B	8	U	C	4	3-4
Expressman	Gaelic Dancer/Olden Times	C+	8	C	C+	5	3-4
Exuberant	What a Pleasure/Beau Purple	C	5	D	F	4	2-3
Fairway Phantom	What a Pleasure/Native Dancer	C	8	F	C	4	3-4
Fairy King*	Northern Dancer/Bold Reason	L	3	L	M	4	2-3
Faliraki	Prince Tenderfoot/Super Sam	C	8	C	F	4	3-4
Falstaff	Lyphard/Sir Ivor	D	7	C	D	4	2-3
Fappavalley*	Fappiano/Road at Sea	M	7	U	M	4	2-3
Fappiano	Mr. Prospector/Dr. Fager	B-	1	C	A	3*	3-4
Faraway Island*	Banquet Table/Our Native	U	10	M	M	2	3-4
Farma Way^	Marfa/Diplomat Way	U	5	U	U	1	2-3
Far North	Northern Dancer/Victoria Park	D	5*	C	C	2	3-4
Far Out East	Raja Baba/Ambehaving	C	7	C	C	4	2-3
Fast	Bold Bidder/Commanding II	D	8	B-	C	4	3-4
Fast Account*	Private Account/Fleet Nasrullah	D	9	C	T	3	3-4
Faster Than Sound*	Five Star Flight/Turn & Count	F	6	U	L	5	2-3
Fast Forward*	Pleasant Colony/Raise a Native	M	8	U	U	4	3-4
Fast Gold	Mr. Prospector/Ack Ack	D	6	B-	F	4	3-4
Fast Passer	Buckpasser/Double Jay	F	10	F	F	4	3-4
Fast Play*	Seattle Slew/Buckpasser	U	7	U	M	2	2-3
Fatih	Icecapade/Graustark	D	7	B-	C	4	3-4
Feel the Power	Raise a Native/Best Turn	B+	8	U	D	5	3-4
Ferdinand	Nijinsky II/Double Jay	F	6	C	C	2	3-4
Festin^	Mat Boy/Con Brio	U	5	U	T	1	3-4

Stallion	Sire/Broodmare Sire	FTS	CL	Trf	OT	SI	AB
Festive*	Damascus/Buckpasser	U	7	U	U	2	3-4
Feu d'Enfer*	Tentam/Victoria Park	M	9	U	L	4	2-3
Fiery Best^	Sunny North/Diplomat Way	L	7	M	M	4	2-3
Fiestero	Balconaje/Sun Prince	C	9	B	F	4	3-4
Fifth Marine	Hoist the Flag/Princequillo	D	6	D	F	4	3-4
Fifty Six Ina Row	Dimaggio/Dusty Canyon	D	9	F	F	5	3-4
Fighting Fit	Full Pocket/Nilo	D	7	D	C	4	3-4
Fight Over	Grey Dawn II/Raise a Native	C	7	D	C	4	2-3
Fire Dancer	Northern Dancer/Native Charger	D	6	D	F	4	2-3
Fire Maker^	Fire Dancer/Raise a Native	M	9	U	M	4	2-3
First Albert	Tudor Grey/Groton	C	7	D	F	4	3-4
First Patriot*	Salutely/Road at Sea	U	7	U	U	2	2-3
Fit to Fight	Chieftain/One Count	C	5	F	C+	3	3-4
Five Star Flight	Top Command/Eddie Schmidt	C	8	C	C	4	3-4
Flashy Image	Dancer's Image/Good Shot	D	8	D	F	5	3-4
Flashy Mac	Minnesota Mac/Cagire II	F	9	U	C	4	2-3
Fleet Mel	Lucky Mel/Fleet Nasrullah	M	•5	M	M	4	2-3
Fleet Nasrullah	Nasrullah/Count Fleet	C	3	C	B	4*	2-3
Fleet Tempo	Fleet Allied/Bright tiny	C	8	D	C	4	3-4
Fleet Twist	Fleet Nasrullah/Prince John	D	8	F	C	3	3-4
Flickering*	Summing/Damascus	U	9	U	M	4	3-4
Floating Reserve*	Olden Times/Quack	C	6	M	L	2	3-4
Floriano	Turn and Count/No Robbery	F	7	C	A	2	3-4
Florida Sunshine^	Alydar/Quadrangle	U	8	U	L	3	3-4
Flout*	Bold Bidder/Ben Lomond	U	9	M	M	4	3-4
Fluorescent Light	Herbager/Ribot	F	7	C	C	I	3-4
Fly By Night	Vaguely Noble/Dark Star	U	9	M	M	2	3-4
Flying Paster	Gummo/Acroterion	C+	4	D	C+	3	2-3
Flying Pidgeon	Upper Case/Minnesota Mac	F	9	C	F	2	3-4
Flying Target	Ack Ack/Nigromante	F	10	D	F	4	3-4
Flying Victor*	Flying Paster/Sir Ivor	C+	7	U	M	4	2-3
Fobby Forbes*	Bold Forbes/Round Table	F	9	C	F	4	3-4
Foligno*	Foolish Pleasure/In Reality	U	8	M	M	4	3-4
Follow the Drum*	Northern Dancer/Lord Durham	L	8	M	M	4	2-3
Folk's Pride*	Turn to/The Doge	A	7	D	F	5	2-3
Foolish Pleasure	What a Pleasure/Tom Fool	D	6	C	D	4	3-4
Fools Dance	Rattle Dancer/Tom Fool	F	7	U	C	4	3-4
Fool the Experts*	Crafty Drone/Zip Pocket	L	8	M	M	5	2-3
Forbidden Pleasure	Foolish Pleasure/Court Martial	C	10	F	F	4	3-4
Forceten	Forli/On and On	U	7	C	B	4	3-4
Foreign Survivor^	Danzig/Drone	M	6	L	M	3	2-3

Stallion	Sire/Broodmare Sire	FTS	CL	Trf	OT	SI	AB
Foretake	Forli/Traffic Judge	D	8	C	F	I	3-4
Forever Silver^	Silver Buck/Correlation	U	5	L	M	I	3-4
Forli	Aristophanes/Advocate	C	3	B	C	I*	3-4
Forlion	Forli/On On	D	7	D	C	3	3-4
Forli Winds	Forli/Windy Sea	D	7	D	D	4	3-4
Forsythe Boy*	Nodouble/Quadrangle	C	6	F	A+	4	2-3
Fort Calgary	Bold Joey/Calgary Brook	F	8	D	F	4	3-4
For The Moment	What a Pleasure/Tulyar	C	8	C	C	4	3-4
Fortunate Moment*	For the Moment/Restless Wind	M	7	M	M	4	2-3
Fortunate Prospect	Northern Prospect/Lucky Debonair	C	5	D	C	4	2-3
Forty Niner	Mr. Prospector/Tom Rolfe	A	I	B-	C+	4*	2-3
Fountain of Gold	Mr. Prospector/Distinctive	B-	7	D	C	4	2-3
Foyt	Raise a Native/On and On	D	7	D	F	4	3-4
Frankly Perfect^	Perrault/Viceregal	U	6	L	U	I	3-4
Fred Astaire	Nijinsky II/Stage Door Johnny	D	7	C	T	I*	2-3
Freezing Rain	It's Freezing/Well Mannered	A	7	U	C	4	3-4
French Colonial	Tom Rolfe/Sir Gaylord	D	8	C	F	2	3-4
French Legionaire*	Grey Legion/Verbatim	C	10	M	M	4	3-4
Frosty the Snowman^	His Majesty/Diplomat Way	M	7	L	U	2	2-3
Full Choke	Full Pocket/Never Bend	C	8	M	C	4	2-3
Full Intent	In Reality/Never Bend	B	6	F	A	5	3-4
Full of Fools*	Full Out/Secretariat	M	9	M	M	4	2-3
Full Partner	Never Bend/Round Table	C	7	D	C	5	2-3
Full Pocket	Olden Times/Summer Tan	C	3	F	C+	4	2-3
Fulmar	Northern Dancer/Bolero	D	8	D	D	4	3-4
Future Hope	Our Native/Sir Ribot	B	7	D	C	4	2-3
Fuzzbuster	No Robbery/Hasty Road	D	7	D	D	3	3-4
Fuzziano^	Fappiano/Turn and Count	M	7	M	M	3*	2-3
Fuzzy	It's Freezing/Raise a Native	F	8	D	D	4	3-4
Gaelic Christian	Gaelic Dancer/Intentionally	C+	8	F	C+	4	2-3
Ga Hai	Determine/Goyamo	D	8	D	F	4	3-4
Gain	Mississipian/Sir Ribot	C	9	F	F	4	3-4
Galaxy Libra	Wolver Hollow/Exbury	C+	8	C	F	4	3-4
Galaxy Road	Kennedy Road/Nashua	D	8	B-	F	5	3-4
Gallant Best	Best Turn/Gallant Man	C	8	D	C	2	3-4
Gallant Man	Migoli/Mahmoud	C	3	D	C	4*	3-4
Gallapiat	Buckpasser/Sir Gaylord	D	6	D	C	4	2-3
Garthorn	Believe It/Cyane	C+	6	D	A	4	3-4
Gate Dancer	Sovereign Dancer/Bull Lea	D	6	D	D	2	3-4
Gato Del Sol	Cougar/Jacinto	D	8	C	F	2	3-4

Stallion	Sire/Broodmare Sire	FTS	CL	Trf	OT	SI	AB
Gaylord's Carousel*	Lord Gaylord/Be Somebody	C+	9	U	B	4	3-4
Gay Old Blade	Hagley/Laugh Aloud	F	7	F	C	4	3-4
Geiger Counter	Mr. Prospector/Nantallah	C	5	D	C+	5	2-3
General Assembly	Secretariat/Native Dancer	D	6	C	C	3	3-4
General Holme	Noholme II/Count Fleet	C	6	C	D	3*	3-4
General Jimmy*	The Irish Lord/No Prevue	C+	9	U	C+	5	3-4
General Silver^	Silver Buck/Count Fleet	U	8	U	M	2	3-4
Gentleman Gene	Maribeau/Rico Tesio	U	7	M	U	5	3-4
Genuine Guy*	Gummo/Windy Sands	F	7	C	C	4	3-4
Georgia's Lyphard	Lyphard/Secretariat	U	8	M	U	3	3-4
Gilded Age	Tom Rolfe/Bold Ruler	C	8	C	C+	3	3-4
Ginistrelli	Hoist the Flag/Round Table	C	7	F	B	4	2-3
Giuseppe*	Fappiano/Graustark	M	9	U	L	5	2-3
Give Me Strength	Exclusive Native/Minnesota Mac	F	7	C	F	4	3-4
Glide*	Mr. Prospector/Northern Dancer	L	7	M	L	4	2-3
Glitterman^	Dewan/In Reality	L	6	U	M	5	2-3
Globe	Secretariat/Hail to Reason	D	8	B	F	4	3-4
Glorious Flag*	Secretariat/Hoist the Flag	U	8	M	M	4	3-4
Go and Go^	Be My Guest/Alleged	U	6	L	M	I	2-3
Gold Alert	Mr. Prospector/Arts and Letters	D	7	F	C	4	2-3
Gold Crest	Mr. Prospector/Northern Dancer	D	8	D	F	4	2-3
Golden Act	Gummo/Windy Sands	F	7	C	F	3	3-4
Golden Eagle II	Right Royal/Princequillo	C	6	C	C+	4	3-4
Golden Gauntlet*	Golden Eagle II/Dr. Fager	M	8	M	M	4	3-4
Golden Hill*	Tri Jet/Quadrangle	U	9	U	L	3	3-4
Golden Reserve	Sir Ivor/Round Table	D	7	F	C	4	2-3
Goldlust	Mr. Prospector/Dr. Fager	B	6	F	C	4	2-3
Gold Meridian	Seattle Slew/Crimson Satan	D	7	C	C+	4	3-4
Gold Pack^	Relaunch/Mr. Prospector	L	9	M	L	4	2-3
Gold Stage	Mr. Prospector/Cornish Prince	C	6	D	C+	4	2-3
Go Loom*	Loom/Gaelic Gold	C	9	U	A+	5	3-4
Gone West	Mr. Prospector/Secretariat	B	3	A	A	4*	2-3
Gonzales	Vaguely Noble/Dark Star	F	9	D	C	4	3-4
Good Rob	Jim J./Faultless	D	7	U	C	5	3-4
Go Step	Bold Reasoning/Native Dancer	B	8	D	C	4	2-3
Government Program	Secretariat/Northern Dancer	C	9	D	C	4	3-4
Grand Allegiance*	Pledge Allegiance/Bold Joey	M	8	U	M	5	2-3
Grand Jette^	Timeless Moment/Stevward	L	9	U	L	5	2-3
Grand Ruler*	Mr. Prospector/Pied d'Or	D	9	U	M	4	3-4
Gran Zar	Raja Baba/Native Charger	C	8	B-	F	4	3-4
Graustark	Ribot/Alibhai	D	2	C	C+	I	3-4

Stallion	Sire/Broodmare Sire	FTS	CL	Trf	OT	SI	AB
Great Above	Minnesota Mac/Intentionally	C	3*	B-	C	4	<u>2-3</u>
Great Deal*	Caro/Raise a Native	U	7	M	M	3	<u>3-4</u>
Great Gladiator	Timeless Moment/Hawaii	C	5	B-	B	3	3-4
Great Neck	Tentam/Dr. Fager	C+	7	C	D	5	<u>3-4</u>
Great Prospector	Mr. Prospector/Hail to Reason	D	7	C	C	4	3-4
Great View*	Raja Baba/Dead Ahead	M	7	U	M	5	2-3
Green Dancer	Nijinsky II/Val de Loir	C	4	C	D	I	3-4
Green Desert <u>c</u>	Danzig/Sir Ivor	C+	6	C	C	4*	<u>2-3</u>
Green Forest <u>c</u>	Shecky Green/The Axe II	C+	6	B	C	4*	3-4
Gregorian	Graustark/Dedicate	C	7	C	C	2	<u>3-4</u>
Greinton	Green Dancer/High Top	F	7	C	C+	I	<u>3-4</u>
Grenfall	Graustark/Swaps	C	7	C	D	4	3-4
Grey Adorn	Grey Dawn II/Bolero	C	8	U	F	I	<u>3-4</u>
Grey Dawn II	Herbager/Mahmoud	D	3	C	D	3*	3-4
Groovy	Norcliffe/Restless Wind	C	5	A	D	4*	2-3
Groshawk	Graustark/Jester	C	6	F	C	4	3-4
Groton	Nashua/Bimelech	C	6	F	C+	4	2-3
Ground Zero*	Silent Screen/Tudor Grey	L	9	U	U	5	2-3
Grub*	Mr. Prospector/Never Bend	M	8	U	L	4	3-4
Guilty Conscience	Court Ruling/Gallant Man	D	8	C	C	2	3-4
Guadalupe Peak^	Mr. Prospector/Roman Line	M	8	M	M	4	2-3
Gulch	Mr. Prospector/Rambunctious	C	5	C	F	4*	2-3
Gumboy	Gummo/Anyoldtime	C	8	F	F	4	<u>2-3</u>
Gummo	Fleet Nasrullah/Determine	C+	4	C	B	4	2-3
Gun Captain	Ack Ack/Crimson Satan	M	9	M	M	4	3-4
Habitat	Sir Gaylord/Occupy	B	I	B	M	4*	<u>2-3</u>
Habitonia*	Habitony/Tudor Minstrel	B+	7	M	F	<u>5</u>	2-3
Habitony	Habitat/Gallant Man	C	7	D	C	4	3-4
Hagley	Olden Times/Jet Action	C	4	F	A	4	3-4
Hail Bold King	Bold Bidder/Hail to Reason	D	8	B-	C+	3	3-4
Hail Emperor	Graustark/Bold Ruler	D	7	F	D	4	3-4
Hail the Pirates	Hail to Reason/Niccolo Dell'arca	D	5	C	F	2	3-4
Hail to Buck*	Silver Buck/Sea Bird	M	9	U	M	3	3-4
Half a Year*	Riverman/Northern Dancer	U	7	L	M	3	2-3
Halo <u>c</u>	Hail to Reason/Cosmic Bomb	C	3	C	C	I*	3-4
Hamza^	Northern Dancer/Val de Loir	U	8	M	U	I	3-4
Hansel^	Woodman/Dancing Count	L	4	U	M	3*	2-3
Happyasalark Tomas^	Wonder Lark/Mr. Washington	M	8	L	M	3	3-4
Happy Escort*	Unconscious/Raise a Native	F	I0	F	F	4	<u>3-4</u>
Harriman^	Lord Gaylord/Restless Native	M	8	U	M	4	3-4

Stallion	Sire/Broodmare Sire	FTS	CL	Trf	OT	SI	AB
Harvard Man	Crimson Satan/Swaps	F	6	C	F	4	2-3
Hasty Spring.	Spring Double/One Count	F	8	D	F	4	3-4
Hatchet Man	The Axe II/Tom Fool	C	3	D	C	I*	3-4
Hawaii	Utrillo II/Mehrail	D	7	B	C	3*	3-4
Hawkin's Special	Great Sun/Nasomo	C	7	C	F	5	3-4
Hawkster^	Silver Hawk/Chieftain	U	5	L	M	I	2-3
Hay Hello*	Halo/Hoist the Flag	C	9	U	C	2	2-3
Heavenly Plain^	Mount Hagen/Prince Regent	U	9	L	M	4	3-4
Heavy Bidder	Bold Bidder/Double Jay	F	10	U	F	5	3-4
Hello Gorgeous	Mr. Prospector/Jet Jewel	C	8	D	F	4	3-4
Henbane*	Alydar/Chateaugay	U	10	U	L	5	3-4
Herat	Northern Dancer/Damascus	D	8	C	F	2	3-4
Here Comes Red*	Peace Corps/Steward	F	10	U	F	5	3-4
Hero's Honor	Northern Dancer/Graustark	C	4	B-	C	I*	2-3
Hey Rob	Tisab/Gaelic Dancer	C	8	F	A	5	2-3
High Brite	Best Turn/Forli	C	6	D	C+	4	2-3
High Counsel	Apalachee/Court Martial	C	7	C	F	4	2-3
High Echelon	Native Charger/Princequillo	C+	7	B	F	4*	3-4
High Gold	Mr. Prospector/Irish Lancer	C	8	D	D	4	3-4
Highest Honor	Kenmare/Riverman	C	3	C	C	2	3-4
High Honors	Graustark/Hail to Reason	F	8	F	D	4	3-4
Highland Blade	Damascus/Misty Flight	D	6	C	C	2	2-3
Highland Park	Raise a Native/Olden Times	C	8	D	D	4	2-3
Highland Ruckus^	Bold Rukus/Victoria Park	M	8	U	M	4	2-3
High LIne	High Hat/Chanteur	C	3	B	M	I	3-4
High Top	Derring Do/Vimy	D	3	C	M	I	3-4
High Tribute	Prince John/War Relic	D	7	C	F	4	3-4
Hilal	Royal and Regal/Whistling Wind	D	8	B	C	4	3-4
His Majesty	Ribot/Alibhai	D	4	C	C	I	3-4
Historically	Raise a Native/Dead Ahead	FD	8	C	C	5	3-4
Ho Choy	Crozier/Northern Dancer	U	8	M	M	4	3-4
Hizaam*	Thatch/Tin Whistle	M	9	M	M	4	2-3
Hoedown's Day	Bargain Day/Dance Lesson	C	9	D	T	4	3-4
Hoist the Flag	Tom Rolfe/War Admiral	C	I	C	C	4	2-3
Hoist the Silver	Hoist the Flag/Prince John	D	8	C	C	4	3-4
Hold Your Peace	Speak John/Eight Thirty	C	3	D	C	3	2-3
Hollywood Brat	Cannonade/Vaguely Noble	D	8	C	A	4	3-4
Holme on Top	Noholme II/Federal Hill	C	8	F	C	4	2-3
Holy War	Damascus/Bold Ruler	C	9	D	F	5	3-4
Home at Last^	Quadratic/Ack Ack	U	6	U	T	I	3-4
Homebuilder*	Mr. Prospector/Vaguely Noble	M	8	M	M	2	2-3

Stallion	Sire/Broodmare Sire	FTS	CL	Trf	OT	SI	AB
Honest Note	Honest Pleasure/Native charger	M	9	U	L	4	3-4
Honest Pleasure	What a Pleasure/Tulyar	D	6	D	C	3	3-4
Honey Jay	Double Jay/Roman	C+	5	C	F	4	2-3
Honeyland*	Stop the Music/Drone	F	8	U	F	4	3-4
Hooched	Danzig/Tom Rolfe	C	7	C	F	4	3-4
Hookano	Silver Shark/Sailor	F	9	D	C	4	3-4
Hopeful Word	Verbatim/Mt. Hope	D	8	D	C+	4	2-3
Horatius	Proudest Roman/Cohoes	D	6	C	C	4	3-4
Horse Flash^	Stay for Lunch/Spectacular Bid	M	7	M	M	5	2-3
Hostage	Nijinsky II/Val de Loir	D	6	D	C	4	3-4
Hot Cop*	Policeman/Noholme II	M	8	M	M	5	3-4
Hot Oil	Damascus/The Axe II	D	7	D	F	4	3-4
Housebuster^	Mt. Livermore/Great Above	L	3	U	L	4	2-3
Houston^	Seattle Slew/Quadrangle	M	5	U	L	3	3-4
Huckster	Mr. Prospector/Sir Ivor	D	7	F	F	4	3-4
Huddle Up*	Sir Ivor/Never Bend	M	6	U	M	5	3-4
Hula Blaze	Hula Chief/Handsome Boy	B	9	U	F	4	2-3
Humbaba*	Raja Baba/Crimson Satan	L	8	M	L	5	2-3
Hurontario	Halo/Buckpasser	U	8	U	U	4	3-4
Hyannis Port	Kennedy Road/Novarullah	D	8	C	F	4	3-4
Hyperborean	Icecapade/Prince John	D	8	D	C	2	3-4
Iades*	Shirley Heights/Sir Gaylord	M	7	U	M	4	3-4
I Am the Game	Lord Gaylord/Dead Ahead	C	8	F	C+	4	3-4
Iam the Iceman^	Pirate's Bounty/Bold Hitter	M	8	U	M	4	2-3
Ice Age	Icecapade/Warfare	F	7	F	A+	4	3-4
Icecapade	Nearctic/Native Dancer	C	2	B	C+	4*	2-3
Ice Power^	Icecapade/Lyphard	M	9	L	M	3	2-3
Idabel^	Mr. Prospector/Briartic	M	7	M	M	3	3-4
Idaho's Majesty	His Majesty/Raise a Native	C+	9	M	M	4	3-4
I Enclose*	Cormorant/No Robbery	B	6	B-	A	4	3-4
Illuminate	Majestic Light/Forli	D	8	F	C	3	3-4
Illustrious	Round Table/Nasrullah	D	5	D	C	4	3-4
Image of Greatness f	Secretariat/Intentionally	D	5	C	C	2	3-4
I'ma Hell Raiser	Raise a Native/I'm for More	B+	7	B-	C+	5	3-4
I'm Glad	Liloy/Idle Hour	D	8	D	D	4	3-4
Imperial Falcon	Northern Dancer/Herbager	D	6	C	F	I*	3-4
Imperial Fling	Northern Dancer/Buckpasser	C	8	C	C	4	3-4
Imperial Guard	Northern Dancer/Victoria Park	D	10	F	F	4	3-4
Implore*	Sovereign Dancer/Herbager	C	8	U	C	4	2-3
Implosion*	Explodent/Diplomat Way	C+	10	L	M	5	3-4

Stallion	Sire/Broodmare Sire	FTS	CL	Trf	OT	SI	AB
Impressability*	Our Native/Bold Ruler	C+	8	U	F	5	3-4
Impressive	Court Martial/Ambiorix	C+	5	B	D	4	2-3
Imp Society	Barrera/Promised Land	D	7	C	C	3	2-3
Incinderator*	Northern Dancer/Gummo	M	7	L	M	4	3-4
Inflated Ego	Ego Eye/Big Brave	M	10	U	M	5	3-4
In From Dixie	Isgala/Grey Dawn II	C	9	U	D	4	3-4
Inherent Star	Pia Star/Dumpty Humpty	B	10	FF	B	4	3-4
In Reality	Intentionally/Rough'n Tumble	B	1	C	A+	4*	2-3
Instrument Landing	Grey Dawn II/Bold Lad	D	8	B-	F	4	3-4
Interco	Intrepid Hero/Majestic Prince	B	6	A-	C	4	3-4
Interdicto	Grey Dawn II/Ahoy	C	6	B	F	4	3-4
In the Swing*	Blushing Groom/Grey Dawn II	M	8	M	M	4	2-3
In the Woodpile	Raise a Native/Prince John	C+	8	C	C	4	3-4
Intimidation*	Sauceboat/Model Fool	B+	9	C	A	4	2-3
In Tissar	Roberto/Bold Bidder	D	8	F	C	5	3-4
In Totality	In Reality/Spy Song	C+	8	F	C+	5	2-3
Inverness Drive	Crozier/Court Martial	D	5	D	C	4	3-4
Irish Bard*	Tudor Melody/Red God	M	9	M	M	4	3-4
Irish Castle	Bold Ruler/Tulyar	D	5	C	C	3	3-4
Irish Conn	Buckpasser/Victorian Era	B	10	U	F	4	3-4
Irish Dreamin^	The Irish Lord/Any Time Now	M	8	U	L	5	2-3
Irish Open*	Irish Tower/Buckpasser	L	6	U	L	4	2-3
Irish River c	Riverman/Klairon	C	4	C	C	3*	3-4
Irish Stronghold	Bold Ruler/Tulyar	D	7	D	F	4	3-4
Irish Sur*	Surreal/Irish Dude	U	9	U	U	4	3-4
Irish Swords*	Raja Baba/Graustark	C+	9	U	U	4	3-4
Irish Tower	Irish Castle/Loom	B	5	D	C+	4	2-3
Iron	Mr. Prospector/Buckpasser	D	7	D	D	4	3-4
Iron Constitution	Iron Ruler/Hail to Reason	D	5	D	B	4	3-4
Iron Courage^	Caro/Fleet Nasrullah	U	7	L	M	2	3-4
Iron Ruler	Never Bend/Mahmoud	D	8	D	C	4	3-4
Island Champ*	Dancing Champ/Search for Gold	C+	8	F	F	3	3-4
Island Sultan	What a Pleasure/Prince John	D	8	F	F	4	3-4
Island Whirl	Pago Pago/Your Alibhai	C	6	C	C	4	3-4
It's Acedemic^	Sauce Boat/Arts and Letters	U	6	U	M	2	3-4
It's Freezing	TV Commercial/Arctic Prince	B-	4*	C	C+	4	2-3
It's the One	Dewan/Hawaii	F	8	F	C	4	3-4
Jacques Who	Grey Dawn II/Nantallah	C	6	D	F	3	3-4
Jade Hunter	Mr. Prospector/Pharly	B	4	A-	BT	4*	2-3
Jaklin Klugman	Orbit Ruler/Promised Land	B-	7	D	C	4	3-4
Jammed Gold*	Kentucky Gold/Mito	U	6	U	L	4	3-4

Stallion	Sire/Broodmare Sire	FTS	CL	Trf	OT	SI	AB
Jane's Dilemma^	Master Derby/Hilarious	U	9	U	M	4	2-3
Jan's Kinsman	Kinsman Hope/Smart	F	10	U	F	3	3-4
Java Gold	Key to the Mint/Nijinsky II	C	5	C	C	2	3-4
Jazzing Around*	Stop the Music/Jacinto	C+	8	M	C	4	2-3
J. Burns	Bold Hour/Double Jay	C	8	F	C	5	3-4
Jeblar	Alydar/Lucky Debonair	B	6	C	F	4	2-3
Jeff's Companion*	Noholme II/Vaguely Noble	B+	9	F	C	4	2-3
Jeloso*	Raise a Native/Nijinsky II	M	8	M	M	5	2-3
Jerimi Johnson	Barachois/Bolinas Boy	D	9	C	C	5	2-3
Jig Time	Native Dancer/Case Ace	C	4	B-	C+	4	2-3
Jihad	Damascus/Bold Ruler	F	8	U	D	4	3-4
Jim French	Graustark/Tom Fool	D	2	F	F	2	3-4
Jitano*	Rattle Dancer/Halter	U	9	M	U	4	3-4
Jitterbug Chief*	Sovereign Dancer/Prove It	U	10	M	M	4	3-4
Joanie's Chief	Ack Ack/Fleet Host	C	7	D	C	4	3-4
John Alden	Speak John/Nashua	D	7	C	D	2	3-4
Johnny Pro	Semi Pro/Mr. Music	F	8	U	F	5	3-4
John's Gold	Bold Bidder/Buckpasser	F	7	D	C	4	3-4
Jolly Johu	Restless Native/Gallant Man	F	6	D	F	4	3-4
Jon Ian^	Our Michael/Catallus	U	10	U	M	5	3-4
J.O. Tobin	Never Bend/Hillary	C	7	C	T	4	3-4
Judge Lex	Judge Kilday/Citation	C	9	F	D	4	3-4
Judge Smells	In Reality/Best Turn	D	6	F	C+	4	2-3
Jump Over the Moon	Vertex/Sadair	C	8	B-	D	4	3-4
Junction	Never Bend/Royal Union	D	8	D	D	4	3-4
Jungle Blade c	Blade/Jungle Road	C	8	D	D	2	3-4
Just a Tab^	Pass the Tab/Faraway Son	M	8	U	M	2	2-3
Just Plain Tuff	Don-Ce-Sar/Nell's Boy	F	10	U	F	4	3-4
Just Right Classi	Cornish Prince/Blue Prince	M	6	U	L	4	2-3
Just Right Mike	Mastadoon/Windy Sands	C	8	F	D	5	3-4
Just the Time	Advocator/Never Bend	C	8	A-	C	4	2-3
Kaintuck*	Kentuckian/Rainy Lake	D	8	U	F	4	3-4
Kala Native	Exclusive Native/Kalamoun	D	8	A	F	4	3-4
Kaldoun	Caro/Le Haar	C	2	B	C	1	3-4
Kalim*	Hotfoot/Lyphard	F	7	L	F	4	3-4
Katowice^	Danzig/Prince John	L	8	L	M	4	2-3
Kenmare	Kalamoun/Milesian	U	3	L	M	4*	3-4
Kennedy Road	Victoria Park/Nearctic	C	7	C	C	4	3-4
Kentucky Cookin*	Best Turn/Tim Tam	M	10	M	A	4	2-3
Kentucky Jazz^	Dixieland Band/Restless Wind	M	8	M	L	3	2-3

Stallion	Sire/Broodmare Sire	FTS	CL	Trf	OT	SI	AB
Key to the Kingdom	Bold Ruler/Princequillo	D	6	C	F	3	3-4
Key to the Mint	Graustark/Princequillo	C	3	D	C	2	3-4
Kibe	Raja Baba/Frosty Mr.	D	8	U	C	5	3-4
Killarney Road*	His Majesty/Damascus	U	6	M	M	4	2-3
King Alphonse*	Ship Leave/Bull Page	M	7	U	C	4	2-3
King Concorde	Super Concorde/Damascus	D	8	F	F	4	3-4
King of Kings	Vaguely Noble/Shut Out	D	8	D	C	2	3-4
King of the North	Northern Dancer/Sensitivo	D	7	B-	C+	2	3-4
King's Bishop	Round Table/Fleet Nasrullah	C	4	B	B	4	3-4
King's Nest^	Rollicking/No Robbery	L	7	M	L	5	2-3
King Pellinore	Round Table/Nantallah	D	7	C	C	4	3-4
Kleven*	Alydar/Tudor Minstrel	C+	7	F	C	3	3-4
Knight*	Mr. Prospector/Jacinto	M	8	U	L	5	2-3
Knightly Rapport*	Inverness Drive/Round Table	L	9	U	M	5	2-3
Knights Choice	Drum Fire/Turn To	B	6	D	C+	5	2-3
Known Fact	In Reality/Tim Tam	B+	5	C	C	4*	2-3
Know Your Aces	Call Me Prince/Nahar	U	9	M	M	4	3-4
Kodiak	Run For Nurse/Mongo	M	8	U	M	3	3-4
Kokand*	Mr. Prospector/Nijinsky II	M	8	M	M	4	2-3
Koluctoo Bay	Creme Dela Creme/Double Jay	D	7	D	C	2	3-4
Kona Tenor	In Reality/Native Dancer	D	6	F	C	5	2-3
Kris	Sharpen Up/Reliance	C	2	B	C	I	3-4
Kris S.	Roberto/ Princequillo	C	3	C	C+	2	3-4
Kyle's Our Man^	In Reality/Graustark	M	6	U	L	4	2-3
Lac Ouimet^	Pleasant Colony/Northfields	U	6	L	M	I	3-4
Lahab*	The Minstrel/Buckpasser	M	10	U	U	4	3-4
Land of Believe	Believe It/First Landing	F	10	U	L	4	3-4
Laomedonte	Raise a Native/Toulouse Lautrec	D	8	D	U	4	3-4
La Saboteur*	Seattle Slew/Exclusive Native	B	8	M	L	5	2-3
Launch a Dream^	Relaunch/Chieftain	M	7	U	L	4	2-3
Launch a Leader*	Relaunch/Caro	M	7	M	L	2	3-4
Launching^	Relaunch/Cougar	U	8	M	L	3	3-4
Lashkari	Mill Reef/Right Royal	D	4	L	U	I	3-4
Last Tycoon	Try My Best/Mill Reef	M	5	L	M	4*	3-4
Late Act	Stage Door Johnny/Tom Fool	D	7	C	C+	2	3-4
Law Society	Alleged/Boldnesian	U	4	L	M	I	3-4
Lazaz^	Blushing Groom/Shoemaker	U	8	L	L	I	3-4
Leading Hour*	Bold Hour/Mr. Leader	U	8	U	L	4	3-4
Lear Fan	Roberto/Lt. Stevens	B	4	B	C+	3*	2-3
Le Danseur	Lord Durham/Dancer's Image	D	6	D	C	4	3-4
Leematt	Turn to Reason/Bull Briar	D	7	F	F	4	3-4

Stallion	Sire/Broodmare Sire	FTS	CL	Trf	OT	SI	AB
Le Fabuleux	Wild Risk/Verso II	F	2	B-	T	I	<u>3-4</u>
Legal Prospector	Mr. Prospector/First Landing	C	7	U	L	4	2-3
Lejoli	Cornish Prince/Sir Ivor	C	6	F	C	4	3-4
Lemhi Gold <u>c</u>	Vaguely Noble/Candy Spots	F	7	D	F	3	<u>3-4</u>
L'Emigrant	The Minstrel/Vaguely Noble	F	5	C	C	2	3-4
L'Empire	The Minstrel/Vaguely Noble	U	8	U	M	4	3-4
L'Enjoleur	Buckpasser/Northern Dancer	D	5	D	D	2	2-3
Leo Castelli*	Sovereign Dancer/Raise a Native	M	5	M	M	3	<u>2-3</u>
Leroy S.	Honest Pleasure/Dr. Fager	C	8	D	D	4	3-4
Les Apres	Riverman/Cadmus	D	8	C	F	4	<u>3-4</u>
Lieutenant's Lark*	Lt. Stevens/Knightly Dawn	U	7	M	F	4	3-4
Light Idea*	MajesticLight/Hail to Reason	M	7	M	U	4	<u>2-3</u>
Light Years	Explodent/Sunny	F	9	U	C	4	<u>2-3</u>
Lil Fappi*	Fappiano/What a Pleasure	L	6	U	L	4	<u>2-3</u>
Liloy	Bold Bidder/Spy Song	C	7	C	C	4	<u>3-4</u>
Lil Tyler*	Halo/TV Lark	F	8	C	F	4	3-4
Lines of Power	Raise a Native/Bold Ruler	C	6	C	C+	4	3-4
Linkage	Hoist the Flag/Cyane	D	6	C	C	4	3-4
Litchi	Jacinto/Graustark	F	9	U	U	4	3-4
Literati^	Nureyev/Bleep-Bleep	M	7	L	U	3	3-4
Little Current	Sea Bird/My Babu	D	5	C	T	3	3-4
Little Missouri	Cox's Ridge/Jacinto	C	5	F	F	4	<u>2-3</u>
Little Secreto^	Secreto/Dike	M	10	M	M	4	3-4
Lively One^	Halo/The Axe II	M	5	U	M	2	2-3
L'Natural	Raise a Native/Seaneen	C	7	B-	C	5	3-4
Local Talent*	Northern Dancer/Vaguely Noble	M	6	<u>L</u>	U	2	<u>2-3</u>
Lode*	Mr. Prospector/Sir Ivor	D	8	U	M	4	2-3
Loft	Roberto/What a Pleasure	C	7	M	C	<u>4</u>	3-4
Lomond	Northern Dancer/Poker	C	7	C	L	2	3-4
London Bells	Nijinsky II/Raise a Native	C+	7	D	F	4	3-4
London Company	Tom Rolfe/Bolero	F	7	D	D	4	3-4
Loom	Swoon's Son/Beau Pere	B	7	F	C+	5	2-3
Loose*	Bailjumper/Misty Flight	M	10	M	M	4	3-4
Loose Cannon	Nijinsky II/Graustark	F	8	C	D	<u>4</u>	3-4
Lord at War	General/Con Brio	C	3	B	BT	2	2-3
Lord Avie	Lord Gaylord/Gallant Man	C	5	C	C	3	3-4
Lord Carlos*	Lord Gaylord/Bold Legend	B	6	U	B	5	<u>2-3</u>
Lord Chilly*	Lord Durham/Nearctic	L	9	U	L	4	<u>2-3</u>
Lord Double Gate*	Lord Gaylord/Groton	U	8	M	L	4	<u>3-4</u>
Lord Gaylord	Sir Gaylord/Ambiorix	C	5	B	C+	2	3-4

Stallion	Sire/Broodmare Sire	FTS	CL	Trf	OT	SI	AB
Lord of All	Seattle Slew/Restless Native	F	8	C	C	3	3-4
Lord of the Apes	Alydar/Jungle Cove	M	7	U	M	4	2-3
Lord of the Night*	Lord Avie/Northern Dancer	M	9	M	M	3	2-3
Lord of the Sea	Torsion/Cyclotron	C	8	C	F	5	2-3
Lord Rebeau	Maribeau/Call Over	C	7	D	F	3	2-3
Lord Treasurer	Key to the Mint/Sir Gaylord	F	8	A	C+	4	3-4
Lost Code	Codex/Ack Ack	C	4	B-	C	3	2-3
Lost Opportunity^	Mr. Prospector/Buckpasser	M	7	U	M	3	3-4
Lothario	Nashua/Tom Fool	B-	7	F	C	4	3-4
Lot o' Gold	Lothario/Young Emperor	D	7	C	D	4	3-4
Louisiana Slew	Seattle Slew/New Policy	C	8	F	F	4	3-4
Loustrous Bid*	Illustrious/Cannonade	C	6	U	A+	5	2-3
Love That Mac*	Great Above/Irish Ruler	M	10	U	M	5	3-4
Loyal Double^	Nodouble/Royal Ascot	U	8	L	M	2	3-4
Loyal Pal*	Caro/Sailor	U	7	M	U	3	3-4
Lt Flag	Delta Flag/Lt Stevens	F	10	U	M	4	3-4
Lt Stevens	Nantallah/Gold Bridge	B	3	C	B	5	2-3
Lucky North	Northern Dancer/Olden Times	D	5	B-	C	4*	2-3
Lucky Prospect	Mr. Prospector/What Luck	F	9	U	F	5	3-4
Lucky So N' So^	Alydar/Nashua	L	7	U	M	4	2-3
Luthier	Klairon/Cranach	M	2	L	U	1	3-4
Lustra^	Danzig/Buckpasser	M	7	M	M	4	2-3
Lyphaness	Lyphard/Ridan	C+	9	U	M	4	2-3
Lyphard	Northern Dancer/Court Martial	B-	3	B	F	1*	3-4
Lyphard's Wish	Lyphard/Sensitivo	D	5	C	F	3*	3-4
Lypheor	Lyphard/Sing Sing	U	3	L	M	4*	3-4
Lytrump^	Lypheor/Bold Bidder	M	9	L	M	4	3-4
Macarthur Park	First Balcony/Conjure	D	7	F	C	3	3-4
Machiavellian^	Mr. Prospector/Halo	L	3	U	M	5	2-3
Macho Hombre	Raise a Native/Sir Gaylord	C	8	D	F	4	3-4
Magesterial	Northern Dancer/Bold Lad	D	6	C	C	4	2-3
Magic North	North Sea/Impressive	D	9	D	F	4	3-4
Magic Rascal*	Raise a Native/Ack Ack	M	10	U	M	4	3-4
Magloire*	Exceller/Round Table	M	7	M	M	4	3-4
Maheras	Walker's/Ambiopoise	C+	8	D	C	4	3-4
Main Debut	Seattle Slew/Bold and Brave	C+	8	D	C	4	2-3
Majestic Light	Majestic Prince/Ribot	C	4*	C	F	1*	3-4
Majestic Reason*	Majestic Prince/Hail to Reason	M	8	M	M	2	3-4
Majestic Shore*	Majestic Prince/Windy Sands	C	8	D	C	4	3-4
Majestic Venture*	Majestic Prince/Illustrious	C	8	F	C	4	3-4
Majesty's Prince	His Majesty/Tom Fool	D	7	C	C	1*	3-4

Stallion	Sire/Broodmare Sire	FTS	CL	Trf	OT	SI	AB
Major Moran*	Colonel Moran/Irish Ruler	M	6	M	U	2	2-3
Malagra*	Majestic Light/Vice Regal	M	8	M	U	5	2-3
Malinowski	Sir Ivor/Traffic Judge	F	9	D	C	2	3-4
Mamaison	Verbatim/Groton	C	7	D	F	4	3-4
Mambo	Northern Dancer/Bold Ruler	C+	8	C	C	4	3-4
Manatash Ridge*	Seattle Slew/Crewman	U	8	M	M	4	2-3
Mane Minister^	Deputy Minister/In Reality	U	6	U	M	2	3-4
Mangaki*	Olympiad King/Isle of Greece	C+	9	L	C	4	3-4
Manila	Lyphard/Le Fabuleux	D	5	C	C+	1	3-4
Man of Vision*	Exclusive Native/Warfare	U	9	M	U	4	3-4
Manzotti*	Nijinsky II/Tom Rolfe	L	8	M	M	3	2-3
Marcellini	Never Bend/Pretense	D	8	A	F	4	2-3
Marfa	Foolish Pleasure/Stratmat	D	6	C	F	3	3-4
Marine Brass	Fifth Marine/In Reality	C	7	C	C	4	3-4
Marine Patrol	Sail On Sail On/Rasper II	F	8	F	C+	4	3-4
Mari's Book	Northern Dancer/Maribeau	C+	6	B-	C	3	3-4
Mark in the Sky*	Graustark/Hasty Road	F	10	U	C	3	3-4
Mark of Nobility*	Roberto/On Your Mark II	B	7	M	C	4	2-3
Marsayas*	Damascus/Round Table	B+	6	U	C	4	2-3
Marshua's Dancer	Raise a Native/Nashua	C	5	C	C	4	3-4
Martial Law*	Mr. Leader/Round Table	U	8	U	T	2	3-4
Masked Dancer	Nijinsky II/Spy Song	D	6	D	F	2	3-4
Master Derby	Dust Commander/Royal Coinage	D	7	D	D	4	3-4
Masterful Advocate*	Torsion/Grey Dawn II	U	9	U	L	4	3-4
Mat-Boy	Matun/Pastiche	U	4	U	T	1	3-4
Matchlite*	Clever Trick/Sir Ivor	F	8	M	C+	3	2-3
Matsadoon	Damascus/Prince John	C	6	D	C	4	3-4
Matter of Honor^	Conquistador Cielo/Venetian Jester	M	8	M	L	4	2-3
Maudlin	Foolish Pleasure/Dr. Fager	C	5	B-	C	3	2-3
Maui Lypheor Jack	Maui Lypheor/In Zeal	U	10	U	M	4	3-4
Mawstuff^	Known Fact/Dancer's Image	L	7	M	M	4	3-4
Maxistar*	Pia Star/To Market	B	9	A	C	4	3-4
McCann	Forli/Better Bee	F	7	B-	F	4	3-4
M Double M	Nodouble/New Policy	D	8	C	D	4	3-4
Meadowlake	Hold Your Peace/Raise a Native	A	4	C	A+	4	2-3
Media Starguest*	Be My Guest/Sea-Bird	M	9	M	U	1	2-3
Medieval Man	Noholme II/Tim Tam	C	5	C	F	4	2-3
Megaturn	Best Turn/Vertex	D	7	F	F	3	2-3
Mehmet	His Majesty/Swaps	C	7	D	C	4	3-4
Melodisk*	Alydar/One for All	U	9	U	M	3	3-4

Stallion	Sire/Broodmare Sire	FTS	CL	Trf	OT	SI	AB
Menderes*	Noholme II/Baybrook	F	8	C	A+	4	2-3
Meneval	Le Fabuleux/Nashua	U	7	M	M	4	3-4
Mercedes Won^	Air Forbes Won/Roman Line	M	6	M	L	3	2-3
Meritable	Round Table/Bold Ruler	C	8	C	D	4	2-3
Mertzon	Mr. Prospector/Damascus	C	8	D	C	4	2-3
Meshach	Icecapade/Prince John	D	9	C	F	4	3-4
Messenger of Song	Envoy/Tudor Minstrel	C	8	C	T	4	3-4
Metfield^	Seattle Slew/Hail to Reason	M	6	M	L	4	3-4
Mickey McGuire	TV Lark/Determine	M	5	U	L	4	2-3
Midway Circle*	Alydar/Round Table	C	8	M	C	3	3-4
Mighty Adversary	Mr. Redoy/New Charger	C	7	D	C	4	2-3
Mighty Appealing	Valid Appeal/Roi Dagobert	C+	5	C	C+	4	2-3
Mighty Courageous*	Caro/Seaneen	U	10	M	M	3	3-4
Mike Fogerty	Royal and Regal/Sea Bird	C	8	C	F	4	3-4
Mileage Miser	Brazen Brother/Beau Prince	F	10	U	C	4	3-4
Mill Native*	Exclusive Native/Mill Reef	U	8	M	M	3	3-4
Mill Reef	Never Bend/Princequillo	U	1	L	M	1	3-4
Mining	Mr. Prospector/Buckpasser	B+	4	D	A+	4	2-3
Minneapple*	Riverman/Umbrella Fella	U	8	M	U	4	2-3
Minnesota Mac	Rough'n Tumble/Mustang	B	4	B	B	4	2-3
Minshaanshu Amad	Northern Dancer/Chieftain	C+	8	C	F	3	3-4
Mi Selecto^	Explodent/Ribot	U	6	L	M	2	3-4
Mister Frisky^	Marsayas/Highest Tide	M	6	U	L	3	2-3
Mister Wonderful^	Mummy's Pet/Petingo	U	7	L	M	2	3-4
Miswaki	Mr. Prospector/Buckpasser	C	4	C	C	4*	3-4
Miswaki Gold*	Miswaki/Lt. Stevens	L	8	U	M	5	2-3
Mogambo	Mr. Prospector/Rainy Lake	D	7	C	F	4	3-4
Mokhieba	Damascus/Royal Vale	D	8	U	T	4	3-4
Mombo Jumbo	Jungle Savage/Iron Peg	C+	8	C	C+	4	3-4
Moment of Hope	Timeless Moment/Baffle	D	7	C	F	3	3-4
Monetary Gift	Gold and Myrrh/Dancing Dervish	B	7	F	C+	4	2-3
Mongo's Image	Mongo/Major Portion	D	8	F	C	5	2-3
Monsieur Champlain	Cutlass/Good and Plenty	F	8	F	F	4	3-4
Moon Up T. C.*	Sharpen Up/Roberto	M	9	L	M	2	3-4
Morning Bob	Blushing Groom/The Axe II	D	7	C	C	2	3-4
Morning Charge	Battle Joined/Eight Thirty	F	8	U	D	4	3-4
Moro*	Full Out/Mito	F	6	D	D	4	3-4
Moscow Ballet	Nijinsky II/Cornish Prince	C	7	D	C	4	3-4
Mountain Express	Caro/Warfare	D	10	M	U	3	3-4
Mountain Lure	Rock Talk/To Market	D	8	D	F	4	3-4
Mount Hagen	Bold Bidder/Tom Fool	D	8	D	C	4	3-4

Stallion	Sire/Broodmare Sire	FTS	CL	Trf	OT	SI	AB
Mr. Badger	Mr. Leader/Pago Pago	C+	10	U	F	4	3-4
Mr. Classic*	Mr. Prospector/Board Marker	M	9	U	M	4	2-3
Mr. Crimson Ruler	Secretariat/Crimson Satan	F	8	U	F	5	3-4
Mr. Goldust^	Mr. Prospector/Crepello	M	8	M	M	3*	3-4
Mr. Leader	Hail To Reason/Djeddah	C	4	C	C	3	3-4
Mr. Prospector	Raise a Native/Nashua	A-	1*	C	A	4*	2-3
Mr. Redoy	Grey Dawn II/Pocket Ruler	C	7	D	D	4	3-4
Mt. Livermore	Blushing Groom/Crimson Satan	B+	4	D	B	4	2-3
Mt. Magazine*	Mr. Prospector/Damascus	U	9	U	L	4	3-4
Much Fine Gold*	L'Natural/Drone	M	10	U	M	5	3-4
Mugatea*	Hold Your Peace/Olden Times	C	8	C	F	5	3-4
Murrtheblurr	Torsion/Cornish Prince	C	8	B	F	5	2-3
Music Prince*	Stop the Music/Call Me Prince	F	7	M	F	3	3-4
Muttering	Drone/Gamin'	D	8	F	F	4	3-4
My Favorite Moment*	Timeless Moment/Dragante	L	10	M	L	4	2-3
My Gallant	Gallant Man/Nashua	C	4	D	C	3	2-3
My Habitony	Habitony/Dogoon	F	9	D	C	4	3-4
My Prince Charming^	Sir Wimborne/Stage Door Johnny	U	8	M	L	3	3-4
Mythical Ruler	Ruritania/Victorian Era	C	7	D	C+	2	3-4
Nabeel Dancer^	Northern Dancer/Foolish Pleasure	U	6	L	M	5	3-4
Naevus	Mr. Prospector/Bold Lad	C	6	C	A	4	2-3
Nain Bleu	Lyphard/Tanerko	D	7	C	F	4	3-4
Naked Sky	Al Hattab/Forli	D	7	C	C	2	2-3
Nalees Man	Gallant Man/Nashua	D	7	B	C	4	2-3
Nantequos	Tom Rolfe/Nantallah	D	7	C	C	4	3-4
Nashwan^	Blushing Groom/Bustino	M	2	L	L	1	2-3
Naskra	Nasram/Le Haar	C	3*	D	B	2	3-4
Nasty and Bold	Naskra/Boldnesian	C+	4	C	C+	4	3-4
Nataraja	Nijinsky II/Admiral's Voyage	F	8	B-	F	4	3-4
Nathan Detroit*	Drone/Forli	D	10	M	U	4	2-3
Native Charger	Native Dancer/Heliopolis	C	4	C	C	4	3-4
Native Host	Raise a Native/Sword Dancer	D	8	D	C	5	3-4
Native Prospector	Mr. Prospector/Carlemont	D	7	C	F	4	3-4
Native Royalty	Raise a Native/Nasrullah	D	5*	C	D	4	3-4
Native Tactics	George Navonod/Delta Judge	D	8	B-	C	5	2-3
Native Uproar	Raise a Native/Never Bend	C	7	B	C	4	3-4
Native Wizard*	In Reality/Northern Dancer	M	8	M	L	4	3-4
Navajo	Grey Dawn II/Double Jay	D	5*	C	F	3	3-4
Negotiated Account*	Private Account/Hoist the Flag	U	8	U	T	3	3-4
Nepal	Raja Baba/Grey Dawn	F	8	D	F	4	3-4

Stallion	Sire/Broodmare Sire	FTS	CL	Trf	OT	SI	AB
Never Bend	Nasrullah/Djeddah	B	I	B	B	3*	2-3
Never Hi	Never Bend/Polynesian	F	8	F	F	5	2-3
Never Tabled	Never Bend/Round Table	C	7	C	B	5	3-4
New Circle*	Circle/Bald Eagle	U	10	M	U	4	2-3
Nice Pirate	Hail the Pirates/Graustark	U	8	L	U	4	3-4
Nicholas^	Danzig/Tom Rolfe	L	4	L	L	5	2-3
Night Mover	Cutlass/Nantallah	C+	8	U	L	4	3-4
Night Shift	Northern Dancer/Chop Chop	D	7	C	C	4	2-3
Nijinsky II	Northern Dancer/Bull Page	C+	I*	B	C	I*	3-4
Nijinsky's Secret c	Nijinsky II/Raise a Native	F	8	D	F	3	3-4
Nisswa^	Irish River/Mr. Prospector	M	8	M	M	3	3-4
No Bend	Never Bend/Traffic Judge	C	8	F	F	4	3-4
Noble Assembly*	Secretariat/Damascus	U	7	U	M	4	3-4
Noble Dancer	Prince de Galles/Sing Sing	C	7	C	C	4	3-4
Noble Monk*	African Sky/Mandamus	C+	8	A	A	5	3-4
Noble Nashua	Nashua/Vaguely	D	5*	D	D	4	3-4
Noble Saint	Vaguely Noble/Santa Claus	D	8	C	D	2	3-4
Nodouble	Noholme II/Double Jay	C	2*	B	C+	3*	3-4
Noholme II	Star Kingdom/	B	3*	B	B	4*	2-3
Noholme Way	Noholme II/Fleet Nasrullah	C	8	C	C	3	3-4
No House Call	Dr. Fager/Olden Times	D	8	F	B	4	3-4
No Louder	Nodouble/Quadrangle	C	5	A-	C	2	2-3
No Points*	Miswaki/Amerigo	M	8	M	M	4	3-4
Norcliffe	Buckpasser/Northern Dancer	D	7	C	C+	4	3-4
Nordic Prince	Nearctic/Hill Prince	C	7	D	C	4	3-4
No Robbery	Swaps/Bimelech	C	3*	C	C	2	3-4
Norquestor^	Conquistador Cielo/Northern Dancer	U	6	M	L	2	3-4
Northern Baby	Northern Dancer/Round Table	D	5	C	C	I	3-4
Northern Baron	Northern Jove/Young Emperor	C	8	D	D	5	3-4
Northern Birdie^	Proud Birdie/Northcliffe	U	8	M	M	4	3-4
Northern Dancer	Nearctic/Native Dancer	C	I	A	C	4*	2-3
Northern Flagship*	Northern Dancer/Raise a Native	M	7	M	M	4	2-3
Northern Fling	Northern Dancer/Hasty Road	C	5*	B	C	I	3-4
Northern Horizon	Northern Dancer/Court Harwell	D	8	U	C	4	2-3
Northern Ice*	Fire Dancer/Vertex	F	9	D	C	4	3-4
Northern Jove	Northern Dancer/Sun Again	D	5	D	F	4	3-4
Northern Magus	Olden Times/Exclusive Native	D	8	F	C	4	2-3
Northern Majesty	His Majesty/Northern Dancer	D	7	C	F	4	3-4
Northern Passage	Northern Dancer/Cyane	M	9	M	L	4	3-4
Northern Prospect	Mr.Prospector/Northern Dancer	C+	4	D	D	4	2-3
Northern Raja	Raja Baba/Northern Dancer	F	7	B	F	4	3-4

Stallion	Sire/Broodmare Sire	FTS	CL	Trf	OT	SI	AB
Northern Ringer	Northern Dancer/Round Table	F	8	C	F	4	3-4
Northern Score*	Northern Dancer/Halo	L	8	M	M	4	2-3
Northern Supremo	Northern Dancer/Habitat	D	8	C	C	4	3-4
Northern Wolf^	Wolf Power/Northern Fling	M	7	M	M	2	2-3
Northfields	Northern Dancer/Occupy	D	3	C	F	1	3-4
Northjet	Northfields/Fortino	F	7	C	F	4*	3-4
North Pole	Northern Dancer/Canadian Champ	B	7	D	C+	5	3-4
North Prospect/	Mr.Prospector/Northern Dancer	C	8	D	F	4	3-4
Northrop/	Northern Dancer/Warfare	C+	7	B-	T	4	3-4
Northwest Passage	Nearctic/Native Dancer	C	9	C	C	4	3-4
No Sale George	Raise a Native/Jacinto	B	7	C	A	4	3-4
Nostalgia	Silent Screen/Herbager	F	8	C	F	2	3-4
Nostalgia's Star*	Nostalgia/Big Spruce	U	9	U	M	4	3-4
Nostrum	Dr. Fager/Native Dancer	D	5	D	C+	4	3-4
Notebook*	Well Decorated/Tom Rolfe	M	7	U	M	4	2-3
Not Surprised*	His Majesty/Never Bend	U	8	M	U	3	2-3
Novel Nashua^	Noble Nashua/Prince John	U	9	U	M	4	3-4
Nureyev	Northern Dancer/Forli	C	3	B	F	3*	2-3
O Big Al	Don B./Traffic Judge	C	8	C	B	4	3-4
Obligato/	Northern Dancer/In Reality	B	2	C	F	4	3-4
Obraztsovy	His Majesty/Nashua	C	8	C	F	4	3-4
Ocala Slew^	Seattle Slew/Diplomat Way	M	8	M	L	4	3-4
Ocean Trick*	Clever Trick/Sham	B	8	U	C+	4	2-3
Ogygian c	Damascus/Francis S.	C+	5	D	F	4	3-4
Oh Say	Hoist the Flag/Cyane	C	5	D	C	4	3-4
Old Broadway*	Broadway Forli/Hail to Reason	M	7	M	U	4	2-3
Olden Times	Relic/Djebel	C	4	D	B	3	2-3
Ole'^	Danzig/Verbatim	M	7	L	L	4	2-3
Ole Bob Bowers	Prince Blessed/Bull Lea	D	6	B	C	4	3-4
Olympic Native*	Raise a Native/Olympiad King	M	8	M	M	5	3-4
Once Wild^	Baldski/What a Pleasure	M	6	U	U	4	2-3
One For All	Northern Dancer/Princequillo	D	7	B	D	2	3-4
One More Slew	Seattle Slew/Caro	F	8	D	F	2	3-4
On to Glory	Bold Lad/Native Dancer	C	6	D	F	4*	2-3
Only Dreamin*	Raise a Native/Prince John	U	10	U	M	5	3-4
Opening Verse^	The Minstrel/Grey Dawn II	M	5	M	U	2	2-3
Opposite Abstract^	Alydar/Bold Commander	U	9	L	M	2	3-4
Oraibi^	Forli/Northern Dancer	U	7	M	M	2	3-4
Orbit Dancer c	Northern Dancer/Gun Shot	D	7	C	C	4	2-3
Orbit Ruler	Our Rulia/Royal Orbit	U	7	U	M	5	3-4

Stallion	Sire/Broodmare Sire	FTS	CL	Trf	OT	SI	AB
Orbit Scene^	Orbit Dancer/Pachuto	L	7	L	U	3	2-3
Order^	Damascus/Reviewer	M	8	U	T	3	2-3
Ormonte*	Danzig/Nodouble	M	8	L	L	4	3-4
Orono*	Codex/Gallant Man	L	8	M	U	4	2-3
Our Bold Landing	Bold Forbes/First Landing	C	8	U	F	5	3-4
Our Captain Willie	Little Current/Ribot	D	8	L	F	3	3-4
Our Hero	Bold Ruler/Aristophanes	M	8	M	M	4	3-4
Our Liberty	Raise a Native/Round Table	B+	7	D	F	4	2-3
Our Michael f	Bolero/Tribe	D	6	D	F	4	2-3
Our Native	Exclusive Native/Crafty Admiral	D	4	C	D	2	2-3
Our Talisman	Cornish Prince/Arrogate	C+	8	A	F	4	3-4
Out of the East	Gummo/Windy Sands	C	7	C	D	4	3-4
Overskate	Nodouble/Speak John	D	4	C	D	4	3-4
Owens Troupe^	For the Moment/Drone	M	9	U	M	3	3-4
Pachuto	Lt. Stevens/Summer Tan	B	10	F	F	4	3-4
Pac Mania	Ramsinga/Boldnesian	M	8	U	M	5	3-4
Pair of Deuces	Nodouble/Your Alibhai	D	8	C	C	2	3-4
Palace Music	The Minstrel/Prince John	C	6	D	F	3	3-4
Palace Panther*	Crystal Palace/Val de Loir	U	8	B-	U	1	3-4
Palmers Tex	Vertex/Ambehaving	F	10	B-	F	4	3-4
Pancho Villa	Secretariat/Crimson Satan	B	6	C	F	4	2-3
Pappagallo	Caro/Hard Sauce	D	8	C	C	4	3-4
Pappa Riccio	Nashua/Raise a Native	D	6	F	C	4	2-3
Paramount Jet*	Tri Jet/Quadrangle	M	7	U	L	4	2-3
Parfaitement	Halo/The Axe II	C	6	C	T	4	3-4
Paris Dust	Dust Commander/Ambiorix	C	8	C	D	4	2-3
Parlay Me^	Irish Tower/Native Charger	L	7	U	L	4	2-3
Parochial^	Mehmet/Hail to Reason	M	7	U	M	2	3-4
Parramon	Lyphard/Cyane	M	9	M	U	3	3-4
Partez	Quack/Tim Tam	C+	8	D	F	4	2-3
Pas de Cheval	Nijinsky II/Bold Ruler	F	9	B-	F	4	3-4
Pas Seul	Northern Dancer/Bold Ruler	C	7	B	C+	4	3-4
Pass'n Raise	Raise a Native/Alibhai	F	8	U	U	5	2-3
Pass the Glass	Buckpasser/Amerigo	B	7	C	C	4	3-4
Pass the Line	Pas Seul/Roman Line	D	8	C	F	2	3-4
Pass the Tab	Al Hattab/Gray Phantom	C	7	D	D	4	2-3
Patch of Sun*	Sunny Clime/Illustrious	M	10	U	L	5	3-4
Patriotically*	Hoist the Flag/Amerigo	C	8	B	A	4	2-3
Pauper Prince*	Majestic Prince/Round Table	C+	8	B-	U	4	2-3
Peace Arch	Hoist the Flag/Restless Native	B	9	M	C	4	3-4
Pencil Point	Sharpen Up/Hill Clown	C	8	C	C	4	3-4

Stallion	Sire/Broodmare Sire	FTS	CL	Trf	OT	SI	AB
Pentelicus^	Fappiano/In Reality	L	7	U	L	5	2-3
Persevered	Affirmed/Never Bend	F	9	F	F	4	3-4
Persian Bold	Bold Lad/Relko	F	8	C	F	1	3-4
Persian Emperor	Damascus/Lt. Stevens	D	8	F	F	4	2-3
Personal Flag	Private Account/Hoist the Flag	C	6	C	T	1*	3-4
Persuasive Leader*	Mr. Leader/Speak John	D	10	U	M	4	3-4
Peterhof	The Minstrel/Cornish Prince	B	7	D	F	4	3-4
Petersburg^	Danzig/Icecapade	L	7	L	L	4	2-3
Petrone	Prince Taj/Wild Risk	D	7	B-	C	2	3-4
Pharly	Lyphard/Boran	D	4	C	D	1	3-4
Phone Order^	Fappiano/In Reality	L	8	U	L	4	2-3
Phone Trick	Icecapade/Finnegan	B+	4	C	C	3	2-3
Pia Star	Olympia/Mahmoud	C	3	D	B	4	3-4
Piaster	Damascus/Round Table	C	7	C	T	2	3-4
Piker	Mr. Prospector/Iron Ruler	D	7	U	B	4	3-4
Pilgrim	Northern Dancer/Victoria Park	D	7	C	F	2	3-4
Pilot Ship	Hoist the Flag/Sir Gaylord	F	8	D	C	4	3-4
Pirateer	Roberto/Chieftain	D	8	C	C	4	3-4
Pirate's Bounty	Hoist The Flag/Stevward	C	6	C	D	4	2-3
Pirouette	Best Turn/White Gloves II	B	7	F	F	4	3-4
Plain Dealing	Northern Dancer/Intentionally	B	8	D	D	4	2-3
Play Fellow	On the Sly/Run For Nurse	D	6	D	B	2	3-4
Play On	Stop the Music/Impressive	C	7	B-	D	4	3-4
Pleasant Colony c	His Majesty/Sunrise Flight	C+	1	C	C+	1	3-4
Pleasure Bent	What a Pleasure/Rasper II	B+	7	F	A	5	2-3
Plugged Nickle	Key to the Mint/Buckpasser	C	6	F	C+	2	3-4
Pluie's Sylvester	Verbatim/Lt. Stevens	D	6	U	D	4	2-3
Pocketful in Vail	Full Pocket/Jungle Road	C	8	C	D	4	3-4
Pocket Park	Verbatim/Jet Pilot	F	8	D	F	4	3-4
Poison Ivory	Sir Ivor/Northern Dancer	C	8	D	F	3	3-4
Poleax	The Axe II/Ambiorix	D	6	D	B	3	3-4
Pole Position	Draft Card/Wallet Lifter	C	8	U	C+	4	2-3
Poles Apart*	Danzig/Cyane	D	6	U	C	2	2-3
Police Car	Nearctic/Menetrier	D	7	B-	D	4	2-3
Polish Navy	Danzig/Tatan	D	6	B-	F	3*	3-4
Polish Numbers^	Danzig/Buckpasser	L	6	M	M	4	2-3
Political Ambition*	Kirtling/Round Table	M	7	L	M	3	3-4
Polynesian Flyer*	Flying Lark/Best Dancer	D	10	M	M	4	3-4
Polynesian Ruler	Manifesto/Polynesian	F	9	U	F	5	3-4
Posen^	Danzig/Best Turn	L	7	M	L	3	2-3

Stallion	Sire/Broodmare Sire	FTS	CL	Trf	OT	SI	AB
Position Leader*	Pole Position/Mr. Leader	M	7	U	T	4	2-3
Positiveness	Exclusive Native/Alcibiades II	C	7	D	C	5	3-4
Positive Step*	Mr. Leader/Prince Dare	M	9	U	T	4	3-4
Potentiate*	FoolishPleasure/NorthrnDancer	U	7	M	U	4	3-4
Powder Horn	Tom Rolfe/My Babu	C	7	B	F	4	3-4
Premiership	Exclusive Native/Never Bend	C	6	D	C+	4	2-3
Present Value^	Halo/Vice Regal	M	6	M	M	3	3-4
Pretense	Endeavor II/Hyperion	C	6	B	T	4	3-4
Prince Card	Prince John/Royal Charger	C+	9	C	B	4	3-4
Prince Don B	Don B/Grey Dawn II	D	7	F	B	3	3-4
Prince John	Princequillo/Count Fleet	C	I	B-	C+	I	3-4
Princely Pleasure	What a Pleasure/Prince John	C	6	C	C	4	3-4
Princely Ruler*	Foolish Pleasure/Prince John	C	9	D	B	4	3-4
Princely Verdict	Prince John/Hail to Reason	F	8	F	F	4	3-4
Prince of Fame^	Fappiano/Secretariat	L	7	M	L	4	2-3
Prince Valid*	Valid Appeal/Klem	M	7	U	L	4	3-4
Private Account c	Damascus/Buckpasser	C+	2	C	T	2	3-4
Private Terms*	Private Account/Bold Ruler	L	4	U	T	2	2-3
Private Thoughts	Pretense/Baghdad	D	6	B	T	4	3-4
Prized^	Kris S/My Dad George	U	5	L	M	I	2-3
Prize Ring	Buckpasser/Turn-to	B	8	F	C	4	3-4
Probable*	Codex/Intentionally	F	8	U	C	4	3-4
Procida	Mr. Prospector/Distinctive	D	7	C	F	3	3-4
Pro Consul	Vice Regent/Bing	D	7	D	F	3	3-4
Productivity	Windy Sands/Fleet Nasrullah	C	8	M	U	4	3-4
Proof	Believe It/Court Martial	C	8	D	F	4	3-4
Proper Reality*	In Reality/Nodouble	U	6	U	M	3	2-3
Prospective Star	Mr. Prospector/Distinctive	D	8	B-	C	4	3-4
Prospect North/	Mr.Prospector/Northern Dancer	D	9	F	C	4	3-4
Prospector's Bid*	Mr. Prospector/Crimson Satan	U	9	U	L	3	3-4
Prospector's Gamble^	Crafty Prospector/Sunny South	L	6	M	L	3	2-3
Prospector's Halo*	Gold Stage/Halo	C	7	U	M	4	2-3
Prosper Fager*	Mr. Prospector/Dr. Fager	C	7	U	F	5	2-3
Prosperous	Mr. Prospector/Sadair	C	7	D	C+	4	2-3
Proud Appeal	Valid Appeal/Proudest Roman	L	3	M	L	3	2-3
Proud Birdie c	Proud Clarion/Bolero	D	5	C	C	2	2-3
Proudest Doon	Matsadoon/Nail	U	9	U	M	4	3-4
Proudest Duke*	Regal and Royal/Proudest Roman	U	9	M	M	4	3-4
Proudest Roman	Never Bend/Roman	C+	3	C	B	4	2-3
Proud Northern*	Northern Jove/Jaipur	C	9	U	C	5	3-4
Proud Truth	Graustark/Summer Tan	C	6	C	C+	I	3-4

Stallion	Sire/Broodmare Sire	FTS	CL	Trf	OT	SI	AB
Prove Out	Graustark/Bold Venture	F	7	D	C	3	3-4
Providential	Run the Gantlet/Primera	D	8	C	F	1	3-4
Publicity	Bold Hitter Terrang	D	7	B	C	4	2-3
Pulsate*	Explodent/Nashua	U	9	M	M	2	3-4
Puntivo*	Restivo/Beau Gar	C	7	F	F	4	3-4
Purdue King^	Dimaggio/Truxton Fair	U	7	U	M	4	2-3
Pursuit f	Never Bend/Prince John	D	8	B-	F	4	3-4
P Vik	Petrone/Viking Spirit	F	9	D	C	3	3-4
Pyrite*	Mr. Prospector/Northern Dancer	U	9	U	L	4	3-4
Quack	TV Lark/Princequillo	D	6*	C	F	2	3-4
Quadratic	Quadrangle/Quibu	F	6*	D	C	4	2-3
Queen City Lad	Olden Times/Royal Union	B+	6	D	D	4	3-4
Quiet American^	Fappiano/Dr. Fager	M	5	U	L	2	3-4
Qui Native	Exclusive Native/Francis S	C	7	B-	D	3	3-4
Racing Star^	Baldski/Imasmartee	M	8	M	U	4	2-3
Raconteur	The Minstrel/Stage Door Johnny	C	7	B-	F	4	3-4
Raft	Nodouble/Round Table	U	6	L	M	2	3-4
Rahy	Blushing Groom/Halo	B+	3	B	M	3	2-3
Rainbow Quest	Blushing Groom/Herbager	C	3	B	L	1	3-4
Rainbows First*	Star Nasrullah/Semi Pro	M	9	U	M	5	2-3
Raise a Bid	Raise a Native/Double Jay	D	4*	F	D	4	2-3
Raise a Cup	Raise a Native/Nashua	C	4*	C	B	4	3-4
Raise a Man	Raise a Native/Delta Judge	C	6	C	F	4	3-4
Raise a Native	Native Dancer/Case Ace	C+	4	D	B	4	3-4
Raised Socially	Raise a Native/Never Bend	C	5*	D	C	4	2-3
Raise Your Glass	Raise a Native/Barbizon	L	7	U	L	5	2-3
Rajab	Jaipur/Princequillo	C	6	C	C	4	2-3
Raja Baba	Bold Ruler/My Babu	C	3	C	A	5	3-4
Raja Native	Raja Baba/Native Charger	C	8	F	F	5	3-4
Raja's Best Boy*	Raja Baba/Icecapade	D	8	F	A+	4	2-3
Raja's Revenge*	Raja Baba/Northern Dancer	F	7	C	C+	4	3-4
Rambunctious	Rasper II/The Solicitor II	B	4*	F	C+	5	2-3
Ramirez	T V Lark/Dedicate	D	9	C	F	4	3-4
Ramplett^	Fappiano/Crimson Satan	M	8	U	L	4	2-3
Rampage	Northern Baby/Vaguely Noble	D	8	D	D	2	3-4
Rare Brick	Rare Performer/Mr. Brick	D	7	C	C	4	2-3
Rare Performer	Mr. Prospector/Better Self	C+	7	C	C	4	3-4
Reach for More*	I'm for More/Villamor	B+	7	U	L	5	2-3
Reading Room	Ribot/Shut Out	F	8	D	F	3	3-4
Real Courage	In Reality/Key to the Mint	D	7	F	C	4	2-3

Stallion	Sire/Broodmare Sire	FTS	CL	Trf	OT	SI	AB
Reality and Reason	In Reality/Boldnesian	D	8	M	A	5	3-4
Really Secret*	In Reality/Bold Ambition	M	7	U	F	4	3-4
Reap	Royal Ski/Vent du Nord	M	9	M	L	3	3-4
Record	Delta Oil/Gallant Romeo	M	9	U	M	4	3-4
Rectory	King's Bishop/First Landing	F	8	C	F	3	3-4
Recusant	His Majesty/Cornish Prince	D	6	F	C	3	3-4
Red Attack	Alydar/Buckpasser	D	7	C	C	2	3-4
Red Clay Country^	Master Derby/Carry Back	U	9	U	U	4	3-4
Red Ransom^	Roberto/Damascus	L	5	L	M	4*	2-3
Red Ryder	Raise a Native/Nashua	U	6*	U	M	4	3-4
Red Wing Bold	Boldnesian/Rocky Royale	B+	7	A	C	5	2-3
Reflected Glory	Jester/Palistinian	C	7	D	C+	4	3-4
Regal and Royal	Vaguely Noble/Native Dancer	C	4*	C	D	3	2-3
Regalberto	Roberto/Roi Dagobert	D	9	C	D	3	3-4
Regal Classic*	Vice Regent/No Double	U	6	U	M	2	2-3
Regal Embrace	Vice Regent/Nentego	D	4	F	F	4	3-4
Regal Intention^	Vice Regent/Tentam	M	6	M	M	2	2-3
Regal Remark	Northern Dancer/Speak John	B+	7	M	C	4	2-3
Regal Search*	Mr.Prospector/Dr. Fager	C	7	U	L	4	2-3
Reinvested	Irish Castle/Crafty Admiral	F	9	D	F	3	3-4
Relaunch	In Reality/The Axe II	B	3	C	A+	4*	2-3
Relaunch a Tune*	Relaunch/Don B	M	7	M	L	4	2-3
Religiously*	Alleged/Admiral's Voyage	U	10	U	M	2	3-4
Restivo	Restless Native/Palestine	C	6	F	D	4	2-3
Restless Con^	Restless Native/Wallet Lifter	M	6	L	M	2	2-3
Restless Native	Native Dancer/Bull Lea	C	5	C	C	4*	2-3
Retsina Run*	Windy Sands/Pia Star	L	10	M	L	4	3-4
Reve Du*	Gallant Best/Fleet Nasrullah	M	8	U	L	4	2-3
Reviewer	Bold Ruler/Hasty Road	C+	3	D	B	4	2-3
Rex Imperator	King Emperor/Imbros	F	7	C	C	4	3-4
Rexson	Bold Bidder/Turn-to	D	4	F	C+	4	2-3
Rexson's Hope	Rexson/Abe's Hope	F	8	B	F	4	3-4
Ribot	Tenerani/El Greco	C	1	C	B	1	3-4
Rich Cream	Creme dela Creme/Turn-to	C	5*	C	F	4	3-4
Rich Doctor*	Doc Scott J/Rich Gift	M	10	U	C	4	3-4
Ridgewood High*	Walker's/Curragh King	M	9	U	M	5	3-4
Right Con*	Beau Buck/Unconscious	M	9	U	M	4	2-3
Rinoso	Noble Sun/Ridan	A	8	B	D	5	2-3
Rio Carmelo	Riverman/Alycidon	F	8	C	F	3	3-4
Risen Star	Secretariat/His Majesty	F	7	D	F	1	3-4
Rising Market	To Market/War Admiral	C	6	F	C	4	3-4

Stallion	Sire/Broodmare Sire	FTS	CL	Trf	OT	SI	AB
Riva Ridge	First Landing/Heliopolis	C	3	F	C+	1	3-4
Riverman	Never Bend/Prince John	C	3	B	F	1	3-4
River of Kings*	Sassafras/Sovereign Path	U	8	M	M	3	3-4
River Prospect*	Riverman/Pretense	U	10	L	M	2	3-4
Roan Drone	Drone/John's Joy	F	8	U	F	4	3-4
Roanoke^	Pleasant Colony/Sea-Bird	U	6	M	M	2	2-3
Roar of the Crowd	Hold Your Peace/Time Tested	M	8	U	M	5	3-4
Roberto	Hail to Reason/Nashua	D	1	A	C+	1	2-3
Rob An Plunder^	Pirate's Bounty/Reflected Glory	M	7	U	M	4	2-3
Robellino	Roberto/Pronto	D	7	L	U	2	3-4
Roberto Reason*	Roberto/Princequillo	U	9	M	U	3	3-4
Rock Lives	Rock Talk/Joust	L	7	U	L	5	2-3
Rock Point^	Believe It/Hoist the Flag	U	8	M	M	4	3-4
Rock Royalty	Native Royalty/Fleet Nasrullah	D	8	D	F	5	3-4
Rock Talk	Rasper II/Polynesian	C	3*	F	C+	1	3-4
Roi Dagobert	Sicambre/Cranach	D	4	C	C	3	3-4
Rolfson	Tom Rolfe/Olden Times	F	8	D	F	3	3-4
Rollicking	Rambunctious/Martins Rullah	C	3*	D	C+	4	2-3
Rollin on Over	Brent's Prince/In the Pocket	D	8	D	F	5	2-3
Rolls Aly*	Alydar/Traffic Mark	C+	8	M	L	2	3-4
Roman Bend*	Proudest Roman/Grey Dawn II	U	9	M	U	3	3-4
Roman Diplomat*	Roberto/Chieftain	C	7	A	F	4	2-3
Roman Majesty	His Majesty/Majestic Prince	C	8	D	F	1	3-4
Roman Reasoning	Bold Reasoning/Roman	C+	7	F	C	4	3-4
Romantic Lead	Silent Screen/Sir Gaylord	D	8	D	C	5	3-4
Romeo c	T.V. Lark/Tiger Wander	D	7	D	C	4	3-4
Roo Art	Buckaroo/Ribot	F	7	D	C	3	2-3
Rougemont	Seattle Slew/Bold Bidder	B	9	U	B	5	2-3
Rough Iron	Nodouble/TV Lark	C	8	U	C	4	3-4
Rough Pearl*	Tom Rolfe/Ruffled Feathers	U	7	M	M	4	3-4
Round Table	Princequillo/Knight's Daughter	C	1	A	D	4	2-3
Roving Minstrel*	Diplomat Way/Poona II	U	10	M	U	5	3-4
Roxbury Park*	Mr. Prospector/Filberto	F	9	U	L	5	3-4
Royal And Regal	Vaguely Noble/Native Dancer	D	5*	C	C	3	3-4
Royal Chocolate	Amber Morn/Prince John	D	7	D	F	4	3-4
Royal Pennant*	Raja Baba/Hoist The Flag	M	10	U	U	5	3-4
Royal Roberto	Roberto/Royal Note	B+	6	C	D	4	3-4
Royal Value*	Regal Embrace/Vice Regal	M	8	M	M	4	3-4
Ruffinal	Tom Rolfe/Nasrullah	U	7	U	M	2	3-4
Ruhlmann^	Mr. Leader/Chieftain	U	5	U	M	2	3-4

Stallion	Sire/Broodmare Sire	FTS	CL	Trf	OT	SI	AB
Rumbo	Ruffinal/Windy Sands	D	8	D	F	4	3-4
Runaway Groom	Blushing Groom/Call the Witness	C	6	C	C	2	2-3
Run Johnny Run*	Graustark/Round Table	M	8	U	U	3	3-4
Run of Luck	Coursing/Berseem	D	6	C	C	4	3-4
Run the Gantlet	Tom Rolfe/First Landing	D	6*	C	C	1	3-4
Ruthie's Native	Native Royalty/Citation	C	6	B	F	4	3-4
Ryeko	Wajima/The Pie King	U	10	U	M	5	3-4
Sabona*	Exclusive Native/Hail to Reason	L	7	L	M	2	2-3
Sadler's Wells	Northern Dancer/Bold Reason	C+	1	B	D	1	3-4
Sagace	Luthier/Chaparral	U	5	L	M	1	3-4
Salem Drive^	Darby Creek Road/Northfields	U	7	L	M	1	3-4
Salem End Road	Big Burn/Hail the Prince	M	7	U	A	4	2-3
Salt Marsh	Tom Rolfe/Sailor	C	5	U	M	1	3-4
Salutely f	Hoist the Flag/Amerigo	D	5*	F	D	2	3-4
Same Direction	Vice Regent/Sallymont	D	7	F	F	4	3-4
Santiago Peak	Alydar/Secretariat	C	10	F	F	4	3-4
Saratoga Six	Alydar/Irish Castle	B+	5	D	A+	4	2-3
Sarawak	Northern Dancer/Bold Bidder	D	8	C	F	5	3-4
Saros c	Sassafras/Floribunda	D	7	C	C+	4	3-4
Sassafras	Sheshoon/Ratification	D	4	C	F	3	3-4
Satan's Flame	Crimson Satan/Kentucky Pride	U	9	M	L	5	3-4
Sauceboat	Key to the Mint/My Babu	D	6	D	C	3	2-3
Saunders	Nijinsky II/Princequillo	D	7	A	D	1	2-3
Savings*	Buckfinder/Star Kingdom	M	6	M	M	2	2-3
Sawbones	The Axe II/Tom Fool	D	4	D	C	1	3-4
Scarlet Ibis*	Cormorant/Cornish Prince	M	8	U	L	4	2-3
Schaufuss	Nureyev/Gallant Romeo	U	10	M	U	3	3-4
School Hero*	Crozier/Noholme II	B	7	M	M	5	2-3
Score Twenty Four	Golden Eagle II/Traffic Judge	M	10	U	M	3	3-4
Screen King	Silent Screen/Nashua	D	7	D	F	3	3-4
Sea Aglo	Sea Bird/Royal Serenade	B+	9	F	C	4	2-3
Sea-Bird	Dan Cupid/Sicambre	D	3	D	C	1	3-4
Seafood	Proud Clarion/Crewman	U	8	U	M	4	3-4
Search For Gold	Raise a Native/Nashua	F	7	C	C	4	3-4
Seattle Battle*	Seattle Slew/Raise a Native	M	9	M	L	4	3-4
Seattle Dancer	Nijinsky II/Poker	C	8	B-	F	2	2-3
Seattle Knight^	Seattle Slew/Prince John	M	9	M	L	5	2-3
Seattle Slew	Bold Reasoning/Poker	C	1	B-	B	3*	3-4
Seattle Song	Seattle Slew/Prince Blessed	F	6	B-	D	2	3-4
Secretariat	Bold Ruler/Princequillo	D	3	B	BT	2	3-4
Secretary of War	Secretariat/Dunce	C	8	D	F	4	2-3

Stallion	Sire/Broodmare Sire	FTS	CL	Trf	OT	SI	AB
Secret Claim*	Mr. Prospector/Secretariat	M	7	L	M	3	3-4
Secret Hello*	Private Account/Silent Screen	M	6	U	T	2	2-3
Secreto	Northern Dancer/Secretariat	C	7	C	C	2	3-4
Secret Prince	Cornish Prince/Lanvin	F	8	F	C	4	3-4
Secret Slew^	Seattle Slew/Ack Ack	M	8	L	L	3	3-4
Seeking The Gold*	Mr. Prospector/Buckpasser	A	3	M	U	4*	2-3
Sejm^	Danzig/Cyane	L	7	M	L	3	2-3
Selous Scout	Effervescing/Roman Line	C	6	D	C	3	3-4
Semillero	Proposal/Sertorious	C	8	D	F	4	3-4
Semi Northern	Northern Dancer/Semi-Pro	D	8	C	F	2	2-3
Sensitive Prince	Majestic Prince/Sensitivo	D	7	D	C	2	3-4
Septieme Ciel^	Seattle Slew/Green Dancer	M	6	L	L	3	3-4
Set Free*	Majestic Prince/Pago Pago	U	8	M	F	4	2-3
Settlement Day*	Buckpasser/Swaps	D	7	D	F	3	3-4
Sewickley^	Star de Naskra/Dr. Fager	L	6	U	M	5	2-3
Sezyou f	Valid Appeal/Dr. Fager	D	7	D	F	4	2-3
Shadeed	Nijinsky II/Damascus	C	7	D	F	3*	3-4
Shahrastani	Nijinsky II/Thatch	C	4	B-	F	1	3-4
Sham	Pretense/Princequillo	D	4	C	D	3	3-4
Shananie	In Reality/Cohoes	C	6	C	B	4	2-3
Shanekite	Hoist Bar/Any Old Time	B+	7	C	C	5	2-3
Shareef Dancer	Northern Dancer/Sir Ivor	M	6	L	M	3*	3-4
Sharif	Damascus/Bold Ruler	C	9	F	D	5	3-4
Sharpen Up	Atan/Rockefella	C	3	B-	C	1*	2-3
Sharper One	Drum Fire/Drone	C	8	F	A+	5	2-3
Sharp Terdankim*	Sharpen Up/Raise a Native	U	9	M	M	4	3-4
Sharrood	Caro/Cougar II	D	5	B-	C	1	3-4
Shawklit Won*	Air Forbes Won/Groshawk	U	8	M	M	4	3-4
Shecky Greene	Noholme II/Model Cadet	B	5	D	C+	5	2-3
Shelter Half	Tentam/Sir Gaylord	C+	6	C	C	4	2-3
Shenadoah River	Dewan/Sea O Erin	C	7	C	D	5	2-3
Shergar's Best*	Shergar/Secretariat	C	8	B-	F	4	2-3
Shernazar	Busted/Val de Loir	D	4	C	U	1	3-4
Shimatoree	Marshua Dancer/Tudor Minstrel	C	7	D	B	4	3-4
Shipping Magnate	What Luck/Royal Dorimar	C	8	C	F	5	3-4
Shirley Heights	Mill Reef/Hardicanute	D	4	B	C	1	3-4
Shirley's Champion	Noholme II/Have Tux	B	7	B-	C	5	3-4
Shot Gun Scott^	Exuberant/Rambunctious	M	9	U	U	4	2-3
Show Dancer	Sovereign Dancer/Sir Gaylord	D	7	B-	F	4	2-3
Show'em Slew*	Seattle Slew/Spanish Riddle	U	8	M	M	4	3-4

Stallion	Sire/Broodmare Sire	FTS	CL	Trf	OT	SI	AB
Siberian Express	Caro/Warfare	D	7	B-	B	2	3-4
Sicilian Law	Wardlaw/Bronzerullah	M	8	U	M	5	3-4
Silent Cal	Hold Your Peace/Arrogate	F	7	B-	F	4	3-4
Silent Fox	Exclusive Native/Crafty Admiral	D	8	B-	F	4	3-4
Silent King*	Screen King/Jet Action	D	9	M	F	5	2-3
Silent Screen	Prince John/Better Self	C	6	C	C	3	3-4
Silk or Satin	Impressive/Olden Times	C	10	B-	C+	5	3-4
Silver Buck c	Buckpasser/Hail to Reason	D	6	D	C	2	2-3
Silver Deputy*	Deputy Minister/Mr. Prospector	D	7	F	C	4	3-4
Silver Ghost	Mr. Prospector/Halo	B	6	F	B	4	2-3
Silver Hawk	Roberto/Amerigo	C	5	A	F	1*	2-3
Silver Nitrate*	Vitriolic/Native Charger	C	9	M	M	4	3-4
Silveyville	Petrone/Successor	C+	9	C	C	3	3-4
Simply Majestic*	Majestic Light/King Emperor	M	6	L	U	3	2-3
Singular	Nodouble/Cohoes	C	7	C	D	4	3-4
Sir Dancer*	Jig Dancer/Linmold	U	9	U	C	4	3-4
Sir Harry Lewis*	Alleged/Mr. Prospector	F	8	M	M	3	3-4
Sir Ivor	Sir Gaylord/Mr. Trouble	D	4	C	C	1*	3-4
Sir Jinsky	Nijinsky II/Tim Tam	D	7	D	F	4	3-4
Sir Maxmillion*	Macarthur Park/Bobby's Legacy	M	10	M	M	4	3-4
Sir Session	Sir Ivor/Hoist the Flag	C	8	C	U	4	2-3
Sir Sizzling Jim	Jim J./Tudorka	D	7	U	C	5	3-4
Sir Spectator*	Sir Ivor/Crewman	M	10	U	U	4	3-4
Sirtaki*	Raja Baba/Nijinsky II	U	10	U	M	4	3-4
Sir Wimborne	Sir Ivor/Tom Fool	M	4	U	M	4	3-4
Sitzmark	J.O. Tobin/Dunce	B-	8	D	F	4	2-3
Siyah Kalem	Mr. Prospector/Graustark	F	8	F	M	4	3-4
Skip Trial	Bailjumper/Promised Land	C	5	D	F	3	2-3
Sky Command*	Relaunch/Kinsman	L	8	M	L	5	2-3
Skywalker	Relaunch/Boldnesian	B-	4	C	A	3*	3-4
Slewacide	Seattle Slew/Buckpasser	C+	8	B	C	3	3-4
Slew City Slew*	Seattle Slew/Berkeley Prince	M	8	M	M	2	2-3
Slewdledo	Seattle Slew/Cyane	A	8	M	T	4	3-4
Slewdonza^	Seattle Slew/Bold Bidder	U	9	L	L	2	2-3
Slew Express	Seattle Slew/L'Enjoleur	U	9	M	M	3	3-4
Slew Machine	Seattle Slew/Sassafras	F	8	B-	C	3	3-4
Slew o'Gold	Seattle Slew/Buckpasser	C	6	C	C	2	3-4
Slewpy	Seattle Slew/Prince John	D	5	C	C	4	3-4
Slew's Folly	Seattle Slew/Tom Fool	C	8	M	F	4	3-4
Slew's Royalty	Seattle Slew/Swaps	B+	5	C	C+	5	3-4
Slew the Bride*	Seattle Slew/Al Hattab	C+	9	M	C	3	3-4

Stallion	Sire/Broodmare Sire	FTS	CL	Trf	OT	SI	AB
Slew the Coup	Seattle Slew/Buckpasser	C	7	D	C	4	2-3
Slew the Knight^	Seattle Slew/Prince John	U	7	M	L	3	3-4
Slewvescent^	Seattle Slew/Herbager	U	7	M	M	2	3-4
Slick*	Alydar/Jacinto	U	7	U	L	4	2-3
Slip Anchor (Eng)	Shirley Heights/Birkhahn	D	4	B	C	1	3-4
Smarten	Cyane/Quibu	C	4	C	C	3*	2-3
Smart Style	Foolish Pleasure/Quadrangle	D	7	D	D	4	3-4
Smart Talk	Speak John/Gallant Man	C+	8	F	C+	4	2-3
Smile	In Reality/Boldnesian	D	7	A-	B	4	2-3
Snake Oil Man	Raise a Native/Tom Fool	B+	9	U	A	4	2-3
Snar^	Sir Ivor/Olden Times	U	9	M	M	2	3-4
Snow Chief	Reflected Glory/Snow Sporting	D	8	F	F	4	3-4
Sobat	Relaunch/Graustark	M	8	M	L	2	3-4
Soda Springs	Herbager/Bold Ruler	F	9	U	F	4	3-4
Solar City	Northern Dancer/Halo	D	5	D	C+	4	3-4
Somethingfabulous	Northern Dancer/Princequillo	C	6*	C	D	4	3-4
Something Lucky*	Somethingfabulous/Lucky Fleet	L	7	M	L	3	2-3
Son Ange	Raise a Native/Tom Fool	D	5*	F	F	4	3-4
Song of Delta*	Sing Along/Great Day	D	9	U	C	5	3-4
Sonny's Solo Halo*	Halo/Solo Landing	C	9	M	U	4	2-3
Son of Briartic	Briarctic/Round Table	C+	6	A-	C	4*	2-3
Southern Sultan*	Stage Door Johnny/Hail to Reason	D	8	B-	C	1	3-4
Sovereign Dancer	Northern Dancer/Bold Ruler	C	4	C	C	3*	2-3
Sovereign Don*	Sovereign Dancer/Hold Your Peace	C	6	D	F	4	2-3
Sovereign Exchange	Sovereign Dancer/Best Turn	U	10	M	M	4	3-4
Sovereign Ruler*	Sovereign Dancer/Roi Dagobert	U	9	U	U	4	3-4
Sovereignty	Affirmed/Never Bend	D	8	M	C	3	2-3
Soviet Lad*	Nureyev/Val de Loir	U	7	L	U	3*	2-3
Soviet Star*	Nureyev/Venture	M	3	L	U	4*	2-3
Soy Numero Uno	Damascus/Crafty Admiral	D	7*	F	M	5	2-3
Space Cup^	Saratoga Six/Run For Nurse	L	10	U	L	4	2-3
Space Station^	Mr. Prospector/Graustark	F	7	U	A	2	3-4
Spanish Drums	Top Command/Drone	D	7	C	F	4	2-3
Spanish Way	Roberto/Olympia	F	8	C	T	3	3-4
Spare Card*	Paris Dust/Mito	U	9	U	M	4	2-3
Sparkly*	Halo/Amerigo	M	8	U	M	4	2-3
Speak John	Prince John/Tornado	C	3	B	B	4	3-4
Speak the Verb	Verbatim/Kauai King	M	8	M	M	4	2-3
Spectacular Bid	Bold Bidder/Promised Land	D	6	C	C+	2	3-4
Spectacular Round*	Spectacular Bid/Never Bend	U	10	M	L	3	3-4
Spellbound	Lyphard's Wish/Elocutionist	U	7	M	U	4	3-4

Stallion	Sire/Broodmare Sire	FTS	CL	Trf	OT	SI	AB
Spend a Buck	Buckaroo/Speak John	C	5	C	C	2	2-4
Spicy Monarch^	Believe the Queen/Sham	U	10	U	M	3	3-4
Spirited Boy*	Monetary Crisis/Manifesto	M	10	U	U	5	2-3
Spirit Rock	Selari/Jet Pilot	F	8	C	D	4	3-4
Splendid Courage	Raise a Native/Gallant Man	C	7	C	B	4	3-4
Spook Dance*	Duck Dance/Double Hitch	L	9	U	U	5	2-3
Sportful	Crozier/Quibu	M	7	M	L	4	2-3
Sportin' Life	Nijinsky II/Round Table	D	7	C	F	2	3-4
Spotter Bay*	Coastal/Iron Ruler	C	10	U	L	5	3-4
Spread the Rumor	Forli/Damascus	C+	8	B-	F	4*	3-4
Spring Double	Double Jay/Hyperion	U	3	U	M	3	3-4
Springhill	Sallust/Trouville	A	8	D	C	5	2-3
Spy Signal	Hoist the Flag/Swaps	F	7	F	C	4	3-4
Squad Car*	Police Car/Winning Shot	M	8	L	M	4	2-3
S.S. Hot Sauce	Sauceboat/Le Fabuleux	U	7	U	M	4	2-3
Stacked Pack	Majestic Light/Buckpasser	C	8	D	F	4	3-4
Staff Riot*	Staff Writer/Tom Rolfe	M	9	M	U	3	3-4
Staff Writer	Northern Dancer/Swaps	C	5	C	D	4	2-3
Stage Door Johnny	Prince John/Ballymoss	F	5	A	C	1	3-4
Stage Door Key*	Stage DoorJohnny/Arts and Letters	U	9	M	M	1	3-4
Stalwart	Hoist The Flag/Iron Ruler	C	5	C	C	4	2-3
Standstead	Gummo/First Landing	C	8	D	D	5	3-4
Star Choice	In Reality/Tirreno	F	7	C	C	3	3-4
Star de Naskra	Naskra/Clandestine	B	5	D	C	4	3-4
Starfields	Northfields/Hugh Lupus	F	9	U	C	2	3-4
Star Gallant	My Gallant/Bold Hitter	D	7	C	C	4	3-4
State Dinner	Buckpasser/Barbizon	D	6	D	C	2	3-4
Stately Don	Nureyev/La Fabuleux	F	7	C	F	2	3-4
Staunch Avenger	Staunchness/Revoked	C	7	F	C	5	2-3
Stay the Course^	Majestic Light/Indian Hemp	C	6	C	F	2	2-3
Steady Growth	Briartic/Crepello	D	5	C	F	3*	3-4
Steinlen^	Habitat/Jim French	U	4	L	U	2	3-4
Stonewalk	Knightly Manner/Besomer	D	7	C	F	4	3-4
Stop the Music	Hail to Reason/Tom Fool	D	5	D	C	4	3-4
Storm Bird	Northern Dancer/New Providence	B	3	C	C	3*	2-3
Storm Cat	Storm Bird/Secretariat	A+	1	B	C	4*	2-3
Stratford	Hoist the Flag/Nantallah	U	10	U	M	4	3-4
Strawberry Road	Whiskey Road/Rich Gift	C	4	C	F	2	3-4
Strike Gold	Mr. Prospector/Roi Dagobert	C+	5	D	C+	5	2-3
Strike the Anvil	Bolinas Boy/Imbros	C	7	C	F	4	3-4
Stutz Blackhawk	Mr. Prospector/Amber Morn	D	7	D	C	3	2-3
Sucha Pleasure	What a Pleasure/Hail to Reason	C	8	D	D	4	3-4

Stallion	Sire/Broodmare Sire	FTS	CL	Trf	OT	SI	AB
Summer Time Guy	Gummo/Crozier	F	6	<u>B</u>	D	4	<u>3-4</u>
Summer Squall^	Storm Bird/Secretariat	L	4	L	M	3*	2-3
Summing	Verbatim/Groton	D	8	D	D	3	<u>3-4</u>
Sunny Clime	In Reality/Newtown Wonder	C	6	C	B	4	2-3
Sunny North	Northern Dancer/In Reality	B	3	C	C	4*	2-3
Sunny's Halo	Halo/Sunny	F	6	D	F	2	2-3
Sun Power	Great Sun/Noble Jay	F	7	C	F	<u>5</u>	2-3
Sunshine Forever*	Roberto/Graustark	M	4	L	U	I	3-4
Superbity	Groshawk/Rough'n Tumble	C	6	B	C	4	2-3
Super Concorde	Bold Reasoning/Primera	C	4	B	D	I*	3-4
Super Moment	Big Spruce/Shantung	D	7	B	F	2	<u>3-4</u>
Superoyale*	Raise a Native/Never Bend	B	7	U	D	5	2-3
Sutter's Prospect*	Mr. Prospector/Nijinsky II	M	8	M	M	4	2-3
Suzanne's Star*	Son Ange/Greek Game	C+	10	U	C	4	2-3
Sweet Candy	Bold and Brave/Blue Prince	C	4	C	C	5	3-4
Swelegant	L'Enjoleur/Mito	F	7	M	C+	5	3-4
Swing Till Dawn	Grey Dawn II/The Axe	C	6	B-	C	3*	<u>3-4</u>
Swiss Trick*	Damascus/Buckpasser	U	10	U	M	3	2-3
Switch Partners	Great sun/Turn-to	F	7	C	C	4	<u>3-4</u>
Swoon*	Secretariat/Sea Hawk	C	8	L	F	3	2-3
Sword Dance*	Nijinsky II/Secretariat	U	9	M	U	I	<u>3-4</u>
Synastry*	Seattle Slew/Nashua	L	7	M	M	4	2-3
Syncopate	Marshua's Dancer/Prince John	C	7	B	F	4	3-4
Table Run	Round Table/Fleet Nasrullah	C	7	D	C	<u>4</u>	3-4
Tabun Bogdo*	Seattle Slew/Caro	U	9	M	L	4	3-4
Tagish	Mr. Prospector/Faraway Son	M	9	M	L	4	2-3
Tahitian King	Luthier/Decathlon	C	7	C	F	<u>4</u>	<u>3-4</u>
Taj Alriyadh*	Seattle Slew/El Relicario	D	8	M	C	4	2-3
Take the Floor	Cornish Prince/Promised Land	D	9	<u>B</u>	F	4	<u>3-4</u>
Talc	Rock Talk/Rash Prince	C	4	C	C	3	3-4
Talinum	Alydar/Riverman	D	7	D	C	3	3-4
Tally Ho the Fox	Never Bend/Hail to Reason	D	8	C	D	3	<u>3-4</u>
Tall Ships	Olden Times/Restless Native	C	10	C	A	I	3-4
Tandem*	Benny Bob/Bold Sultan	L	8	U	M	4	2-3
Tank's Prospect	Mr. Prospector/Pretense	F	7	D	C	4	<u>3-4</u>
Tanthem	Tentam/Nashua	C	4	C	C	4	3-4
Tantoul	Tatan/Hill Prince	D	8	C	F	4	<u>3-4</u>
Tapping Wood*	Roberto/Tudor Melody	F	8	M	C	3	3-4
Tarr Road*	Grey Dawn II/Chieftain	U	8	U	M	4	2-3
Tarsal	Tom Rolfe/Royal Union	C+	8	B	C+	4	3-4

Stallion	Sire/Broodmare Sire	FTS	CL	Trf	OT	SI	AB
Tasso	Fappiano/What a Pleasure	C	7	B-	D	4*	3-4
Taufan	Stop the Music/Sadair	U	8	M	M	3*	3-4
Taxachusetts	Gallant Romeo/Hasty Road	C	7	D	B	4	3-4
Tayfun*	Lord Tomboy/Correlation	L	8	M	M	4	3-4
Taylor	Taylor's Falls/Roman Line	M	9	M	M	4	3-4
Taylor Road	Proudest Roman/Groton	M	8	U	L	4	3-4
Taylor's Falls	In Reality/Newtown Wonder	D	6	B-	C	4	2-3
Taylor's Special	Hawkin's Special/Espea	F	8	C	F	4	3-4
Teddy's Courage	Exclusive Native/Dead Ahead	F	8	B	D	4	3-4
Tejano*	Caro/Exclusive Native	C	6	C	C	3	2-3
Tell	Round Table/Nasrullah	C	6	B	D	4	3-4
Tella Fib	Bold Laddie/Tell	C	8	U	C	4	3-4
Temerity Prince	Cornish Prince/Native Dancer	D	8	F	F	4	3-4
Temperence Hill	Stop the Music/Etonian	C	6	D	T	2	3-4
Ten Gold Pots	Tentam/Search for Gold	C	7	F	C	4	2-3
Tentam	Intentionally/Tim Tam	C+	4	C	B	4	2-3
Text	Speak John/Gunshot	D	8	C	F	4	3-4
Thaliard	Habitat/Sing Sing	M	5	M	M	5	2-3
Thatching	Thatch/Abernant	C+	5	B	C	4*	2-3
That's a Nice	Hey Good Lookin/Palestinian	C	6	B	F	2	3-4
T.H. Bend*	Full Out/Bold Reason	B	9	U	C	4	3-4
Theatrical	Nureyev/Sassafras	B	4	A	F	1*	3-4
The Axe II	Mahmoud/Shut Out	F	1	C	C	1	3-4
The Bart	Le Fabuleux/St Crespin III	F	7	C	F	2	3-4
The Captain	Stevward/Francis S.	D	10	U	L	5	3-4
The Carpenter	Gummo/Fulcrum	D	8	U	BT	4	3-4
The Cool Virginian	Icecapade/Royal Gem II	D	6	D	D	4	3-4
The Corps	Drone/Lt. Stevens	M	10	U	M	4	2-3
The Crowd Roars*	Secretariat/Lt. Stevens	U	10	M	L	4	3-4
The Hague	Transworld/Reform	U	9	U	M	3	3-4
The Irish Lord	Bold Ruler/Double Jay	C	6	D	C+	5	3-4
The Miller	Mill Reef/Sir Ivor	F	8	U	M	4	3-4
The Minstrel	Northern Dancer/Victoria Park	B	4*	C	F	4*	2-3
The Pruner	Herbager/Better Self	D	7	C	F	4	3-4
The Wedding Guest	Hold Your Peace/Olden Times	U	7	M	M	4	3-4
Thirty Eight Places	Nodouble/Dancing Count	F	6	C	C	2	3-4
Thirty Six Red^	Slew o' Gold/Stage Door Johnny	U	5	U	M	1*	3-4
Thorn Dance^	Northern Dancer/Bold Forbes	L	8	L	L	4	2-3
Three Martinis	Third Martini/One Count	C+	6	B-	C	4	2-3
Thunder Puddles	Speak John/Delta Judge	D	6	C	F	1	3-4
Tiffany Ice	Icecapade/Chieftain	C	7	C	C	4	2-3

Stallion	Sire/Broodmare Sire	FTS	CL	Trf	OT	SI	AB
Tilt the Odds^	Tilt Up/Handsome Boy	L	9	U	M	5	2-3
Tilt Up	Olden Times/Tudor Melody	C+	7	F	C	4	2-3
Timber Native^	Hula Chief/American Native	M	10	U	M	5	2-3
Time for a Change c	Damascus/Reviewer	C	5	C	C	4*	3-4
Timeless Moment	Damascus/Native Dancer	B	5	C	C+	4	2-3
Timeless Native	Timeless Moment/Executioner	C	6	C	A	4	2-3
Timely One*	Bold Hour/First Balcony	F	7	U	C	4	3-4
Time to Explode	Explodent/Olden times	C	6	A-	C+	3	2-3
Tim the Tiger	Nashua/Court Martial	C	6	F	F	5	3-4
Tisab	Loom/Nashua	F	8	F	F	4	3-4
Titanic	Alydar/Lt Stevens	D	7	D	F	4	3-4
To-Agori Mou	Tudor Music/Cracksman	U	7	M	M	4	3-4
Toll Key^	Nodouble/Buckpasser	U	8	M	U	3	3-4
Tobin Bronze	Arctic Explorer/Masthead	U	8	U	T	1	3-4
To B. or Not	Don B./Fair Truckle	D	10	D	F	4	3-4
Tolstoy	Nijinsky II/Key to the Mint	F	8	D	F	4	3-4
Tom Rolfe	Ribot/Roman	U	3	M	M	1*	3-4
Tong Po^	Private Account/Hoist the Flag	U	8	U	T	3	3-4
Tonzarun	Arts and Letters/Buckpasser	C	8	D	C	4	3-4
Top Avenger	Staunch Avenger/Dunce	B	7	D	F	5	2-3
Top Command	Bold Ruler/Prince Bio	M	7	M	M	2	3-4
Topsider	Northern Dancer/Round Table	B	3	C	C+	3*	2-3
Top Ville	High Top/Charlottesville	U	2	L	M	1	3-4
Toronto	Sir Ivor/Larkspur	F	7	C	F	1*	3-4
Torquelle	Torsion/Isgala	M	9	U	L	4	3-4
Torsion	Never Bend/Prince John	D	6	DC	C	4	3-4
Total Departure	Greek Answer/Manifesto	C	8	C	F	4	2-3
Totality*	In Reality/Boldnesian	M	7	U	L	5	2-3
Tough Knight	Knights Choice/Capt. Courageous	C+	8	U	C	5	2-3
Track Barron	Buckfinder/Sir Gaylord	C	7	D	F	3	3-4
Track Dance*	Green Dancer/L'Enjoleur	U	8	M	M	3	2-3
Tralos^	Roberto/Briarctic	M	8	L	M	2	2-3
Transworld	Prince John/Hornbeam	D	7	C	F	1	3-4
Trapeze Dancer*	Northern Dancer/Tim Tam	M	9	M	U	3	3-4
Trapp Mountain*	Cox's Ridge/Round Table	F	8	U	U	4	2-3
Travelling Music	Spring Double/Silent Screen	B	7	D	C+	5	3-4
Travelling Victor	Hail to Victory/Travelling Dust	C+	8	U	C	5	2-3
Tree of Knowledge	Dr. Fager/Nasrullah	D	9	F	B	5	3-4
Trempolino	Sharpen Up/Vice Regal	C+	5	C	B	1	2-3
Tricky Creek^	Clever Trick/His Majesty	M	6	L	M	2	2-3
Tricky Tab^	Clever Trick/Al Hattab	M	8	M	M	3	2-3

Stallion	Sire/Broodmare Sire	FTS	CL	Trf	OT	SI	AB
Tridesseus	Northern Dancer/Ribot	F	7	C	F	3*	3-4
Tri Jet	Jester/Olympia	B	3*	C	B	4	2-3
Triocala	Tri Jet/Francis S	C+	7	D	D	4	3-4
Triple Bend	Never Bend/Gun Shot	C	7	F	C	4	3-4
Triple Sec	Tri Jet/Duck Dance	C	8	C	F	4	2-3
Trooper Seven	Table Run/Holandes II	C	9	L	D	4	3-4
Trophy Man^	Strike the Anvil/Tumiga	M	8	M	M	4	2-3
Truce Maker*	Ack Ack/Gallant Man	F	7	B-	A	2	3-4
True Colors	Hoist the Flag/Princequillo	C	7	B	D	4	2-3
Trulo^	Halo/Bold Bidder	M	8	M	M	3	3-4
Truxton King	Bold Ruler/Swaps	F	7	B-	C	4	3-4
Tsunami Slew	Seattle Slew/Barbizon	F	7	C	F	2	3-4
Tunerup	The Pruner/Rocky Royal	D	7	C	C	2	3-4
Turkey Shoot	Seattle Slew/Swaps	D	8	C	F	2	3-4
Turkoman	Alydar/Table Play	C	4	B-	C	1	2-3
TV Alliance*	TV Commercial/Staunchness	L	7	M	M	4	3-4
TV Commercial	TV Lark/Alibhai	C	4	C	C+	2	2-3
Twice Burned	Singh/Buckpasser	C	10	C	C	5	3-4
Twice Worthy	Ambiopoise/Vandale	F	7	B	F	2	3-4
Two Davids	Olden Times/In Reality	C	6	C	C	4	2-3
Two Punch	Mr. Prospector/Grey Dawn II	B	4	F	C+	4*	2-3
Two's A Plenty	Three Martinis/Dark Star	C	6	C	F	2	2-3
Tyrant	Bold Ruler/My Babu	C	6	B	C+	2	2-3
Tyrone Terrific	Lyphard/Bold Ruler	D	10	M	U	5	3-4
Ultimate Pride	Lyphard/Princequillo	D	8	F	F	4	3-4
Unbridled^	Fappiano/Le Fabuleux	M	5	M	L	3*	2-3
Understanding	Promised Land/	M	3	U	T	2	3-4
Under Tack	Crozier/Sailor	C	7	D	C	4	3-4
United Holme	Noholme II/Drawby	D	7	D	D	5	3-4
Uno Roberto	Roberto/Secretariat	C	8	D	C	2	3-4
Unpredictable	Tri Jet/Ambehaving	D	8	C	F	4	2-3
Unreal Zeal	Mr. Prospector/Dr. Fager	B	6	C	B	5	2-3
U.S. Flag	Hoist the Flag/Sir Gaylord	B	7	F	A	3	3-4
Uzi*	Rich Cream/Dr. Fager	F	8	D	B	4	3-4
Unzipped^	Naked Sky/Specialmante	M	9	M	M	5	2-3
Vaal Reef*	Raise a Native/Nashua	B	8	D	C+	4	3-4
Vaguely Noble	Vienna/Nearco	D	4	B	D	1	3-4
Val de Loir	Vieux Manoir/Sunny Boy	D	3	C	C	1	3-4
Val de L'orne	Val de Loir/Armistice	D	4	B	F	1*	3-4
Valdez	Exclusive Native/Graustark	C	4	C	T	4	3-4
Valid Appeal	In Reality/Moslem Chief	B	3	C	C	4	2-3

Stallion	Sire/Broodmare Sire	FTS	CL	Trf	OT	SI	AB
Vanlandingham	Cox's Ridge/Star Envoy	D	7	D	C+	3	3-4
Varick*	Mr.Prospector/Fleet Nasrullah	C	8	C	F	4	2-3
Variety Road*	Kennedy Road/Macarthur Park	D	8	M	U	3	3-4
Vencedor	Flag Rasier/Fulcrum	C	7	C	F	4	3-4
Ventriloquist^	Lyphard/Stage Door Johnny	U	10	L	U	1	3-4
Verbago*	Verbatim/My Babu	U	10	M	M	5	3-4
Verbatim	Speak John/Never Say Die	D	4	D	C	2	2-3
Verge*	Alydar/Damascus	U	7	U	L	4	3-4
Verzy*	Vice Regent/New Providence	M	7	M	M	4	2-3
Verification*	Exceller/Buckpasser	C	10	U	F	4	2-3
Vernon Castle*	Seattle Slew/Prince John	F	9	U	L	2	3-4
Vice Regal	Northern Dancer/Menetrier	C	3	C	C	2	3-4
Vice Regent	Northern Dancer/Menetrier	D	2	C	C	3*	2-3
Victoria Park	Chop Chop/Windfields	C	5	D	D	4	3-4
Victorious	Explodent/MajesticPrince	D	8	D	C	2	3-4
Vigors	Grey Dawn II/El Relicario	F	5	C	C+	1*	3-4
Vilzak*	Green Dancer/Hilarious	M	8	M	M	4	2-3
Vitriolic	Bold Ruler/Ambiorix	C	6	D	C	4	3-4
Vittorioso	Olden Times/Native Charger	C	7	B	F	4	3-4
Wajima	Bold Ruler/Le Haar	C	5	D	C	3	3-4
Walesa^	Danzig/Gallant Man	M	6	M	L	4	2-3
Walker's	Jaipur/Swaps	C	9	D	C	5	2-3
Wander Kind	Hold Your Peace/Bupers	D	9	C	F	4	2-3
Waquoit	Relaunch/Grey Dawn II	D	6	C	C	3*	3-4
War*	Majestic Light/Victoria Park	U	9	M	U	2	2-3
Wardlaw	Decidedly/King of the Tudors	C	5	D	C	2	2-3
Ward Off Trouble*	Wardlaw/Francis S.	U	8	U	M	2	3-4
Warning*	Known Fact/Roberto	B	3	A	M	4	2-3
Wasa	Dust Commander/Intentionally	U	9	U	M	4	3-4
Wassl	Mill Reef/Tudor Melody	D	5	B	M	1	3-4
Water Bank	Naskra/Crewman	C	6	C	C	4	3-4
Water Gate*	Political Coverup/Bargain Ticket	M	9	U	U	4	2-3
Waterway Drive*	Poker/Royal Serenade	U	9	M	M	4	3-4
Water Moccasin*	Topsider/Hawaii	L	8	L	U	4	3-4
Wavering Monarch c	Majestic Light/Buckpasser	C	5	B-	C	3	2-3
Wayne's Crane*	L'Enjoleur/Gallant Romeo	B	7	U	A	4	2-3
W.D. Jacks*	Matsadoon/Young Emperor	L	8	U	M	5	2-3
Well Decorated	Raja Baba/Majestic Prince	D	6	D	C	4	3-4
Weshaam	Fappiano/What a Pleasure	C+	7	F	C+	4	2-3
Westheimer	Blushing Groom/Northern Dancer	C	8	D	F	4	2-3

Stallion	Sire/Broodmare Sire	FTS	CL	Trf	OT	SI	AB
Whadjathink^	Seattle Song/Great Nephew	U	6	M	U	2	3-4
What a Hoist	Hoist the Flag/Cyane	D	8	U	B	4	2-3
What a Pleasure	Bold Ruler/Mahmoud	C	5	C	C	4	3-4
What Luck	Bold Ruler/Double Jay	C	4	D	C+	4	3-4
Wheatley Hall*	Norcliffe/Miracle Hill	U	8	M	U	4	3-4
White Mischief*	Roberto/Intentionally	M	6	L	M	2	2-3
Whitesburg	Crimson Satan/Prince Bio	C	7	C	C	4	3-4
Who's Fleet*	Villamor/Hafiz	M	6	F	F	4	2-3
Who's For Dinner	Native Charger/Intentionally	F	7	D	F	2	3-4
Wild Again	Icecapade/Khaled	C	3	C	C+	4*	2-3
Wild Injun*	Exclusive Native/Warfare	F	10	F	F	5	3-4
Willard Scott*	Roanoke Island/Quadrangle	U	9	M	M	3	3-4
Willow Hour	Bold Hour/Pia Star	D	8	D	C	4	3-4
Will Win	Raise a Native/Our Michael	C+	8	F	D	5	2-3
Wind & Wuthering	No Robbery/John's Joy	D	8	D	D	3*	3-4
Windlord	Secretariat/Pappa Fourway	F	10	U	F	4	3-4
Windy Sands	Your Host/Polynesian	C	5	C	B	4	3-4
Wing Out	Boldnesian/Toulouse Lautrec	C	7	C	D	4	3-4
Wing's Pleasure*	Quiet Pleasure/Emblem II	U	10	U	U	5	3-4
Winning Hit	Bold Ruler/Ambiorix	C	6	D	C	4	2-3
Winrightt*	Distinctive/Sir Khalito	B+	7	U	F	5	2-3
Wintersett*	Gummo/Acroterion	C+	9	U	B	4	3-4
Wise Times	Mr. Leader/He's a Pistol	D	7	D	C	4	3-4
With Approval^	Caro/Buckpasser	M	4	L	M	1	3-4
Wolf Power	Flirting Around/Casabianca	C	5	C	C+	4	3-4
Wollaston*	Lord Gaylord/Le Fabuleux	U	8	U	M	3	3-4
Wonder Lark	Verbatim/TV Lark	D	9	D	F	4	2-3
Woodland Lad	Woodland Pines/Atomic	B	10	B	B	5	3-4
Woodman	Mr. Prospector/Buckpasser	C+	5	C	C	3*	2-3
World Appeal	Valid Appeal/Restless Wind	C+	5	C	C	4	2-3
Worthy Endevor*	Ruken/Port Wine	U	10	U	U	5	3-4
Wynslew*	Seattle Slew/Shantung	U	8	L	L	3*	3-4
Yarnallton Native	Raise a Native/T.V. Lark	F	10	U	C	5	3-4
Yes I'm Blue^	Marfa/No Fooling	U	7	U	U	5	2-3
Yesterdays Hero	Olden Times/Round Table	C	8	D	B	4	2-3
Young Commander	Lt. Stevens/Blue Prince	F	9	D	U	4	3-4
Your Dancer	Gaelic Dancer/The Pie King	F	9	F	F	4	3-4
Yukon	Northern Dancer/Nashua	C	7	C	F	4	3-4
Zafarrancho*	Farnesio/Martinet	U	8	U	M	2	3-4
Zaizoom*	Al Nasr/Turn-to	C	8	M	U	4	2-3
Zamboni	Icecapade/Eddie Schmidt	C	8	C	C+	4	3-4
Zanthe	TV Lark/Turk's Delight	M	6	M	M	4	3-4

Stallion	Sire/Broodmare Sire	FTS	CL	Trf	OT	SI	AB
Zen	Damascus/Tulyar	C	6	C	C	4	3-4
Zevi	Cornish Prince/Carry Back	C+	8	D	C+	5	2-3
Ziad	Key to the Mint/Jacinto	C	7	B-	F	4	3-4
Ziggy's Boy	Danzig/Sunrise Flight	C	6	D	C	4*	2-3
Zilzal*	Nureyev/Le Fabuleux	M	6	L	U	4*	2-3
Zoning	Hoist the Flag/Round Table	U	7	M	M	4	3-4
Zoot Alors	Raise a Native/Saint Crespin	C	8	F	C	5	3-4
Zuppardo's Love*	Icecapade/Jester	M	10	M	M	4	3-4
Zuppardo's Prince	Cornish Prince/Primate	B	7	D	C+	4	2-3

Academy Award	2K	Distinctive Pro	15K	Java Gold	12.5K
Alwuhush	7.5K	Dixie Brass	7.5K	Jazzing Around	4K
Apalachee	5K	Dixieland Band	40K	Jeblar	7.5K
Ascot Knight	6.5K	Dixieland Brass	3K	Judge Smells	3.5K
Batonnier	5K	Dr. Blum	5K	Katowice	2.5K
Beau Genius	4K	Dynaformer	5K	Key to the Mint	10K
Black Tie Affair	15K	Eastern Echo	10K	Kleven	2K
Blushing John	20K	Easy Goer	50K	Known Fact	12.5K
Bold Rukus	15K	El Gran Senor	30K	Kris S.	40K
Bolger	3.5K	Explodent	10K	Leo Castelli	10K
Briartic	3.5K	Falstaff	2.5K	Lively One	3.5K
Broad Brush	12.5K	Fatih	2.5K	Local Talent	5K
Buckaroo	7.5K	Fast Play	7.5K	Lord at War	10K
Bucksplasher	6.5K	Ferdinand	15K	Lost Code	15K
Candi's Gold	3.5K	Festin	2.5K	Lucky North	5 K
Capote	40K	Fit to Fight	10K	Lyphard	75K
Carnivalay	6K	Fly Till Dawn	4K	Majestic Light	20K
Carr de Naskra	6.5K	Fortunate Prospect	7.5K	Manila	10K
Carson City	7.5K	Forty Niner	70K	Marfa	5K
Caveat	7.5K	Fred Astaire	5K	Meadowlake	12.5K
Cherokee Colony	5K	Gate Dancer	8.5K	Mehmet	2.5K
Chief's Crown	15K	Geiger Counter	10K	Mining	15K
Clever Trick	15K	Gone West	35K	Miswaki	30K
Conquistador Cielo	20K	Green Dancer	25K	Moon Up T.C.	1K
Cool Victor	10K	Green Forest	5K	Moscow Ballet	7.5K
Copelan	10K	Groovy	5K	Mr. Prospector	200K
Cormorant	7.5K	Gulch	40K	Mt. Livermore	15K
Cox Ridge	30K	Gumboy	1.5K	Naevus	5K
Cozzene	10K	Habitony	3K	Nasty and Bold	2.5K
Crafty Prospector	20K	Half A Year	3K	Nepal	2.5K
Crusader Sword	5K	Halo	20K	Norquestor	3.5K
Cryptoclearance	25K	Hansel	30K	Northern Baby	12.5K
Cure the Blues	9.5 K	Herat	5K	Northern Flagship	15K
Cutlass Reality	5K	High Brite	5K	Northern Prospect	5K
D'Accord	5K	His Majesty	15K	Northern Score	2.5K
Danzig	175K	Homebuilder	5K	Notebook	2.5K
Danzig Connection	7.5K	Housebuster	20K	Nureyev	125K
Darn That Alarm	10K	Houston	15K	Ogygian	12.5K
Dayjur	50K	Interco	5K	Oh Say	3.5K
Demons Begone	5K	Irish Open	3.5K	Olympic Native	3.5K
Deputy Minister	50K	Irish River	25K	Opening Verse	10K
Desert Wine	7.5K	Irish Tower	7.5K	Order	2K
Devil's Bag	25K	Island Whirl	3K	Pancho Villa	7.5K
Diesis	30K	Jade Hunter	25K	Pentelicus	3K

Personal Hope	7.5K	Slewdledo	2K
Phone Trick	15K	Son of Briartic	7K
Pirate's Bounty	8K	Spectacular Bid	9.5K
Pleasant Colony	60K	Spend A Buck	5K
Premiership	7.5K	Star de Naskra	10K
Present Value	5K	Steinlen	7.5K
Private Account	40K	Stop The Music	10K
Private Terms	7.5K	Storm Bird	45K
Private Reality	10K	Storm Cat	60K
Prospectors Gamble	5K	Strawberry Road	15K
Proud Birdie	5K	Strike Gold	3.5K
Proud Truth	7.5K	Summer Squall	25K
Rahy	30K	Sunny Clime	2.5K
Raise A Man	5K	Sunny North	5K
Rare Brick	3K	Swing Till Dawn	3.5K
Rare Performer	3K	Sword Dance	5K
Red Ransom	7.5K	Synastry	2K
Red Wing Bold	2.5K	Talinum	3.5K
Relaunch	35K	Tasso	5K
Riverman	75K	Tank's Prospect	2K
Roman Diplomat	2.5K	Tejano	5K
Ruhlmann	7.5K	Temperence Hill	2.5K
Runaway Groom	6.5K	Theatrical	30K
Sabona	4K	Thirty Six Red	12K
Saratoga Six	10K	Time For A Change	15K
Saros	3.5K	Top Avenger	3.5K
Seattle Dancer	7.5K	Trapp Mountain	2.5K
Seattle Slew	80K	Trempolino	20K
Seattle Song	5K	Tsunami Slew	3.5K
Secret Claim	2.5K	Turkoman	7.5K
Seeking the Gold	50K	Two Punch	5K
Sejm	2.5K	Unbridled	20K
Septieme Ciel	7,5K	Unreal Zeal	3.5K
Shadeed	10K	Valid Appeal	20K
Shanekite	5K	Vanlandingham	2.5K
Show Dancer	3K	Vice Regent	15K
Silver Deputy	7.5K	Vigors	5K
Silver Hawk	40K	Waquoit	7.5K
Simply Majestic	5K	Wavering Monarch	3.5K
Siyah Kalem	3.5K	Wayne's Crane	3.5K
Skywalker	15K	White Mischief	2K
Slew City Slew	7.5K	Wild Again	35K
Slew o'Gold	15K	With Approval	15K
Slew's Royalty	3K	Wolf Power	7.5K
Slewpy	10K	Woodman	50K
Smarten	7.5K	World Appeal	5K
Smile	3K	Ziggy's Boy	2.5K
Snow Chief	3.5K	Zilzal	60K

Also Available From City Miner

Scared Money, $19.95
by Mark Cramer

Kinky Handicapping: The Path To Promiscuous Profits, $26.95
by Mark Cramer

Bred To Run: The Making Of A Thoroughbred, $27.50
by Mike Helm

A Breed Apart: The Horses And The Players, $22.95
by Mike Helm

Trainer Pattern Pocket Guide, $50.00
by Bill Olmsted

Pedigree Update News (coming in July 1995), $50.00 year - 2 newsletters

Send orders + $4 shipping to:

California residents please add 8.25 % sales tax

City Miner Books

P.O. Box 176

Berkeley, CA 94701